Acidity and Basicity of Solids

NATO ASI Series

Advanced Science Institutes Series

A Series presenting the results of activities sponsored by the NATO Science Committee, which aims at the dissemination of advanced scientific and technological knowledge, with a view to strengthening links between scientific communities.

The Series is published by an international board of publishers in conjunction with the NATO Scientific Affairs Division

A Life Sciences **B Physics**	Plenum Publishing Corporation London and New York
C Mathematical **and Physical Sciences** **D Behavioural and Social Sciences** **E Applied Sciences**	Kluwer Academic Publishers Dordrecht, Boston and London
F Computer and Systems Sciences **G Ecological Sciences** **H Cell Biology** **I Global Environmental Change**	Springer-Verlag Berlin, Heidelberg, New York, London, Paris and Tokyo

NATO-PCO-DATA BASE

The electronic index to the NATO ASI Series provides full bibliographical references (with keywords and/or abstracts) to more than 30000 contributions from international scientists published in all sections of the NATO ASI Series.
Access to the NATO-PCO-DATA BASE is possible in two ways:

– via online FILE 128 (NATO-PCO-DATA BASE) hosted by ESRIN,
Via Galileo Galilei, I-00044 Frascati, Italy.

– via CD-ROM "NATO-PCO-DATA BASE" with user-friendly retrieval software in English, French and German (© WTV GmbH and DATAWARE Technologies Inc. 1989).

The CD-ROM can be ordered through any member of the Board of Publishers or through NATO-PCO, Overijse, Belgium.

Series C: Mathematical and Physical Sciences - Vol. 444

Acidity and Basicity of Solids

Theory, Assessment and Utility

edited by

Jacques Fraissard

Laboratoire de Chimie des Surfaces,
Université Pierre et Marie Curie,
Paris, France

and

Leonidas Petrakis

Department of Applied Science,
Brookhaven National Laboratory,
Upton, New York, U.S.A.

Springer Science+Business Media, B.V.

Proceedings of the NATO Advanced Study Institute on
Acidity and Basicity of Solids: Theory, Assessment and Utility
La Colle sur Loup (Nice), France
June 13–25, 1993

A C.I.P. Catalogue record for this book is available from the Library of Congress.

ISBN 978-94-010-4427-1   ISBN 978-94-011-0986-4 (eBook)
DOI 10.1007/978-94-011-0986-4

Printed on acid-free paper

TABLE OF CONTENTS

vi

PREFACE

The subject of acidity and basicity has enormous economic and technological value while it continues to present significant scientific challenges with prospects for further important technological developments. Historically, technological developments in acidity/basicity have often preceded the scientific understanding of the phenomena involved, certainly in the petroleum industry, a key beneficiarry and user of the concepts of acidity. This process, however, is very expensive and less efficient than developments based on a fundamental understanding of the scientific phenomena involved. This has been recognized over the years and it explains why university, government and industrial laboratories have in the last 50 years devoted large efforts to understanding acidity (and basicity to a lesser extent) so they can gain the technological advantage.

The scientific and technological literature on the subject is truly enormous. There have been some very important articles and books on the subject that have attempted to critically review many individual contributions. During the last few years there have been three developments that led us to organize the Advanced Study Institute on which this volume is based:

a) Significant developments in the theory of acids and bases;

b) Developments in instrumentation that allow the detailed characterization of materials including in-situ conditions relevant to industrial processes;

c) The realization that closer coupling of scientific and technological pursuits can lead to greater scientific understanding and better technology.

The structure of the ASI reflected the coming-together of these three factors. Following a brief historical review of the subject, the ASI and the book focus essentially equally on theory, assessment and utility. A serious effort has been made to address basicity on an equal footing with acidity. Topics covered include : inherent condensed matter acidity and basicity; medium effects; proton transfer mechanisms; superacidity; various spectroscopic characterization techniques; properties; prospects for different types of materials including zeolites, aluminum phosphates, molecular metal-oxygen clusters; various quantum mechanical approaches to calculation of relevant properties; carbocation chemistry; nature of intermediates; preparation of catalysts; industrial applications.

This book is addressed to a wide readership. Specialized workers in the field should find the updated material on several areas of this topic very useful; university teachers could use the material in the book for introductory or graduate courses; and those who have a general interest in the subject should find the overviews offered particularly interesting. There are extensive literature references for further detailed studies.

Many people have contributed to the success of the ASI on which this volume is based. It is a great pleasure to acknowledge the financial support provided by the Scientific Affairs Division of the North Atlantic Treaty Organization; also the National Science Foundation and Dr. Richard Metcalf of the National Science Foundation for the fellowships provided to American students and postdoctorate fellows; and the

International Science Foundation for the support they provided to scientists from Russia and the Ukraine. Many individuals who helped greatly with the organization of the ASI and whose help we acknowledge with gratitude include Drs. M.A. Springuel-Huet, J.L. Bonardet, A. Gedeon, Ms C. Bonmkratz, Mr. La Porte and Ms. P. Taylor.

Jacques Fraissard, Paris, France
Leonidas Petrakis, Port Jefferson, NY, USA
July 1993

HISTORICAL PERSPECTIVES OF ACIDITY AND BASICITY. PART 1: 4000 B.C. - 1800 A.D.

H. SCHOFIELD and J. DWYER
Chemistry Department
University of Manchester Institute of Science and Technology
P.O. Box 88
Sackville Street
Manchester
M60 1QD
United Kingdom

ABSTRACT. The earliest uses and preparations of acids and bases are described. The awareness of the differences between acids and bases as classes of compounds and between different acids and bases is discussed, and the transition to definitions and finally theories concerning the mechanism of their activity and composition are examined. The discoveries are placed in the context of the social and economic times, and the roles played by prominent workers are evaluated. The discoveries of related concepts, such as indicators, are also included.

1. Introduction

The discovery, uses and gradual development of theories concerning acids and bases and their activity goes back to the origins of chemistry. An attempt has been made in this paper to extract and bring together the discoveries and theories that concern acids and bases, and the works of scientists who contributed to these aspects of the history of chemistry. Some important chemists are not featured in cases where they did not make significant contributions in this area, similarly many who are included also made discoveries in other areas. The research was carried out chronologically, therefore this history is presented in roughly chronological order.

2. Origins of the Terms Alkali, Acid and Base

The term alkali is derived from the Arabic *al-qaliy*, meaning calcined ashes of plants, this being derived from the word *qalay* which in turn means "'to roast in a pan".[1] Thus alkali originally referred to potash, or potassium carbonate, and the root therefore also indicates the geographical origin of the term. Alkalis may well have been referred to as such several thousand years before Christ. The word base was not introduced until much later, namely in 1736 when it was first used by Duhamel du Monceau and popularised by Rouelle. The term was defined in the context that "a neutral salt is a product formed by union of an acid with any substance capable of forming a base for it by giving it a concrete form".[1,2] The term acid is derived from the Latin word *acidus*, meaning sour. All terms were in use before any true understanding of their action or composition existed.

3. Earliest Uses of Alkalis and Acids

Most events in the early history of alkalis and acids appear to have been in the Middle East. The historical literature indicates that alkalis were in use before acids, probably as early as 4000 B.C. during the New Stone Age period.[3] The earliest occurrence appears to be in the production of quicklime (CaO), which was carried out by roasting limestone (CaCO$_3$). At this time the principal

1

J. Fraissard and L. Petrakis (eds.), Acidity and Basicity of Solids, 1-12.

use of quicklime was to remove fat and hair from leather, although later (probably around 3000 B.C.) it became widely used for cement production, which is its main use today. The production process was basically the same as that employed today.

The next alkali to be discovered appears to have been potash (K_2CO_3),[3] around 3000 B.C., when the first soap was probably (inadvertently) prepared. A dilute solution of potash was produced by leaching ashes from wood fires, and this was discovered to have cleaning powers. If more ashes were added, the cleansing properties increased, due to the increase in the concentration of the alkali solution. Probably by accident, someone added fat to the solution, and soap was produced. These discoveries were made at around the same time as the invention of the wheel, at a time when the Sumerian civilisation in Mesopotamia was at its peak.

Also in about 3000B.C. there is evidence that glass was being produced, and this requires sodium carbonate. Support for this was obtained by the archaeologist Petrie,[4,5] who discovered a glass bead which has been dated to about this time. Glass was not produced on a large scale until abut 1500 B.C., evidence for this arising from the discovery of a glass factory in Egypt.

The first acid to have been commonly used was probably acetic acid, in the form of vinegar, around 400-300B.C. Acetic acid was used in the production of white lead by reacting it with metallic lead. The white lead produced was used as a pigment. This occurred at the time of the development of Greek Science and philosophy and the preparation was described by the Greek worker Theophrastus in his *Treatise on Stones*, where he states: "lead is placed in an earthen vessel over sharp vinegar, and after it has acquired some thickness of a kind of rust, which it commonly does in about ten days, they open the vessels and scrape it off........what has been scraped off they beat to a powder and boil with water for a long time, and what at last settles to the bottom is white lead".[4] Vinegar is also mentioned in the Bible in the Book of Proverbs: "He who sings with a heavy heart is like one who takes off his clothes on a cold day, and like vinegar on a wound".[6] Theophrastus gives another use in the preparation of mercury in about 400 B.C., which was done by rubbing cinnabar in vinegar.

4. Greek Science

Greek science and philosophy signified the beginnings of a scientific way of thinking, and the first theories of science. Communication was improving, not least due to the ease with which the Greek alphabet could be learnt (compared to say Egyptian hieroglyphics), enabling a greater proportion of the population to be literate. In addition, worldwide communication was improving,

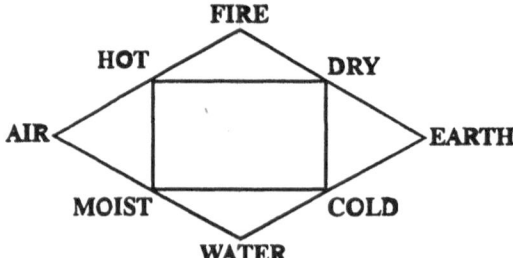

Figure 1 - Aristotle's four elements

which meant that ideas were able to be transmitted from one civilisation or culture to another. Initially the Greeks organised knowledge into natural philosophy, the predecessor of philosophy and science. Greek science had its beginnings in about 450B.C., and the culture dominated for about 1000 years, until Islam became predominant.

The key figure of the times was Aristotle, who brought together the ideas of the earlier thinkers. He developed his theory of the four elements (fire, earth, air and water) and four qualities (hot, dry, cold and moist), each combination of properties being associated with an element (figure 1). All substances were considered to be composed of different combinations of the qualities, and their properties were considered to be the result of an excess of a particular quality; for example hard substances han an excess of the quality "dry", dense substances an excess of "cold". These elements and qualities related to earthly or material substances, and Aristotle later introduced a fifth element "quintessence" which referred to the composition of the heavens.

Another product of Greek science was the atomic theory of Leucippos and Democritos, who believed that one substance, "prime matter", existed as indestructible indivisible atoms.[7] This was not of course the atomic theory recognised today, but a revolutionary way of thinking at the time. The shapes and sizes of atoms changed to give rise to the different properties of substances. Aristotle rejected this atomic theory, and as his work had considerable influence for so long it was many centuries before an atomic theory of any sort reemerged.

Although not producing any major discoveries in the history of acidity and basicity, events in these areas were influenced considerably for many centuries by Greek science.

5. The Rise and Fall of Alchemy

In parallel with the thinking and theories of the Greek natural philosophers were the attempts of the alchemists. The alchemists indeed made contributions to Greek thinking and *vice versa*. There are problems with discovering definite dates of events and facts arising from alchemy because the documentation was often written in unusual language with much appearing to be in a kind of code.[3,4]

There were two types of alchemy, esoteric and exoteric. The former was the religious branch, and was in fact a mix of astrology and religion. It was believed that studying the heavens could help the understanding of events on earth, and therefore it was supposed that studying earthly matters would help the understanding of the heavens: heaven and earth were believed to mirror one another. This belief led to studies of chemical substances and metallurgy. Exoteric alchemy was the worldly form and involved the search for the philosopher's stone, which would transmute metals (principally lead) into gold, and the elixir of life.

Alchemy appears to have been practised worldwide, estimates of dates of its beginnings being 140 B.C. in China, 400 A.D. in the Middle East, 800 A.D. in India and 1100A.D. in Europe, when many of the Arabic works were translated into Latin.

Centuries of failure to produce gold gave rise to scepticism about alchemists and their science. By the end of their period of influence, it was possible to analyse lead by use of acids and it was thus discovered that lead in fact contained some gold, the proportion of which did not increase after attempts at transmutation. Some alchemists kept their faith. After their research had failed in one location, they appear to have wandered from place to place looking for new benefactors to finance their experimentation. After another failure, they moved on again.

The Church was opposed to alchemy as it was perceived to be an attempt at magic and linked with paganism. The Christian world had started a three-hundred year war against anyone suspected of practising magic, and alchemy was banned in most European countries from 1300A.D. onwards.

Alchemy did, however, produce a number of workers with contributions to the history of acidity and basicity mainly, through Islamic science.

6. Islamic Science

Islamic science built on the Greeks' ideas and improved existing processes.

6.1. JABIR (720 - 815 A.D.)

Jabir was the most famous Islamic alchemist. He was a member of the Isma'ilia sect, involved with esoteric alchemy. Jabir is attributed with responsibility for over 2000 books, which in fact were probably written by many different members of the sect. He is believed to have prepared nitric acid, which was probably used for medical purposes and as a solvent. Many of his works were translated into Latin, so dates quoted are often much later, and he appears in these works under the name Geber.

6.2. AL- RAZI (866-920 A.D.)

Al-Razi was the first person to group acids as a class of compounds, describing them as "sour waters". In this class he included sour milk, lemon juice and vinegar.[3] Also his works mention the discovery of caustic alkalis, giving methods for the preparation of NaOH and KOH by treating Na_2CO_3 and K_2CO_3 with slaked lime ($Ca(OH)_2$).
　　During and after this period of Islamic dominance, the centre of activity in scientific research appears to have moved to Western Europe. A prominent worker was Raymond Lull (born in Majorca in 1232), who was a missionary enthusiast, an alchemist and eventually a martyr. He described preparations of aqua regia and nitric acid.

7. Iatrochemistry

The invention of the printing press around 1400 gave rise to a great improvement in communication and a hunger for knowledge as books became widely available.
　　The major area of activity in science at this time centred around medicine and the development of pharmaceutical cures for ailments and diseases, and many discoveries were made in these areas in the 15th and 16th centuries.

7.1. PARACELSUS (1493-1541)

Theophrastus Bombast von Hohenheim, otherwise known as Paracelsus, was born in Switzerland, the son of a German Licentiate in medicine who had migrated there.[9] He taught his son medicine, mineralogy and chemistry. Paracelsus obtained his degree of M.D. from Ferrara in Italy, and became professor of medicine at Basel in 1527. Paracelsus is regarded as the founder of iatrochemistry, chemistry applied to the service of medicine. He was a reformer of medicine, but his work laid the foundations for discoveries concerning the nature of fluids in the human body. Paracelsus believed in Aristotle's four elements, but he thought they appeared in the body as "three principles", salt, sulphur and mercury. Salt was regarded as the principle of "fixity and incombustibility", mercury the principle of "fusibility and volatility" and sulphur the principle of "inflammability", and these were likened to body, soul and spirit. Paracelsus coined the term "archaeus" to mean the spirit in each body organ.

7.2. VAN HELMONT (1579-1644)

The work of Paracelsus led to theories about the acid/alkaline character of bodily fluids which appears to have been first recognised by Van Helmont.[4] He postulated the existence of "ferments" which reacted with the "archaeus" to produce acids or alkalis, dependent on the organ of the body in question: in the stomach for instance acids were produced. The ferments were different in each organ in order that they could perform the different bodily functions, and to some extent actually equate with enzyme activity. Six different fermentations were thought to occur as food passes through the organs of the body.

7.3. DE LE BOE (1614-1672)

Franciscus de le Boe (also known as Sylvius) represents the culmination of iatrochemistry, and came closest to the truth, rejecting both ferments and archeus and believing that bodily functions are determined by the acid/alkaline characters of the bodily fluids.

8. The Discovery of Hydrochloric Acid

The problems of discovering an accurate sequence of events in a historical study of this kind are typified by the case of hydrochloric acid.

Although aqua regia had been known about for some time, as mentioned by Lull, the discovery of the preparation pure hydrochloric acid seems to be a bit of a mystery and there seems to be disagreement among historians. Salzberg[3] states that hydrochloric acid was known by Moorish workers, reported in the influential works *De Aluminibus et Salibus* (Concerning alums and salts) and *Alchemica de Anima* (About the alchemical spirit) in the late 11th or early 12th century. The first work mentions heating sal ammoniac with vitriol to produce the alchemists' "spirit of salt" - hydrochloric acid. Reti[10] states, however, that there is no clear reference to its preparation in the Middle Ages and that HCl appears to have been discovered at the end of the 16th century. In 1845 Kopp in *Geschicte der Chemie,* III, Braunschweig, refers to a fifteenth century chemist Basilius Valentinus who described how aqua caustica could be obtained by heating vitriol and common salt. Kopp later questioned the date and authenticity of the Valentinus writings, and today's historians believe they were written after Paracelsus, not before. There are some earlier reports that probably led to the preparation of hydrochloric acid by Rosetti in a book called "*Plicheto*", published in 1540. Apparently another writer, Porta, described a process of producing an acid by heating salt and bricks together in 1589. The first detailed description of the above process has also be attributed to Glauber, in 1648, Libavius, in 1595, or van Helmont in about 1646 (by heating a mixture of common salt and dried potter's clay). This last method was improved upon by Glauber, who used green vitriol and alum instead of clay. This led to his standard method of preparation in 1658 from salt and oil of vitriol.

Reti claims, however, to have found a fifteenth century manuscript dealing with colours, dying, recipes and the decorative arts, the author or compiler of which is unknown. There is apparently a clear description of the preparation of hydrochloric acid, and its use for softening bones, thus making it what appears to be the first awareness of that acid.

9. Seventeenth Century Theories of Acidity and Basicity

A scientific revolution occurred in the 17th century as workers began to develop theories about scientific events, including the activities of acids and alkalis. Much was known already about the properties of acids and alkalis: acids had been recognised as a class of compounds with certain common properties, most common mineral acids were known along with aqua regia, many poorly defined organic acids were recognised, many alkalis had been prepared. The fundamental reaction

$$acid + alkali \rightarrow salt$$

was known, but no theories as to the activities or the reasons for the properties of acids and bases had been postulated. The theories of the 17th century were mostly mechanical in approach, and became increasingly elaborate.

9.1. GLAUBER

Glauber had a fairly clear idea that salts consist of acid and base, and also of the concept of affinity. In 1648 he explained that when ammonium chloride is heated with zinc oxide the latter combines with the acid because of its greater affinity, and that it lets the ammonia go free. He also recognised the "oppositeness" of acids and bases.

9.2. BOYLE

Robert Boyle also recognised the oppositeness of acids and bases and postulated in 1663 that acids have high solvent power.[1] He did not attempt to provide a mechanical theory as to the activities of acids and alkalis, and in fact condemned the concepts of acidity and alkalinity as examples of the "vulgar chemical doctrine of occult qualities".[11] Boyle also made fundamental contributions to indicator chemistry (see later).

9.3. TACHENIUS

Tachenius arrived at perhaps the most accurate theory when he stated in 1669 that "all salts can be divided into two parts, into alkali and acid".

9.4. ST ANDRE

The first of the elaborate mechanical theories can be attributed to St Andre in 1680. He stated that an acid is "a simple body of pointed shape, which ferments in the presence of alkalis and is the essence of all mixtures. The sharp points of acids fit so perfectly into the holes of the alkalis and fill them so completely that fresh acid encounters no empty pores to arrest its movement; hence the fresh acid reacts so violently that it tears the constituents of these bodies apart".[1,12]

9.5. BERTRAND

Along similar lines, Bertrand in 1683 hypothesised that "an acid is a liquid body composed of small firm pointed particles, slightly resembling fine delicate needles". He described alkalis as consisting of "particles which have between their junctions pores of different structures". Neutralisation apparently occurred when the needles of the acid enter the pores of the alkali.[13]

9.6. LEMERY

Lemery developed a classification of substances into three groups, mineral, vegetable and animal, and also used an atomic theory which put forward the idea that the properties of substances depended on the shapes of their atoms. In 1690 Lemery believed that acids have sharp, spiky atoms which prick the tongue, and their salts form sharp crystals. In precipitation reactions, the spikes of the acid particles break off in the pores of metal atoms and are carried down in the precipitate. According to Lemery metals dissolve in acids because the points of the acids tear apart the particles in the mass of the metal.[1,14]

9.7. NEWTON

Newton had a slightly different concept in his definition, introducing a dynamical aspect to the purely mechanical theories. He put forward the idea of short-range interparticle forces which gave rise to his 1692 definition that acids are substances "with a great attractive force, in which force their activity exists".[1,15]

9.8. SUMMARY OF SEVENTEENTH CENTURY THEORIES

The definitions mentioned above, as well as not referring to acids and alkalis in any chemical sense, were not quantitative. No explanations or definitions were offered about the sizes, shapes or motions of the needles, pores or particles. They did, however, reflect a change in that scientists started to question why events took place, representing a move towards a more inquisitive thinking.

10. Discovery of Indicators

The first discovery of colour changes in plant materials upon exposure to acids and alkalis seems to have been by Pliny in the first century A.D. , who tested for the presence of iron with papyrus soaked in a solution of nutgall, but the most significant discoveries can be attributed to Robert Boyle.[16,17]

Boyle was born at Lismore Castle, Ireland, in 1627. He became a prolific writer, his most famous work being "*The Sceptical Chemist*". Boyle believed in studying chemistry for its own sake, not just in the service of medicine or alchemy. He introduced a rigorous experimental approach to chemistry, and rejected Aristotle's four elements and Paracelsus's three principles.

Reactions leading to colour changes had been observed before, but usually they were irreversible and very specific. Boyle is therefore regarded as the originator of indicator chemistry because of his work on reversible changes. He shared with the alchemists a fascination with colour changes, and studied the effects of acids and alkalis upon a variety of plants, observing that syrup of violets was reddened by acids and made green by alkalis and that reversible colour changes occurred when blue cornflowers, brazilwood, pomegranate buds and blossoms and turmeric amongst others were treated with acids and alkalis. "All acid salts destroy the blewness of the infusion of our wood and all sulphureous salts (both animal and alcalisate) have the vertue of restoring it".[17] He also experimented with cochineal and probably litmus, and was responsible for producing a considerable list of indicators to which no new ones were added until the mid-19th century.

Boyle used the syrup of violets to classify substances into acid, alkali and neutral on the basis of colour change, anything which failed to change the colour he considered neutral. He also classified acids and alkalis according to their chemical reactions, such as whether or not they effervesced with limestone. Boyle used acids and alkalis as analytical reagents in preference to pyrolysis which was the most widely used analytical technique at the time.

10. The Chemical Industry

By about 1700, large scale industrial chemistry was growing rapidly. Living standards were improving and there was increasing demand for soap and glass products which were widely produced in Europe, leading to an increased requirement for alkalis in particular. This was despite there still being little comprehension concerning chemical reactions.

11. Phlogiston

Phlogiston was important in that it influenced chemistry for over one hundred years, although science historians debate whether it helped or hindered chemical progress.[3,4] Phlogiston theory did promote a more "chemical" approach to understanding reactions and properties, and thus represented a progression from the earlier mechanical theories. Phlogiston was used to explain many chemical events, and the theory was modified when the experimental evidence did not match the theory.

11.1. ORIGINS OF PHLOGISTON

Phlogiston was originally postulated by the German Johan Becher (1635-82), who in 1669 put forward his five elements theory. This was derived from Aristotle's four elements theory, Becher's five comprising air, water and three solids which corresponded to the three responses to combustion: vitreous earth, fatty earth and mercurial earth, these bearing a resemblance to Paracelsus's three principles (salt, sulphur and mercury). Despite the similarities, Becher rejected both Aristotle's and Paracelsus's theories. Becher's work was popularised by Stahl (1660-1734), who changed the name of Becher's combustible fatty earth to phlogiston.

11.2. PHLOGISTON IN COMBUSTION REACTIONS

Phlogiston was employed to explain combustion, smelting and reactions of lime. Combustion had always been regarded as the loss of something - Stahl decided it was loss of phlogiston. By definition, substances which burn easily, such as sulphur, wax and charcoal, were considered to be rich in phlogiston. Phlogiston was believed to be a real substance, which transferred from one substance to another. When a substance was heated, it was believed to take in phlogiston from the flame. However, phlogiston had to be ascribed a negative weight in order to account for the fact that when metals burn to give powders this is accompanied by a gain in weight.

11.3. ACIDS AND BASES IN THE CONTEXT OF PHLOGISTON

Stahl "proved" that sulphur was a compound of sulphuric acid and phlogiston because sulphur burnt easily with a flame, thus giving off phlogiston, to give sulphuric acid; similarly phosphorus was a compound of phlogiston and phosphoric acid.

Phlogiston was believed to be the ingredient that made alkalis caustic. When limestone, soda or potash are heated to high temperature, they change to quicklime, caustic soda or caustic potash, because they absorb phlogiston from the fire (reaction 1).[1,18] (When these caustic compounds are left in air they lose some of their causticity because the phlogiston was believed to leak into the air.)

$$CaCO_3 \longrightarrow CaO + CO_2 \ (1)$$
(lime + phlogiston $\longrightarrow$ quicklime)

$$CaO + H_2O \longrightarrow Ca(OH)_2 \ (2)$$
(dissolution of quicklime in water)

$$M_2CO_3 + Ca(OH)_2 \longrightarrow CaCO_3 + 2MOH \ (3)$$
(mild alkali + quicklime (= lime + phlogiston)) $\longrightarrow$ lime + caustic alkali (= mild alkali + phlogiston))

In reaction 2 the quicklime undergoes physical dissolution in water. When the resulting calcium hydroxide solution, already rich in phlogiston absorbed from the heat in reaction 1, is treated with mild alkali, a strong caustic alkali is produced due to the absorption of more phlogiston. Hence an increase in phlogiston was believed to lead to an increase in alkalinity.

Neutralisation was believed to be a redistribution or sharing of phlogiston between a phlogiston-rich alkali and a phlogiston-poor acid.

11.4. PROBLEMS WITH PHLOGISTON

Flaws were apparent in the theory almost from the outset, but always answers were contrived to explain disparate phenomena. For instance, coke and charcoal, which must have been almost pure phlogiston because they burnt almost completely, leaving only a little ash, were not alkaline as would have been expected. It was argued that phlogiston alone was mild and only in combination was it caustic.

12. Joseph Black (1728-99)

Black was a Scottish physician turned chemist. He was born in Bordeaux, the son of a Scottish descendent who owned a wine business in the area. Black became a physician at Glasgow University where he obtained his M.D., and he later moved to Edinburgh where he became professor of chemistry. Black was an unassuming man, modest about his discoveries, who regarded himself principally as a teacher. He was the founder of thermodynamics, discovering heat capacity and latent heat. Black was a believer in phlogiston, although his work contributed to its downfall. He welcomed the views of Lavoisier, and incorporated his work into his lectures. Black was, like Lavoisier, a man with little regard for theories unless they were backed up by experiment. The two met in 1786.

Black's experiments and theories contributed to events in the history of acids and bases. Black discovered "fixed air" (carbon dioxide) in 1754, through his observations of the effervescence which occurred during reactions of acids on limestone.[8] He also differentiated between mild and caustic alkalis, but some of his most significant work resulted from his M.D. thesis which led to his transition from medic to chemist.

In the course of his M.D. research,[3] Black was searching for a solvent to dissolve kidney stones. It was known that caustic alkalis and solutions of quicklime would dissolve the stones, but they also dissolved bladder tissue. He decided to investigate quicklime in more detail, but encountered a problem in that the two professors in his department preferred different types of quicklime, one favouring that from limestone, the other that from cockle or oyster shells. Black decided to be diplomatic and opted to investigate magnesia alba instead. Black soon discovered that magnesia alba did not dissolve kidney stones but exhibited other interesting properties. His findings were reported in 1754 in his dissertation "*On the acid humour arising from food and magnesia alba*", which deals with the acidity of the stomach and the value of magnesia as an antacid. An appendix to the work gives a full explanation of the chemical experiments and the relation between mild and caustic alkalis.

Black investigated the decomposition of magnesia alba on heating:[3,8,19]

$$\text{Magnesia alba} + \text{heat} \longrightarrow \text{calcined magnesia} + \text{fixed air} + \text{water}$$
$$x\text{MgCO}_3 . y\text{Mg(OH)}_2 . z\text{H}_2\text{O} \longrightarrow \text{MgO} + \text{CO}_2 + \text{H}_2\text{O}$$

Black observed that the magnesia lost 7/12 of its weight on heating, which he considered to be too much to be attributable to a gain of phlogiston from the heat alone, but Black was uncertain how much weight loss was due to emission of the gas and how much due to gain in phlogiston. He also observed that the product, MgO, was not strongly alkaline as it did not effervesce in the presence of acids, despite having absorbed phlogiston from the heat. "Of the volatile parts contained in that powder, a small proportion only is water; the rest cannot, it seems be retained in vessels under a visible form.....the volatile matter lost in the calcination of magnesia is mostly air; and hence the calcined magnesia does not emit air or make effervescence when mixed with acids".[8,19] Black repeated the experiment and observed that a gas was being produced. He decided that the gas was probably the same as was evolved when magnesia was reacted with acid. Black measured the difference in weight resulting from the reaction with acid and concluded that the same amount of gas was produced during this reaction as when magnesia alba was heated. Hence the total weight loss was due to evolution of the gas, and apparently none due to gain of phlogiston.

Black also employed affinity tables in his lectures. These are tables showing the relative reactivities of compounds, the top line usually showing compounds with which all substances listed below in the same column will react. Various acids and bases were represented in these tables, and the affinities of different acids for metals could be determined.

13. The Discovery of Oxygen

Oxygen was discovered more or less simultaneously by two scientists, Joseph Priestley (1733-1804) and the Swedish chemist Scheele.[20] Scheele probably made the discovery first, but did not publish his results until 1777, after Priestley in 1774. Priestley was a believer in phlogiston, and concluded that the gas he had discovered was dephlogisticated air, containing little or no phlogiston. It could therefore absorb more phlogiston than ordinary air (hence it supported combustion). Priestley's discoveries had importance for Lavoisier, whom he met in 1795. Scheele was a prolific discoverer of organic acids, preparing oxalic, lactic, uric, citric and malic acids, all between 1774 and 1786.

14. Antoine Lavoisier and the Demise of Phlogiston

Antoine Lavoisier was born in 1743 in Paris, to a relatively affluent home.[21] His father, Jean Antoine, was a lawyer. Lavoisier attended an expensive school, the College Mazarin, and was encouraged by his father to become a lawyer also, but Lavoisier was more interested in natural and scientific matters. Lavoisier's chemistry teacher was apparently not the best, but each lecture was followed by a demonstration, and Lavoisier's demonstrator was Rouelle, a methodical experimenter full of enthusiasm. Rouelle taught Lavoisier the methods of science, to verify facts by experiment and search for new ideas from facts. After leaving school, Lavoisier was pressurised into studying law, but continued to study science in his spare time. Lavoisier was admitted to the Academy of Sciences in 1766 at the early age of 23. He published many articles in the Memoires of the Academy.

Although fairly affluent, to ensure a secure future, Lavoisier bought a share in the Ferme Generale, thus becoming a tax farmer; these people were responsible for collecting taxes for the king. The tax farmers paid for the privilege of collecting taxes, and kept the money collected for themselves. Through this he met his wife, Marie Anne, who became a great help to him as his laboratory assistant and secretary, documenting his results and sketching his laboratory apparatus.

Lavoisier's scientific work signified a fundamental change in scientific thinking, and his work on acids and bases forms the core of anti-phlogistic chemistry. He produced the first truly non-mechanical, chemical theory of acidity.[22,23] In 1772 Lavoisier experimented with phosphorus. He wanted to discover if phosphorus absorbed air when it burnt. He discovered that this was the case, forming "acid spirit of phosphorus", and gaining weight in the process. This weight gain had been observed before, and attributed to a loss of phlogiston. Lavoisier assumed that the phosphorus absorbed air, but he knew that the air absorbed was not the "fixed air" discovered by Black. In 1775 Lavoisier met Priestley, who informed him of his work on different types of air, and in 1775 Lavoisier declared that the air involved with the calcination process was part of ordinary air.

In 1776, Lavoisier concentrated on nitric acid, studying the "air" contained in it, and he stated "I am now in a position to advance affirmatively that not only the air, but the purest part of the air enters into the composition of all acids without exception; that is the substance which constitutes their acidity". In 1779 Lavoisier worked with sulphuric acid, and confirmed his nitric acid results. He referred to this air as the acid principle, the term principle deriving from the old three principles of Paracelsus, namely the sulphur principle responsible for inflammability, stating in 1779 "henceforth I shall designate dephlogisticated or eminently respirable air in a state of combination or fixity by the name of acidifying principle". This he followed by introducing the word oxygen, derived from the Greek *oxus*, meaning sour, and *gennao*, meaning I produce.

Although not factually correct in his assumption that all acids contain oxygen, it should be remembered that most of the acids that Lavoisier encountered did indeed contain oxygen. Lavoisier spent much time repeating the experiments of others, and often was able to explain the results of such experiments. He met Black, and was able to explain scientifically much of Black's experimental work.

At the time of Lavoisier's discoveries in chemistry, the French Revolution was taking place, and considerable resentment was felt for the tax farmers, who were accused of fraud and charging excessive interest rates. Lavoisier's association with this body of people proved to be his downfall, and he was arrested in 1794. He continued to write whilst in prison, but was eventually guillotined in 1795, immediately after his father-in-law.

References

1. Jensen, W.B. "Lewis acid-base concepts: an overview". Chapter 1. Wiley (1980).
2. Duhamel du Monceau, H.L. *Mem. Acad. R. Soc.(Paris)*, 215 (1736).
3. Saltzberg, H.W. "From caveman to chemist: circumstances and achievements". Americn Chemical Society (1991).

4. Partington, J.R. "A short history of chemistry". Macmillan (1937).
5. Multhauf, R.P. "The origins of chemistry". Oldbourne (1966).
6. The Holy Bible: Proverbs (25:20). Revised Standard Version. The Bible Societies (1952).
7. Hudson, J. "The history of chemistry". Macmillan (1992).
8. Partington, J.R. "A history of chemistry". Macmillan (1962).
9. Pachter, H.M. "Paracelsus: magic into science". Collier (1961).
10. Reti, L. "How old is hydrochloric acid?" *Chymia*, **10**, 11-23.
11. Boyle, R. in "The works of the honourable Robert Boyle". Vol. 4, p.284-292. edited by T. Birch. W. Johnston (1772).
12. Andre, F. "Entretiens sur l'acide et sur l'alcali" 2nd ed. Paris (1680).
13. Bertrand, G. "Reflexions nouvelles sur l'acide et sur l'alcali". Lyon (1683).
14. Lemery, N. "Cours de chymie", 7th ed. Paris (1690).
15. Newton, I. "Lexicon technician or an universal English dictionary of arts and sciences", Vol.3. edited by J. Harris. Brown et al. (1710).
16. Boyle, R. "Experiments and considerations touching colours", p.212. London (1664).
17. Baker, A.A. *Chymia*, **9**,147-167.
18. White, J.H. "The history of the phlogiston theory". Edward Arnold (1932).
19. Alembic Club Reprints, Edinburgh, 16.
20. Crane, W.D. "The discoverer of oxygen: Joseph Priestley". Julian Messner (1962).
21. Riedman, S.R. "Antoine Lavoisier: scientist and citizen". Abelard-Schuman (1957).
22. Crosland, M. *Isis*, **64**, 306-325 (1973).
23. Le Grand, H.E. *Ann. Sci.* ,**29**(1), 1-18 (1972).

HISTORICAL PERSPECTIVES OF ACIDITY AND BASICITY. PART II: 1800 A.D. -

J. DWYER and H. SCHOFIELD
Chemistry Department
University of Manchester Institute of Science and Technology
P.O. Box 88
Sackville Street
Manchester
M60 1QD
United Kingdom

ABSTRACT. The development of concepts of acidity and basicity during the past two hundred years is examined. Progression from concepts focusing on chemical composition as outlined originally by Lavoisier to the dissociation of substances in aqueous solution (Arrhenius, Ostwald) and thence to the role of specific ions in aqueous and non-aqueous solutions (Brønsted, Lowry) are discussed. Finally explanations based on electronic features starting with Lewis and progressing to the involvement of quantum mechanics is outlined.

1. Introduction

The latter part of the eighteenth century had witnessed the exciting intellectual development of the enlightenment which was centred largely in France. This period, which saw intense social, political and philosophical activity, also resulted in considerable progress in science, particularly in chemistry. From this time on, the concepts and definitions of acidity and basicity have evolved in concert with the development of experimental techniques and theoretical outlook.

In this paper an attempt is made to follow this evolution and to focus on the key developments and scientists involved.

2. Overview of Developments

From 1800 on, the evolution of definitions and concepts relating to acids and bases can be conveniently summarised as follows.

1800-1850	Definitions in terms of chemical composition
1850-1900	The role of substances capable of generating ions in aqueous solution
1900-1950	(a) A focus on ions in solution both aqueous (e.g. protic acids) and non-aqueous (b) The introduction of electronic effects
1950-	Further development of the role of electronic effects leading to explanations of acidity and basicity in terms of quantum chemistry

3. Acidity and Chemical Composition 1800-1850

As discussed in Part I of this lecture, Lavoisier (1777) played a key role in defining acids and bases by his recognition of the role of chemical composition such that oxygen was regarded as the

13

J. Fraissard and L. Petrakis (eds.), Acidity and Basicity of Solids, 13–31.

acidifying principle.[1] Doubts about the necessity for oxygen existed from the beginning since Berthollet (1779) showed that prussic acid (HCN) contained no oxygen.[2] However, since it was possible to argue that prussic acid was not truly an acid, this was not generally taken as conflicting with Lavoisier's view.

However, when Humphry Davy (1810) demonstrated[3] that muriatic acid (HCl) contained no oxygen, it became more difficult to sustain definitions based on the presence of oxygen. Although Davy originally thought that chlorine was responsible for acidity in muriatic acid, he later suggested that hydrogen might be the acidifying principle and subsequently proposed that acidity depended upon "peculiar combinations of substances" rather than on "peculiar substances". Several other scientists subscribed to this view, but understanding of the mode of combination of substances was completely lacking whereas stoichiometric composition could be determined. Consequently emphasis remained on defining acidity in terms of substances which provided the acidifying principle. Gay Lussac (1814, 1815) grouped several acids (HCl and HI) with prussic acid (HCN) and termed them hydroacids[4] to distinguish them from acids containing oxygen. By 1830, the list of hydroacids included HI, HF, HSCN, H_2S, H_2Se, H_2Te, H_2SiF_6, HBF_4 and in 1838 Liebig[5] proposed that "acids are ......hydrogen compounds in which hydrogen may be replaced by metals".

The above developments by Lavoisier, Davy and others, who related acidity to chemical composition, were greatly facilitated by improvements in methods of chemical analysis which became available in the late eighteenth and early nineteenth centuries. The discovery of the battery by Volta (1801) resulted in extensive research into the effect of electric current on solutions. Davy and later Faraday made detailed studies of electrolysis as did Berzelius who observed (1803) that the electrolysis of aqueous solutions tended to give acids at the negative pole and bases at the positive pole. In addition to introducing ideas concerning "electropositive" and "electronegative" species (these terms are not identical with current usage) Berzelius suggested (1812) that "all chemical reactions are neutralisations of opposite electrical charges". For example, the reaction of calcium oxide (considered as a basic substance with excess positive charge) with carbon dioxide (considered as an acid with excess negative charge) could be written as follows:

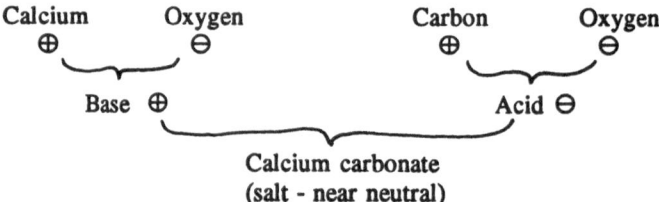

Calcium carbonate
(salt - near neutral)

Combinations were represented in pairs so that the scheme was essentially "dualistic" and in accord with the views of Lavoisier. The "strongly oxidised substance" (CO_2) reacted with the "weakly oxidised substance" (CaO) by sharing oxygen to give a "neutral salt", which was also in accord with the dualistic views of Lavoisier. Berzelius (1819) coined the term "electrochemical dualism" to describe the system above.[6] The system of electrochemical dualism does not appear to have played a significant part in the further development of views of acidity, but Berzelius staunchly defended the role of oxygen as an acidifying principle.

However, gradually the role of hydrogen, particularly replaceable hydrogen, as outlined earlier by Liebig who had worked on organic acids became widely accepted. Acids and bases were then seen as parent compounds from which salts were obtained by substitution rather than as "true" components of the salts. Neutralisation was not addition (dualistic) but substitution (unitary).

4. Ionic Definitions

It was Arrhenius (1877) who developed, along with Ostwald and van't Hoff, the ionic view of acids and bases. The theory was based on the new concept of ionisation in solution. Faraday had shown, previously, that ions were present in aqueous solution during electrolysis but Arrhenius proposed that the ions were present in aqueous solution and were not produced during electrolysis.

Svante Arrhenius (1859-1927) was born near Uppsala in Sweden.[7] He took a first degree in physics at Uppsala University but moved to Stockholm for his doctorate work because, it is suggested, the professor of physics did not encourage independent research. His doctorate involved measurement of aqueous solutions of acids, bases and salts with a view, initially, to estimation of molecular weights of solutes. His doctoral dissertation was not well received and the grade awarded was too low for a staff position at Stockholm. Later the offer of a position by Ostwald resulted in his appointment to the staff at Stockholm. His work and his collaborations with Wilhelm Ostwald, in 1888 at Riga and in 1889 in Leipzig, with Jacobus van't Hoff in 1888 at Amsterdam laid the foundation for the first theory of acidity and basicity which is clearly recognisable to present day scientists. Arrhenius also worked with Kolsrauch (Wurzberg, 1886) and Boltzman (Graz, 1887). Against strong opposition he was finally elected to the Swedish Academy in 1901 and was awarded the Nobel prize in 1903 - the first Swedish scientist to receive this award. He had difficulty achieving a professorship but in 1895, over strong objection he was appointed professor of physics at Stockholm Technical University. Interference by the king who wanted to keep Arrhenius in Sweden resulted in his appointment as first director of the Nobel Institute for physical chemistry, founded by the Swedish Academy of Scientists in 1905. He held the post until his retirement in 1927. He died the same year.

Arrhenius published most of his work and summarised the work of others in 1883 and 1884 but it was not until 1887 that he stated[8]

"an acid is any hydrogen-containing compound which gives hydrogen ions in (aqueous) solution and as a base is any hydroxyl containing compound which gives hydroxyl ions in (aqueous) solution."

Consequently in answer to the long-standing question "why do all (aqueous) acids and bases exhibit common properties?" Arrhenius replied that the acidity of H_2SO_4 , HNO_3 etc. was not a property of the molecule but the property of a single ionic species (H^+) common to all their aqueous solutions. Similarly the basicity of NaOH, KOH etc. was a property of the ion (OH⁻) common to all their aqueous solutions.

This view considerably clarified the concepts of acidity and basicity. The suggestion that electrolytes were dissociated in solution, although not widely accepted immediately, was developed and strengthened by Arrhenius, particularly in his collaborations with Ostwald and van't Hoff. For example van't Hoff had shown that the osmotic pressure (π) of dilute solutions of non-electrolytes followed a simple gas law ($\pi = CRT$) but for electrolytes it was necessary to introduce a factor "i" such that $\pi = iCRT$. For strong electrolytes the value of "i" corresponded to the number of ions in the fully dissociated salt (e.g. $i = 2$ for NaCl, 4 for $LaCl_3$) Arrhenius writing to van't Hoff (April 1887) suggested that "at extreme dilution all salts are completely dissociated.......the degree of dissociation (α) is the ratio of the equivalent conductivity of the solution in question (Λ) to the equivalent conductivity at the most extreme dilution (Λ_0)"

i.e. $\alpha = \Lambda/\Lambda_0$

Consequently, if a fraction (α) of a mole of electrolyte dissociated to give N ions the number of particles in solution (i) is given by

$$i = 1 - \alpha + N\alpha \text{ so that } \alpha = (i\text{-}1)/(N\text{-}1)$$

relating the degree of dissociation to the van't Hoff parameter.

Similarly Ostwald (1888) utilised the suggestion that $\alpha = \Lambda/\Lambda_0$ when applying mass action to the dissociation of an electrolyte (AB) of concentration (C),

$$\begin{array}{ccc} AB & \rightleftharpoons & A + B \\ C(1\text{-}\alpha) & & C\alpha \quad C\alpha \end{array}$$

$$K = C^0\alpha^2/C(1\text{-}\alpha) = C\alpha^2/(1\text{-}\alpha)$$

so that $K = C\Lambda^2/\Lambda_0(\Lambda_0\text{-}\Lambda)$,

the Ostwald dilution law.

Wilhelm Ostwald,[9] often regarded as the founder of physical chemistry, was born in Riga, Latvia in 1853. After studying at the Dorpatsche Hochschule he was appointed professor of chemistry at the University of Riga in 1881. He invited Arrhenius to Riga and following publication of a paper by van't Hoff on the studies of chemical dynamics all three were brought into contact. Ostwald went to Leipzig in 1887 and spent a year at Harvard in 1905. In 1909 he was awarded a Nobel prize. Throughout his life Ostwald was very interested in sketching, painting and music. In later life he became interested in philosophy and, based on his understanding of thermodynamics, produced a happiness formula. During World War 1 he was a pacifist. He died in 1932.

An explanation for the observed constancy for the enthalpy of neutralisation of aqueous acids and bases also followed from the Arrhenius view since, in terms of ionic species the neutralisation of an acid and a base, in aqueous solution, could be written

$$M^+ + OH^- + H^+ + X^- \longrightarrow M^+ + X^- + H_2O$$

the common reaction being

$$H^+ + OH^- \longrightarrow H_2O$$

which had an enthalpy change of around 58 kJ mol^{-1} for several acid/base neutralisations.

In this vein the establishment of appropriate equilibria provided a basis for defining acid strength and resulted in quantitative relationships for dissociation, hydrolysis, buffer solutions, indicator equilibria and later for acid catalysis. the positive features of the Arrhenius-Ostwald-van't Hoff collaboration are summarised below.

	Ionic Dissociation in Water	Key Dissociation Product
1. ACID HB;	$HB \rightleftharpoons H^+ + B^-$	H^+
2. BASE MOH;	$MOH \rightleftharpoons M^+ + OH^-$	OH^-
3. NEUTRALISATION	$H^+ + OH^- \rightleftharpoons H_2O$	
4. ACID STRENGTH	$K_a = [H^+][B^-]/HB$	

5. QUANTITATIVE RELATIONSHIPS FOR:- Dissociation, Hydrolysis, Buffer Solutions, Indicator Equilibria, Acid Catalysis.

There were weaknesses also. For example Arrhenius regarded acids as HCl, H_2SO_4 etc., not as H^+ and bases as NaOH, KOH etc. and not OH^- and there was no clear explanation as to why some H-X substances dissociated in water to give acids and others did not. This was difficult to understand since H_2O was presumed to be inert. It was some years later (1893) when Werner and co-workers[10] emphasised the role of solvation (complexation) in stabilising ions in solution. An additional weakness of the Arrhenius formulation concerned the limitation of acids and bases to aqueous solutions. Moreover, Ostwald[11] alluded to a "crisis of identity" with regard to acid-base concepts when he suggested that neutralisation, when H^+ and OH^- mutually destroy each other to form H_2O, was not, in principle, different from the destruction of say Ni^{2+} by S^{2-} to form NiS. Why then emphasise acid-base reactions? However, as in previous similar situations, the acid-base concepts, because of their utility, survived.

The limitation of Arrhenius's concepts to aqueous solutions was addressed by E.C. Franklin (1905) and others. Franklin's comparison[12] of acid-base reactions in aqueous solution and in liquid ammonia is summarised below.

EXTENSION OF ARRHENIUS IDEAS TO NON-AQUEOUS SOLUTION
E.C. FRANKLIN (1905)

	SOLVENT	
	WATER	LIQUID AMMONIA
AUTOIONISATION	$H^+ + OH^- \rightleftharpoons H_2O$	$NH_4^+ + NH_2^- \rightleftharpoons 2NH_3$
ACID	$HX \rightleftharpoons H^+ + X^-$	$NH_4X \rightleftharpoons NH_4^+ + X^-$
BASE	$MOX \rightleftharpoons M^+ + OH^-$	$MNH_2 \rightleftharpoons M^+ + NH_2^-$

NEUTRALISATION WITH:-

ACID	$M(OH)_2 + 2H^+ \rightleftharpoons M^{2+} + H_2O$	$M(NH_2)_2 + 2NH_4^+ \rightleftharpoons M^{2+} + 4NH_3$
BASE	$M(OH)_2 + 2OH^- \rightleftharpoons M(OH)_4^{2-}$	$M(NH_2)_2 + 2NH_2^- \rightleftharpoons M(NH_2)_4^{2-}$

Further relaxation of the Arrhenius views resulted in so-called "solvent system" definition of acidity. For example G.N. Lewis (1923)[13] suggested that "an acid is a substance that gives off a cation or combines with an anion of the solvent (and) a base.....gives off the anion or combines with the cation". For example, in the case of water and liquid ammonia:

SOLVENT	ACID	SOLVENT CATION
H_2O	$MCl_3 + H_2O \rightleftharpoons M(OH)^{2+} + 3X^- + H^+$	H^+
NH_3	$HBr + NH_3 \rightleftharpoons Br^- + NH_4^+$	NH_4^+

The solvent system proposed by Lewis, although limited to the liquid phase, could clearly be applied to non-protic solvents such as SO_2 for which autoionisation was $2SO_2 \rightleftharpoons SO^{2+} + SO_3^{2-}$. The terms acid and base, as used in solvent system definitions were still focused on substances generating ions rather than the ions themselves.

5. Protonic Definitions of Acidity

Following the development of the Arrhenius/Ostwald views, the notion that hydrogen ion concentration could be used to measure acidity was readily accepted, particularly in regard to protic solvents, such as water, liquid ammonia or alcohols. There was, though, some resistance to the restriction of base strength to hydroxyl ion concentration. In particular, organic chemists tended to regard, as a base, any substance which decreased the concentration of hydrogen ions. In 1920, Langmuir[14] proposed that "acids are substances from whose molecules hydrogen nuclei are readily detached, while bases are substances whose molecules can easily take up hydrogen. The more easily the hydrogen nuclei are given up the stronger the acid.......the greater the tendency to take up hydrogen nuclei the stronger the base".

The protonic views regarding acidity were, however, considerably clarified[15] by Brønsted (1923) and, independently by Lowry[16] in the same year. Johannes N. Brønsted[17] was born in Varde, Denmark, in 1879. He took his doctorate in 1908 and in the same year he was appointed professor of physical chemistry at the University of Copenhagen. His research was largely concerned, initially, with chemical affinity determined from EMF measurements. Later he made solubility studies, worked on the specific interaction of ions, proton concepts of acids and bases and on catalysis. He died in Copenhagen in December 1947. Thomas M. Lowry[17] was born in Bradford, England, in 1874. He was awarded a doctorate from the University of London in 1899 and was appointed to the newly established chair of physical chemistry at Cambridge in 1920. His work was mainly on the application of physical methods, particularly optical rotation, to organic chemistry. He died in Cambridge in November 1936.

Brønsted viewed an "acid as a species that acts as a proton donor. The stronger the donor the stronger the acid". Similarly a "base is a species that acts as a proton acceptor. The stronger the proton acceptor the stronger the base".

The Brønsted-Lowry view of acids and bases is summarised below.

(i) All acid-base reactions were identified as proton transfers. For example an acid HB may react with a base B',

$$HB + B' \rightleftharpoons HB' + B$$

Where HB and B, also HB' and B' were referred to as conjugate acid-base pairs. The equilibrium position in the proton transfer decided the strength of the acids and bases which hold the proton. Weak acids have a strong conjugate base and strong acids a weak one.

ACID	CONJUGATE BASE
WEAK (CH_3COOH)	STRONG (CH_3COO^-)
STRONG (HNO_3)	WEAK (NO_3^-)

(ii) Conjugate pairs identified the role of the solvent in the ionisation of acids:

$$HB \; + \; H_2O \; \rightleftharpoons \; H_3O^+ \; + \; B^-$$

acid   base   acid   base

CONJUGATE

Water acted as a common base towards (Arrhenius) acids. Acid HB would dissociate in water only if B^- was a weaker base than H_2O.

(iii) The definitions were solvent independent and could be applied to proton transfer in gas or solid phases or between phases. Similarly ions as well as pure substances could be included within the definitions:-

ACID	BASE

$$HCO_3^- + OH^- \rightleftharpoons H_2O + CO_3^{2-} \qquad RCOO^- + H_2O \rightleftharpoons RCOOH + OH^-$$

$$NH_4^+ + H_2O \rightleftharpoons NH_3 + H_3O^+ \qquad [Al(H_2O)_5OH]^{2+} + H_3O^+ \rightleftharpoons [Al(H_2O)_6]^{3+} + H_2O$$

(iv) Competitive protonation equilibria using a single reference base (B_{REF}) provided a basis for quantification of acid strength.

$$HB' \; + \; B_{REF} \; \rightleftharpoons \; HB_{REF} \; + \; B' \qquad K'$$

$$HB'' \; + \; B_{REF} \; \rightleftharpoons \; HB_{REF} \; + \; B'' \qquad K''$$

Equilibrium constants K' and K'' provided a measure of the acid strengths for HB' and HB''. In aqueous solution the reference base was water and K', K'' etc. were designated as K_a values (now in common use). Typical examples are given below

ACID	CONJUGATE BASE	K_a
HSO_4^-	SO_4^{2-}	1.2×10^{-2}
H_3PO_4	$H_2PO_4^-$	7.5×10^{-3}
H_2CO_3	HCO_3^-	9.2×10^{-7}
H_2O	OH^-	1.0×10^{-14}

The strength of the acids decreases and the strength of the conjugate bases increases in the order HSO_4^- to H_2O.

(v) In the Brønsted definitions the focus was clearly on proton transfer. This obscured the association of acid-base interactions with neutralisation to form salts and Brønsted acids did not react directly with bases but acted as a source of protons to react with the base. The basis for quantification of acid strength, following from the Brønsted definitions was applied to non-aqueous solutions by Hammett[18] who defined an acidity function (H_0) as follows.

The protonation of a weak base, B, in acid solution

$$B \; + \; H^+ \; \rightleftharpoons \; HB^+$$

20

could be considered in terms of dissociation of the conjugate acid

$$HB^+ \rightleftharpoons H^+ + B$$

such that

$$K_a = a_{H^+} C_B / C_{BH^+} (\gamma_B / \gamma_{BH^+})$$

where a, C and γ referred to activity, concentration and activity coefficient.

Hammett defined a parameter (h_0) as a measure of the proton donating ability of an acid solvent to a base B

$$h_0 = K_a C_{BH^+} / C_B = a_{H^+} \gamma_B / \gamma_{BH^+}$$

Consequently, $\log C_{BH^+} / C_B = pK_a - H_0 \quad (H_0 = -\log h_0)$

which defined the acidity function H_0.

By using a series of structurally similar bases, having different pK_a values, for which the ratio C_{BH^+} / C_B could be measured (e.g. spectroscopically), the value of H_0 for a particular solvent system could be determined and this value, in turn, could be used to find pK_a for other bases for which C_{BH^+} / C_B could be measured.

In practice, when setting up experiments to determine H_0 for acid solutions, it is necessary to use several types of base/indicator to show the independence of H_0 on type of base. Values of H_0 for typical mineral acids are shown in Figure 1.

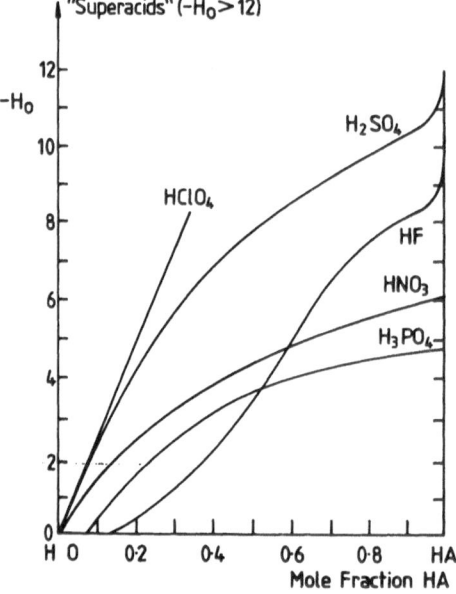

Figure 1 - The H_0 (Hammett) acidity function for acid/water mixtures (redrawn from G.A. Olah, 1973)

6. Ionotropic Definitions

Lux[19a] and Flood,[19b] in 1939, suggested that in oxide melts acids could be conveniently defined as oxide acceptors, and bases defined as oxide donors. Consequently, the reaction

$$BaO + CO_2 \rightleftharpoons Ba^{2+} + CO_3^-$$
$$\text{BASE} \quad \text{ACID} \quad \text{ACID} \quad \text{BASE}$$

involved transfer of an oxide ion, analogous to the proton transfer in Brønsted acid-base reactions. The acids Ba^{2+} and CO_2, which have their conjugate bases BaO and CO_3^{2-}, compete for the (characteristic) oxide anion. Subsequently, Gutmann and Lindqvist (1954)[20] expanded on this and similar views to include the transfer of any characteristic ion such that

base + characteristic cation = acid
acid + characteristic anion = base

The first of these reactions includes the Brønsted definition (cation = H^+) and the second includes the Lux-Flood definition (anion = O^{2-}). By considering both characteristic cations and anions a very general definition, albeit limited to an ionic definition, for both acids and bases can be established.

Usanovich (1939) broadened the definition.[21] An acid was any substance capable of (a) reacting with a base to form a salt, (b) giving up cations or (c) combining with anions or electrons. Similarly, a base could react with acids to form salts, give up anions or electrons, or combine with cations. This very general definition included odd electron transfers so that redox reactions were included in acid-base interactions. Because of this extreme generality, and also perhaps of the focus on ionic phenomena this most general of definitions has not been widely accepted thus far.

7. Electronic Definitions

During the nineteenth century understanding of valence and chemical bonding developed and this has been strongly reflected in views of acidity and basicity.

7.1. LEWIS ACIDS AND BASES

The first of the electronic definitions[13] was due to Gilbert Newton Lewis (1923). Lewis[22] was born in Weymouth, Massachusetts in October 1875. He started formal school in Nebraska at the age of thirteen and completed his first degree, in chemistry, at Harvard. Subsequently he obtained a masters degree and PhD from Harvard and took fellowships in Europe to work with Wilhelm Ostwald and Walther Nernst. He started work on his octet theory as early as 1902.

Lewis based his definitions of acids and bases on the concept of the shared electron pair bond which he introduced[24] in 1916, a concept which was extended and developed by Sidgwick in 1927.[24] Prior to the publication[13] of Lewis's views on acidity (1923), Langmuir, utilising the Lewis concept of shared electron pairs had suggested that, since protons possess no shared electrons, bases acted as electron pair donors to protic acids. However, it was the monograph by Lewis which clarified the electron pair definitions. In this publication, Lewis stated that "a basic substance is one which has a lone pair of electrons which may be used to complete the stable group of another atom..... An acid substance is one which can employ a lone pair from another molecule to complete a stable group of one of its own atoms". Simply, bases donate lone pairs and acids accept lone pairs. This view, expressed in 1923, was not developed by Lewis for fifteen years, and although Sidgwick (1927) pointed out that the Werner coordination compounds fell within the Lewis acid-base concepts and Lapworth, Ingold, Robinson and others introduced the concepts electrophiles and nucleophiles in relation to electron-accepting and electron-donating species, there was little discussion of Lewis acids prior to 1938 when Lewis developed his earlier suggestions and provided supporting experimental data.[25] Lewis proposed the following criteria for recognising acidity and basicity:

(i) Neutralisation is rapid

(ii) Acids or bases displace weaker acids or bases from their compounds

(iii) Acids or bases may be titrated against each other by means of indicator species

(iv) Both acids and bases function as catalysts

Lewis demonstrated that a range of acids including

$$H^+, Ag^+, SO_3, BCl_3, SnCl_4$$

and bases including

$$I^-, SCN^-, S^{2-}, OH^-, (CH_3)_3N, CH_3COCH_3$$

fulfilled the above criteria. Lewis made use of non-aqueous titrations to demonstrate neutralisation and hence expand the notion of acidity. He pointed out that Brønsted bases were also Lewis bases but that acid behaviour was not confined to the proton. In fact the proton becomes one example of the more general Lewis acids.

The concepts of Lewis met with a mixed reception. For example R.P. Bell (1959)[26] concluded that the "Lewis definition of acids does not represent an extension or a generalisation of the older concepts but......the use of the word acid in a fundamentally different sense". Bell argues for a separation of the protic acid definitions ("classical acid-base concepts") from Lewis's views and points out that the "important aspects of classical acid-base concepts are the quantitative relationships to which they lead. No such general quantitative relationship can be envisaged for Lewis acids". He quotes, in support of his opinions Lewis's view that the "relative strengths of acids and bases depend not only on the chosen solvent but also on the particular acid-base used for reference". In contrast to this somewhat critical view, Luder and Zuffant (1948) compare the importance of the Lewis acid-base concepts in chemistry with that of relativity in physics.[27]

7.2. SUPERACIDITY. AN EXTENSION OF THE LEWIS CONCEPTS

More recently the concept of acids has been extended to acids which can accept bonding electron pairs (π or α) as well as non-bonding pairs. The term superacid, first coined by Conant and Hall in 1927 is frequently used to describe such acids.[28] The term superacid was applied by Gillespie (1927) to Brønsted acids stronger than 100% sulphuric acid,[29] that is acids with values of the Hammett function $-H_0 > 12$. Extensive studies of superacid solutions are reported by G. Olah and coworkers.[30] Olah pointed out that the ionisation of Brønsted acids could be increased by the addition of stronger Brønsted acids or of Lewis acids.

(1) Brønsted-Brønsted

$$HA + HB \rightleftharpoons H_2A^+ + B^-$$

(a) $H_2S_2O_7 + H_2SO_4 \rightleftharpoons H_3SO_4^+ + HS_2O_7^-$ (oleum)

(2) Brønsted-Lewis

$$2HA + L \rightleftharpoons H_2A^+ + LA^-$$

(b) $2HSO_3F + SbF_5 \rightleftharpoons H_2SO_3F^+ + SbF_5SO_3F^-$ (magic acid)

Species in solutions (a) and (b) depend upon composition but very strong acids can be produced in this way, as shown by values of the Hammett function in Figure 2. Lewis superacids have been defined as acids stronger than $AlCl_3$. In hydrochloric acid the decreasing order of acid strength is reported to be

$$SbF_5 > AsF_5 > TaF_5 > BF_3 > NbF_5$$

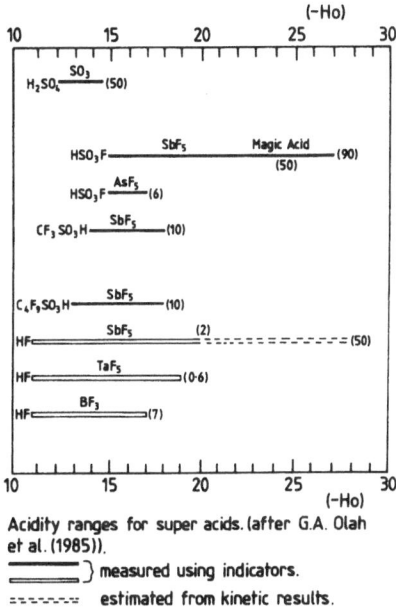

Acidity ranges for super acids. (after G.A. Olah et al. (1985)).

‒‒‒‒‒ } measured using indicators.

‒‒‒‒‒ estimated from kinetic results.

Figure 2 - Acidity ranges for superacids (after G.A. Olah et al., 1985)[30]

8. Extension of Electronic Concepts

8.1. EMPIRICAL APPROACH

The point made by R.P. Bell concerning difficulty in quantifying Lewis acid/base strength had some validity since the relative strength of Lewis acids depended upon the base used for comparison. For cation-ligand reactions in solution the classification of ions and ligands into class (a) and (b) acceptors and class (a) and (b) donors by S. Arhland, J. Chatt and N.R. Davies (1958) considerably clarified the picture.[31] The classification of ligands or metal ions as belonging to type (a) or (b) was based on their observed bonding preferences. Class (a) metal ions included the alkali and alkaline earth metals and also lighter transition metals in higher oxidation states. The tendency for ligands to complex with either class (a) or class (b) metals to form stable complexes could then be represented as follows:

Tendency to form complexes with class (a) metals	Tendency to form complexes with class (b) metals
N > > P > As > Sb	N < < P > As > Sb
O > > S > Se > Te	O < < S < Se ≈ Te
F > Cl > Br > I	F < Cl < Br < I

Thus, class (a) metals (acceptors) tend to form strong complexes with first-row ligand (donor) atoms (e.g. Al^{3+} and F^-) but not with second or higher row atoms. Conversly class (b) metals form weak complexes with first row donor atoms but strong complexes with second row donors (e.g. Pd^{2+} and R_3P). Pearson, below, suggested the terms hard and soft to describe members of class (a) and (b).

Approaches to the quantification of Lewis acid-base interactions was made using empirical correlations. R.S. Drago and B. Wayland proposed equations to predict the enthalpy of interaction of acids and bases in poorly coordinating solvents or in the gas phase.[32] The equation involved two parameters (E and C) which characterised the acid (A) and the base (B). This resulted in the four parameter equation:

$$\Delta H_{AB} = E_A E_B + C_A C_B \qquad (1)$$

where the E and C parameters measured the ability of the acid and base to participate in electrostatic (E) and covalent (C) bonding respectively. By fixing values of E and C parameters for reference acids and bases a set of relative E and C parameters for other acids and bases could be established and used predictively. The form of equation (1), which was justified by reference to R.S. Mulliken's quantum mechanical approach to adducts, was found to be inadequate for strongly interactive systems and, subsequently, an expression for prediction of ΔH_{AB} in these systems was proposed by A.P. Marks and R.S. Drago.[33]

Equation (1) above may be considered as a modified form of a linear free energy relationship. Such relationships, the first of which was suggested by Brønsted, have been widely used by organic chemists. In one such study, R.G. Pearson collaborated with J.O. Edwards on the use of the Edwards equation, a linear free energy relationship, to correlate rates of nucleophilic displacement reactions.[34] Pearson followed this study by considering[35] the general acid-base reaction

$$A + B: \rightleftharpoons A:B \qquad (2)$$

Accepting that strong acids and bases form stable complexes, Pearson argued that, if Lewis acids and bases shared a unique order for strength then the equilibrium constant for equation (2) would be related to strength factors for the acid (S_A) and also base (S_B). For example,

$$\log K = S_A S_B$$

where S_A and S_B would be modified from the intrinsic (gas phase) strength by appropriate solvent corrections. In the absence of a unique order for Lewis strength, Pearson wrote,

$$\log K = S_A S_B + \sigma_A \sigma_B$$

where σ_A and σ_B measured characteristics of the acid and base which differed from their strength parameters. These he called softness parameters and classified a range of acids and bases as either hard, soft or borderline. Some examples are given below.

HARD AND SOFT ACIDS AND BASES (PEARSON, 1963)

	HARD	SOFT	BORDERLINE
BASES	H_2O, OH^-, F^-, ROH, R_2O, NH_3	R_2S, RSH, I^-, SCN^-, R_3P, H^-	C_6H_5N, NO_3^-, N_2
ACIDS	H^+, Li^+, Ca^{2+} Al^{3+}, Sc^{3+}	Cu^+, Hg^+, CH_3Hg^+ Pd^{2+}, I_2, Br_2	Fe^{3+}, Sn^{2+} $B(CH_3)_3$, SO_2

In keeping with other suggestions regarding acid-base concepts the HSAB principle was not always well received, and Pearson writes[36] that by 1969 "I was heartily sick of the subject and resolved to give no more lectures (on HSAB)...also.....never to write on the subject again". As with the classification into types (a) and (b) by Ahrland, Chatt and Davies, Pearson observed that like preferred like. He proposed the HSAB principle which stated that hard acids prefer hard bases and soft acids prefer soft bases. For example the displacement of a soft cation by a hard cation depends on the nature of the anion:

REPLACEMENT OF SOFT CATION BY HARD CATION

	HARD	SOFT
ACIDS	H^+	CH_3Hg^+
BASES	OH^-	S^{2-}

			K
(a)	H^+ + CH_3HgOH	$\rightleftharpoons$ H_2O + CH_3Hg^+	$10^{6.3}$
(b)	H^+ + CH_3HgS^-	$\rightleftharpoons$ HS^- + CH_3Hg^+	$10^{-8.4}$

When the soft cation is bound to a hard anion it is readily displaced by a hard cation (a) but not when it is bound to a soft anion.

The characteristics of hardness and softness were outlined. Typically the properties of soft donor atoms included high polarisability and low electronegativity. They were easily oxidised and had empty low-lying orbitals. Hard donor atoms had low polarisability, high electronegativity, were hard to reduce and had empty orbitals of high energy.

8.2. QUANTUM MECHANICAL APPROACH

Two quantum mechanical approaches have been used to provide a theoretical basis for HSAB. The first approach by G. Klopman and R.F. Hudson[37] used perturbation molecular orbital theory. In the simplest case where the acid-base interaction

$$A + B: \longrightarrow A:B$$

is dominated by the frontier orbitals of the donor and acceptor atoms, that is by the HOMO of atom "r" on the base and the LUMO of atom "s" on the acid the interaction energy consists of three terms two of which contribute to the attractive energy (ΔE_{ATT}) and the third represents a first order repulsive term. For the simplest case above

$$\Delta E_{ATT} = \frac{Q_r Q_s}{\epsilon R_{rs}} + \frac{2(C_r^{HOMO} C_s^{LUMO} \beta_{rs})^2}{E_{(HOMO)BASE} - E_{(LUMO)ACID}}$$

The term $Q_r Q_s / \epsilon R_{rs}$ is a first order Coulombic term determined by charges (Q_r, Q_s) and the distance of approach (R_{rs}) of the interacting atoms (ϵ is the appropriate dielectric). The second term is the second order perturbation representing an orbital term, with coefficients "C" and resonance integral "β", corresponding to the HOMO-LUMO interaction. This second term (orbital term) is small when $E_{HOMO} - E_{LUMO}$ is large so that chemical attack is dominated by the charge term and is "charge controlled". Conversely when $E_{HOMO} - E_{LUMO}$ is small the orbital term is large and chemical interaction is dominated by "orbital control". Although this approach was developed in connection with the chemical reactivity it could be applied to the stability of compounds and was used in this context in regard to HSAB.

Whereas this approach did not provide a satisfactory definition of hardness and softness it did focus attention on the influence of the HOMO-LUMO gap on the extent of covalence which came to be identified, quantitatively, with soft-soft interactions, as indicated in Figure 3.

SOFT-SOFT INTERACTION   ### HARD-HARD INTERACTION

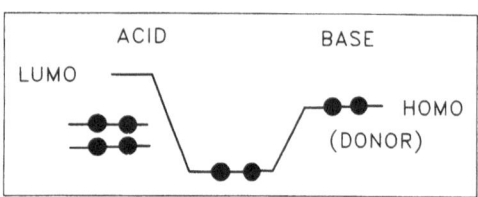 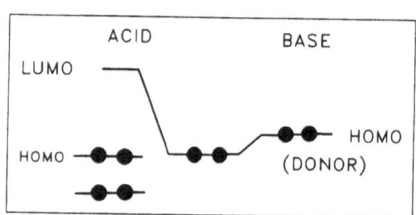

Strong Covalency by Donation   Weak Covalency by Donation

HOMO (BASE) → LUMO (ACID)   HOMO (BASE) → LUMO (ACID)

Figure 3 - Influence of HOMO-LUMO gap in acid-base interactions

A second quantum mechanical approach to HSAB utilised density functional theory.[38] Starting with the representation of the electronic energy of an atom or molecule (M) as a function of the number of electrons (N) in the species (Figure 4).

Density functional theory defined the electronic chemical potential (μ) as

$$\frac{\delta E}{\delta N}_V = \mu$$

where V is a (constant) potential. Also it was shown that the electronegativity (χ) could be defined by μ ($\chi = -\mu$) and was constant throughout the molecule, which justified the electronegativity equalisation principle of Sanderson.[39] Moreover, density functional theory identified the absolute hardness (η) of a system as

$$\eta = \frac{\delta^2 E}{\delta N^2} = \frac{\delta\mu}{\delta N}$$

and the global softness as $s = 1/2\eta$.

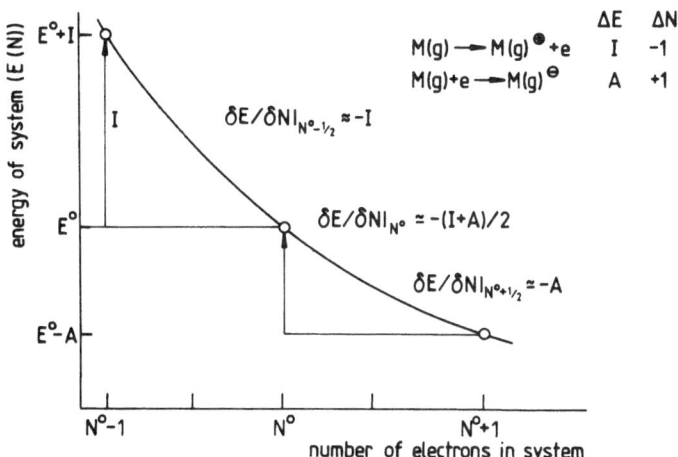

Figure 4 - Dependence of energy of a system on number of electrons

Finite difference approximations gave

$$\left.\frac{\delta E}{\delta N}\right|_{N0} \simeq -\frac{(I + A)}{2} = \mu = -\chi$$

and

$$\left.\frac{\delta^2 E}{\delta N^2}\right|_{N0} \simeq -A + I = 2\eta$$

and the transfer of charge δN to M was written

$$\left.\frac{\delta E}{\delta N}\right|_{N0} \simeq \left.\frac{\delta E}{\delta N}\right|_{N0} + \left.\frac{\delta^2 E}{\delta N^2}\right|_{N0}\delta N = \mu + 2\eta\delta N$$

Consequently, when isolated species A and B, with electronic chemical potentials μ_A^0 and μ_B^0, react by transfer of charge δN from B to A to give the adduct AB with electronic chemical potential $\mu = \mu_A = \mu_B$ then[40]

$$\mu_A = \mu_A^0 + 2\eta_A\delta N, \qquad \mu_B = \mu_B^0 - 2\eta\delta N$$

and

$$\delta N = \frac{\mu_B^0 - \mu_A^0}{2(\eta_A + \eta_B)} = \frac{\chi_A^0 - \chi_B^0}{2(\eta_A + \eta_B)}$$

This suggested that charge transfer was facilitated by differences in atom electronegativities and was resisted by the sum of hardnesses. Using this simplified approach Pearson related the density functional theory approach to MO theory.[41] Accepting that ionisation energy and electron affinity could be related to the energies of the HOMO and LUMO respectively (Kloopman) the relationship of electronegativity and hardness was represented by Pearson as in Figure 5 which refers to a molecule with ionisation energy and electron affinity of +10 and -2eV respectively.

This very simple approach related hardness (η) and softness ($s = 1/2\eta$), concepts which initially were very loosely defined in HSAB, to the HOMO-LUMO gap via a simple MO energy diagram.

Further impact of DFT on acidity and basicity, including solid acids, was provided recently by the development of the electronegativity equalisation method (EEM) of W.J. Mortier. This approach (discussed in detail in the present proceedings) used a spherical atom approximation to represent the charge-dependent electronegativity (χ_α) of an atom in a molecule or crystal as

$$\chi_\alpha = \chi_\alpha^0 + \Delta\chi_\alpha + 2(\eta_\alpha + \Delta\eta_\alpha)q_\alpha + \Sigma_{\beta \neq \alpha} q_\beta / R_{\alpha\beta}$$

This expression provided corrections to electronegativity ($\Delta\chi_\alpha$) and hardness ($\Delta\eta$), and included the influence of the external potential due to surrounding charges (q_β), which arose as a result of confining the atom within a molecule.[42]

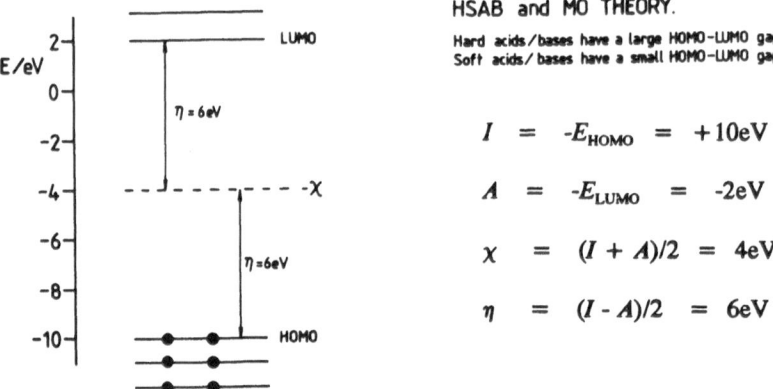

HSAB and MO THEORY.

Hard acids/bases have a large HOMO-LUMO gap.
Soft acids/bases have a small HOMO-LUMO gap.

$$I = -E_{HOMO} = +10eV$$

$$A = -E_{LUMO} = -2eV$$

$$\chi = (I + A)/2 = 4eV$$

$$\eta = (I - A)/2 = 6eV$$

Figure 5 - HOMO-LUMO gap and hardness

Several solid-state properties were estimated using the EEM formulation including acid-base properties.

Finally quantum mechanics has provided a means for accurate calculation of intrinsic acidities as defined by the standard enthalpy change for the reaction:

$$HA_{(g)} \rightarrow H^+_{(g)} + A^-_{(g)}$$

For example, molecular orbital calculations[43] at the G2 level of theory gave very good agreement with experimental results for gas phase acidities as shown below.

GAS-PHASE ACIDITIES
ΔH^0_{298}

ACID (HA)	CALCULATED	EXPERIMENT
CH_4	1751	1744 ± 7
H_2O	1632	1635
HF	1552	1554 ± 1
HCl	1400	1395 ± 1, 1396 ± 9
HBr	1358	1354, 1349 ± 9

Solvation effects result in considerable modification of acid dissociation and progress has been made in calculating such effects using combined quantum mechanical and molecular mechanics or Monte Carlo simulations. This aspect is further discussed by K.N. Houk in the present proceedings.

9. Summary

During the last two hundred years, concepts of acidity and basicity have evolved from a focus on chemical composition and "acidifying principles" to representation of acids and bases in terms of ionic equilibria in aqueous, and later non-aqueous, solutions and latterly to descriptions in terms of chemical bonding theory.

At all stages in this evolution, the definitions and concepts of acids and bases reflected the interests of and the experimental techniques and theoretical understanding available to the scientists of the period. During the latter part of the nineteenth century the growing understanding of chemical bonding and the advent of computational facilities located the concepts of acidity and basicity firmly within the electronic theory of bonding. No doubt this trend will continue and since computational methods can, in principle, be applied to any reactions it is likely that there will be yet another "crisis of identity" in the definitions of acids and bases. Nevertheless, it seems, at least to the present authors, that the concepts will be retained for a long time to come. Primarily this is because, as at all other periods in the evolution of science, there is still utility in the concepts of acids and bases both in teaching chemistry and in the comprehension of a wide range of related phenomena with the associated predictive quality which this gives to the scientist. More trivially, there is within the human psyche an historic desire to see paired phenomena, particularly reflecting opposite features or values, and acids and bases fall into this category. According to Doris Kolbe[44]

> The base has two electrons it can share
> The acid can accommodate a pair
> Their mutual attraction
> Leads to an interaction
> And both are neutralised by the affair.

Acknowledgement

The writers would like to single out the texts by J.R. Partington, "A History of Chemistry", Macmillan (1962) and by W.B. Jensen "The Lewis Acid-Base Concepts", Wiley (1980) which were of considerable help in preparing this lecture.

References

1. Part 1 of this paper.
2. Berthollet, C.L. *Mem. Acad. Sci. (Paris)*, 148 (1787).
3. Le Grand, H.E. *Ann. Sci.*, **31**, 213 (1974).
4. Gay-Lussac, H.E. *Ann. Chim. (Paris)*, **91**, 5 (1814); **95**, 136 (1915).
5. Liebig, J. *Ann. Chem.*, **26**, 113 (1838); Klickstein, H.S. (ed.) "A sourcebook in chemistry 1400-1900". Harvard Univ. Press (1952).
6. Berzeleus, J.J. "Essai sur la théorie des proportions chimiques et sur l'influence chimique de l'électricité". Paris (1819)
7. Kauffman, G.B. *J. Chem. Ed.* **65**(5), 437.
8. Arrhenius, S. *Z. Physik. Chem.* **1**, 631 (1887); "Foundations of the theory of dilute solutions", Alembic Club Reprint No. 19, Livingstone, Edinburgh, 43 (1991).
9. Wall, F.E. Paper to the Division of History of Chemistry 111th ACS, Atlantic City, NJ (1947).
10. Werner, A. *Z. Anorg. Chem.* **3**, 267 (1893).
11. Ostwald, W. "The principles of inorganic chemistry" 4th ed. London:Macmillan (1914).
12. Franklin, E.C. *J. Am. Chem. Soc.* **42**, 274 (1920).
13. Lewis, G.N. "Valence and the structure of atoms and molecules". New York:The Catalog Company (1923).
14. Langmuir, I. *J. Am. Chem. Soc.* **42**, 274 (1920).
15. Brønsted, J.N. *Rec.Trav. Chim. Pays-Bas*, **42**, 718 (1923).
16. Lowry, T. *Chem. Ind. (London)*,**42**, 43, 1048 (1923).
17. Orna, M.G. *J. Chem. Ed.*, **65**(1), 28 (1988).
18. Hammett, L.P. and Deyrup, A.J. *J. Am. Chem. Soc.* **59**, 2721 (1932).
19a. Lux, H. *Z. Electrochem.* **45**, 303 (1939).
19b. Flood, H. and Förland, T. *Acta Chem.Scand.* **1**, 592, 781 (1954).
20. Guttman, V. and Lindqvist, I. *Z. Phys. Chem.* **203**, 250 (1954).
21. Usanovich, M. *Zh. Obsch. Khim.* **9**, 182 (1939).
22. Tierman, N.F. *J. Chem. Ed.* **62**(7), 569 (1985).
23. Lewis, G.N. *J. Am. Chem. Soc.* **38**, 762 (1916).
24. Sidgwick, N.V. "The electronic theory of valence". Oxford:Clarendon Press (1927).
25. Lewis, G.N. *J. Franklin Inst.*, **226**, 293 (1938).
26. Bell, R.P. "The proton in chemistry", Methuen (1959).
27. Luder, W.F. and Zuffanti, S. "The electronic theory of acids and bases". New York: Wiley (1946).
28. Hall, N.F. and Conant, J.B. *J. Am. Chem. Soc.* **49**, 3047 (1927).
29. Gillespie, R.J. and Peel, T.E. *Adv. Phys. Org. Chem.* **9**, 1 (1972); *J. Am. Chem. Soc.* **95**,5173 (1973).
30. Olah, G.A., Surya Prakash, G.K. and Sommer, J. "Superacids". Wiley (1985).
31. Arhland, S., Chatt, J. and Davies, N.R. *Q. Rev. Chem. Soc.* **12**, 265 (1958).
32. Drago, R.S. and Wayland, B. *J. Am. Chem. Soc.* **87**, 3751 (1975).
33. Marks, A.P. and Drago, R.S. *J. Am. Chem. Soc.* **97**, 3324 (1975).

34. Edwards, J.O. and Pearson, R.G. *J. Am. Chem. Soc.* **84**, 16 (1962).
35. Pearson, R.G. *J. Am. Chem. Soc.* **85**, 3333 (1963*)*; *Chem. Br.* **3**, 103 (1967); *J. Chem. Ed.* **45**, 581, 643 (1968).
36. Pearson, R.G. "Hard and soft acids and bases". Strondsburg, PA: Hutchinson and Ross (1973).
37. Klopman, G. and Hudson, R.F. *Tetrahedron Lett.* **12**, 1103 (1967); *Theor. Chim. Acta* **8**, 165 (1967); Klopman, G. *J.Am. Chem. Soc.* **90**, 223 (1968).
38. Parr, R.G., Donelly, R.A., Levy, M. and Palke, W.E. *J. Chem. Phys.*, **68**, 3801 (1978).
39. Sanderson, R.G. *Science*, **114**, 670 (1951); "Polar covalence". New York:Academic (1983).
40. Pearson, R.G. *J. Am. Chem. Soc.* **107**, 6801 (1985).
41. Pearson, R.G. *Proc. Nat. Acad. Sci. USA* **83**, 8440 (1986).
42. Mortier, W.J., Ghosh, S.K. and Shankar, S. *J. Am. Chem. Soc.* **108**, 4315 (1986); Paper in the present proceedings.
43. Smith, B.J. and Radom, L. *J. Phys. Chem.* **95**, 1059 (1991).
44. Kolbe, D. *J. Chem. Ed.* **7**, 461 (1978).

THEORY OF ACIDS AND BASES
I. RELATIONSHIPS BETWEEN STRUCTURE AND ACIDITY AND BASICITY
II. THEORY OF MEDIUM EFFECTS ON ACIDITY AND BASICITY

K. N. HOUK
Department of Chemistry and Biochemistry
University of California, Los Angeles
Los Angeles, CA 90024-1569
USA

I. Relationships Between Structure and Acidity and Basicity

These lectures provide the theoretical foundation for the NATO Workshop study of acid catalysis in the solid state. We begin with the basic definitions and theoretical concepts which allow us to understand acids and bases, from both an equilibrium and kinetic point of view. The goal is to provide a conceptual framework, the tools to understand the relevant theoretical literature, and the background to predict experimental conditions to effect a desired result. We then study medium effects - solution and solid-state - which influence acidities and basicities. Third, we discuss the rates of proton transfer and kinetics of acid-catalyzed reactions.

Brønsted acids are proton donors. The strength of Brønsted acids is generally referenced to pH in dilute aqueous solution.

$$pH = -\log [H+] = -\log \frac{K_A[HA]}{[A-]}$$

J. Fraissard and L. Petrakis (eds.), Acidity and Basicity of Solids, 33–51.
© 1994 *Kluwer Academic Publishers.*

The strength of a Brønsted acid in solution is dependent upon bond strengths and the stabilization of the acid and conjugate base, both inherent and stabilized by the medium.

$$\Delta G(g)$$

$$HA\ (gas) \rightleftharpoons H^+(gas) + A^-(gas)$$

$$\Big\downarrow \Delta G(S) \qquad \Big\downarrow \Delta G(S) \qquad \Big\downarrow \Delta G(S)$$

$$\Delta G°(s)$$
$$HA\ (S) \rightleftharpoons H^+(S) + A^-(S)$$

These medium effects are enormous, and we must keep them in mind whenever discussing acidity and acid catalysis. Much of my lecture concentrates on this.

Lewis acids are electron-pair acceptors, while Lewis bases are electron-pair donors. The strengths of Lewis acids are also profound functions of medium. Take for example, the following association constants:

$$BF_3 + H_2CO \rightleftharpoons H_2CO\cdot BF_3 \qquad \Delta H = -18.1\ kcal/mol$$

$$THF + BF_3 \rightleftharpoons THF\cdot BF_3 \qquad \Delta H = -21.5\ kcal/mol$$

$$THF\cdot BH_3 + H_2CO \rightleftharpoons THF + H_2CO\cdot BF_3 \qquad \Delta H = +3.4\ kcal/mol$$

The enormously exothermic coordination of BF_3 to formaldehyde becomes endothermic in THF.

We will begin with a discussion of gas-phase acidities and basicities. We then turn to theory, and finally to medium effects.

Gas Phase Acidity

Ion cyclotron resonance measurements of equilibria in the gas phase have permitted measurements of inherent acidities.[1-3]

Gas Phase Acidities (Data from Ref. 2)

Acid	ΔH (kcal/mol)
MeOH	373
EtOH	371
^{i}PrOH	367
^{t}BuOH	366
PhOH	344
MeCO$_2$H	344
CF$_3$CO$_2$H	318
CH$_4$	409
H$_2$S	347
HF	365
HCl	327
HBr	315
HI	306
MeCN	361
MeCHO	360
NH$_4^+$	196
MeCO$_2$H$_2^+$	190
H$_3$O$^+$	162

A great deal of data is now available for gas phase acidities,[1,2] and for comparisons with solutions,[3] especially DMSO from the work of Bordwell.[4]

Early on in the study of gas phase acidities, it was noted that medium effects have a profound effect on acidity orders. Good examples can be found in the alcohols. In the gas phase, acidity increases with the size of the alkyl group. It is 7 kcal/mol easier to deprotonate tert-butanol then methanol. The alkyl groups are polarizable and stabilize anions. The opposite order is found in solution - tert-butanol is much less acidic, that is tert-butoxide is more basic - because the large alkyl group hinders solvation of the alkoxide.

The Hammett Acidity Scale[5]

Although pH is a very useful concept in dilute aqueous solution, it becomes less so in concentrated solution, where the activity coefficient may deviate significantly from the value of 1. In 1932, Hammett and Deyrup devised an acidity scale based upon a series of aniline indicators.

$$H_o = pK_{BH+} - \log \frac{[BH^+]}{[B]}$$

If basicity is known and the ratio of the protonated and unprotonated species can be measured, then H_o can be determined. This is a very useful way to quantitate the ability of a strong acid to protonate bases.

Superacids

The H_o of 100% H_2SO_4 is -11.9. The quantity for HF is -11.0. Gillespie classified all stronger acids, those with $H_o < -11.9$ as "superacids".[6] Some common superacids are $HClO_4$ (13.0), HSO_3F (-15.6), CF_3SO_3H (-14.6). Even stronger acids can be made by combinations of Brønsted and Lewis acids.[7] For example:

$$HCl + AlCl_3 \longrightarrow H^+ AlCl_4^-$$
$$HF + BF_3 \longrightarrow H^+ BF_4^-$$

have H_0 of -15 to -16. Magic acid, HSO_3F - SbF_5 has an H_0 of about -25. These acids have been used by George Olah to revolutionize our understanding of cationic intermediates in organic reactions.[7] They all involve the coordination of the basic lone pairs of a Brønsted acid.

Hard and Soft Acids and Bases

Thirty years ago, Pearson defined a new classification of Lewis acids and bases or electrophiles and nucleophiles.[8] He defined _hard_ and _soft_ acids and bases - qualitative concepts which were given theoretical description by Klopman.[9] A large selection of these is given below.

Bases (nucleophiles)	_Acids (electrophiles)_
Hard	_Hard_
H_2O, OH^-, F^-	H^+, Li^+, Na^+, K^+
$CH_2CO_2^-$, PO_4^{3-}, SO_4^{2-}	Be^{2+}, Mg^{2+}, Ca^{2+}
Cl^-, CO_3^{2-}, ClO_4^-, NO_3^-	Al^{3+}, Ga^{3+}
ROH, RO^-, R_2O	Cr^{3+}, Co^{3+}, Fe^{3+}
NH_3, RNH_2, N_2H_4	CH_3Sn^{3+}
	Si^{4+}, Ti^{4+}
	Ce^{3+}, Sn^{4+}
	$(CH_3)_2Sn^{2+}$
	$BeMe_2$, BF_3, $B(OR)_3$
	$Al(CH_3)_3$, $AlCl_3$, AlH_3
	RPO_2^+, $ROPO_2^+$
	RSO_2^+, $ROSO_2^+$, SO_3
	I^{7+}, I^{5+}, Cl^{7+}, Cr^{6+}
	RCO^+, CO_2, NC^+
	HX (hydrogen-bonding molecules)
Borderline	_Borderline_
$C_6H_5NH_2$, C_5H_5N, N_3^-, Br^-, NO_2^-,	Fe^{2+}, Co^{2+}, Ni^{2+}, Cu^{2+}, Zn^{2+}, Pb^{2+}, Sn^{2+},
SO_3^{2-}	$B(CH_3)_3$, SO_2, NO^+, R_3C^+, $C_6H_5^+$
Soft	_Soft_
R_2S, RSH, RS^-	Cu^+, Ag^+, Au^+, Tl^+, Hg^+
I^-, SCN^-, $S_2O_3^{2-}$	Pd^{2+}, Cd^{2+}, Pt^{2+}, Hg^{2+}, CH_3Hg^+,
R_3P, R_3As, $(RO)_3P$	$Co(CN)_5^{2-}$
CN^-, RNC, CO	Tl^{3+}, $Tl(CH_3)_3$, BH_3
C_2H_4, C_6H_6	RS^+, RSe^+, RTe^+
H^-, R^-	I^+, Br^+, HO^+, RO^+
	I_2, Br_2, ICN, etc.
	Trinitrobenzene, etc.
	Chloranil, quinones, etc.
	Tetracyanoethene, etc.
	O, Cl, Br, I, N, $RO^.$, $RO_2^.$
	M^0 (metal atoms)
	Bulk metals
	CH_2, carbenes

Theory of Hard and Soft Acids and Bases

Soon after Pearson established the concept of hard and soft acids and bases, Klopman provided a theoretical framework, which also provides a useful general theory by which to understand acids and bases more broadly.[9] The theory was developed with second-order perturbation theory which Salem, Fukui, and Hoffmann used extensively for other types of reactions.

Acid-base interactions can be divided into core, electrostatic, and orbital interactions. Core interactions refer to repulsion between electrons in filled orbitals. Electrostatic interactions refer to the electrostatic attraction between opposite charges, either the full charge of an ion, or a partial charge of a polar molecule.

$$\Delta E = - \sum (q_a + q_b)\beta_{ab}S_{ab} + \sum_{k<1} \frac{q_k q_1}{\varepsilon r_{kl}} + \sum^{occ} \sum^{unocc} - \sum^{occ} \sum^{unocc} \frac{2(\sum_{ab}c_{ra}c_{sb})^2}{E_r - E_s}$$

$$\text{core} \qquad\qquad \text{electrostatic} \qquad\qquad \text{overlap}$$

Orbital interactions are a consequence of the overlap of orbitals on two molecules. When two orbitals interact, the extent of mixing and energy charge depends on overlap.

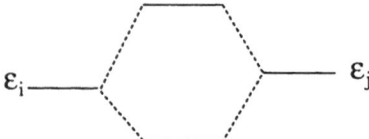

Whether or not the interaction is repulsive or attractive depends upon the electron occupation. Repulsive interactions occur when both of A and B are occupied:

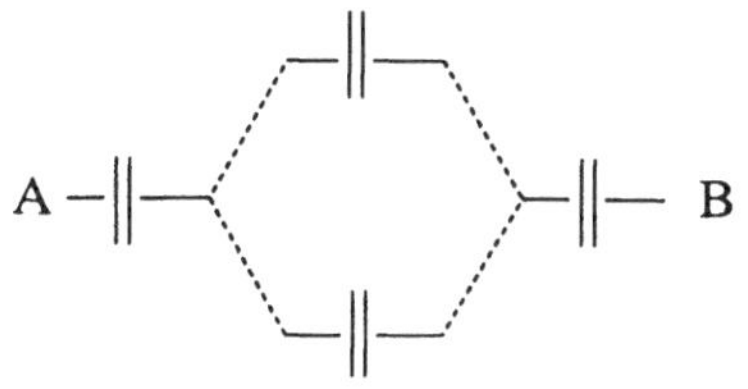

The orbital mixing results in destabilization, closed-shell repulsion.

Stabilizing interactions involve filled orbitals on one molecule and vacant on the other. This is shown schematically below:

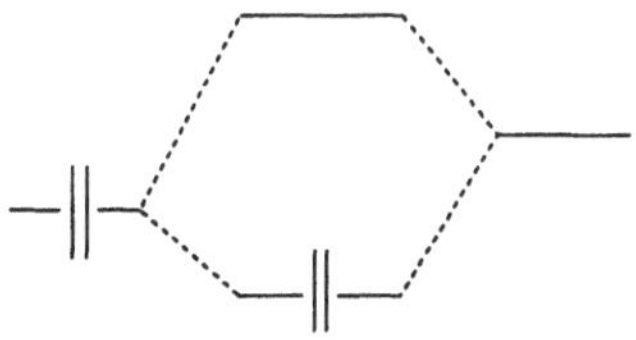

This two-electron interaction is a fundamental aspect of bonding, and it is worth spending some time discussing it in detail. The extent of interaction, $\Delta\varepsilon$, between any two orbitals is, according to second-order perturbation theory, proportional to

$$\Delta\varepsilon = - H_{ij}^2/(\varepsilon_i - \varepsilon_j)$$

where ε_i and ε_j are the orbital energies before interaction, $\Delta\varepsilon$ is the charge in orbital energy, and H_{ij} is the interaction integral,

$$H_{ij} = \int \phi_i H \phi_j d\tau$$

The closer in energy the orbitals, the stronger the interaction. H_{ij} is roughly proportional to the overlap between ϕ_i and ϕ_j.

Klopman showed that some acid-base interactions involve mainly electrostatic interactions. These are called "charge-controlled" reactions and have the same characteristics as Pearson's hard-acids and hard bases.

Charge-controlled interactions (hard-hard):

      high charge

      IP of base is high

      EA of acid is low

      small size

      well-solvated

      form ionic bonds

The other extreme involves cases where orbital interactions dominate. These are known as "frontier-controlled" interactions, or soft acid-soft base interactions.

Frontier-controlled interactions [soft-soft]:

      low charge

      IP of base is low

      EA of acid is high

      large size

      poorly-solvated

      form covalent bonds

Recently, Parr proposed a different theoretical definition of hardness based upon density functional theory. Density functional theory is an *ab initio* quantum mechanical method which involves direct computation of the charge density. From this, all properties, including the energy, of a molecule can be calculated. Here the "absolute hardness" of a fragment is calculated.[10]

$$\eta_s = \frac{1}{2}\left(\frac{\delta^2 E}{\delta N^2}\right)_z$$

In this formula, z is the total number of charges in the system. N is the number of electrons. This becomes approximately:

$$n_s = 1/2(I_s - A_s)$$

That is, the hardness is related to the difference between the ionization potential and the electron affinity. According to Koopmans' Theorem, $I_s = -\varepsilon_{HOMO}$ and $A_s = -\varepsilon_{LUMO}$, so that $\eta_s = 1/2(\varepsilon_{LUMO} - \varepsilon_{HOMO})$. The larger this gap, the harder the fragment. When small, the system has a low hardness, i.e., it is soft.

Theoretical Calculation of Acidity-An Example

Several years ago, we employed *ab initio* quantum mechanical calculations to study an interesting case of acidity in organic molecules.[11] The example reveals something about the subtleties of acidity and of the ability of modern quantum mechanical calculations to reproduce experimental acidities.

Meldrum's acid, a dilactone, is considerably more acidic than the corresponding diester, malonic ester.[11]

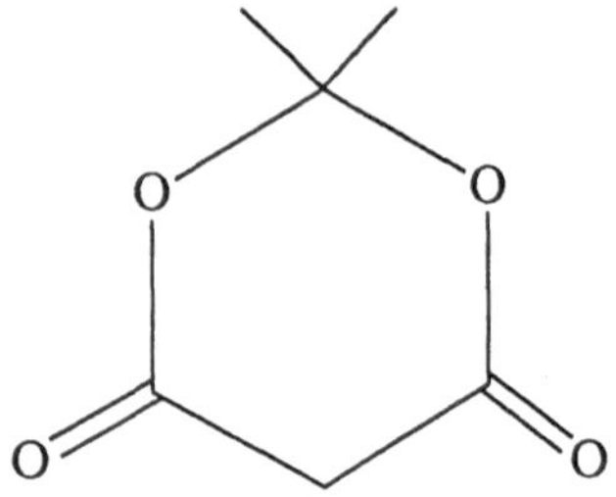

pK$_a$(DMSO) = 7.3

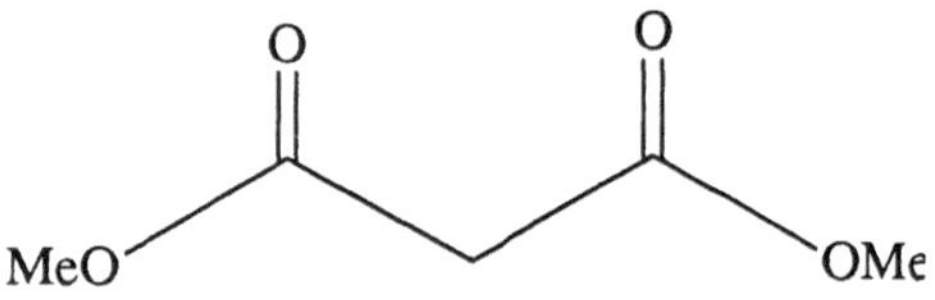

pK$_a$(DMSO) = 15.9

We, and at the same time, Wiberg and Laidig, studied the origin of this effect.[11] Calculations involved *ab initio* optimization of methyl acetate in *syn* and *anti* conformations, as well as the anions formed by removal of the protons shown.

Formula B

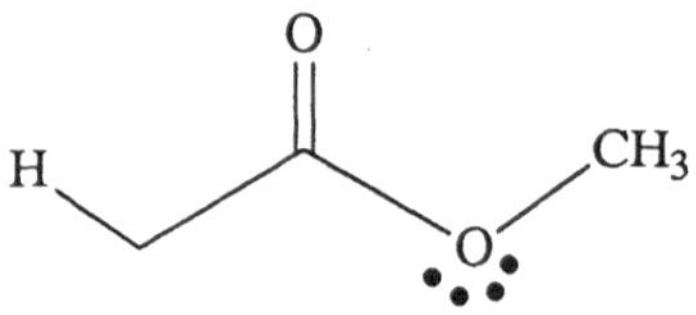

syn

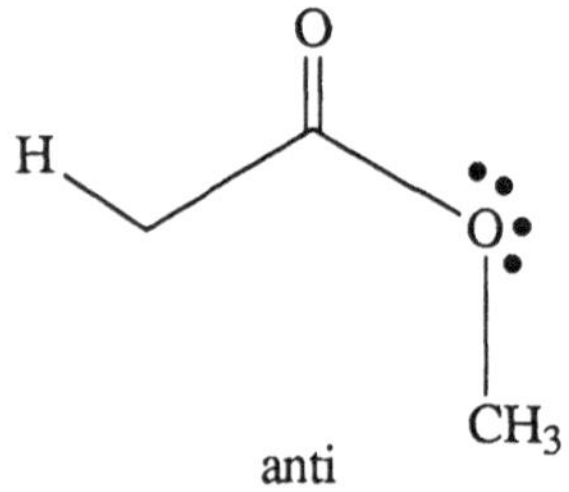

anti

Deprotonation of the normal *syn* methyl acetate is more difficult by about 5 kcal/mol than deprotonation of the higher energy *anti* conformer. In the former, electrostatic repulsion is increased much more than the latter. Constraining two esters to be *anti*, as in Meldrum's acid, decreases the protonation energy by about 10 kcal/mol.[11] This is a substantial electrostatic effect and signals the importance of electrostatic interactions on acidities.

II. Theory of Medium Effects on Acidity and Basicity

The Influence of Solvent on Acidity.

Methods have been developed for the theoretical study of solvation. In the case of carboxylic acids, we can directly compare measurements for the gas phase to acidities in the solution phase. Jorgensen has studied such systems using Monte Carlo techniques, and several examples will be discussed here.

Jorgensen and Briggs have developed a theoretical procedure to calculate pK_as of organic molecules in aqueous solution.[12] The method begins with *ab initio* calculations of deprotonation energies and entropies. The 6-31+G* basis set is used for geometry optimizations, and the energies are carried out at these geometries with MP3/6-31+G* calculations. For simple acids, CH_3CH_3, CH_3NH_2, CH_3OH, CH_3SH and CH_3CN, the gas phase acidities are within a few kcal/mol of those calculated at this level.[11]

Ab initio calculations are also used to calculate interaction of neutrals and anions with individual water molecules. An empirical OPLS force field is parameterized to closely reproduce these interaction energies. The OPLS force field gives the energy of interaction between molecules as:

$$\Delta E_{ab} = \sum_i^{\text{on a}} \sum_j^{\text{on b}} (q_i q_j e^2 / r_{ij} + A_{ij}/r_{ij}^{12} - C_{ij}/r_{ij}^6)$$

The solution of the neutral and the anion are then computed in water using Monte Carlo simulations and statistical perturbation theory. The differences in acidities can be calculated quite accurately as compared to experimental values.[12]

This procedure is quite time consuming, although it has been applied to calculate a variety of solution acidities. Alternate methods are under investigation, such as that reported by Gao.[13] This hybrid QM-MM approach permits the solute to polarize in the presence of the solvent, a factor which is quite important for polar solvents.

Solution simulations have been applied to the Meldrum's acid problem discussed earlier.[14] Whereas *syn* methyl acetate is 8.8 kcal/mol more stable than *anti* in the gas phase, this difference drops to 6.1 kcal/mol in acetonitrile and 5.8 kcal/mol in water.[14] Wiberg and Wong obtained a value of 5.2 kcal/mol for acetonitrile, using an Onsager cavity model.[15]

The anions of *syn* and *anti* methyl acetate differ in energy by only 2.8 kcal/mol.[11] This drops to 0.5 kcal/mol in water. Consequently, the 6 kcal/mol greater acidity of *anti* methyl acetate in the gas phase drops to 5.3 kcal/mol in water.[14]

Solid State Acidity

The calculation of properties of acids and bases in the solid state, or indeed any other properties, is a much larger and more complex problem than calculations on isolated molecules. The general approach is to compute small finite model systems at a high level, or to use relatively approximate methods, but include as much as possible of the system, using periodic boundary conditions. Many of the pioneers in this area are participants in this NATO Institute. I will give some examples of the calculation of solid state acidities as an introduction here.

There have been a variety of studies of the acidity of zeolites.[16-20] These provide a good introduction to the methods which can be applied to the study of acidity in the solid state. Early calculations by Mortier *et al.* on acidic sites in zeolites provide evidence of the dramatic effect which solid state interactions can have on functional group acidity.[16] *Ab initio* calculations with the 3-21G basis set were applied to study the energy of deprotonation of free silanol.

$$H_3SiOH \longrightarrow H_3SiO^- + H^+ \qquad \Delta E = 390 \text{ kcal/mol}$$

This is very similar to the value calculated for deprotonation of water at this level. In zeolites, acidic sites are the result of substitution of aluminum in the silicate framework.

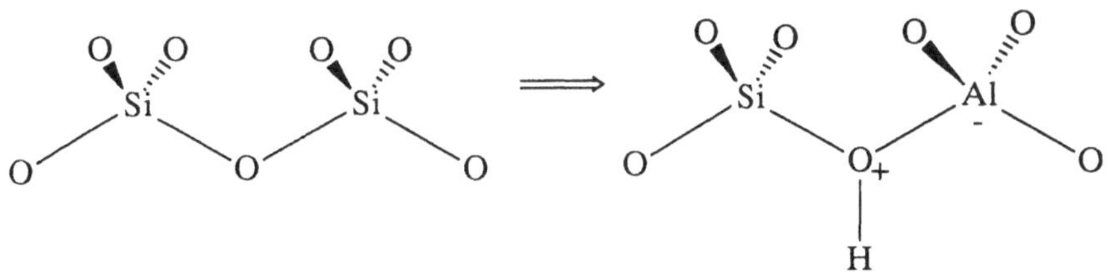

When a model for this is calculated, the following deprotonation energy is obtained:

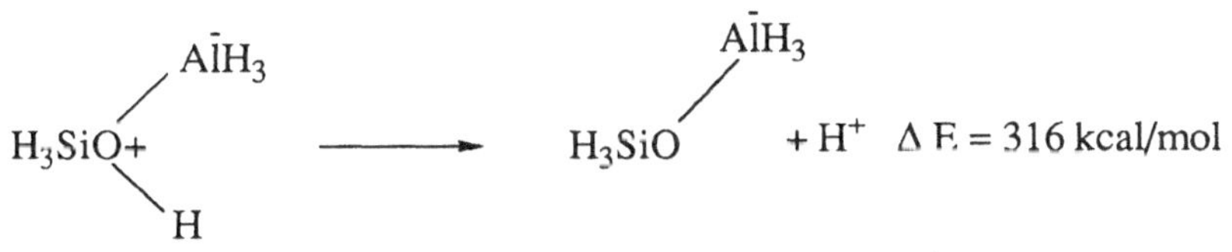

This represents an enormous decrease of 74 kcal/mol in the energy to deprotonate the oxygen. Clearly, the coordination of a silanol oxygen with the Lewis acidic Al causes a huge increase in acidity, related closely to the formation of superacids by complexation of a Brønsted acid-acting as a Lewis base - to a Lewis acid.

The 3-21G level is not very accurate for such deprotonation energies, but the large influence of coordination will certainly be found in the much more accurate calculation which will be carried out in the future.

Some higher-level *ab initio* calculations of this type have already been reported.[18] Curtiss and coworkers studied a number of clusters containing the unit shown above. They first tested methods including RHF/3-21G calculations through G2 calculations.[18] The G2 method devised by Pople includes correlation energy corrections and very large basis set calculations. Some comparisons with different methods for simple systems are given below.

<u>Proton affinities (kcal/mol)</u>

	3-21G	MP2/6-31G*	G2
HO$^-$	450	429	381
H$_3$SiO$^-$	391	368	363
H$_3$SiOAlH$_3^-$	324	313	310

With the G2 method, which is said to be accurate to +/- 2 kcal/mol, the model zeolite site is 53 kcal/mol more acidic than silanol, as compared to 67 kcal/mol by 3-21G.

A variety of larger clusters were also calculated, but only at the 3-21G level in most cases.

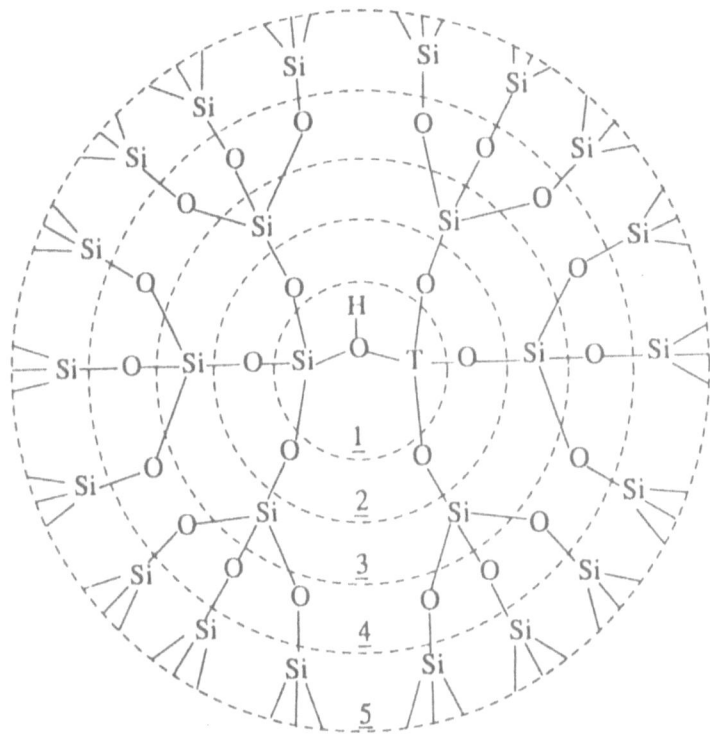

Compared to the PA of 324 kcal/mol for the simplest model, an increase in the size of the cluster causes the PA to vary from 300-350 kcal/mol, depending on size and structure. Electrostatic effects were involved to explain the oscillating results, with oxygens increasing acidity and silicons decreasing acidities. This effect has a simple analog in the Meldrum's acid results noted earlier.

A variety of other calculations of zeolite acidities have been performed with other methods, such as inclusion of the Madelung potential - that is electrostatic effects on the acidic site.[19] These were performed in an investigation of hydrogen-bonded and hydrogen-transferred forms of zeolites plus ammonia.

The hardness and softness of various zeolite cluster models have been investigated with semi-empirical calculations. These calculations predict that an increase in the Si/Al ratio increases the zeolite hardness, due to an increase in the LUMO energy.[20]

Proton Transfer Theory

Brønsted acid or base catalysis may occur by specific or general acid or base catalysis. Specific acid catalysis is a term which applies to reactions in which the proton is the only acid appearing in the rate expression:

$$rate = k \, [H^+][reactants]$$

This indicates a situation where a fast and reversible protonation by one or more acids present is followed by a slower, rate-determining reaction of the protonated substrate. Specific base catalysis, similarly, involves rate determining reaction of a conjugate base formed by fast reversible deprotonation of a reactant.

General acid catalysis applies to reactions which have rates dependent upon the concentrations of individual different Brønsted acids in solution. Kinetics of this type involve proton transfer simultaneous with reaction, or slow proton transfer followed by rapid reaction.

For relatively simple organic systems, proton transfer from hydroxylic acids or other acids where the acidic proton is attacked to a heteroatom are relatively fast. These "Eigen acids" have very fast proton transfer rates. For example, proton transfer from water to a second water is about 10^8 faster than proton transfer from carbon to oxygen.

Scheiner has reported a number of *ab initio* computational studies of proton transfer in a variety of systems.[21] Simple model studies, such as the transfer of a proton from water to hydroxide, or from ammonium to ammonia, have quite low barriers of 4.3 kcal/mol and 2.7 kcal/mol, respectively. The barrier increases rather dramatically if the distances between the heavy atoms are made larger than the ideal. For example, the water-hydroxide H-bond length is 2.64Å. When this length is increased to 2.95Å, the barrier increases to 18 kcal/mol. Similar results are obtained for a large number of single proton transfers. The origin of this effect is clear from the theory of double-well potentials. If dissociation of a X-H bond is represented by a Morse potenteal, and if two of these are superimposed, the net may give either a single-well potential or a double-well potential with a high barrier.[21] The closer the heavy atoms, the lower the barrier. Rapid proton tunneling through the barrier, or even a single-well potential may result. When the heavy atoms are for apart, the barrier is high.

Another interesting aspect of proton transfers which may influence proton transfer in the solid state is the role of angular constraints. For example, when the N-H-N angle is varied, the proton-transfer potential energy increases. A change of the angle for 180° to 140° increases the barrier from 12 kcal/mol to 33 kcal/mol for an NN distance of 2.95 Å.[21]

An interesting application of these ideas is in Scheiner's study of intramolecular proton transfer in protonated diamines.[22] With *ab initio* calculations and the 6-31G* basis set, the NHN proton transfer angle in the transition state and the activation energies for 1,3-, 1,4-, and 1,5- diamines are 111° and 34 kcal/mol, 125° and 11 kcal/mol, and 149° and 6 kcal/mol, respectively. These systems may model accessible geometries in some solid state systems, such as the pores of zeolites.

Summary. This brief review of acidity in gas, liquid, and the solid state, and various theoretical studies, provides a backdrop for the intensive study of solid state acidity from both experimental and theoretical viewpoints.

50

References

1. R. W. Taft, Progr. Phys. Org. Chem., *14* 247 (1983).

2. J. E. Bartmess and R. T. McIver, "Gas Phase Ion Chemistry," M. T. Bowers, Ed., Academic, New York, 1979; Vol. 2, Chapter 11.

3. (a) R. W. Taft and F. G. Bordwell, Accounts Chem. Research, *21*, 463 (1988). and references therein; (b) G. Klopman, J. Am. Chem. Soc., *90*, 223 (1968).

4. F. G. Bordwell, Accounts Chem. Research, *21*, 456 (1988) and references therein.

5. C. D. Ritchie, "Physical Organic Chemistry", Marcel Dkker, New York, 1979, pp. 165-169.

6. R. J. Gillespie, Accounts Chem. Research, *1*, 202 (1968).

7. G. A. Olah, G. K. S. Prakash, J. Sommer, Science, *206*, 13 (1979).

8. R. G. Pearson, J. Am. Chem. Soc., *85*, 3533 (`963).

9. G. Klopman, J. Am. Chem. Soc., *90*, 223 (1968).

10. Parr, R.G.; Pearson, J. Am. Chem. Soc., *105*, 7512 (1983).

11. X. Wang and K. N. Houk, J. Am. Chem. Soc., *110*, 1870 (1988). See also K. B. Wiberg and K. E. Laidig, *ibid*, *110*, 1872 (1988).

12. W. L. Jorgensen and J. M. Briggs, J. Am. Chem. Soc., 111, *4190* (1989).

13. J. Gao and J. J. Pavelites, J. Am. Chem. Soc., *114*, 1912 (1992).

14. J. D. Evanseck, K. N. Houk, J. M. Briggs, and W. L. Jorgensen, submitted for publication.

15. K. B. Wiberg and M. W. Wong, J. Am. Chem. Soc., *115*, 1078 (1993).

16. J. Sauer, "Modeling of Structure and Reactivity in Zeolites," 1992, C. R. A. Catlow, Ed., Academic Press, London, 1992, Chapter 8.

17. W. J. Mortier, J. Sauer, J. A. Lercher and H. Noller, J. Phys. Chem., *88*, 905 (1984).

18. H. V. Brand, L. A. Curtiss, and L. E. Iton, J. Phys. Chem. *96*, 7725 (1992).

19. M. Allavena, K. Seiti, E. Kassab, G. Ferenczy, and J. G. Ángyún, Chem. Phys. Lett., *168*, 461 (1990).

20. A. Corma, G. Sastre, R. Viruela, and C. Zicovich-Wilson, J. Catalysis, *136*, 521 (1992).

21. S. Scheiner, Accounts Chem. Research, *18*, 174 (1985).

22. X. Duan and S. Scheiner, J. Am. Chem. Soc., *114*, 5849 (1992).

QUANTUM MECHANICAL STUDY OF ACID BASE REACTIONS IN THE SOLID STATE.

M. ALLAVENA
C.N.R.S. - Dynamique des Interactions Moléculaires
Université Pierre et Marie Curie - Tour 22 -
Case Courrier 137
4, place Jussieu
75252 PARIS Cedex 05
France.

Introduction.

The main goal of this lecture is to show how Quantum Chemistry methods can be adapted to the description of chemical reactions occurring in solid media, with special attention to acido-basic reactions in zeolites.

Quantum Chemistry offers both a theoretical framework and accurate tools to interpret molecular structures and interaction processes, but its applicability is restricted to isolated systems containing a finite number of atoms, though, as we will show later some of these methods have been extended to systems with translational symmetry.

Nevertheless it has been a current practice to use Quantum Chemistry resources for investigating chemical reactions which take place in condensed phase. The basic assumption which allows surpassing the strict domain of validity of these methods was the commonly accepted belief that chemical reactions were essentially local and that the host medium was merely a passive "reservoir" which may eventually accept energy but does not influence significantly the reactants.

More and more accurate comparisons with experimental data have clearly shown that this assumption underestimate significant contributions originating in what is usually called "environmental effects" or more traditionally, in chemistry, "solvent effect". These are now well identified, and are, for example, responsable for frequency shifts and band shape deformations in optical spectroscopies or chemical shifts in N.M.R. It has also been shown that they may be used to monitor the rates of many chemical reactions in liquid or solid states.

Presently in order to account for these reactions in situ the quantum chemists are developing various strategies which can be classified in three main directions.

a) The local approach, which consists of assuming that the reaction considered concerns a limited number of atoms or molecules of the medium. This small number

J. Fraissard and L. Petrakis (eds.), Acidity and Basicity of Solids, 53–93.
© 1994 *Kluwer Academic Publishers.*

of particles (atoms or molecules) constitutes a subsystem which may be isolated from the rest of the bulk and can be treated as an isolated molecular complex which will be designated later as the "cluster model".*

b) Extensions of the previous model are numerous and may be broadly designated as "embedding models". The common process adopted here is to reinsert the isolated subsystem (cluster) into the original bulk and define some physical interactions which couple cluster and bulk. The coupling is treated by perturbative methods.

c) The development of quantum mechanical methods able to treat systems containing a very high number of particles is in progress. Preliminary results have already been obtained on special systems like periodic crystals, but these methods apply only to perfect samples, and cannot take account of systems containing interacting species.**

Some aspects of these problems will be discussed in the course which is divided in four chapters.

In the first chapter, a brief review of basic concepts and standard tools of Quantum Chemistry will be given. Next, in chap. 2, it will be shown that these tools are not appropriate for the treatment of infinite media and the system of interest has to be represented by some size reduced model. Rules for the construction of the cluster model and related problems will be discussed.

The third chapter will be devoted to the application of these principles to the analysis of acido-basic reactions in solid media, especially zeolites. Examples from work in our laboratory or taken from recent literature will serve to illustrate the performance of Quantum Chemistry in this domain.

Finally, in chapter 4, the deficiencies of the cluster model will be examined, in particular the effect of the neglect of long range electrostatic forces. Embedding techniques, especially in the framework of Reaction Field Theory will be discussed. Others possibilities, like periodic Hartree-Fock theory will also be presented. In all cases, the quality of the methods will be tested on relevant examples.

* The reader should be aware that the word "cluster" may have two different meanings. The most traditional significance is related to a large domain of physics devoted to the study of small aggregates of atoms and the detection of specific properties which may be different from bulk properties [1].
In the present discussion, the cluster is considered as a model which should reproduce the local properties of the bulk. In our case the cluster should mimic some part of the bulk.

* * Molecular Dynamics based on classical Mechanics or Monte Carlo method may give informations on microscopic behavior of large systems of particles but ignore chemical reactions.
New possibilities will be offered by more elaborate techniques like the Car-Parinello methods [2], but no application has been yet attempted with systems like zeolites.

Chapter I - <u>Basic concepts and methods of Quantum Chemistry (3)</u>.

I.1.-<u>Generalities</u>

The basic concepts of Quantum Mechanics (3) will not be discussed within the context of this course and it will be admitted that fundamental principles, Schrödinger's equation and general properties of its solutions, the wave function, are familiar to the reader. Our purpose is only to recall how this formalism applies to molecular systems (4-6).

Before doing any mathematical derivations, it is important to remark that molecular systems are made of a mixture of two sets of particles having very different properties : the nuclei with heavy masses and positive charges and the electrons, much lighter and negatively charged. The second point of importance is that isolated molecular systems, like atomic systems under the same conditions, are stable systems in time, which means, using the formalism of Quantum Mechanics, that the hamiltonian operator associated to these systems is time independent. So that, in the general Schrödinger's equation :

$$i\hbar \partial \Psi_v (Xxt)/\partial t = H (X x,t) \Psi_v (X,x,t) \tag{1}$$

where X and x represent sets of coordinates associated to the nuclei and the electrons. If $H (X, x,t) = H (X,x)$, the time dependant solution has the simple form :

$$\Psi_v(X,x, t) = \Psi_v (X x O)e^{-(i/\hbar)H.t} \tag{2}$$

The system is said to be in a stationary state v and will stay for ever in the same state unless submitted to some disturbance. This is the case when the system is submitted to electromagnetic radiation (optical spectroscopy for example) or to collisions with other particles.

In simple cases, time dependant perturbation theory permits calculating probabilities of transition from state v to state v' under the time dependant disturbance. This aspect of the problem will not be considered here, only steady states will be retained, which are the eigenstates of the time independant Schrödinger's equation :

$$H(X,x)\Psi_v(X,x) = E \Psi_v(X, x) \tag{3}$$

An essential step in the resolution of equation (3) is to take advantage of the mass discrepancy pointed out previously between nuclei and electrons since this must have a direct consequence on the dynamics of each of these particles. These considerations are at the origin of a fundamental theorem of molecular Physics, the Born Oppenheimer theorem, which states, in classical language, that motions of nuclei and electrons may be examined separately. It is worth mentioning at this stage and because it will be used later on in this course, that separability has become a ubiquitous principle in Quantum Chemistry. It is as the basis of separation of slow and fast motion, R (A....B) and r (A-H), in hydrogen bonded complex A-H....B, or of rotational and vibrational modes. It will also be invoked with separate core and valence electrons and in many other situations. Separability is not a rigorous principle and it is often supported by arguments comparing magnitudes (energies, frequencies,...). A direct consequence of the Born Oppenheimer principle is the separation of equation (3) into a set of two coupled equations:

$$[T^e + V(X,x)]\, \Psi^e_n(X,x) = E^e_n(X)\, \Psi^e_n(X,x) \qquad (4)$$

$$[T^N + E^e_n(X)]\, \Psi^{(N)}_{nm}(X) = E^{(N)}_{nm}\, \Psi^N_{nm}(X) \qquad (5)$$

T^e and T^N are the kinetic energy operators associated with electrons and nuclei. The first equation is the electronic equation from which the eigenfunction Ψ^e_n and the eigenvalue $E^e_n(X)$ may be deduced for a given position of the nuclei.

The second equation, the nuclear equation, signifies that the nuclei are moving on a potential energy surface which is deduced point by point from the electronic energy calculated with the help of the first equation. Resolution of this equation would tell us how nuclei redistribute on the potential energy surface in the course for example of a bimolecular reactions :

$$A\text{-}B + C \rightarrow A + B\text{-}C \qquad (6)$$

From these remarks it may be easily seen that the Born-Oppenheimer approximation has led Quantum Chemistry to develop along a unique direction whose principal steps may be summarized as follows.

a) Solve equation (4) for a given position of the nuclei.

b) Repeat the calculation as many times as it is necessary in order to build the hypersurface $E^e_n(X)$.

c) Explore the hypersurface in order to determine the extrema : minima, maxima or saddle point, and then deduce the most stable geometric configuration of the system.

d) Resolution of equation (5) is undertaken in general to study the stationary motions of the nuclei and determine the eigen frequencies of the system.

If it is assumed that nuclei may be assimilited to classical particles, resolution of Hamiltonian equations*, using the nuclear operator $H^N = T^N + E^e_n(X)$, permit following the dynamics of the nuclei on the potential energy surface $E^e_n(X)$.

In ordrer to estimate the interest of theoretical investigations and also the limits of their validity, more attention should be paid to point a), i-e, to methods used for the resolution of eq. (4). Only very essential steps will be recalled here, for any detailed demonstrations, readers should refer to basic text books.

In general, the resolution of equation (4) can be decomposed in two steps :

1) Find a specific form of the $\Psi^e_n(X,x)$ function.

2) Apply the variational principle.

The objective of step 1) is rather easily reached by imposing requirements using the Aufbau principle (shell model), or spatial and spin symmetries (electrons are Fermions) or by using chemical intuition.

From these principles it is possible to deduce that the total wave function Ψ^e must be constructed with molecular orbital (M.O.) functions, each of the M.O. being associated with 2, 1 or eventually O electron (s). Starting from the independent model approximation (electrons move independently from each other in the field of the nuclei) and taking account of symmetry constraints, it can be shown that the total wave function φ^e should have the form (Slater determinant).

$$\Psi^e = \frac{1}{\sqrt{N!}} \begin{vmatrix} \varphi_1(1) \ \varphi_2(1) \ \varphi_3(1)... \\ \\ \varphi_1(2) \ \varphi_2(2) \ \varphi_3(2)... \\ \\ .... \end{vmatrix} \tag{7}$$

The wave function Ψ^e has been written here for a closed shell system containing $N = 2n$ electrons. With each orbital φ_i are associated two electrons, one with spin function α, and the other one with spin function β. Furthermore, a specific form may be attributed to the M.O. φ_i thanks to the L.C.A.O. approximation assuming that an M.O. may be represented by a Linear Combination of Atomic Orbitals. This is a rather intuitive statement based on the fact that there must exist some close similarity between the molecular system considered and its constituent atoms.

---

* If X_i is one fo the nuclear coordinate and P_i its conjugate momentum, the set of Hamilton equations is : $dX_i/dt = \partial H/\partial P_i$ and $-dP_i/dt = \partial H/\partial X_i$. For further details on classical Mechanics see : H. Goldstein : Classical Mechanics - Addison Wesley ed., (1980).

$$\varphi_i = \sum a_{ij} \chi_j \tag{8}$$

where χ_i is an atomic orbital

Step 2) provides the tool which permits determining effectively the solution of equation (4). The variational principle states that if E^e_n is stationary, then the Ψ functions are eigenfunctions of H. Practically this means that for any linearly independent variations of the Ψ's ($\Psi \rightarrow \Psi + \Delta\Psi$) the corresponding variation of E^e_n should be zero ($\Delta E = 0$). The procedure may be applied to any type of trial functions.

I.2.-Hückel and the Hartree-Fock family

These general principles are now illustrated in two specific cases.

a) Case of a wave function which is a product of P functions given by expansion(8). In this case the unknown coefficients a_{ij} are solution of a system of P linear equations:

$$\sum a_{ij} (H_{ij} - \epsilon S_{ij}) = 0 \qquad\qquad j = 1, 2...P \tag{9}$$

accompanied with the compatibility condition (secular equation) :

$$\det |H_{ij} - \epsilon S_{ij}| = 0 \tag{10}$$

when restricted to π valence electrons and with the following approximations : $H_{ii} = \alpha$, $H_{ij} = \beta$ if i and j are adjacents, O otherwise, and $S_{ij} = \delta_{ij}$ (δ is the Dirac function), the Hückel Theory (H.T.) is obtained. For a closed shell system, the total energy of the system is given by $E = \sum n_i \epsilon_i$ (n_i = 2, 1, 0 following occupation of the M.O. and ϵ_i the energy associated to φ_i). Reformulation of this original Theory to include σ and π valence electrons has been proposed by R.J.Hoffmann, and has given rise to the familiar extended Hückel Theory (EHT).

b) When the Slater type function is used as trial function, the variational method leads to the famous Hartree-Fock (H.F.) equations :

$$F(\varphi_i)\, \varphi_i = \epsilon_i\, \varphi_i \tag{11}$$

it is important to remark that each φ_i is an H.F. orbital associated with the ith electron moving in the average field created by its N-1 partners, and on the other hand, that the operator F (whose exact form is of no importance for our discussion) is constructed with these H.F. orbitals. As a consequence, equation (11) is not an eigenvalue problem, this equation is not linear and can only be solved by iterative procedure. Starting with an original guess of $\varphi^{(0)}$, the Fock operator (F) is constructed and equations (11) are solved, this leads to a new set of M.O., said $\varphi^{(1)}$ and the procedure is pursued (iterated) up to order n, when variations of calculated quantities (energy,...) are less than some fixed threshold. At this stage, self consistency is reached (the procedure is often referred to as self consistent field (S.C.F.) method). Resolution of Hartree Fock equations is a formidable task for molecular systems. Solutions exist for atomic systems and a few very simple molecules. For the rest, it is only in 1951 that an approximate solution of the Hartree Fock equations has been proposed by C.J. Roothaan [7]. It consisted of resolving the set of equation (11) within the framework of the L.C.A.O. approximation. Under these conditions, for the set of differential equations (11) is substituted a set of linear algebraic equations :

$$\sum c_{ij}(F_{ij} - \varepsilon S_{ij}) = O \qquad\qquad i = 1,...N \qquad (12)$$

where the c_{ij} are the coefficients of the expansion of φ_i on the atomic basis set functions, F_{ij} are the matrix elements of the Fock operator and S_{ij} the overlap matrix elements. The structural similarity of equations (12) and (9) should not be misleading. The Fock matrix elements F_{ij} are rigorously calculated and contain the bielectronic Coulomb and Exchange integrals (J_{ij} and K_{ij}) whereas the H_{ij} are only parameters.Furthermore, as with the original Hartree Fock equations, the Fock operator must be calculated iteratively starting from a set of initial guess of C_{ij}.In the Hartree Fock method the total energy is $E = 2\Sigma\varepsilon I + \Sigma(J_{ij} - K_{ij})$ (for a closed shell system).

Though the Hartree Fock procedure following Roothaan method can be completely worked out without any more mathematical approximations (these are "ab initio" methods depending only on first principles) it is important to underline that the solution found is not an exact solution of the Schrödinger's equation (use of variational method allows predicting that energies found are always above the exact value) and this essentially for two reasons*.

A third one is the neglect of relativistic effects, which can be significant for systems containing heavy metal atoms.

One is intrinsic to the method and is a direct consequence of the neglect of interelectronic correlation. As already pointed out, in the Hartree Fock method, for electron i all the others electrons appear as an averaged field. An exact treatment should consider all individual interactions. Correlation energy is defined as the difference between exact and H.F. energy :

$$E_{corr} = E_{EXACT} - E_{H.F.} \qquad (13)$$

The second reason invoked is a practical one and is related to the L.C.A.O. approximation. In effect, both terms of equation (8) are mathematically identical if, and only if, the expansion is infinite. For obvious reasons, this requirement is never satisfied and the error introduced by the use of truncated basis set functions is a subject of great concern. The choice of a good basis set functions (compatible with the computer resources available) is more a question of practice than rigorous rules. The question raised by use of troncated basis set functions, is a central problem for all ab initio calculations, and control the quality of the results. Here again, two factors play a key role : the nature of the AO's and number of orbitals used. The best choice of AO's are the atomic orbitals calculated by Hartree Fock method for the atoms (H.F.O.). Unfortunately these functions are obtained numerically and are not very useful for traditional calculations. Next, are the Slater orbitals (S.T.O), usually two S.T.O. are asked to represent one H.F.O. (This is called a double zeta (DZ) calculation). But calculations of polycentric integrals are very tedious with S.T.O. For this reason, most calculations now use the Gaussian orbital (G.T.O.) which are very easy to manipulate for integral calculations. But the number of G functions to introduce is usually higher than the number of S.T.O. In Table -1- we have summarized some of the most familiar labels designating basis set and their signification.

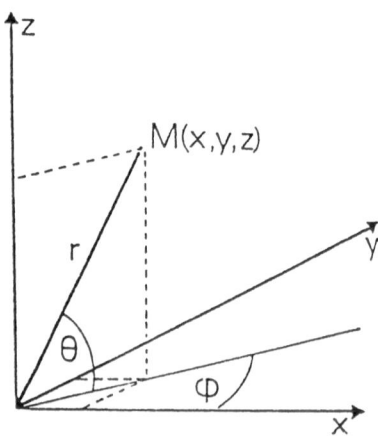

Fig.-1- Polar coordinates (r,θ,φ) and cartesian coordinates (xyz) used as arguments in S.T.O. and G.T.O. functions and in spherical harmonic functions Y_{lm}

a) Type of orbitals (T.O.).

Slater   (S.T.O.) : $Y_{lm} (\theta, \varphi) \, p^{n-1} \, EXP \, (-\zeta r)$ $= \Psi^{(S)}$ notations are

Gaussian (G.T.O.) : $X^l \, y^m \, z^n \, EXP \, (-\alpha r)$ $= \Psi^{(G)}$ indicated on Fig. -1-

---

b) Expansion of a molecular orbital (M.O.).

$\Phi = \sum a_i \, \Psi_i^{(S)}$

a_i

$= \sum b_i \, \Psi_i^{(G)}$

b_i varied coefficients

c_i

$= \sum c_i \left[\sum d_{il} \, \Psi_e^{(G)} \right]$

d_{il} fixed coefficients

combinations of Ψ_G with fixed coefficients (d_i) are called "contractions".

---

c) Minimal basis set signifies that n_i (i = 1, 2 or 3) = number of core + valence, atomic orbitals.

Double zeta (D.Z.) : two   STO (or GTO) are used to represent one A.O.
Triple zeta (T.Z.) : three STO (or GTO) are used to represent one A.O.

---

d) STO 3G : one Slater orbital is represented by three Gaussian functions.

4 31G, 5 31G, ...K MNG, signifies that contractions containing 4, 5,, ...L $\Psi^{(G)}$ are used to represent core orbitals and that two contractions containing respectively 3, 1, ...$\Psi^{(G)}$ are used to represent valence orbitals.

---

e) Polarization functions :

L MNG (*) = L MNG + polarization functions on heavy atoms
L MNG (**) = L MNG (*) + polarisation functions (type p) on hyddrogene.

---

f) Diffuse functions

G.T.O. with small parameters (α) are used to represent Rydberg states or anions.

Table 1 - a) Atomic orbitals   b) The introduction of contracted functions allows to reducing the number of coefficients to vary c) The spacial extension of electronic density represented by $\Psi^*\Psi$ depends on the magnitude of exponant α (cf. a)) to a small α correspond a density diffuse in space, on the contrary a large α leads to a very condensed density. Polarization and diffuse functions correspond to high and small coefficient respectively.

The Hartree Fock method is one of the basic tools of Quantum Chemistry and is the source of many standard "ab initio" codes like : GAUSSIAN [8], HONDO [9], MONSTERGAUSS [10],...

The development of the ab initio methods and the increase of their domain of application is rather recent and has closely followed the rapid burst of computer technologies.

Before, or simultaneously to the development of ab initio methods, the Hartree Fock scheme has been used in the production of the so called semi-empirical methods. A large number of them can be more or less easily deduced from the Hartree Fock methods by introducing ad hoc approximations such as (11) :

i) treating only valence electrons and ignoring core electrons.

i i) reducing the computer time spent for integral calculations by the simple device of estimating these integrals, or replacing them with parameters fitted on experimental data.

When using these approximate methods, one should keep in mind that a) their objective is not to compute absolute values but trends. b) they are not general purpose methods, but according to the type of experimental data used as reference, they are often dedicated to the calculation of specific quantities.

The CNDO (Complete Neglect of Differential Overlap) method is one of the most famous semiempirical methods used, it gives rather good results for structural parameter determination.

Others closely related methods should be mentionned, they are the MNDO (Modified Neglect of Diatomic overlap), INDO (Intermediate Neglect of Differential Overlap).

I.3.-Beyond Hartree-Fock.

As previously pointed out, the Hartree-Fock method suffers from a serious defect, an incorrect treatment of the electronic correlation. Though the H.F. wave function introduces some correlation (Fermi correlation) since the determinant form prevents two electrons with identical spin being in the same orbital, the correlation due to Coulombic repulsion is underestimated. The consequences are numerous : dissociation limit is badly reproduced, structure of compounds where correlation is important like transition metals is also erroneous, long range interaction which depends on van der Waals forces cannot be treated either since dispersion energy has, as origin, intersystem correlations.

The magnitude of correlation effects represents a few percent of the total energy in most systems, and it is quite justified to use perturbative treatments to recuperate this missing contribution. In order to satisfy internal requirements of Perturbation Theory, the calculations are performed by taking as zero order hamiltonian the Fock operator. This procedure, called Moller-Plesset (M.P.) perturbation theory is now implemented in some of the ab initio codes and designated as MPn correction according if perturbation expansion has been extended up to order n. An alternative to the perturbation method is the application of the variational method to a trial wave function developed over a basis of ground and excited state functions.

$$\Psi = \sum c_i \Phi_i \qquad (14)$$

The c_i are coefficients and the Φ_i are, in general, determinental functions constructed with the full set of φ orbitals (occupied and unoccupied) extracted from the H.-F. calculation for the ground state.

This is the Configuration Interations (C.I.) method. Given the set of Φ_i, the coefficients are solutions of a linear system of equations whose compatibility condition has the form of the secular equation (9) or (12).The procedure is straightforward but raise many technical difficulties. A full C.I. calculation, i-e. including all possible Φ_i functions, is practically unfeasible and most C.I. calculations use a restricted number of Φ_i selected on various criteria.

Directly related to the C.I. method, is the multiconfigurational method (MCSCF) where the coefficients C_i and orbitals φ_i are simultaneously varied.

I.4.- Conclusion.

This very brief review did not intend to give a full account of the present state of Quantum Chemistry. Many more methods are now available for solving the electronic equation (4). In recent years new developments have been prolific among uncorrelated and correlated methods. Inclusion of pseudo potentials have allowed Hartree-Fock methods to treat molecular systems containing heavier atoms [12]. New tracks have been explored trying to avoid the truncated basis set problem. Such are the Density Functional Method (D.F.T.) where calculations are initiated from a given electronic density function [13] or Quantum Monte Carlo (Q.M.C.) method based on similarity between equation (4) and diffusion equation [14].

Chapter II - Modelization of active sites.

II.1.-The favorite domain of application of Quantum Chemistry methods.

In principle there are no limitations as to the size of the system whose wave function can be determined from Schrödinger's equation.

But practical limits are imposed by mathematical difficulties or computer facilities. Even with microscopic systems the volume of data to treat may be enormous. For a system like $(V_{10} O_{28})^{6-}$ (decavanadate ion) containing 460 electrons, an ab initio calculation at H.F.(15) level requires calculating of the order of 430 millions of bielectronic integrals.

Then it can be seen by these examples that the size of the system which can be handled by Quantum Chemistry is rather small. On the other hand there is actually some sort of inverse law between the size of the system treated and expected accuracy of the results. Effectively, a current practice is to reduce the volume of calculation by using small truncated basis set functions, a procedure leading in general to results of lower quality. In practice, with respect to possibilities of calculations, the molecular systems may be classified in three categories :

i) very small systems containing a few electrons, like H_2, LiH,...
i i) systems containing less than about 50 electrons.
i i i) systems containing more than 50 electrons.

The first category contains few representatives of interest and in general may be viewed as test cases for exploring the performances of new methods. Calculations may be applied up to their extreme limits with accuracies comparable to experimental values. In certain cases high quality ab initio results even challenge experimental values.

The second category concerns a much larger class of systems. Good accuracy may be expected if large basis set functions are used and electronic correlation taken into account. At H.F. level, accuracy on structural parameters may be of the order of 0.1Å for bond length and 2 or 3° for bond angles. Harmonic frequencies are generally too high by a few percent and dissociation limit uncertain. For these two last quantities, correction for electronic correlation may greatly improve the results.

Concerning energetics a precision less than 4 or 5 Kcal/mole is difficult to achieve. Here again the role of electronic correlation is decisive, especially when comparing energies of systems with different electronic distributions (determination of ionisation potentials, U.V. transitions) and is indispensable if dealing with van der Waals complexes.

Systems containing from 50 electrons to a maximum of about 1000 electrons fill the third category. Here accuracy estimations are uncertain and depends a great deal on the quality of the basis set functions and of correction for correlation effect.

II.2.-Cluster model

From the above discussion it is seen that only molecular systems containing a finite number of atoms can be treated at a reasonable level of accuracy using Quantum Chemistry methods. As a consequence, when treating atomic or molecular interactions in solids, it is out of the question to deal with the overall system (bulk). First, we should make certain that the problem considered involves, as a priority, local interactions. Then, conjugating chemical intuition and material constraints, determine what are the particles which interact significantly. These particles will constitute a subsystem which can be treated by Quantum Chemistry as an isolated system. This is the essence of the "cluster" model, and it is on the construction of this cluster and the validity of the model we would like to focus attention now. Fig. 2 illustrates a situation where the cluster model may be a good starting model for a first order study of catalytic reaction[16-17].

Considering only solid state media, the separation between an active region (the cluster) and a passive one, later ignored, is not trivial and may create serious problems which may affect the validity of the results deduced from the cluster model. As it will result from the forthcoming discussion it is convenient to distinguish at least three types of host media which can be classified according to the dominant type of bonding: i) van der Waals or hydrogen bonding, ii) covalent bonding and iii) ionic system.

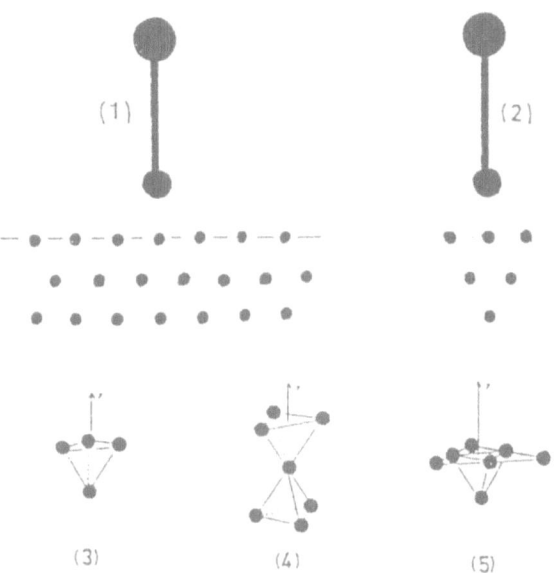

Fig.-2- (1) Adsorption of a diatomic molecule on a surface
(2) Cluster model : the surface is simulated by a small set of atoms.
(3), (4) and (5) depict a few possible arrangements [16].

66

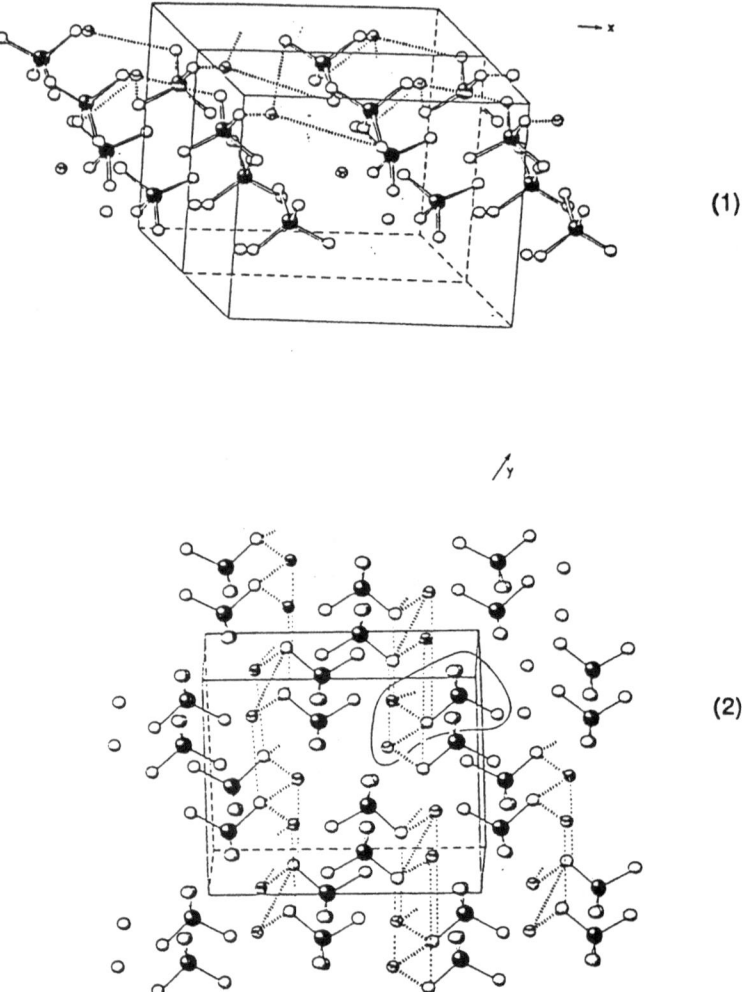

(1)

(2)

Fig.-3- Schematic representation of the crystal structure of hydronium perchlorate
(1)hydrogen bonding pattern along the X-axis, dashed lines represent
possible orientations of hydrogens in the hydronium ion (the proton are not
shown).
(2)hydrogen bonded pattern along the y-axis is indicated with dashed lines,
and broken lines connect atoms on the same plane. The quantum system
selected to study proton translocations is surrounded with a continuous
line(18).

By host media of the first category we mean complex systems constituted of subunits (small aggregates) weakly bonded. A typical example is the perchlorate of oxonium Fig.-3- where subunits H_2O and $HClO_4$ are hydrogen bonded along chains. Lateral interactions between the chains are of van der Waals type. Observation of high ionic conductivity has been an incitation to research a model for proton translocation. Transfer along the chain has been evoked as a possibility and simple Quantum Chemistry calculations have tried to support this hypothesis. An elementary step in this mechanism is proton jump via interaction with the ClO_4^- groups. The potential energy-barrier which controls this motion may be derived from calculations performed on a cluster made of $H_2O...H...Cl O_4$. As a matter of fact this cluster is a natural subunit of the crystal and no arbitrary partitioning is needed.

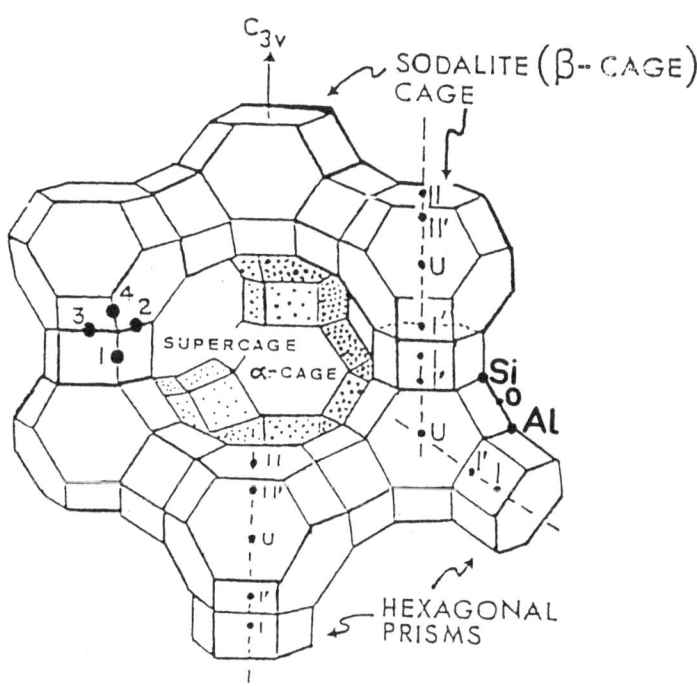

Fig.-4- Zeolite structure, Si-O-Al bridge are indicated. Following the rule of Lowenstein no two Al atoms can be consecutive.

In the case of covalent solids where atoms are selectively bonded, the isolation of a subsystem necessarily requires the breaking of several convalent bonds and the redistribution of the bonding electrons between the cluster and the rest of the solid considered as a host medium. The net result is the formation of a highly charged anion, an unsatisfactory situation which may enhance irrelevant interactions. Charge neutrality for the cluster must be imposed even artificially and to reach this objective the dangling bonds must be "healed" by saturating the bonds with appropriate monovalent atoms.

The case of acido basic reactions in zeolites is a good illustration of these problems. Zeolites (19-20) are aluminosilicate crystallin solids made of three dimensional arrangements of SiO_4 and AlO_4 tetrahedral sub units. An array of positive ions (H^+, Na^+, Mg^{++},...) insures electrical neutrality of the overall system. Fig.4 depicts the framework of a special type of zeolite, the Faujasite containing Si-OH-Al bridge where the OH group is a Brönstëd acid site* is located.

Characterization of Brönstedt acidity in zeolites has become in the last 10 years a popular subject of investigation for quantum chemists. The majority of these calculations use the cluster model, and their analysis offers an excellent occasion to verify the validity of this model.

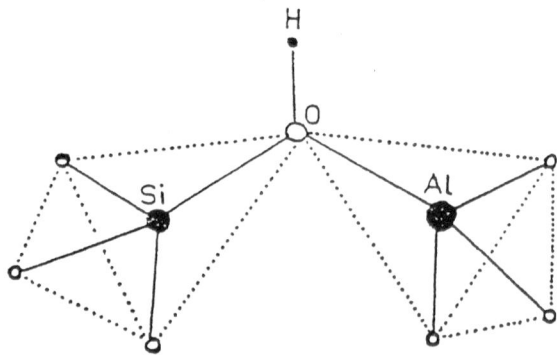

Fig.-5-A two tetrahedral cluster for
simulating acidic site in Faujasite.

---

*　Lewis acid sites may also be present in these zeolites, but their formation is related to the presence of a three coordinated Al atom on the surface. Localisation of Lewis sites in zeolites are still controversial (21).

Most ab initio calculations have started with the minimum cluster made with Si-O-Al bridge. Fig.-5- illustrates a situation where six covalent Si-O and Al-O bonds have been removed. A very common practice consists of saturating the bonds with Hydrogen atoms. The resulting cluster is a system of two tetrahedra H_3 Si O and H_3 Al O sharing an oxygen atom. This H_3 Si OH Al H_3 cluster is often designated as ZOH.

Naturally, the minimal cluster able to simulate the zeolite environment of the acid- OH is the two tetrahedral model. Increasing the size of the cluster to have a more realistic environment is a permanent ambition and now an increasing number of authors try to use models containing 3, 4, or more tetrahedra, with dangling bonds saturated either by H atoms or (OH) groups.

Recently developed computer facilities now allow treating very large systems. Test calculations have been performed on 24-tetrahedral model terminated by OH groups ($Si_{24} O_{60} H_{24}$).[22].

Semi-empirical methods allow exploring large cluster models but above all introduce some flexibility in the choice of terminal atoms. Pseudo atoms with appropriate parameters may be chosen to simulate, better than by using H atom, the interrupted bonding.[23]. Criteria to define the properties of these pseudo atoms are however missing, they could be tentatively deduced from fitting quantities (charges for example) calculated both in bulk and cluster, if calculated quantities in bulk are known.

Alumina systems are good cases to illustrate some of the difficulties which appears when dealing with ionic type crystals. In particular the case of corundum ($\alpha-Al_2 O_3$) where each Al atom is hexacoordinated to six oxygen atoms (Fig.-6-) may well

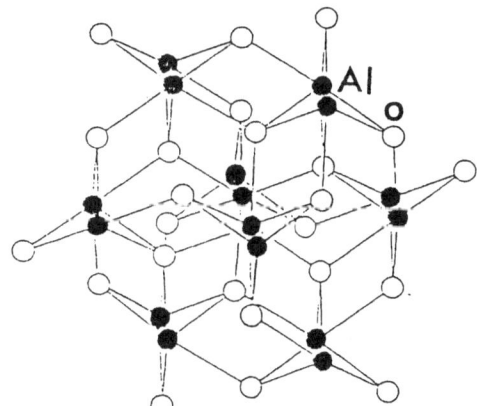

Fig.-6- Structure of α-$Al_2 O_3$.

illustrate the problem. Calculations have been performed in order to examine surface reactivity. The problems considered are very similar to the ones examined in zeolites and concerns characterization of Brönstedt or Lewis acid sites or dissociation of molecular species (H_2O, NH_3,....) by adsorption on these sites. Construction of small clusters surrounding these sites presents some difficulties due to the fact that tetracoordinated or hexacoordinated Al are transfering fractions of charges (3/4 or 1/2 if we use Pauling's rule)(24) towards their oxygen partners. The fictive terminal atoms should then compensate fractional charges and this cannot be realised with standard atoms whose valence numbers are integers. In the context of semi empirical methods pseudo atoms with specific parameters can be defined to overcome this difficulty.(23). The case of reactions on surface Alumina also offer delicate problems concerning the structure of the surface itself. Different models of surface have been proposed (25-26) but there are still some incertainties on the exact location of Brönstedt or Lewis acid sites.

Kawakami and al. (27) have proposed several models and some procedure to build corresponding clusters. Their basic principle, to avoid the problem of fractional charges compensation, is to construct clusters from electrically neutral fragments like Al(OH)$_3$, (HO)$_2$AlOAl (HO)$_2$, H$_2$0. Fig.-7-. show an example of such cluster to study Brönstedt acidity. The major disavantage of the method is that it yields large clusters and requires using small basis set functions (STO 3G)to perform the calculations.

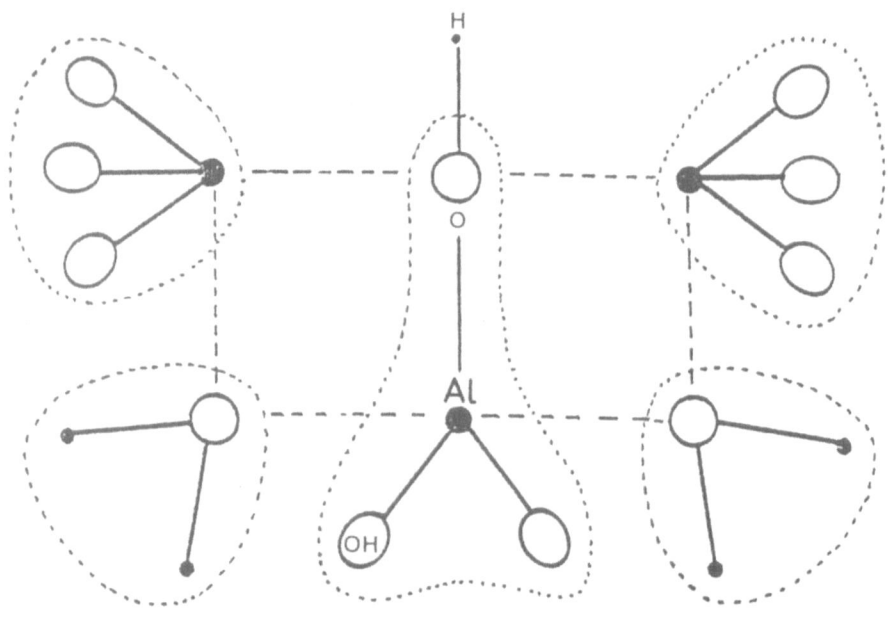

Fig.-7- Cluster model for the study of Brönstedt acidity on central
oxygen. Basic units are Al (OH)$_3$ and OH$_2$. Al atom is
hexacoordinated. Dangling bonds are saturated by H.

II.3 - Conclusion

In this chapter we have insisted on several problems which occur when constructing cluster models. But whatever be the type of site we are investigating and the number of atoms we think convenient to retain around the active site, the use of the cluster model raises a series of questions.

i) are the calculated quantities dependent on the size of the cluster ?

ii) should the geometry of the cluster be kept identical to the configuration the atoms had in the original medium or be recalculated ?

iii) does the model lead only to a first order approximation and is it necessary to take into account long range forces having their origins in the neglected host medium.

Probably no general answers are now available to these questions, but the accumulation of work using this model permits at least comparing series of results at different levels of sophistication and for a well identified physical problem, the Brönstedt acidity in zeolites.

Analysis of these results will be presented in the next chapter.

Chapter III - <u>Applications of cluster models to the characterization of Brönstedt acidity in zeolites.</u>

III.1.-<u>Compared quantities</u>

According to Brönsted, acidity is defined as the strength of attachment of the proton. The proton detachment can be simulated by two different procedures :

i)   the proton can be removed to infinity by a thought experiment, which gives rises to a unimolecular reaction : $ZOH \rightarrow ZO^- + H^+$

i i)   a base molecule can approach the acid site and proton transfer may result from acido-base interaction, in that case the reaction is bimolecular : $ZOH + B \rightarrow ZO^- + BH^+$
        the first reaction is usually designated as the intrinsic acidity and the latter as relative acidity. Analysis of the results will be presented separately.

The quantities most frequently calculated are the following

i)   deprotonation energy Dp.   In the case of intrinsic acidity, Dp is the analogue of dissociation energy.

$$D_P = E(ZOH) - E(ZO^-) \qquad (15)$$

Relative acidity may also be characterized by the energetic difference between hydrogen bonded and zwitterionic complexes.

$$D = E(ZOH...B) - E(ZO^-...BH^+) \qquad (16)$$

but to be complete, ,the reaction path should be followed and the height of the energetic barrier (position of transition state) determined.

ii)   vibrational frequency γ_{OH} of the OH group, deduced from second derivative of the electronic energy with respect to distance D(O-H).

iii) electric charge q_H on the proton (Mulliken definition)

iiii) bending angle Si O Al

        geometry of the ZOH (or ZOH + B) cluster is sometimes recalculated to adjust to each new position of the proton (partial or full optimization) or is kept fixed when proton is moving.

III.2.-<u>Analysis of the results</u>

Some of the points discussed previously will be illustrated here by comparing a few results chosen from the recent litterature. The prospect is to examine how the quantities listed above, Dp, γ_{OH}, q_H and α, are dependent on the size of the cluster, the truncated basis set and correlation effect. In the abscence of a systematic study varying each parameter independently of the others, only partial results may be given.

a) Intrinsic acidity
Deprotonation energy is the basic information - its dependence on cluster size and truncated basis set are given on table I.

	3-21G	6-31G	6-31G+D	6-31G(**)	6-31G(**)+D
H_3 Si OH Al H_3					
a	1354.75	1358.72			
b	1350.42	1354.31			
c	1355.59	1359.83			
d	1356.28	1360.03			
$(OH)_3$Si OH Al$(OH)_3$		1384.22	1376.16	1411.12	1404.79

Table-1-Deprotonation energy (Intrinsic acidity) in KJ/m. Cases a) and b) correspond to structures where the Si-O and Al-O bonds which are contained in the Si O Al plane point both in (or are both opposite) to the O-H direction. c) and d) cases designate a similar situation but with opposite Si-O and Al-O bonds.(28).

In zeolite the maximum value of size effect is of 30 KJ/m, whereas truncated basis set effects reach a maximum (20 KJ/M) when passing from 6-31G to 6-31G**+D, and is rather small (4KJ/m) when improving the basis set from 3-21G to 6-31G.
Both effects seems affect acidity in the same direction, a larger cluster or a richer basis set lowers the acidity.
To obtain stability in calculated geometrical parameters requires at least double zeta (D.Z.) quality basis set functions and even triple zeta (T.Z.) basis set functions with two d-functions, as stated by Nicholas and al. (29-30).
Extensive calculations have been recently presented by Brand and al. (31) for

evaluation of acidity strength in ZSM-5 zeolites. The originality of the calculation was to explore a large variety of clusters. Starting with : the traditional H_3 Si O T H_3 (T = Al or Si) cluster, designated as (1) four other structures are built by successive addition of new shells of neighbors. One obtain successively H_6Si T O_7 (2), $H_{18}Si_7$ T O_7 (3), $H_{18}Si_7T$ O_{25}(4) and H_{54} Si_{25} T O_{25}(5). Size effect may then be studied on a large scale. One of the results is presented in Fig.-8- where original results have been summarized for the case T = Al. The interesting result is that proton affinity does not converge towards some limit when the size of the cluster is increasing, but has an oscillatory behaviour. The authors suggest that electrostatic effects are dominant, so that a shell of positively charged Si decreases the proton affinity whereas an inverse effect is expected when one adds a new shell of negatively charged oxygen atoms.

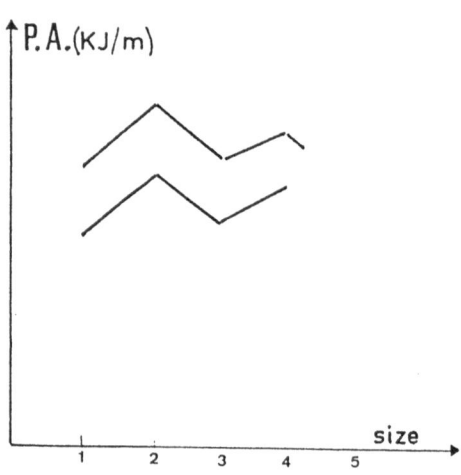

Fig.-8- Evolution of proton affinity
with cluster size (31)

As it could be expected the increase of proton affinity found here when passing from cluster (1) to cluster (2) is in agreement with the decrease of acidity depicted in Table-1-
The effect of electronic correlation is also reported by Teunissen and. al. (32) for a few clusters, they are small, at least in that case.

	Al $(OH)_2$ H_2	Al $(OH)_3$ H_2	H_3Si OH Al H_3	NH_4^+
SCF	1379	1364	1359	907
SCF + MP2	1377	1356	1358	905

Table -2- Effect of electronic correlation.

The charge q_H on the acidic proton is often considered as indicative of the degree of acidity, and it is easily obtained as a by-product of any quantum chemistry calculations in so far as one accepts Mulliken's approximation. However this information should be used with care for at least two reasons : a) the Mulliken definition of charge rests on an arbitrary partition of the total charge attributed to a bond (anyway there is no exact definition of an atomic charge in a molecule, but some partitioning, like Bader's (33) may seems to have better justification), b) charges are very sensitive to the quality of the basis set functions adopted and in particular to the introduction of polarization functions. Table -3- depicts variation of Mulliken charges on cluster size and extension of basis set (28).

		3-21G	6-31G	6-31G+D	6-31G(**)	6-31G(**)+D
H_3 Si OH Al H_3	O	- 0.93	- 1.10			
	H	0.46	0.50			
	Al	0.82	0.90			
	Si	1.20	1.21			
$(OH)_3$ Si OH Al $(OH)_3$	O		- 1.08	- 1.50	- 0.97	- 1.40
	H		0.50	0.54	0.52	0.57
	Al		2.05	2.02	1.46	1.46
	Si		2.39	2.76	1.72	2.10

Table-3- Net charges distribution on H_3 Si OH Al H_3 and (OH_3) Si OH Al $(OH)_3$ complexes.(28).

Comparison between poor quality ab initio and semi-empirical methods may be of interest. Some data extracted from a work of Senchenya and al.(34), are given in table-4- and show that trends are equally well reproduced by ab initio STO 3G and CNDO/2 calculations.

α angle	110	130	150	180
STO 3G	0.3074	0.2947	0.2794	0.2509
CNDO/2	0.1959	0.1871	0.1782	0.1659

Table-4- Charges on acidic hydrogen (q_H)calculated for various Si O Al = α angle with ab initio (STO 3G) and semiempirical method (CNDO/2)(34). (model cluster is $(OH)_3$ Si OH Al $(OH)_3$)

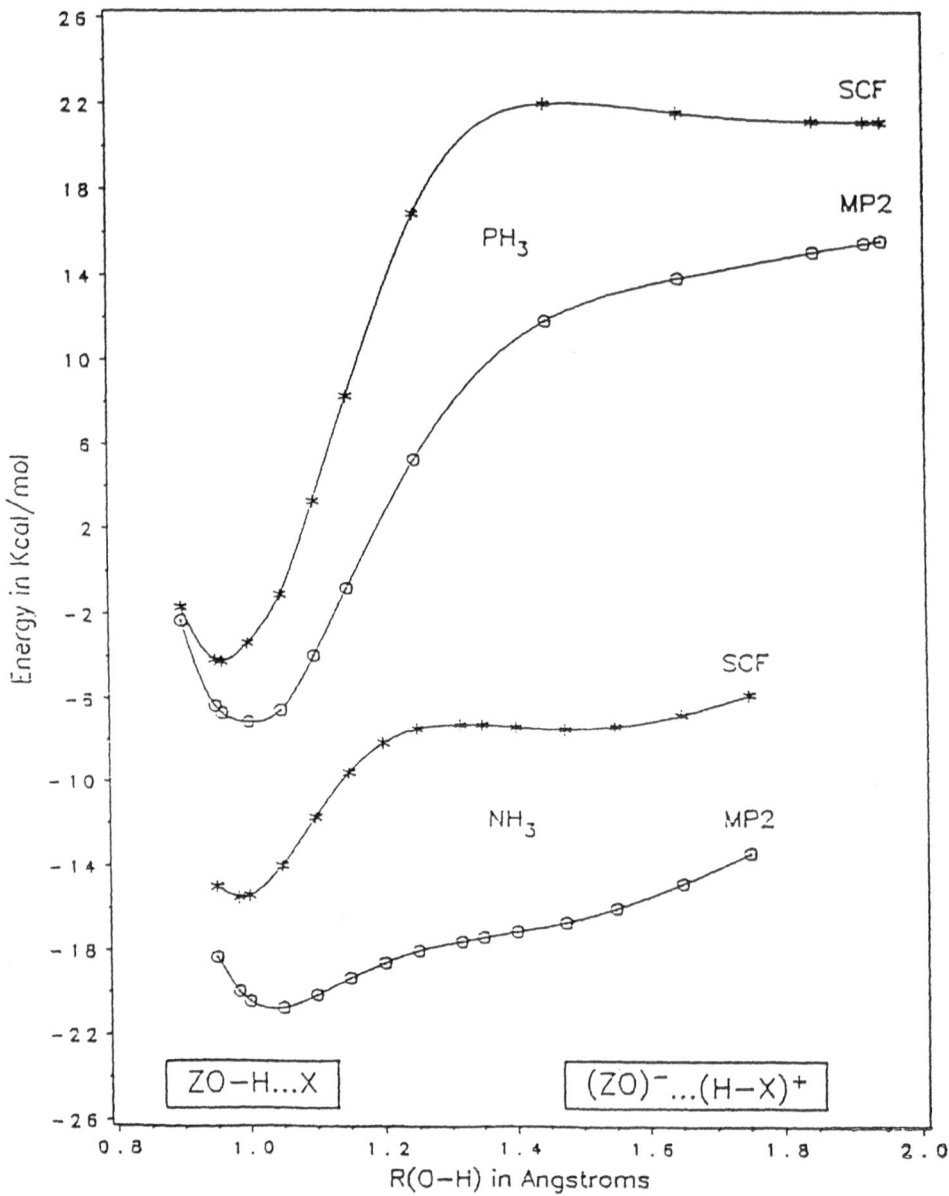

Fig.-9- Energy of interaction of NH$_3$ (PH$_3$) with the cluster model H$_3$ Si O Al H$_3$
versus O-H distance.(35).

b) Relative acidity

Ab initio calculations of effective interaction between ZOH cluster and base species (H_2O, H_2S, NH_3,....) approaching the acid site along the OH direction have all failed to reproduce the proton transfer. The H-bonded complex is always the more stable, and this statement is almost independent of the quality of the calculation. Results obtained by recent calculations on the ZOH + NH_3 reaction with large basis set and correlation correction are depicted on Fig.-9- and confirms this statement[*].

To explain this repeated failure two main reasons have been evoked :

i) solvent effect, a problem we shall discuss in the next chapter

i i) the possibility of multiple bonding between proton donor and acceptor, or of coadsorption.

This last alternative has reoriented the research in what is the best model cluster. Instead of increasing systematically the size by adding more shells of atoms arround a unique acid site, new models are investigated, offering several sites of interaction. A model cluster simulating double bonding is shown on Fig.-10- (35). Calculations performed with cluster models offering possibilities of formation of two or triple bonds (36-37) between cluster and base species (H_2O, NH_3,....) have demonstrated that the occurrence of basic oxygen sites play a much more decisive role than the number of shells of atoms surrounding active sites. This is confirmed by the results given in Table -5- where absorption energy of the NH_4 cation with two active site clusters of various sizes is reported for calculations at SCF and SCF + MP_2 level

	$(OH)_2 Al H_2$ (32) (a)	$(OH)_3 Al H$ (32) (b)	$(O Si H_3)_2 Al H_2$ (35) (c)
SCF	-110	-112	- 95
SCF + MP2	-141	-139	- 92,5

Table-5- Adsorption energy of NH_4^+ with various clusters .
SCF calculations for cases a) and b) are at 6-311 + G (d,p)/STO3G level and for case c) at 6-31G (d) level. Energies are given in Kj/mol.

[*] H.V. Brand and al. (31) are the only authors to have reported recently that at H.F. level they could reproduce proton transfer in the ZOH + NH_3 reaction. This may be an artifact of calculation since the results have been obtained with small 321G basis set and with fixed crystal related geometries. It is nevertheless an interesting result showing how geometry relaxation may be important.

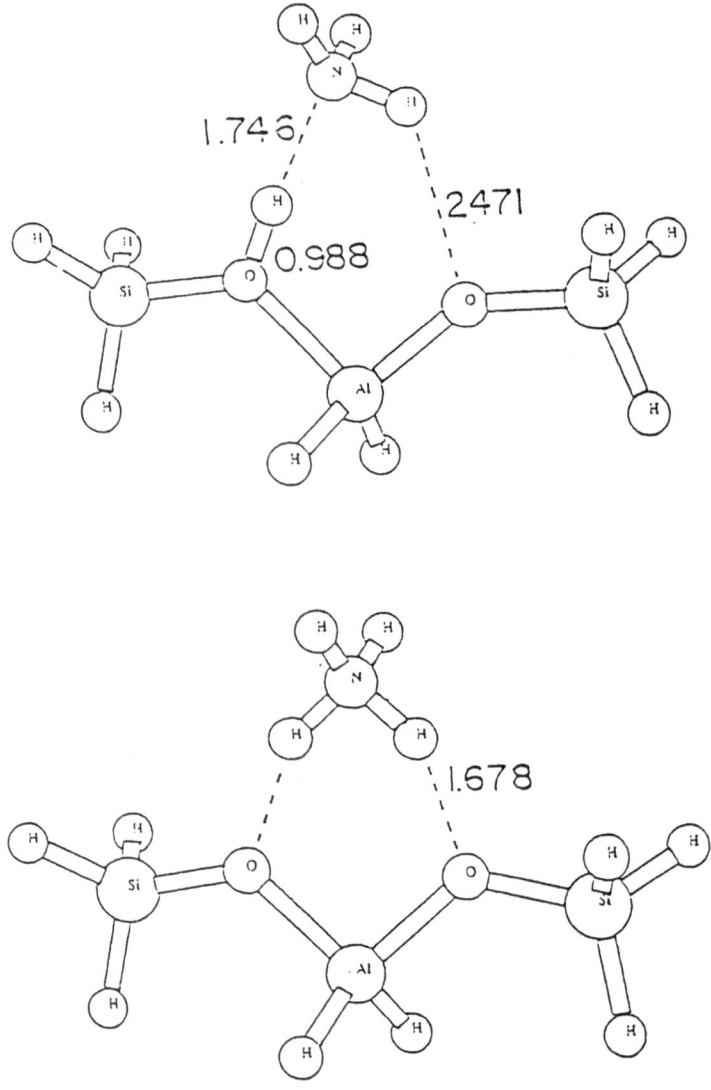

Fig.-10-Interaction of NH₃ with a two active site model cluster H₃ Si OH AlH₂ O SiH₃ a) original structure where nitrogen atom is directed toward the hydroxyl group b) After full optimization formation of a bidented complex. Lengths are given in Å (35).

A further example can be found in a recent work of Pelmenschikov (38) on the interaction of water with OH bonds belonging to $Si(OH)_4$ cluster.

Some of his results are reported in table -6-. They show that a decisive step is the extention of monodented to multidented complex, the magnitude of the effect being competitive, in that case, with correlation energy.

	6-31G*	6-31G**//6-31G**	MP2/6-31G**//6-31G*
1	26.4	25.9	30.4
2	37.3	36.5	47.1
3	37.8	36.9	48.5

Table-6- Adsorbtion energy of H_2O on $Si(OH)_4$ cluster for mono, bi and tridented complex. Energies in Kj/m.(38).

III.3.-Conclusion

The few examples presented here demonstrate that theoretical prediction depends on a delicate balance between different competitive effects. Calculation should start with a good quality basis set and correction for correlation effect cannot be neglected. But above all, full optimization should be complete in order to reveal all the possible mechanisms of interaction which may combine for the achievement of the reaction. The model chosen plays a major role.

It should also be emphasized that in this problem of acido-basic reaction, the P.A. of the base species is a major factor. Comparison of NH_3 and PH_3 shows that lowering the P.A. by 7% is sufficient to prevent any proton transfert in the reaction ZOH + PH_3 as it is illustrated on Fig.-9-.

The works mentioned in this chapter have more methodological than chemical interest. The goal of further developments should be the interpretation of Aromatics fragmentation following proton transfer, and the eventual formation of carbocations. Tentative calculations in that direction are still rare.(39-42).

Chapter IV - Non local effects.

This part will be devoted to one of the most frequent criticisms directed at the cluster model, i.e. the neglect of long range interactions. This problem is somewhat comparable to the frequently discussed question of solvent effects in liquid state but has two specific features. 1) the subsystem (cluster) embedded in the solid has an artificial frontier built to neutralise the cluster but which may leave dangling bonds on the bulk side. 2) the medium benefits from translational symmetry.

Two aspects of the problem will be presented :

a) How to complement the results obtained from the cluster model using an embedding method. The procedure has been often used in amorphous systems by surrounding the cluster by a continuous medium usually characterised by a dielectric constant. More attention will be paid to the Reaction Field Theory (R.F.T.) which in its simplified version allows evaluating long range electrostatic effects. Essential features of the theory will be presented as well as its implementation in the framework of the H.F. Theory.

Application of the R.F.T. requires computational techniques familiar to solid state Physicists, but above all raises questions concerning the definition of charges in the bulk system.

b) Comparison of H.F. calculations on cluster and bulk systems.

Taking advantage of translational symmetry it has been demonstrated that the Schrödinger equation may be solved for periodic systems.

Examples of applications of embedding and periodic Hartree-Fock calculations to proton transfer in zeolites will be presented.

IV.1.-Generalities

Simulation of solvent effects by immersion of the cluster in a continuum medium is a current practice since the pionneering work of Born [43-45]. Besides, the original procedure has been substantially improved, in particular in the adjustment of the borders of the cavity on the shape of the embedded molecule (or complex). [46-48]. A mixed model has also been worked out which consists of treating quantum mechanically the original cluster augmented by one or two shells of solvation and plunge that super system into the continuum medium. These procedures suffer from at least two major defects, a)results depend on parameters such as dielectric constant or radius of cavity b)structural specificity of the medium has been lost. They appear to be better adjusted to reproduce averaged effects of liquids solvents, than the specific interactions which may be expected from organised media. None of these methods seems able to reproduce the supposed effect of the variation of Si/Al ratio on acidic properties in zeolites.

A simple procedure to reproduce local environment in ionic type crystals is very similar to the strategy followed in the mixed model previously mentioned. It consist of replacing nearest neighbor atoms by point charges and performing the H.F. calculation on the cluster and its point charge environment. A few hundreds point charges may be introduced without complicating too much the H.F. calculation since the added charges introduce only supplementary monoelectronic integrals.This "dressed" cluster could eventually be embedded in a continuum.(49).

IV.2.-The Reaction Field Theory (R.F.T.).(50-54)

The object of the R.F.T. is to calculate the mutual influence resulting from electrostatic interactions which cluster and host medium exert on one another. It must also be recalled that solvent effects are usually weak and consequently that theoretical tools used are borrowed from perturbation theory. Furthermore, expansions are often limited to linear terms. This signifies that whatever be the system considered, the hamiltonian operator will always be separable :

$$H = H_C(X,x) + H_M(R,r) + V(X,x,R,r) \qquad (17)$$

where H_C, H_M and V are operators associated with the cluster and the medium and V represents the interaction between variables of both systems. X, R, x, r are sets of variables (nuclear and electronic) defined in both systems.

In conformity with the method followed for the treatment of isolated systems, the calculations are performed within the framework of the Born Oppenheimer approximation where wave functions are calculated keeping nuclei at fixed positions. It is furthermore assumed that wave functions of cluster and medium are not overlapping, i.-e. there is no exchange of electrons between the two systems, so that the total wave function may be written :

$$\Psi = \Psi_C \Psi_M \qquad (18)$$

Averaging electronic variables of the medium and ignoring medium energy, we end with the following expression for the perturbed energy of the cluster :

$$E_C = <\Psi_C H_C \Psi_C> + <\Psi_C \Psi_M V\Psi_M \Psi_C> \qquad (19)$$

If the host medium is a crystal, the Ψ_M may be approximated by Bloch's function. But in most cases and as it was stated earlier, the determination of Ψ_M is out of the question. Practically the averaged V operator is replaced by a multipolar expansion deduced from the Coulombic interaction of two non overlapping charge distributions.

$$V = \int dr\, \rho(r) \int T(rq)\, \Omega(q)\, dq \qquad (20)$$

where $\rho(r)$ and $\Omega(q)$ are charge density operators for the cluster and the medium and r, q, vectors which denote points inside and outside of the cluster. T(r,q) represent the expansion of the inverse distance : (Fig.-11-)

$$T(rq) = |r - q|^{-1} \qquad (21)$$

Further development requires explicit expression of the charge density operators $\rho(r)$ and $\Omega(q)$. Retaining the first two terms of the expansion of T(r,q), V may be written as :

$$V = \int dr\, \rho(r) \int dq\, [\Omega(q)\, T_1(rq) + P(q)T_2(rq)] \qquad (22)$$

$$= \int \rho(r)dr\, [VMA_D(r) + V_{POL}(r)] \qquad (23)$$

with :

$$T_1 = |r - q|^{-1} \quad \text{and} \quad T_2 = -\nabla_r |r - q|^{-1} \qquad (24)$$

It can be shown that the V_{POL} operator reduces to :

$$V_{POL} = -\vec{\mu}\, \tilde{g} < \Psi\, \vec{\mu}\, \Psi> \qquad (25)$$

where $\vec{\mu}$ is the dipole moment associated to the cluster and $\tilde{g}$ a coupling tensor. And V_{MAD} can be obtained from a calculation similar to the one applied for the determination of the Madelung constant in ionic crystals, if it is admitted that :

$$\Omega(q) = \sum_{\alpha} Q_{\alpha}\, \delta(q - q_{\alpha}) \qquad (26)$$

then :

$$V_{MAD} = \int \frac{\Omega(q)dq}{|r - q|} = \sum_{\alpha} \frac{Q_{\alpha}}{|r - q_{\alpha}|} \qquad (27)$$

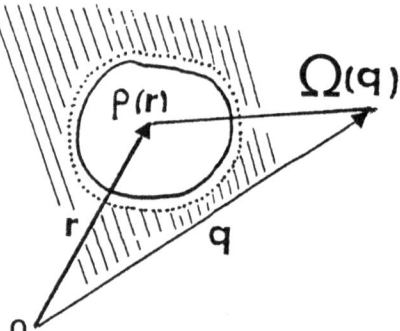

Fig.-11- vector systems for points in the cluster (r) and in the bulk(q).

Addition of V to the original hamiltonian operator leads to a non linear Schrödinger's equation :

$$[H + V(\Psi)]\, \Psi = E\Psi \qquad (28)$$

Implementation of the H + V operator in the Hartree Fock operator would lead to calculation of the matrix elements of V. Further simplifications may be introduced by assuming that the cluster density of charge $\rho(r)$ may be represented by a Dirac distribution of charges over the atoms of the cluster.

$$\rho(r) = \sum Q(r)\, \delta(r\text{-}r_S) \tag{29}$$

and in these conditions V (r) becomes :

$$V = \sum_S Q(r_S)\, V_{MAD}(r_S) \tag{30}$$

where $Q(r_S)$ is the charge operateur associated with atom S. The host medium contribution to the energy is then reduced to the calculation of $V_{MAD}(r_S)$, i.-e., the estimation of the Coulombic potential created by the host medium at each atom site in the cluster. Given a charge distribution (formula 29) the calculation of $V_{MAD}(r)$ may be rather easily performed with the help of the Ewald procedure (55-56).

Difficulties arise concerning the definition of charges in the overall system considered. Fig.-12- illustrates the situation : The region of interest, consisting of atoms A_c, is selected around a point r_0 and the dangling bonds of the cluster atoms are saturated by H atoms. The cluster thus obtained is designated as "zone I". The bulk (zone III) is separable from zone I by an intermediate region, "zone II" or frontier region. In this region are found the frontier atoms A_F, originally bonded to the cluster atoms A_c and which are now replaced by the terminal H atoms. The problem is now to determine the charges associated with these atoms. It has been demonstrated (57) that constraints may be imposed on these charges. Effectively the partition just described is an arbitrary mathematical device imposed only by the necessity of building a closed shell system for our model cluster. Calculated quantities should be independent of this partition.

84

In particular, calculation of the Madelung potential at any point of the cluster should give the same result prior to or after partitioning. Let us then consider the original over all system and calculate the Madelung Potential at some point r_0 :

$$V_{MAD}(r_0) = \sum_k q_k / |r_0 - r_k| \qquad (31)$$

where r_k and q_k are the position vectors and charge of the Kth atom in the crystal and N is the number of atoms in the crystal ($N \rightarrow \infty$).

After partitioning, the same potential may be written :

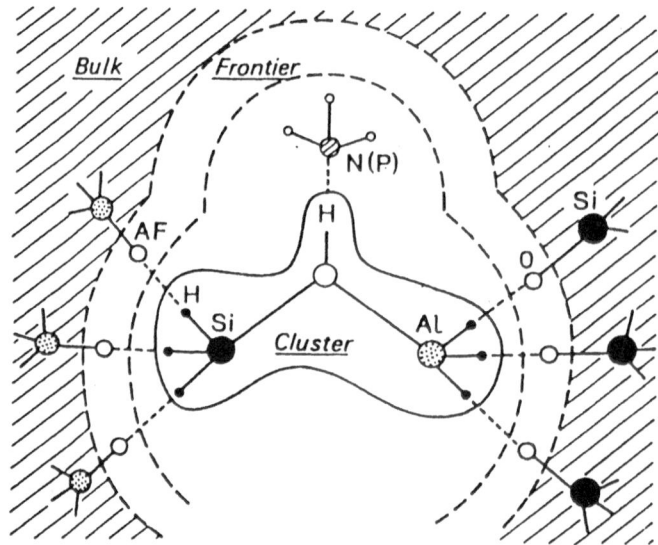

Fig.-12- Partition of charges in the zeolite sample : cluster, frontier and bulk.

$$V(r) = \sum_C^{N_C} \frac{q_C}{|r_0 - r_C|} \; + \; \sum_F^{N_F} \frac{q_F}{|r_0 - r_F|} \; + \; \sum_B^{N-(N_C+N_F)} \frac{q_B}{|r_0 - r_B|} \qquad (32)$$

where N_C, and N_F are respectively the number of atoms in the cluster (including terminal H atoms), and the frontier region. N is the total number of atoms contained in the zeolitic sample.

As for the charges : q_C are the Mulliken charges deduced from quantum calculation, q_M are identical to the charges used in the calculation of V_{MAD} they are deduced from semi empirical Density Functional methods (58-59). And finally the q_F may be determined by the condition :

$$V(r_0) = V_{MAD}(r_0)$$

This leads in principle to a set of linear equations for the n unknown q_F charges.

	E	
	isolated cluster	embedded cluster
ZOH...NH$_3$	-1115.17796	-1115.54074
	(3.1)	(50.2)
ZO$^-$...NH$_4$$^+$	-1115.18295	-1115.62070

Table-7- Stabilization energies of the covalent and ionic nitrogen complex when isolated and embedded in a distribution of charges. Energies are in au (numbers in parentheses are energy differences in kcal/mol) (60).

This embedding procedure has been applied to the interaction of NH$_3$ with (OH)$_3$ Si OH Al (OH)$_3$ cluster(60) and it was shown (table-7-) that monodented ZO$^-$...NH$_4$$^+$ complex is stabilized by the electrostatic potential. Stabilization energy is 50.2Kcal/mole with respect to hydrogen bonded complex.

Though much remains to be done in order to refine the representation of the environment, the present results demonstrate the extreme importance of embedding effects.

IV.3.-The quantum chemistry methods for periodic systems.

Development of the embedding method is not yet completed and improved results should be expected in the near future. However new opportunities should be considered with the development of Quantum Mechanical methods applicable to periodic systems.

The extension of Hartree Fock methods to periodic systems has been known a long time (61-64) but direct application to problems of Quantum Chemistry have been developed in particular by Andre and al. using the L.C.A.O. approximation and above all by Dovesi and al. who are at the origin of an ab initio program for the periodic Hartree-Fock method. This method follows very closely the successive steps of standard Hartree Fock, but orbitals must satisfy symmetry requirements as formulated by Bloch Theorem :

$$\Phi_K (r + R_j) = e^{ikR_j} \Phi_K(r) \tag{33}$$

and to the M.O.

$$\varphi_k = \sum C_{ki} \chi_i \tag{34}$$

are now substituted Bloch orbitals :

$$\varphi_{kn} = \sum_{J}^{N} \sum_{I} e^{ikR_j} C_{knI} \, \chi_I (r-R_j) \tag{35}$$

where K is the Kth level in band n, and N is the number of unit cells. The c_{knI} coefficients are the solution of a set of linear equations with compatibility conditions :

$$\det \left| \sum e^{ikR_h} (F_{kI} - \varepsilon S_{kI}^h) \right| = O \tag{36}$$

The procedure is only apparently simple and arise numerical calculation problems among which are i) the presence of interaction potential containing summations over all the vectors of the direct lattice, with as a consequence the recourse to truncation of infinite sums. ii) calculation of integrals in K-space (momentum space).

The perspectives offered by the development of quantum chemistry method applicable to periodic systems are extremely interesting but their actual possibilities have some limitations.

i) electronic correlation cannot be taken into account.

i i) the number of atoms which may be included in the unit cell is still limited and this restricts the possibility of application.

i i i) impurities which break the symmetry cannot be introduced unless the impurity is introduced into the unit cell and infinitely repeated by symmetry operations.

Concerning points i) and ii) solutions will be found probably in a near future, but the impurity problem (iii) will create more difficulties. Presently, a practical procedure consists of introducing the impurity into an arbitrarily defined unit cell, in order to reintroduce periodicity. But the major defect of this artifact is to introduce an infinite member of impurities into the system. The resulting mutual interaction between these impurities may be minimized by choosing a large unit cell, a decision which make problem ii) more complicated.

To our knowledge, application of these procedures to the case of impurities contained in zeolite systems has not yet been published. But the method has been largely applied for structural studies of simpler crystals and in particular for the description and analysis of phase transitions (65).

Structural studies offer the possibility of comparing results obtained from periodic H.F. applied to bulk systems with traditional quantum chemistry calculations using cluster models.

Interested readers may consult both calculations performed on $\alpha-Al_2O_3$ (corundum), one with periodic HF (66) and the other one with a cluster model using an ab initio density functional method (67). Comparison of electronic density maps exhibits significant difference in the region of the oxygens, which may be indicative of artefacts due to cluster representation.

Conclusion

Through in principle there are no restrictions in the application of traditional quantum chemistry methods to systems containing a very large number of particles (bulk sample), practically these applications are limited to rather small molecular systems including a finite number of electrons.

As a consequence, in order to study chemical reactions occurring in solid media, it should be assumed that these reactions are essentially local and involve only a small number of particles surrounding the reactants. Reactants and their nearest neighbors constitute a subsystem often designated as a "cluster" and the quantum chemical treatment of the cluster should provide complete information on the reaction in situ.

The simulation of the physical problem by the cluster model is not trivial and it has been shown by taking examples of catalytic reactions in zeolites that calculated results may depend on the nature of the cluster.

Besides truncated basis sets functions problems and correlation effect which are common defects in all H.F. calculations, the cluster model may introduce two sources of error originating in a) the structure of the cluster (size, number of active sites). b) the relaxation of the cluster following motions of the reactants.

Moreover, proton transfer leading to the formation of ionic complexes, may be taken into account either by using cluster models and allowing formation of multidented complexes or by embedding monodented complex in media simulated by an electrostatic field. This plurality of interpretations added to the fact that energetics differences between conformations are small (less thant 3 or 4 Kcal/mole) suggest that the reactions are probably conditioned by effects whose energetic magnitudes are very close.

Future developments will rely on progress in computer facilities (parallel processing) which will permit the treatment of giant clusters (more than several thousands electrons) and also on methodological progress in the treatment of environmental effects which could be combined with a periodic H.F. method, this latter method providing a zero order wave function indispensable for the application of perturbation theory to the treatment of impurity problem in crystals.

Finally, a complete prospective should not omit the possibilities of Molecular Dynamics (68-72), though presently it is restricted in most cases to physabsorbtion or diffusion of particles rather than reactive processes.

Acknowledgements :

It is a pleasure to thank Dr. E. Evleth for rereading the manuscript and for helpful discussion.

eferences

1 - J. Davenas, P. Rabette "Contributions of Cluster Physics to Material Science and Technology", Nijhoff eds., Amsterdam, (1986).

2 - R. Car, M. Parrinello, Phys. Rev. Lett., $\underline{55}$, 2471, (1985).

3 - Readings for updating theoretical background may extend from simple text books for graduate students to more elaborate works like : C. Cohen-Tannoudji, B. Diu, F. Laloë, "Quantum Mechanics", Wiley ed.

4 - A. Szabo, Neil S. Ostlund, "Modern Quantum Chemistry", Mac Millan ed., (1982).

5 - J.-L. Rivail, "Elements de Chimie Quantique", Intereditions/ed. du C.N.R.S., (1989).

6 - J. Simons, J. Phys. Chem., $\underline{95}$, 1017, (1991).

7 - C.C.J. Roothaan, Rev. Mod. Phys., $\underline{23}$, 69, (1951).

8 - Gaussian 92, Revision B, M. J. Frisch, G.W. Trucks, M. Head-Gordon, P.M.W. Gill, M.W. Wong, J.B. Foresman, B.G. Johson, H.B. Schlegel, M.A. Robb, E.S. Replogle, R. Gomperts, J.L. Andres, K. Raghavachari, J.S. Binkley, C. Gonzalez, R.L. Martin, D.J. Fox, D.J. Defrees, J. Baker, J.J.P. Stewart, and J.A. Pople, Gaussian, Inc. Pittsburgh PA, (1992).

9 - M. Dupuis, J. Rys., H.F. King, Quantum Chemistry Programm Exchange, $\underline{11}$, 336, (1977), $\underline{11}$, 338, (1977), $\underline{13}$, 401, (1981).

10 - M.R. Peterson, R.A. Poirier, Programm MONSTERGAUSS, Univ. of Toronto, (1981).

11 - J.A. Pople, D.L. Beveridge, "Approximate Molecular Orbital Theory", Mc.Graw Hill ed., (1970).

12 - J.D. Weeks, A. Hazi, S.T. Rice, Adv. Chem. Phys., $\underline{16}$, 283, (1969).

13 - R.G. Parr, "Density Functional Methods in Physics", R.M. Dreizler and J. Da Providencia, Eds. Plenum Publ. Corp., New-York, (1985).

14 - M. Caffarel, P. Claverie, J. Chem. Phys., $\underline{88}$, 1088, (1988).

15 - J.-Y. Kempf, M.M. Rohmer, J.-M. Poblet, C. Bo, M. Bénard, J.A.C.S., $\underline{114}$, 1136, (1992).

16 - V. Russier, D.R. Salahub, C. Mijoule, Phys. Rev. B., 42, 5046, (1990).

17 - E.L. Muetterties, T.N. Rhodin, E. Band, C.F. Brucker, W.R. Pretzer, Chem. Rev., 79, 91, (1979).

18 - J. Angyan, M. Allavena, M. Picard, A. Potier, O. Tapia, J. Chem. Phys., 77, 4723, (1982).

19 - D.W. Breck, "Zeolite Molecular Sieves", Wiley ed., N.-Y., (1975).

20 - "Modeling of structure and reactivity in zeolites", C.R.A. Catlow ed., Academic Press, (1992).

21 - A.S. Medin,Y.Yu. Borovkov,V.B.Kazansky,A.G. Pelmentschikov,G.M. Zhidomirov Zeolites, 10, 668, (1990).

22 - R. Ahlrichs, M. Bär, M. Häser, C. Kölmel, J. Sauer, Chem. Phys. Lett., 164, 199, (1989).

23 - M.B. Fleisher, L.O. Golender, M.V. Shimanskaya, React. Kinet. Catal., 24, 25, (1984).

24 - L. Pauling "The Nature of the Chemical Bond", Cornell Univ. Press, p. 335, (1948).

25 - H. Knözinger, P. Ratnuzami, Catal. Rev. Sci. Eng., 17, 31, (1978).

26 - J. B. Peri, J. Phys. Chem., 69, 220, (1965).

27 - H. Kawakami, S. Yoshida, J. Chem. Soc., Faraday Trans., 81, 1117, (1985), idem, 81, 1129, (1985).

28 - E. Kassab, K. Seiti, M. Allavena, J. Phys. Chem., 92, 6705, (1988).

29 - J.B. Nicholas, R.E. Winans, R.J. Harrison, L.E. Iton, L.A. Curtiss, A.J. Hopfinger, J. Phys. Chem., 96, 10247, (1992).

30 - J. Sauer, J. Phys. Chem., 91, 2315, (1987).

31 - H.V. Brand, L.A. Curtiss, L.E. Iton, J. Phys. Chem., 96, 7725, (1992).

32 - E.H. Teunissen, R.A. van Santen, F.B. van Duijneveldt, A.P.J. Jansen, J. Phys. Chem, 97, 203, (1993).

33 - R.F.W. Bader, "Atoms in Molecules", Clarendon Press, Oxford, (1990).

34 - I.N. Senchenya, V.B. Kazansky, S. Beran, J. Phys. Chem., 90, 4857, (1986).

35 - E. Kassab , J. Fouquet, M. Allavena, E. Evleth, J. Phys. Chem., (accepted for publication).

36 - E.H. Teunissen, F.B. van Duijneveldt, R.A. van Santem, J. Phys. Chem., 96, 366, (1992).

37 - J. Sauer, C.M. Kölmel, J.-R. Hill, R. Ahlrichs, Chem. Phys. Lett., 164, 193, (1989).

38 - A.G. Pelmenschikov, G. Morosi, A. Gamba, J. Phys. Chem., 96, 2241, (1992).

39 - S. Beran, P. Jiru, L. Kubelkova, J. Mol. Catal., 12, 341, (1981).

40 - S. Beran, P. Jiru, L. Kubelkova, J. Mol. Catal., 16, 299, (1982).

41 - V.B. Kazansky, Acc. Chem. Res., 24, 379, (1991).

42 - L.R. Sierra, E. Kassab, E.M. Evleth, J. Phys. Chem., 97, 641, (1993).

43 - M. Born, Z. Phys., 1, 45, (1920).

44 - J.L. Pascual-Ahuir, E. Silla, J. Comp. Chem., 11, 1047, (1990).

45 - E. Silla, I. Tunon, J.L. Pascual-Ahuir, J. Comp. Chem., 12, 1077, (1991).

46 - J. Langlet, P. Claverie, J. Caillet, A. Pullman, Phys. Chem., 92, 1617, (1988).

47 - J.-L. Rivail, D. Rinaldi, Chem. Phys., 18, 233, (1976).

48 - F.J. Olivares del Valle, R. Bonaccorsi, R. Cammi, J. Tomasi, J. Mol. Struct., 230, 295, (1991).

49 - R. Vetrivel, R.A. Catlow, E.A. Colbourn, Proc. Roy. Soc., London, A417, 81, (1988).

50 - L. Onsager, J. Am. Chem. Soc., 58, 1486, (1936).

51 - S. Yomosa, J. Phys. Soc. Japan, 35, 1738, (1973) ; 36, 1655, (1974) ; 44, 602, (1978).

52 - O. Tapia, O. Goscinski, Mol. Phys., 29, 1653, (1975).

53 - O. Tapia, in "Molecular interactions", H. Ratajczak and W.J. Orville-Thomas, eds., 3, (Wiley, Chichester, U.K.,1982), chap. 2.

54 - O. Tapia, J. Mol. Struct., 226, 59, (1991).

55 - P.P. Ewald, Ann. Phys., 64, 253, (1921).

56 - F. Bertaut, J. Phys. Rad., 13, 499, (1952).

57 - M. Allavena, K. Seiti, E. Kassab, Gy. Ferenczy, J.G. Angyan, Chem. Phys. Lett., 168, 461, (1990).

58 - K. van Genechten, W.J. Mortier, P. Geerlings, J. Chem. Phys., 86, 5063, (1987).

59 - K. van Genechten, W.J. Mortier, P. Geerlings, J. Chem. Soc. Chem., Comm. 1278, (1986).

60 - E. Kassab, K. Seiti, M. Allavena, J. Phys. Chem., 95, 9425, (1991).

61 - J.-M. André, L. Gouverneur, G. Leroy, Int. J. Quant. Chem., 1, 427, 451, (1967).

62 - A. Zunger, Phys. Rev., B17, 626, (1978).

63 - C. Pisani, R. Dovesi, Int. J. Quant. Chem., 17, 501, (1980).

64 - R. Dovesi, C. Pisani, C. Roetti, Int. J. Quant. Chem., 17, 517, (1980).

65 - B. Silvi, L.-H. Jolly, Ph. d'Arco, J. Mol. Structure, 260, 1, (1992).

66 - L. Salasco, R. Dovesi, R. Orlando, M. Causa, V.R. Saunders, Mol. Phys., 72, 267, (1991).

67 - S. Nagel, J. Phys. C., 18, 3673, (1985).

68 - W.F. van Gunsteren, H.J.C. Berendsen, Angew. Chem. Int. Ed., Engl., 29, 992, (1990).

69 - P. Demontis, G.B. Suffritti, A. Alberto, S. Quartieri, E.S. Fois, A. Gamba, Gazzetta Chim. Ital., 116, 459, (1986).

70 - L. Leherte, G.C. Lie, K.N. Swamy, E. Clementi, E.G. Derouane, J.-M. André, Chem. Phys. Lett., 145, 237, (1988).

71 - K.-P. Schröder, J. Sauer, M. Lestre, C.R.A. Catlow, J.M. Thomas, Chem. Phys. Lett., 188, 320, (1992).

72 - G.J. Kramer, R.A. van Santem, J. Am. Chem. Soc., 115, 2887, (1993).

ACIDITY AND BASICITY CONCEPTS: THE DENSITY FUNCTIONAL VIEWPOINT BASED ON AN ATOMS-IN-MOLECULES SENSITIVITY ANALYSIS - A TUTORIAL

B.G. Baekelandt, G.O.A. Janssens, H. Toufar, W.J. Mortier, R.A. Schoonheydt
Centrum voor Oppervlaktescheikunde en Katalyse, K.U.Leuven, Kardinaal Mercierlaan 92, B-3001 Heverlee (Belgium)

ABSTRACT: A compilation is given of a complete set of sensitivity coefficients, developed in the frame of the charge sensitivity analysis and the electronegativity equalization method (EEM). The relevance for chemical reactivity in general, and for acidity and basicity in particular, is demonstrated. The algorithms to calculate these sensitivity coefficients are outlined. Their physical significance is discussed and illustrated by examples.

Introduction

All aspects of this tutorial have been adequately discussed and referenced in a series of (review) papers to which reference is primarily made [1-5]. This tutorial merely serves to bring the attention of the reader to the many possible applications.

1. Definition of Acidity and Basicity: dv Perturbation

Density Functional Theory (DFT) provides a framework for studying the electronic structure of matter and has been shown to be especially suited to elucidate chemical concepts such as electronegativity and chemical hardness [6]. Based on the Hamiltonian operator and the Hohenberg and Kohn theorems [7], we can safely state that a system is completely determined by the total number of electrons (N) and the external potential (v) to which the electrons are subject to (usually the nuclear charges in a given disposition).

A corollary to this is that we know at the same time exactly how we can change a system of electrons and nuclei: this can only be done by interfering with the number of electrons (dN-perturbation) or with the external potential (dv-perturbations).

According to Hohenberg and Kohn, perturbations can be in the total number of electrons (dN) or in the external potential (dv): these are the only possibilities to change a system.

In this context acid-base reactions typically involve dv-perturbations. The proton transfer (Broensted acidity) and the initial polarization of reactants by ions (or charges) (Lewis acidity) do not change the number of electrons of the system. Changes in chemical composition and charge (electron) transfer reactions, are typical examples of dN perturbations (cf metal catalysis). In radical mechanisms, both N and v are changing.

95

J. Fraissard and L. Petrakis (eds.), Acidity and Basicity of Solids, 95–126.

Although we will mainly concentrate on dv perturbations, the dN sensitivities will be included as well for completeness.

Acidity and basicity (and reactivity in general), cannot be defined as intrinsic properties in the absence of a reaction partner: mutual perturbation of the reacting molecules influences the 'intrinsic' reactivities drastically. The sensitivity analysis as presented here allows the computation of density relaxation effects, *i.e.* the adaptation of the electronic charge distribution of a reactant to the perturbing (electrostatic) potential produced by the other reactant (at an atomic resolution).

Finally it should be noticed that, to be effective, a perturbation in v must be a potential gradient:

$$E'[\Psi] = \int \Psi^* (\hat{H} + C)\Psi d\tau = E[\Psi] + C\int \Psi^*\Psi d\tau = E + C$$

Adding a constant to the original external potential (Hamiltonian) does not affect the wave function, but only the energy level.

2. Complete Set of Sensitivities

The "sensitivity" of a system is defined as the response of one of its properties (*e.g.* energy, electron density, electronegativity, hardness) to a perturbation in N or v, i.e. the derivative with respect to N or v [8]. Up to second order, the derivatives of the total energy are as follows [9]:

$$E[N, v]$$

$$\left(\frac{\partial E}{\partial N}\right)_v = \mu = -\chi \qquad \left(\frac{\delta E}{\delta v(\bar{r})}\right)_N = \rho(\bar{r})$$

$$\left(\frac{\partial^2 E}{\partial N^2}\right)_v = \eta \qquad \left(\frac{\delta^2 E}{\delta v(\bar{r})\partial N}\right) = f(\bar{r}) = \left(\frac{\delta^2 E}{\partial N \delta v(\bar{r})}\right) \qquad \left(\frac{\delta^2 E}{\delta v(\bar{r}_1)\delta v(\bar{r}_2)}\right)_N = p(1,2)$$

$$\parallel \qquad\qquad \parallel \qquad\qquad\qquad \parallel \qquad\qquad\qquad \parallel$$

$$-\left(\frac{\partial \chi}{\partial N}\right)_v \qquad \left(\frac{\delta \mu}{\delta v(\bar{r})}\right)_N \qquad \left(\frac{\partial \rho(\bar{r})}{\partial N}\right)_v \qquad \left(\frac{\delta \rho(\bar{r}_1)}{\delta v(\bar{r}_2)}\right)_N$$

μ denotes the electronic chemical potential (minus the electronegativity χ), ρ the electron density distribution function, η the absolute global hardness, f the Fukui function and p the two-variable linear response function (response kernel).

The analysis can be carried out using several degrees of detail; we have chosen an atomic resolution in which local properties ($\bar{r}$-dependent functions) are integrated over the atomic region. For instance:

$$\int_\alpha f(\bar{r})dr = f_\alpha \qquad \text{and} \qquad \int_\alpha \rho(\bar{r})dr = N_\alpha$$

are respectively the Fukui function and number of electrons on atom α.

Defining the 'atomic region' is only necessary for parametrization of the method. The essential message is that the set of sensitivity coefficients defined above is complete, *i.e.* we do not need to consider any other variable or quantity to describe the response of a system to perturbations in N or v.

3. Electronegativity Equalization Method (EEM)[1-5,9-12]

The Density Functional expression for the total molecular electronic energy [6]

$$E[\rho] = F[\rho] + \int \rho(\bar{r}).v(\bar{r}).d\bar{r} \qquad (1)$$

is explicitly introduced in the EEM formalism (using a spherical-atom approximation) as a sum of atomic contributions [11]:

$$E[q] = F^*[q] + \sum_\alpha^n \frac{1}{2}q_\alpha \cdot \phi_\alpha \qquad (2)$$

All local quantities are defined as atomic quantities and correspond to an integration of the density functional analogue over the atomic region: $N_\alpha = Z_\alpha - q_\alpha$ (with Z_α and q_α the atomic number and charge respectively) is the number of electrons on atom α, that is, an integration of ρ over the atomic region of atom α; ϕ_α is the external electrostatic potential at atom α:

$$\phi_\alpha = k \sum_{\beta \neq \alpha}^n \frac{q_\beta}{R_{\alpha\beta}}$$

$R_{\alpha\beta}$ is the internuclear separation between atoms α and β and k is a conversion factor (1 e.s.u. = 14.4 eV for R expressed in Å); $F[\rho] = T[\rho] + V_{ee}[\rho]$, with T and V_{ee} the electronic kinetic and the electron repulsion energy respectively. Because an explicit expression for $F[\rho]$ has not yet been derived, it is introduced as a second order expansion of the energy (for each atom):

$$F^*[q] = \sum_\alpha^n \left\{ E_\alpha^* + \left(\frac{dE_\alpha}{dq_\alpha}\right)q_\alpha + \frac{1}{2}\left(\frac{d^2E_\alpha}{dq_\alpha^2}\right)q_\alpha^2 \right\} = \sum_\alpha^n \left\{ E_\alpha^* + \chi_\alpha^* q_\alpha + \eta_\alpha^* q_\alpha^2 \right\} \quad (3)$$

One can thus write:

$$E[q] = \sum_{\alpha}^{n} \left\{ E_{\alpha}^{*} + \chi_{\alpha}^{*} q_{\alpha} + \eta_{\alpha}^{*} q_{\alpha}^{2} + \frac{1}{2} q_{\alpha} \cdot \phi_{\alpha} \right\} \qquad (4)$$

The expansion coefficients $(E_{\alpha}^{*}, \chi_{\alpha}^{*}, \eta_{\alpha}^{*})$ can be calibrated for several atom types so as to reproduce ab initio (STO-3G) data. Eq. 4 allows the calculation of the derivatives of E with respect to N and ϕ, as compiled in Table 1 [9,12]; matrix notations are shown in the examples. In the EEM scheme, each set of derivatives can be computed directly through solving of a set of (n+1) linear equations in (n+1) unknowns (n=number of atoms in a "molecule"), i.e. with an equation for each atom-in-a-molecule, plus one normalization condition). The quantities that can be obtained in such a way are given together with the set of equations in Table 1. Since ϕ is a function of q and R one can also envisage these respective derivatives.

All first and second order partial derivatives of E with respect to N, ϕ (q and R), their inverses and their mutual relations will be discussed with their relevance for chemical reactivity, and especially acidity and basicity, in mind. Each time some historical background as well as the new Density Functional interpretation of the quantities is given. The relevance for acidity/basicity is briefly touched upon, and a numerical example for water, illustrating the calculation method, is given presented in the insets throughout the text.

Table 1: EEM equations and definitions

$$E[q] = \sum_{\alpha}^{n} \left\{ E_{\alpha}^{*} + \chi_{\alpha}^{*} q_{\alpha} + \eta_{\alpha}^{*} q_{\alpha}^{2} + \frac{k}{2} \sum_{\beta \neq \alpha}^{n} \frac{q_{\alpha} \cdot q_{\beta}}{R_{\alpha\beta}} \right\} \qquad \sum_{\alpha}^{n} E_{\alpha} = E \qquad (I)$$

$$\chi = -\left(\frac{dE}{dN} \right)_{v} = \left(\frac{dE}{dq_{\alpha}} \right)_{v} = -\mu \qquad \phi_{\alpha} = k \sum_{\beta \neq \alpha}^{n} \frac{q_{\beta}}{R_{\alpha\beta}}$$

$$\chi_{\alpha} = \left\{ \chi_{\alpha}^{*} + 2\eta_{\alpha}^{*} q_{\alpha} + k \sum_{\beta \neq \alpha}^{n} \frac{q_{\beta}}{R_{\alpha\beta}} \right\} = \chi \qquad \sum_{\alpha}^{n} q_{\alpha} = Q_{m} \qquad (II)$$

$$\overline{\eta} = \left(\frac{d^2 E}{dN^2} \right)_{v} = -\left(\frac{d\overline{\chi}}{dN} \right)_{v} \qquad f_{\alpha} = -\left(\frac{d^2 E}{dNd\phi_{\alpha}} \right) = -\left(\frac{dq_{\alpha}}{dN} \right)_{v} \quad \sum_{\alpha}^{n} -\left(\frac{dq_{\alpha}}{dN} \right)_{v} = \left(\frac{dN}{dN} \right)$$

$$\eta_{\alpha} = \left\{ 2\eta_{\alpha}^{*} f_{\alpha} + k \sum_{\beta \neq \alpha}^{n} \frac{f_{\beta}}{R_{\alpha\beta}} \right\} = \overline{\eta} \qquad \sum_{\alpha}^{n} f_{\alpha} = 1 \qquad (III)$$

$$f_{\alpha} = -\left(\frac{d^2 E}{d\phi_{\alpha} dN} \right) = \left(\frac{d\overline{\chi}}{d\phi_{\alpha}} \right)_{N} \qquad p(\beta,\alpha) = \left(\frac{d^2 E}{d\phi_{\alpha} d\phi_{\beta}} \right)_{N} = \left(\frac{dq_{\beta}}{d\phi_{\alpha}} \right)_{N}$$

$$f_\alpha = \left\{ (1) + 2\eta_\alpha^* p(\beta,\alpha) + k \sum_{\gamma \neq \beta}^n \frac{p(\gamma,\alpha)}{R_{\gamma\beta}} \right\} \qquad\qquad \sum_\beta^n p(\beta,\alpha) = 0 \qquad (IV)$$

$$\left(\frac{d\overline{\chi}}{d\overline{\chi}}\right) = 1 \qquad\qquad \left(\frac{dq_\alpha}{d\overline{\chi}}\right)_v = s_\alpha \qquad\qquad \left(\frac{dN}{d\overline{\chi}}\right)_v = S$$

$$1 = \left\{ 2\eta_\alpha^* s_\alpha + k \sum_{\beta \neq \alpha}^n \frac{s_\beta}{R_{\alpha\beta}} \right\} \qquad\qquad \sum_\alpha^n s_\alpha = S \qquad (V)$$

$$\left(\frac{d\overline{\chi}}{d\phi_\alpha}\right)_{\overline{\chi}} = 0 \qquad\qquad \left(\frac{dq_\beta}{d\phi_\alpha}\right)_{\overline{\chi}} = s(\beta,\alpha) \qquad\qquad \left(\frac{dN}{d\phi_\alpha}\right)_{\overline{\chi}} = s_\alpha$$

$$0 = \left\{ (1) + 2\eta_\alpha^* s(\beta,\alpha) + k \sum_{\gamma \neq \beta}^n \frac{s(\gamma,\alpha)}{R_{\gamma\beta}} \right\} \qquad\qquad \sum_\beta^n s(\beta,\alpha) = s_\alpha \qquad (VI)$$

The respective equations can be solved for the following quantities:

I	:	E
II	:	$q_\alpha, q_\beta, \ldots, q_n$, and $\overline{\chi}$
III	:	$f_\alpha, f_\beta, \ldots, f_n$, and $\overline{\eta}$
IV	:	$p(\alpha,\alpha), p(\beta,\alpha), \ldots, p(n,\alpha)$, and f_α
V	:	$s_\alpha, s_\beta, \ldots, s_n$, and S
VI	:	$s(\alpha,\alpha), s(\beta,\alpha), \ldots, s(n,\alpha)$, and s_α

EXAMPLE: H_2O. INPUT PARAMETERS

The geometry used for water, is the AMPAC/AM1 optimized structure. The Cartesian coordinates are shown below. The EEM parameters are also listed.

	x	y	z	χ^*	η^*
O	0.00000	0.00000	0.00000	8.50000	11.08287
H_1	0.96130	0.00000	0.00000	4.40877	13.77324
H_2	-0.2249	0.93460	0.00000	4.40877	13.77324

4. First Order Derivatives of E

In the EEM formalism, the first order derivatives (the average electronegativity and the atomic charges) are directly calculated for the AIM electronegativity equalization for a closed system (Table 1, eq II).

EXAMPLE : H_2O. EQUATION II, CHARGES AND ELECTRONEGATIVITY

$$
\begin{pmatrix}
2\eta_O^* & \dfrac{k}{R_{OH_1}} & \dfrac{k}{R_{OH_2}} & -1 \\[2ex]
\dfrac{k}{R_{H_1O}} & 2\eta_H^* & \dfrac{k}{R_{H_1H_2}} & -1 \\[2ex]
\dfrac{k}{R_{H_2O}} & \dfrac{k}{R_{H_2H_1}} & 2\eta_H^* & -1 \\[2ex]
1 & 1 & 1 & 0
\end{pmatrix}
\cdot
\begin{pmatrix}
q_O \\[1ex] q_{H_1} \\[1ex] q_{H_2} \\[1ex] \overline{\chi}
\end{pmatrix}
=
\begin{pmatrix}
-\chi_O^* \\[1ex] -\chi_H^* \\[1ex] -\chi_H^* \\[1ex] Q_m
\end{pmatrix}
\Rightarrow
\quad
\begin{aligned}
q_O &= -0.38069 \\
q_{H_1} &= +0.19034 \\
q_{H_2} &= +0.19045 \\
\overline{\chi} &= +5.76444
\end{aligned}
$$

Q_m is the total molecular charge. In this case (neutral water molecule) it is zero.

4.1. $[dE/dN] = [dE/dq] = \mu = -\chi$: THE AVERAGE ELECTRONEGATIVITY

The first order derivative of E with respect to N (or q) has been identified as the electronic chemical potential or minus the electronegativity. The electronegativity concept is often used as an intuitive tool for rationalizing not only Lewis acidity and basicity (electron acceptor and donor capacity) but also Broensted acidity. In particular, the Sanderson electronegativity is widely used for estimating proton charges in hydroxyl groups. They are related to the OH bond ionicity, the bond strength and the deprotonation energy for homogeneous series of compounds (such as bridging OH groups in zeolites) [13,14].

The electronegativity is an overall (global) property of a system, and in those models (such as Sanderson's) where a linear relation exists between the average electronegativity and the atomic charges, there is no more information in the thus calculated charges than in the average electronegativity itself. These express (weighted average) changes in the chemical composition. Geometry-dependent information cannot be extracted from such models.

In density functional theory, μ is the Lagrange multiplier associated with the energy minimization of a closed system. As a physico-chemical property, it is related to the compactness of the electron cloud and the Fermi level. In reactions between two systems (such as atoms), electronegativity differences mainly control the direction of the electron flow.

4.2. $[dE / d\phi_\alpha] = q_\alpha$: THE ATOMIC CHARGES

The derivative of the energy with respect to the electrostatic potential is the atomic charge. Notice that the charge distribution is a finite approximation to the electron density distribution function, which contains all information about a chemical system (just as the wave function in quantum theory). Charges are therefore widely used for rationalizing trends in chemical reactivities. We will not go into details and refer to section 9 for some illustrative examples, especially for the solid state (zeolites).

The AIM electronegativity is geometry-dependent (equation II) and it does *e.g.* matter for its basicity in zeolites in which structure or crystallographic position the oxygen is located. Apart from their equilibrium values, the charge sensitivities (*i.e.* responses to changes) due to electrophilic or nucleophilic attack, are well known tools to explain and predict site-selectivities (for example in electrophilic aromatic substitution). This is particularly the case when one has charge control, *i.e.* electrostatic interactions are the dominating forces at the initial stage of the reaction.

EXAMPLE: H_2O. EQUILIBRIUM CHARGE DISTRIBUTION

One can demonstrate that the charge distribution obtained from EEM is the equilibrium one, *i.e.* corresponds to minimum energy. This is demonstrated for a water molecule by taking other charge distributions (still corresponding to zero global molecular charge, while keeping the molecule symmetric: symmetrical internal polarization) and calculating the energy with equation I. The results are shown below. Since the molecule is symmetric and neutral, only the H charge is listed.

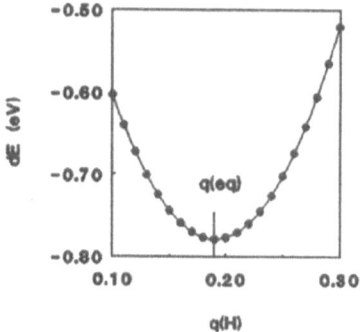

Figure 1: Energy versus proton charge upon symmetrical internal polarization

4.3. $[dE/dR] = 0$. REMARK: VALIDITY CRITERIUM OF THE EEM FORMALISM

The AIM electronegativity equalization as used to derive the EEM is for fixed geometry, dR=0 or at equilibrium geometry (dE/dR=0). In a diatomic molecule, $E(N_\alpha, N_\beta, R)$, we have that

$$dE = \left(\frac{\partial E}{\partial N_\alpha}\right)_{R,N_\beta} dN_\alpha + \left(\frac{\partial E}{\partial N_\beta}\right)_{R,N_\alpha} dN_\beta + \left(\frac{\partial E}{\partial R}\right)_{N_\alpha,N_\beta} dR$$

which for $dN_\alpha = -dN_\beta$, and at equilibrium (dE=0), reduces to

$$\left(\frac{\partial E}{\partial N_\alpha}\right)_{R,N_\beta} = \chi_\alpha = \chi_\beta = \left(\frac{\partial E}{\partial N_\beta}\right)_{R,N_\alpha} \text{ for } \left(\frac{\partial E}{\partial R}\right) = 0 \text{ (i.e. } R=R_e) \text{ or } dR=0 \text{ (i.e. } R=\text{constant)}$$

However, since bond length variation rules are attractive for the chemist for interpretative purposes, one can include dR-sensitivities as well: infinitesimal displacements (*i.e.* sensitivities at the given configuration, indicating the trend only, but this is already very valuable information). Further elaboration of R-dependencies will be given in section 5.5 and 6.

5. Second Order Derivatives of E

5.1. $[d^2E/dN^2] = -[d\chi/dN] = \eta$: THE HARDNESS

The second order derivative of the electronic energy with respect to the number of electrons has been identified with the chemical hardness [15]. Hardness generally refers to the resistance to changes; chemical hardness is the resistance against electron transfer (changes in the number of electrons) or electron displacements (fluctuations). While electronegativity mainly determines the <u>direction</u> of the electron flow, the <u>amount</u> is mainly determined by the hardness. Equation III shows that the average effective hardness is also, as is the electronegativity, a structure and composition dependent property. It is a global quantity of the molecular system (and is equalized). In the molecular orbital interpretation it can be considered as the energy (HOMO-LUMO) gap between the frontier orbitals [16], and is also associated with polarizability. Its inverse is the global softness.

Hardness came in the chemical vocabulary due to Pearson, who also formulated the HSAB principle [17]: "Hard Acids prefer to coordinate to Hard Bases and Soft Acids prefer to coordinate to Soft Bases". Since then, many empirical rules could be tested on their merit with the hardness as a strong conceptual basis. In '83 hardness was given a firm theoretical basis, rooted in DFT [15]. The implications for the theory of acidity and basicity are evident already from the HSAB principle.

EXAMPLE : H_2O: HARDNESS AND FUKUI FUNCTION

The matrix notation of eq III for water is:

$$\begin{pmatrix} 2\eta_O^* & \dfrac{k}{R_{OH_1}} & \dfrac{k}{R_{OH_2}} & -1 \\[2mm] \dfrac{k}{R_{H_1O}} & 2\eta_H^* & \dfrac{k}{R_{H_1H_2}} & -1 \\[2mm] \dfrac{k}{R_{H_2O}} & \dfrac{k}{R_{H_2H_1}} & 2\eta_H^* & -1 \\[2mm] 1 & 1 & 1 & 0 \end{pmatrix} \cdot \begin{pmatrix} f_O \\ f_{H_1} \\ f_{H_2} \\ \overline{\eta} \end{pmatrix} = \begin{pmatrix} 0 \\ 0 \\ 0 \\ 1 \end{pmatrix} => \quad \begin{array}{l} f_O = +0.33136 \\ f_{H_1} = +0.33432 \\ f_{H_2} = +0.33432 \\ \overline{\eta} = 17.36096 \end{array}$$

5.2. $\left[d^2E / dNd\phi_\alpha\right] = -\left[dq_\alpha / dN\right]_v = f_\alpha = -\left[d\chi / d\phi_\alpha\right]_N = \left[d^2E / d\phi_\alpha dN\right]$:   FUKUI FUNCTION

The mixed second derivative of E is the Fukui function. It is a local property measuring the charge shift on an atom upon changing the total number of electrons in the molecule (or change in global χ upon a local change of the electrostatic potential at atom α); Electrodynamically, the Fukui function can be considered as a charge capacity and in terms of molecular orbital theory it is the electron density of the frontier orbitals [18]. Formally, it is a relative measure of local softness (since it is normalized and integrates to one). Absolute local softnesses can be obtained by dividing f_α by the global hardness: $s_\alpha = f_\alpha/\eta$, or by equations V and VI. The Fukui function plays a key role in the frontier electron theory of Fukui [19], where it is identified as a reactivity index (for orbital controlled reactions). The principle that then formally generates the frontier electron theory is the postulate of Parr and Yang [6]: "Of two different sites with generally similar dispositions of reacting with a given reagent, the reagent prefers the one which, on the reagent's approach is associated with the maximum response of the system's electronic chemical potential". In short: $|d\mu|$ big is good ! Since a general relation for $d\mu$ can be written:

$$d\mu = \eta dN + \int f(\overline{r}).dv(\overline{r}).d\overline{r}$$

it is seen that the Fukui function is particularly suitable for probing different sites within a molecule: a large value of f corresponds to a large $|d\mu|$ and indicates a high reactivity of the site. In the EEM formalism, the largest atomic f value for positively charged atoms will be the preferred locus for a nucleophilic attack, and the reverse, for negatively charged atoms, the preferred site for an electrophilic attack.

EXAMPLE: H_2O. HARDNESS EQUALIZATION

In this illustration the hardness equalization is demonstrated for a water molecule. We look for $\bar{\eta}$, given $\eta_O = f(f_H)$ and $\eta_H = f(f_H)$. Via EEM one finds:

$$\eta_O = 2\eta_O^* + \left[(-4\eta_O^*) + \frac{2k}{R_{OH}}\right].f_H \text{ and } \eta_H = \frac{k}{R_{OH}} + \left[2\eta_H^* - \frac{2k}{R_{OH}} + \frac{k}{R_{HH}}\right].f_H$$

$$\eta_O = 22.16574 - 14.64076.f_H \quad \text{and} \quad \eta_H = 14.84500 + 7.27491.f_H$$

These lines are drawn in Figure 2. The intersection gives for the two atom types $\eta_H = \eta_O = \bar{\eta} = 17.27512$ (hardness equalization) and the corresponding equilibrium Fukui function $f_H = 0.33402$.

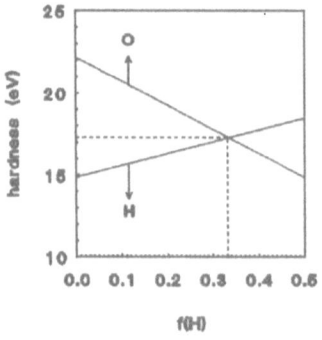

Figure 2: Hardness equalization principle

5.3. $\left[d^2E / d\phi_\alpha d\phi_\beta\right]_N = \left[dq_\alpha / d\phi_\beta\right]_N = \left[dq_\beta / d\phi_\alpha\right]_N = p(\alpha,\beta)$: THE TWO-VARIABLE LINEAR RESPONSE FUNCTION

$p(\alpha,\beta)$ is called the linear response function (or response kernel) and describes the mutual relation between the potential perturbation at one atom ($d\phi_\alpha$) and the charge response at another (or the same) atom (dq_β or dq_α). The $d\phi$-perturbation is defined at constant global number of electrons and is thus a dv perturbation. This also means that (initial) polarizations (e.g. at Lewis sites) as well as proton jumps (H^+, $dN=0$) can be considered. Because the response kernel describes an internal redistribution of electrons, it is not surprising that is can be used to calculate atomic polarizabilities [20]: the induced dipole moment (dq/dR) upon applying an electric field $(d\phi/dR)$ originates from an internal charge separation (dq):

$$\frac{dq}{dR} = \frac{dq}{d\phi} \cdot \frac{d\phi}{dR}$$

It has also been recognized that the problem of molecular vibration (constant N) can be expressed in terms of the density change $\Delta\rho(r)$ when nuclei are displaced by small amounts from their equilibrium positions, with a change in nuclear potential $\Delta v(r)$ [21]:

$$\Delta\rho(\bar{r}) = \int p(\bar{r},\bar{r}') . \Delta v(\bar{r}') . d\bar{r}'$$

This gives prospects for using the atomic equivalent $p(\alpha,\beta)$ of the linear response function to study nuclear vibrations, as will be pointed out later. The relevance of the response kernel and its relation with nuclear vibrations becomes clear in view of the RRKM theory (of unimolecular reactions). In this theory, bond breaking is achieved by harmonic osscilators via resonance with a catalytic surface [22,23]. Notice that, in the atomic resolution, $p(\alpha,\beta)$ can be written for each pair of atoms and as such defines a symmetrical square matrix of dimension n (n=number of atoms in the molecule): $\mathbf{p} = \{p(\alpha,\beta)\}$.

EXAMPLE : H_2O. LINEAR RESPONSE AND FUKUI FUNCTION

For a perturbation at H_1 the set of equations of type IV can be written as:

$$\begin{pmatrix} 2\eta_O^* & \dfrac{k}{R_{OH_1}} & \dfrac{k}{R_{OH_2}} & -1 \\ \dfrac{k}{R_{H_1O}} & 2\eta_H^* & \dfrac{k}{R_{H_1H_2}} & -1 \\ \dfrac{k}{R_{H_2O}} & \dfrac{k}{R_{H_2H_1}} & 2\eta_H^* & -1 \\ 1 & 1 & 1 & 0 \end{pmatrix} \begin{pmatrix} p(O,H_1) \\ p(H_1,H_1) \\ p(H_2,H_1) \\ f_{H_1} \end{pmatrix} = \begin{pmatrix} 0 \\ -1 \\ 0 \\ 0 \end{pmatrix} \Rightarrow \begin{array}{l} p(O,H_1) = +0.04652 \\ p(H_1,H_1) = -0.05102 \\ p(H_2,H_1) = +0.00450 \\ f_{H_1} = +0.33432 \end{array}$$

As such the perturbations can be done at O and H_2 as well, and the total $\mathbf{p}$-matrix becomes:

$$\mathbf{p} = \begin{pmatrix} -0.09305 & +0.04652 & +0.04652 \\ +0.04652 & -0.05102 & +0.00450 \\ +0.04653 & +0.00450 & -0.05102 \end{pmatrix}$$

5.4. $\left[d^2E/dq_\alpha dq_\beta\right] = \left[d\chi_\alpha / dq_\beta\right] = \eta(\alpha,\beta)$: HARDNESS KERNEL

To describe the mutual interactions between atoms upon changing the global number of electrons ($dN \neq 0$), the hardness matrix has been defined [24,25]. By taking equation II one directly writes:

$$\eta(\alpha,\alpha) = \left(\frac{d\chi_\alpha}{dq_\alpha}\right) = 2\eta_\alpha^* \qquad \text{and} \qquad \eta(\alpha,\beta) = \left(\frac{d\chi_\alpha}{dq_\beta}\right) = \frac{k}{R_{\alpha\beta}}$$

for the diagonal and off-diagonal terms respectively. These elements are called the hardness kernels. Their inverses are the softness kernels:

$$\left\{\eta(\alpha,\beta)\right\}^{-1} = \left\{s(\alpha,\beta)\right\} = \left\{\left(\frac{dq_\alpha}{d\chi_\beta}\right)\right\}.$$

Notice that the hardness and softness matrices are defined in a closed atom representation, so that the (local) electronegativity is not equalized during the perturbation.
As in the response matrix, the hardness matrix $\eta = \left\{\eta(\alpha,\beta)\right\}$ describes strongly coupled local charge fluctuations, now as a response to a global dN (instead of a local $d\phi$). The hardness matrix can be interpreted as an electron repulsion matrix (by making reference to the PPP-theory [26]). Diagonal and off-diagonal elements are the one- and two-center electron repulsion integrals respectively. Atoms at large distance have weak coupling.

EXAMPLE : H_2O. HARDNESS MATRIX

The hardness matrix used for the calculations in the previous sections is:

$$\eta = \begin{pmatrix} 2\eta_O^* & \dfrac{k}{R_{OH_1}} & \dfrac{k}{R_{OH_2}} \\ \dfrac{k}{R_{H_1O}} & 2\eta_H^* & \dfrac{k}{R_{H_1H_2}} \\ \dfrac{k}{R_{H_2O}} & \dfrac{k}{R_{H_2H_1}} & 2\eta_H^* \end{pmatrix} = \begin{pmatrix} 22.16574 & 14.97970 & 14.98001 \\ 14.97970 & 27.54648 & 9.53547 \\ 14.98001 & 9.53547 & 27.54648 \end{pmatrix}$$

Notice that it is strongly coupled (i.e. not diagonally dominated).
The inverse of the hardness matrix is the softness matrix:

$$\sigma = \begin{pmatrix} +0.09937 & -0.04014 & -0.04014 \\ -0.04014 & +0.05746 & +0.00194 \\ -0.04014 & +0.00194 & +0.05746 \end{pmatrix}$$

5.5. $\left[d^2E / dR^2\right] = H(i, j)$

A model force field is being developed within EEM [27], which defines the rigid bond stretch Hessian $\mathbf{H} = \{H(i, j)\}$. Rigid means that charges are kept constant during the dR perturbation (vibration); bond stretch means that angles are fixed and only pure stretch vibrations are considered. Apart from a few studies on bendings, Broensted acidity has mainly been investigated by means of stretching O-H frequencies. This is understandable in view of the hypothetical process in which the proton dissociates from the oxygen (homolytically or heterolytically). Relevance will be stressed later from the RRKM point of view.

EXAMPLE : H_2O. FORCE MATRIX (BOND STRETCH HESSIAN)

$$\mathbf{H} = \begin{pmatrix} \dfrac{d^2E}{dR_1^2} & \dfrac{d^2E}{dR_2 dR_1} \\ \dfrac{d^2E}{dR_1 dR_2} & \dfrac{d^2E}{dR_2^2} \end{pmatrix} = \begin{pmatrix} 42.59526 & -1.30842 \\ -1.30842 & 42.59825 \end{pmatrix}$$

Notice that the bond stretch Hessian for water is dominated by the diagonal elements. However, the off-diagonal terms will be shown to determine the right sequence for the force constants of the normal modes of vibration.

---

6. Alternative Representations of the Matrices: Eigensolutions

The last three paragraphs (3-5) introduce matrices, describing the strongly coupled mutual interactions between either atoms or bonds in a molecule. By a transformation of the axes these have alternative representations [24,25]. For example for the dN perturbations of a diatomic molecule one can define the sensitivities:

$$\left(\frac{dE}{dN_1}\right) \quad \text{and} \quad \left(\frac{dE}{dN_2}\right)$$

but also the alternative (pure polarization and pure charge transfer sensitivities):

$$\left(\frac{dE}{d(N_1 - N_2)}\right) \quad \text{and} \quad \left(\frac{dE}{d(N_1 + N_2)}\right)$$

The transformation is schematically represented below.

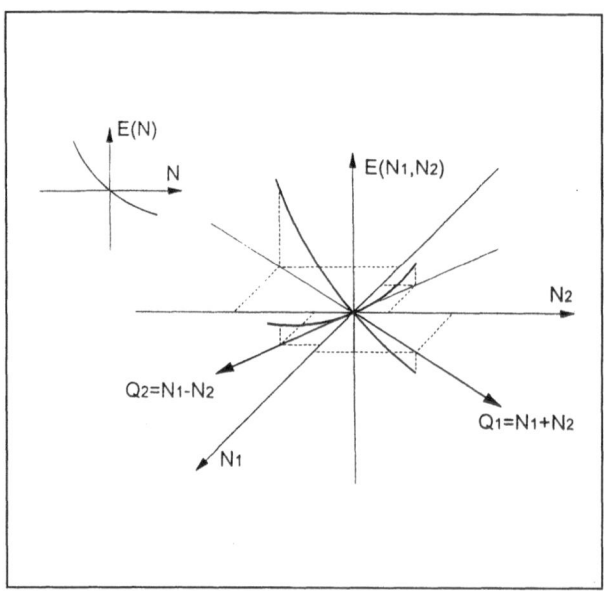

In general, such transformation can be realized by an orthogonal linear unitary transformation (*i.e.* the eigensolutions) providing eigenvectors and eigenvalues of the matrix (principal axes representation). The respective eigenproblems for the three matrices are shown below:

$$U^t.\eta.U=h \qquad O^t.p.O=r \qquad Q^t.H.Q=k$$

$$U^t.U=1 \qquad O^t.O=1 \qquad Q^t.Q=1$$

η, p, and H are the hardness matrix, response matrix and force matrix (bond stretch Hessian) respectively. The accompanying eigenvalues are the principal hardnesses, principal responses and molecular harmonic force constants (of stretch vibration) respectively. These represent the difficulty with which a certain normal mode takes place.

The eigenvectors of vibration, Q, represent collective nuclear displacements. The bond stretch Hessian generates only the (n-1) pure stretch modes. These can also be selected from quantum-mechanical calculations (as will be pointed out later: see section 10).

The eigenvectors of electron displacements, U and O, represent independent (decoupled) collective electron fluctuations in a molecule [24,25,27], with perturbations at variable and constant N respectively. A pattern of independent electron displacements (normal mode) is called a charge transfer or polarization channel. The hardness matrix provides n normal modes; the p-matrix provides (n-1) normal modes: the last one is not allowed because of linear dependencies (zero eigenvalue).

EXAMPLE : H_2O. EIGENSOLUTIONS OF MATRICES

The changes in atomic charges are represented as white and black circles, corresponding to an electron inflow and outflow respectively. The radius of the circle is proportional to $|U_{ij}|$. For the nuclear vibrations, the stretches and shrinkages are represented by arrows.

$h = \begin{pmatrix} 7.16465 & 18.01101 & 52.08303 \end{pmatrix}$

$U = \begin{pmatrix} -0.81611 & 0.00000 & +0.57790 \\ +0.40862 & +0.70711 & +0.57707 \\ +0.40865 & -0.70711 & +0.57708 \end{pmatrix}$

$r = \begin{pmatrix} -0.13957 & -0.05552 & 0.00000 \end{pmatrix}$

$O = \begin{pmatrix} -0.81650 & 0.00000 & +0.57735 \\ +0.40825 & +0.70711 & +0.57735 \\ +0.40825 & -0.70711 & +0.57735 \end{pmatrix}$

$k = \begin{pmatrix} 41.28834 & 43.90518 \end{pmatrix}$

$Q = \begin{pmatrix} +0.70711 & -0.70711 \\ +0.70711 & +0.70711 \end{pmatrix}$

7. Relations between Sensitivities

7.1. MAXWELL RELATIONS [8]

In the above compilation, use has been made of several Maxwell relations linking the sensitivities expressed in different ways (related via Legendre transformations):

$$p(\alpha,\beta) - \left(\frac{dq_\alpha}{d\phi_\beta} \right)_N - \left(\frac{dq_\beta}{d\phi_\alpha} \right)_N$$

$$f_\alpha = -\left(\frac{dq_\alpha}{dN}\right)_v = \left(\frac{d\overline{\chi}}{d\phi_\alpha}\right)_N$$

$$s_\alpha = \left(\frac{dq_\alpha}{d\overline{\chi}}\right)_v = \left(\frac{dN}{d\phi_\alpha}\right)_{\overline{\chi}}$$

These Maxwell relations have more than mathematical significance. It is for example of interest to see that the response function is symmetric with respect to the indices. The Fukui function can be described, either by a local response to a global perturbation, or, a global response to a local perturbation. Both descriptions are with the independent variables N and ϕ, and with reversed constraints. Since the Fukui function determines site selectivity (in orbital controlled reactions), it is clear that the implications of this Maxwell relation for catalysis are important: the same 'activation' (e.g. an increase of f) can be achieved either by a dN or a dϕ perturbation.

Apart form the interpretation of s_α as absolute local softness (by considering it as a fraction f_α of the global softness: $[-dq_\alpha/dN].[dN/d\chi]=f_\alpha.S$), it is seen that it can also be defined as the dN response of the system at constant χ. This is only possible when the system is coupled to an external electron reservoir (cf. metal catalysis), that can exchange electrons with it. This leads to theoretical evidence for the relation between softness and reactivity from the ensemble theory, where it is shown that large softness and large number fluctuations go together [28]:

$$S = \frac{1}{kT}\left(\langle N^2\rangle - \langle N\rangle^2\right)$$

Similarly, local softness can be written as:

$$s(\overline{r}) = f(\overline{r}).S = \frac{1}{kT}\left(\langle\rho(\overline{r}).N\rangle - \langle N\rangle.\langle\rho(\overline{r})\rangle\right)$$

In this statistical mechanical interpretation, local softness measures local fluctuations in the local electron density. Low-energy fluctuations have been shown to determine site selectivity for metals in chemisorption and catalysis [38]. For metals, local softness is the local density of states at the Fermi level, while global softness is the total density of states.

7.2. CLOSURE RELATION: dN=0

Notice the similarity between the softness kernels and response kernels:

$$s(\alpha,\beta) = -\left(\frac{dq_\beta}{d\phi_\alpha}\right)_{\overline{\chi}} \qquad\qquad p(\alpha,\beta) = \left(\frac{dq_\beta}{d\phi_\alpha}\right)_N$$

Only the limiting conditions determine the difference. For very hard systems the condition N=constant is automatically fullfilled and the condition for the softness kernel equals that of the response kernel. This can also be derived from another point of view. An exact relation between $p(\alpha,\beta)$ and $s(\alpha,\beta)$ can be written [29]:

$$p(\alpha,\beta) = -s(\alpha,\beta) + f_\alpha . f_\beta . S$$

For S=0 (*i.e.* for dN=0, or infinite hardness)

$$p(\alpha,\beta) = -s(\alpha,\beta)$$

In general for hard systems, *i.e.* S small (*e.g.* small organic molecules) one may write:

$$p(\alpha,\beta) \cong -s(\alpha,\beta)$$

7.3. RELATION BETWEEN ELECTRON POPULATION AND VIBRATIONAL NORMAL MODES [27,30]

On an intuitive basis one can expect a relation between the p-matrix and the bond stretch Hessian:
- both perturbations $d\phi$ and dR occur at constant N
- equal dimensionality of eigensolutions: n-1
- intuitively each charge density reorganization is accompanied by a bond length variation and *vice versa*.
The expected equivalence between nuclear vibrations and electron vibrations can schematically be represented as follows:

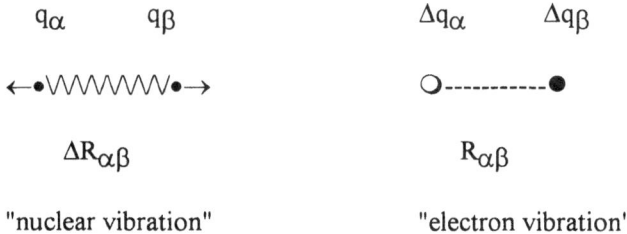

One can envisage the same analogy in the local as well as in the normal picture. This resulted in the construction of "translator matrices", translating local and normal modes of the electron population space into that of the nuclear position space ([27], in preparation).
Instead of using this formal approach, the assignments (from nuclear vibrations into polarization channels) have been done intuitively up to now: first the (n-1) normal modes of dominating stretch vibration are selected from a list containing all (3n-6) vibrations, calculated from semi-empirical quantummechanical methods; then these are related to the (n-1) independent polarization channels, on the basis of large localized contributions from atomic charge shifts (in the electron population space), sharing the bond which is most

affected in a given normal vibration. Group theoretical arguments, such as (anti-)symmetry and degeneracies are helpful in the assignments. Finally, some typical patterns of conjugated electronic and nuclear displacements can be identified and can be used to relate the patterns, *e.g.* in functional groups (*cf.* group vibrations).

To amplify the impact of our statement it will be shown that not only the patterns, but also the physical quantities associated with them (*e.g.* energetic aspects) are related [30]. For a series of small organic molecules for instance, the relation between the force constants of nuclear vibrations and the principal responses of internal charge displacements (polarization) will be demonstrated and discussed.

EXAMPLE : H_2O. ELECTRONIC AND NUCLEAR VIBRATIONS

In the case of water it is evident that one links to one-another the symmetric and anti-symmetric modes of both electronic and nuclear vibration respectively. In section 6 we anticipated the above-mentioned analogy by ligning up the conjugated patterns with each-other (we refer to the example in section 6). Notice that the physical properties associated with the normal modes are also related: analogous dipole moment changes can be observed, and the sequence of the force constants is the same:

$$k_{symmetric} \quad < \quad k_{asymmetric}$$
$$h_{symmetric} \quad < \quad h_{asymmetric}$$

The above concept is vital for the theory of chemical reactivity, since it allows one to interpret polarizational perturbations (the usual language of the chemist) in terms of bond breaking mechanisms. These are elementary steps in chemical reactions and provide guidelines to direct (bond breaking) activity and selectivity (*cf.* Gutmann rules).

8. Rules of Thumb

In section 9 we review some applications which have been carried out within the EEM formalism. We mainly focus on solid state problems, especially zeolites. We often follow a line of thought which is briefly summarized by the following rules of thumb. These can be derived from the sensitivity analysis [4].

(i) 'local softness is dominantly positive'

(ii) 'local softness is additive'

(iii) 'there is an inverse relation between the softness of an atom-in-a-molecule (s_α) and the softness of its environment (S_α)'

Schematically one can write:

(i) $s_\alpha \geq 0$

The first rule cannot be mathematically (formally) derived. It is based on the outcome of the calculations and intuition (it is intuitively reasonable to presume that for a $dN>0$, all atoms will share the extra electrons, and none would end up with less). Negative values for s_α are very rare, and if they occur they are generally small.

(ii) $\sum_{\alpha=1}^{n} s_\alpha = S$

The second rule is a consequence of the normalization condition for the Fukui function and of the definition of absolute local and global softnesses:

$$\sum_{\alpha=1}^{n} f_\alpha = 1 \qquad\qquad s_\alpha = f_\alpha / \overline{\eta} \qquad\qquad S = 1/\overline{\eta}$$

(iii) $s_\alpha \propto -S_\alpha$

This rule directly follows from eq V (the sum of softnesses is a zero-sum game) by assuming that the external softness potential is written as:

$$S_\alpha = k \sum_{\beta \neq \alpha}^{n} \frac{s_\beta}{R_{\alpha\beta}}$$

9. Applications: First Order Derivatives

9.1 $d\phi$-SENSITIVITIES

One of the first successes of the EEM formalism applied to solids is the calculation of the charge distribution in various zeolite type structures (with hypothetical SiO_2 composition) and dense silica polymorphs [10]. An advantage is that long-range effects can be fully taken into account, by performing Ewald type summations in reciprocal space. Various charge-structure relationships could be rationalized and demonstrated. In the dense parts of the structure for example, *i.e.* at high local density, an increased charge separation (ionicity) is observed in accordance with eq II. This has direct consequences for the negative charges (basicities) of the oxygens *e.g.* which are position dependent.

Next, the charge distribution in zeolite Faujasite was followed as a function of the chemical composition [4]. The results are shown in Figure 3, where these are plotted against the experimentally observed average chemical shifts. We observe five groups of silicon atoms according to the number of aluminum atoms in the second coordination sphere [Si(nAl)]. The correlation with the ^{29}Si NMR chemical shifts is excellent and agrees qualitatively with the general relation derived from shielding theory.

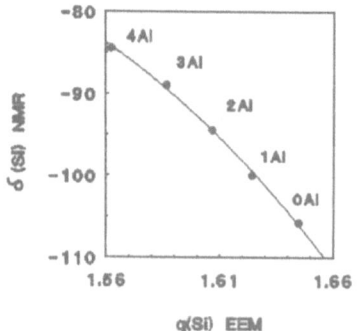

Figure 3: Silicon NMR chemical shift versus Si charge as calculated from EEM.

Finally the charge distribution and other properties of bridging hydroxyls in zeolites have been studied [31]. By calculating energy differences (between crystallographically different proton sites) from EEM charges, the proton distribution could be obtained from Boltzmann's law and was in good agreement with experimental literature data in that protons prefer oxygens of the supercages, large channels and channel intersections.

9.2. dN-SENSITIVITIES

The structural features of the zeolites are not only reflected in the well predicted charge distribution, but also in a variation of the average effective electronegativity with structure types (even for a pure SiO_2 compostion) [10]. This is due to an adequate equalization of electronegativity which explicitly takes into account the external (electrostatic) potential contributions at each crystallographic site. The results for different dense silica polymorphs are shown in Figure 4 and are plotted against the respective refractive indices. As pointed out, the electronegativity is a fundamental constant of a system, and contains significant information on the shape ("compactness") of the electron cloud. Since the electronegativity directly relates to the strength with which the electrons are held into place (upon interference with electromagnetic radiation), a nearly perfect correlation with the refractive index is found. It was also demonstrated that there exists an inverse relation between the average electronegativity and the framework density, with the open zeolite structures being characterized by a high electronegativity (Figure 5). Given the same chemical composition, the structure type (framework density) must have a direct influence on the acidity/basicity (probably more so for Lewis acidity/basicity).

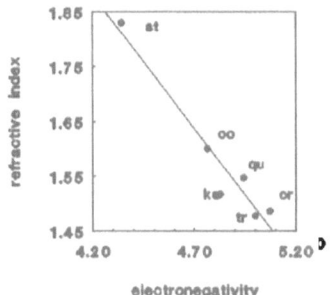

Figure 4: Variation of the average framework electronegativity with the refractive index n at λ=589 μm for some dense silica polymorphs: stishovite (st), coesite (co), quartz (qu), tridymite (tr), keatite (ke), and cristoballite (cr).

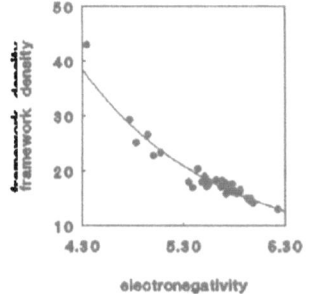

Figure 5: Variation of χ with the global framework density (number of Si atoms per nm^3) for the compounds of Figure 4 and for some zeolite-type frameworks with hypothetical SiO_2 composition.

9.3. dR-SENSITIVITIES

A possible application of the zero gradient (dE/dR=0) would be an energy minimization as a function of R, and thus a geometry optimization. However, the force field developed, is designed for vibrational analysis, and should be used to define, rather than to obtain, zero forces at equilibrium. As such the EEM formalism is complementary to molecular mechanics simulations in which an equilibrium geometry or configuration is obtained as a function of a fixed charge distribution, via energy minimization. In EEM, an equilibrium charge distribution is obtained as a function of a fixed geometry, through an electronegativity equalization procedure (both principles are of course equivalent).

Schematically one can write:

MMM	EEM
E_{min}	χ_{equal}
$R_e(q)$	$q_e(R)$

10. Applications: Second Order Derivatives.

In this section we demonstrate the meaning of the second order sensitivity coefficients with a typical example, namely the perturbation (dϕ and dN) of the zeolite framework by adsorbed cations, and its relation with the intrinsic framework properties. The effects observed are rationalized with the rules of thumb of section 8, and their possible implications for acidity/basicity are outlined.

10.1 dϕ -SENSITIVITIES

10.1.1. *Charge Perturbation Sphere.* We consider the perturbation of the framework by a charge compensatin cation (Na^+). The charge rearrangements in the Faujasite framework are calculated as a function of the distance from the locus of the perturbation (Figure 6). The cation is placed in the center of the six-ring of a hexagonal prism (site I'). No charge transfer between the lattice and the cation is considered and the framework is rigid.

The perturbation is compensated for locally by means of polarization effects in the lattice. The positive potential of the cation is balanced by the negative potential induced in the lattice. It is seen that the relaxational correction are restricted to a perturbation sphere of radius 7-8 Å. This confirms one of the hypotheses about the cation distribution in zeolites [32]: it was supposed that cations generate their own site energy by attracting negative charges around themselves. An increased Al substitution (harder) gradually decreases the extent of this perturbation sphere.

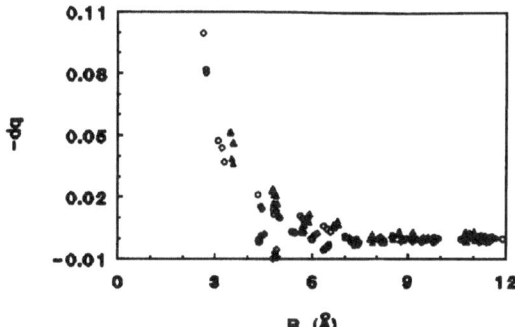

Figure 6: Perturbation sphere of a cation (Na^+ at site I' in Faujasite; Si/Al=3): charge shifts on all framework atoms versus distance (in Å) from the cation (O: ° ; T: ▲).

10.1.2. *Perturbation Potential versus Softness Potential.* Intuitively the softness potential S_α has been identified with the ability to perturb the environment. S_α indeed contains the Fukui function which is a measure for the charge shifts on the atoms upon perturbation. The electrostatic potential reflects the <u>charges</u> in the surroundings, while the softness potential measures <u>charge shifts</u> (or sensitivities). Essentially:

$$\phi_\alpha = \sum \frac{q}{R} \qquad \text{and} \qquad S_\alpha = \sum \frac{dq}{R}$$

To clarify the physical meaning of these expressions, the calculations carried out on real perturbations of the lattice, were reationalized in terms of the above relations; the changes in external potential should correlate with the softness potential.

In Faujasite type zeolites, the most probable cation sites (I, I' and II) are located on the threefold axis A Na^+ cation was displaced along this axis (whereby each time the framework charges were recalculated) and at regular distances the electrostatic potential was calculated (v2). The same was done without any cation present (v1). The difference v1-v2 (the perturbation potential dv) is due to the charge shifts on the lattice, caused by the cation. On the same axis the softness potential was calculated. The results are represented in Figures 7 and 8. It is seen that the perturbation potential and the softness potential exhibit the same maxima and minima. For a maximum in the softness potential the framework relaxes most easily, *i.e.* low energy fluctuations take place.

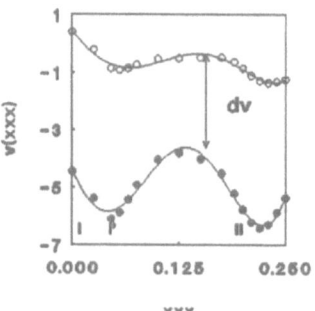

Figure 7: Potential along the trigonal axis of Faujasite with (v2) and without (v1) cation present; the difference (dv) is the perturbation potential.

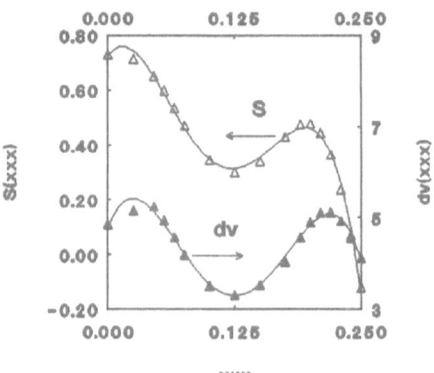

Figure 8: Correlation between the softness potential S(xxx) and the perturbation potential dv(xxx) along the trigonal axis of Faujasite (xxx=fractional coordinate).

10.2 dN-SENSITIVITIES

10.2.1 *Intrinsic Framework Properties*. The intrinsic electronic properties of 30 zeolite structure types with hypothetical SiO_2 composition have been estimated by EEM. The structure and connectivity dependent local softnesses (Fukui functions) for all crystallographically different atom types were calculated. The results are summarized in Figure 9, giving the local softness s as a function of the local topological density d. The latter is defined as $d_\alpha = \sum_{\beta \neq \alpha} 1/R_{\alpha\beta}$, and measures the number of atoms at their respective distances from a central point or atom. It allows to formally distinguish between (atoms in) dense and open parts of a structure.

With the rules of section 8 we are in a position to rationalize the trend observed in Figure 9, namely the inverse relation between the local topological density of an atom and its local softness: since in all cases the local softness is positive, and since local softness is additive, each atom in the environment of a central atom will contribute to generate a positive softness potential at that atom; because of the inverse relation between the softness of an atom in a molecule and its environment, we can expect an increased local softness upon decreasing the local density. Therefore, atoms in open parts of the crystal are predicted to be the softest because they have the hardest environment. Thus, in zeolites, the softest oxygens are located in the walls of the large channels and cavities; these are the most basic. Also the Si/Al ratio comes into play. Al atoms (since they are hard), soften the neighbouring oxygens. This is confirmed by several experimental data, showing an increased basicity upon increasing the Al-content (catalysis, XPS O(1s) binding energies).

The data also reveal that intrinsically, T-atoms are softer than oxygens. This means that T-atoms are the electron reservoirs in such systems, which agrees with the electrodynamical and statistical mechanical interpretation for local softness (charge capacitance and low energy density fluctuations respectively) and confirms previous work concerning the charge delocalization in zeolites.

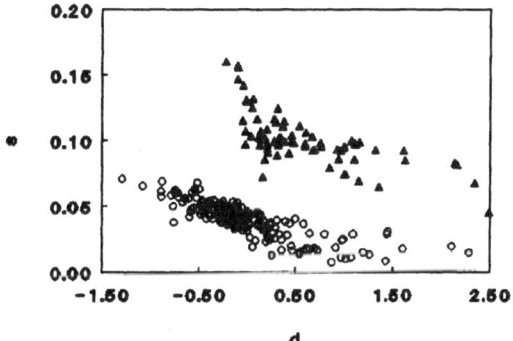

Figure 9: Local softness versus local topological density for O-and T-atoms in 30 zeolite structure types (O: ° ; T: ▲)

10.2.2. *Charge Perturbation Sphere.* Returning now to the sensitivity analysis of the cation, and investigating the perturbation sphere in detail, it can be seen that the charge shifts for the different atom types follow the order $dq_T > dq_O$, in full agreement with the order of sensitivities or local softnesses: $s_T > s_O$: T-atom charges are most flexible, while the oxygen electron populations are rather rigid.

However, upon increasing the Al-content (increasing the global hardness), the neighbouring oxygens are softened and these exhibit more pronounced charge shifts as is shown in Figure 10. This polarization effect directly influences the effective charges on the oxygens and their basicity. Together with the cations they will be responsible for the sorption characteristics of the surface.

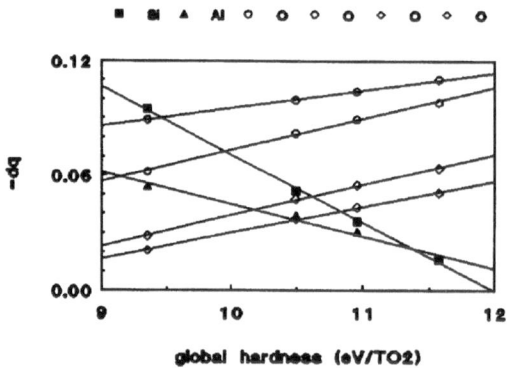

Figure 10: Charge shifts (on some atoms of the perturbation sphere in the neighbourhood of the perturbing cation) as a function of the global hardness. Upon increasing the T-atom hardness, the charge shifts on the oxygen atoms increase, reflecting their increased softness.

10.2.3. *Local Softness versus Local Charge Transfer Properties.* Global charge transfer properties in zeolites and oxides have been rationalized in terms of the (optical) electronegativity concept [33]: the 'electron attractive power' of the oxygens governs the ligand to metal charge transfer. The frequency of the band maximum in the electronic spectra has indeed been shown to be related to the average electronegativity [34]. However, in this way structural effects cannot be distinguished and within one structure one cannot separate the different components of the band envelope, due to different locations of the transition metal ion (TMI) in the structure.

Detailed theoretical analyses of the spectroscopic data of Cu^{2+} ions in several zeolites, allowed to assign all signals and to distinguish between different sites within one structure [35,36]. The difference in the spectroscopic parameters so obtained has a physical significance for the coordinative bond and reflects the local charge transfer properties of the TMI complex. As is obvious by now the local softnesses also reflect local charge transfer properties. We thus expect the spectroscopic and softness parameters to correlate. This is illustrated by the following example. For Cu^{2+} on different sites in some zeolites, the EPR parameters are plotted against the local softnesses of the oxygens of the first coordination sphere in Figure 11. The general trend is that $g_{//}$ decreases and $A_{//}$ increases with increasing softness of the oxygens. This is expected for an increasing covalency of the Cu-O bond [37]. Thus increasing the local softness of the oxygens also means increasing the covalent character of the coordinative bond, a clear manifestation of the HSAB principle (since, generally, TMI's are known to be soft). An increased oxygen softness, which can *e.g.* be induced by a nearby hard Al-atom, reflects an increased electron donor capacity (basicity), *i.e.* the easiness with which dN-perturbations take place and the amount of charge transfer during the perturbation.

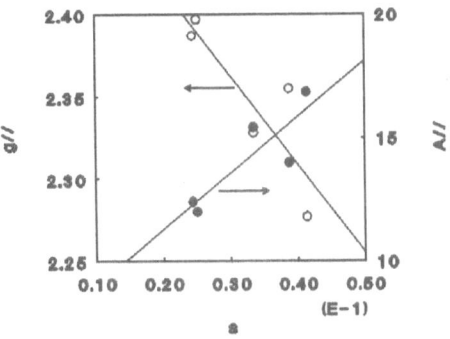

Figure 11: Correlation between the EPR parameters ($g_{//}$ and $A_{//}$) of Cu^{2+} at different sites in various zeolites, and the softness of the oxygens in the first coordination sphere.

10.3 dR-SENSITIVITIES

10.3.1. *Mapping of the Patterns.* To illustrate the one-to-one relation between the polarization patterns and the normal modes of nuclear stretch vibrations, calculations have been performed on a series of small (mainly organic) molecules RX, with R= H-, CH_3-, C_2H_5-, C_3H_7-, C_4H_9- C_5H_{11}- and X= -CH_3, -OH, -NH_2, -OR. Equilibrium geometries and vibrational data were obtained from the semi-empirical quantummechanical code AMPAC, with AM1 Hamiltonian. The principal hardnesses and responses were calculated by the EEM formalism. For 3 simple cases, the (n-1) eigenvectors of stretch vibration and the (n-1) patterns of independent electron redistribution are represented schematically in Table 2. Clearly the vibrational modes can be predicted from the polarization channels. To know whether or not a bond is affected (contributes to a mode), one divides the molecule in two parts by breaking that bond. If $\sum_{i}^{n} O_{i\alpha} = 0$ for α belonging to one part, no electron transfer (polarization) takes place over the bond, and as such there is no bond length variation. Degeneracies are convincingly reproduced.

Table 2: Relation between electron and nuclear vibrations for H_2O, CH_4 and CH_3OH.

r_γ (eV)	O_γ	Q_i	k_i(mdyne/ Å)	ν_i (cm^{-1})
H_2O				
-0.13688			6.56	3584.95
-0.05516			7.56	3505.73
CH_4				
r_γ (eV)	O_γ	Q_i	k_i(mdyne/ Å)	ν_i (cm^{-1})
-0.24853			3.07	3215.43
-0.05099			6.04	3102.51
-0.05098			6.04	3103.49
-0.05098			6.04	3104.70

		CH$_3$OH		
r_γ (eV)	O_γ	Q_i	k_i(mdyne/ Å)	ν_i (cm^{-1})
-0.22611			3.48	3149.00
-0.10418			6.97	3503.42
-0.05083			5.85	3056.52
-0.05073			5.99	3074.22
-0.03635			8.14	1361.48

124

10.3.2. *Energetic Aspects.* In Figure 12 the principal responses are plotted against the stretch force constants. It is seen that these correlate well. This means that the difficulty of electron displacements (principal response or hardness) reflects the difficulty of nuclear displacements. It is remarkable that the 'functional groups' (O-H, N-H, C-N, C-O) fall out of range (lower group in the figure): electronic fluctuations in functional groups are easier than expected on the basis of their force constants. For all other bonds, the relation seems to be universal for the different chain lenghts as well as for the different types of molecules.

This constatation is very important for the theory of reactivity and catalysis: it has been argued (in the spirit of the RRKM theory of unimolecular reactions) that the role of a catalyst is, *inter alia,* to supply energy by vibrational resonance directly into that vibrational mode of the reacting molecule that most closely represents that distorsion (perturbation) of the molecule which is critically needed for the reaction to proceed [22,23]. In view of the equivalence just established, this is what the 'unlocking' of a polarization channel is all about.

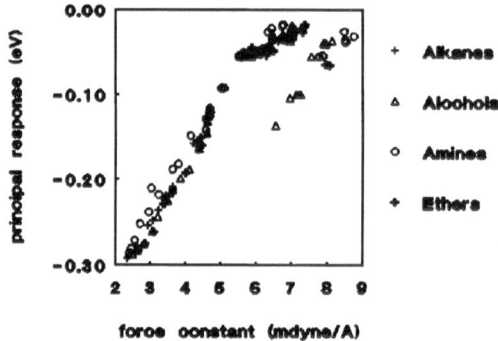

Figure 12: Correlation between principal responses (approximately minus inverse hardnesses) and force constants for the bond stretches in a series of organic molecules.

11. Acknowledgements

BGB thanks the Research Council of the Catholic University of Leuven (Onderzoeksraad) for a Postdoctoral Mandate (PDM 93/57); GOAJ thanks the Institute for Scientific Research in Industry and Agriculture (I.W.O.N.L); HT is grateful to the European Community for a Postdoctoral Mandate (Human Capital and Mobility Program); the authors acknowledge financial support from the Belgian State Secretariat for Scientific Research in the form of a Concerted Research Action (G.O.A.).

12. References

1. Mortier, W.J.; *Structure and Bonding: Electronegativity*, **66:** 125 (1987).
2. Mortier, W.J.; in *"Theoretical Aspects of Heterogeneous Catalysis"* (J.B. Moffat, Ed.), Van Nostrand Reinhold Catalysis Series, New York, pp. 135-159 (1990).
3. Mortier, W.J.; *Academie voor Wetenschappen, Letteren en Schone Kunsten van België*, Jaargang **52**, Nr. 1 (1990).
4. Baekelandt, B.G., Mortier, W.J., Schoonheydt, R.A.; in *"Modelling of structure and Reactivity in Zeolites"* (R.A.Catlow, Ed.), Chapter **8**, pp. 157-182 (1992).
5. Baekelandt, B.G., Mortier, W.J., Schoonheydt, R.A.; in *"Structure and Bonding: Chemical Hardness"* (K.D. Sen, Ed.), Volume **80**, pp. 187-227 (1993).
6. Parr, R.G., Yang, W.; *"Density Functional Theory of Atoms and Molecules"*, The Internat. Ser. Monographs. Chem. **15**, Oxford Univ. Press: New York (1989).
7. Hohenberg, P., Kohn, W.; *Phys. Rev. Ser.*, **136B:** 864 (1964).
8. Nalewajski, R.F., Parr, R.G.; *J. Chem. Phys.*, **77:** 399 (1982).
9. Baekelandt, B.G.; *Ph.D. Thesis*, **#224**, K.U. Leuven, Faculty of Agronomy (nov. 1992).
10. Van Genechten, K.; *Ph.D. Thesis*, **#160**, K.U. Leuven, Faculty of Agronomy (nov. 1987).
11. Mortier, W.J., Ghosh, S.K., Shankar, S.; *J. Am. Chem. Soc.*, **108:** 4315 (1986).
12. Baekelandt, B.G., Lievens, J.L., Mortier, W.J., Schoonheydt, R.A.; *J. Am. Chem. Soc.*, **113:** 6730 (1991).
13. Jacobs, P.A.; *Catal. Rev. - Sci. Eng.*, **24:** 415 (1982).
14. Jacobs, P.A., Mortier, W.J.; *Zeolites*, **2:** 226 (1982).
15. Parr, R.G., Pearson, R.G.; *J. Am. Chem. Soc.*, **105:** 7512 (1983).
16. Pearson, R.G.; Proc. *Natl. Acad. Sci. USA*, **83:** 8440 (1986).
17. Pearson, R.G.; *J. Am. Chem. Soc.*, **85:** 3533 (1963).
18. Parr, R.G., Yang, W.; *J. Am. Chem. Soc.*, **106:** 4049 (1984).
19. Fukui, K.; *Science (Washington D.C.)*, **218:** 747 (1982).
20. Vela, A., Gazquez, J.; *J. Am. Chem. Soc.*, **112:** 1490 (1990).
21. Handler, G.S., March, N.H.; *J. Chem. Phys.*, **63(1):** 438 (1975).
22. Larsson, R.; in *"Catalysis Today"* Elsevier Science Publishers B.V. Amsterdam, **3:** 387 (1988).
23. Larsson, R.; *Chemica Scripta*, **12:** 78 (1977).
24. Nalewajski, R.F., Korchowiec, J., Zhou, Z.; *Int J. Quant. Chem.: Quant. Chem. Symp.*, **92:** 349 (1988).
25. Nalewajski, R.F.; in *"Structure and Bonding: Chemical Hardness"* (K.D. Sen, Ed.), Volume **80**, pp. 115-186 (1993), and references therein.
26. Pariser, R., Parr, R.G.; *J. Chem. Phys.*, **21:** 767 (1953); Pople, J.A.; *Trans Faraday Soc.*, **49:** 1375 (1953).
27. Baekelandt, B.G., Janssens, G.O.A., Toufar, H., Mortier, W.J., Schoonheydt, R.A. and Nalewajski, R.F.; in preparation
28. Yang, W., Parr, R.G.; *Proc. Natl. Acad. Sci. USA*, **82:** 6723 (1985).
29. Berkowitz, M., Parr, R.G.; *J. Chem. Phys.*, **88:** 2554 (1988).

30. Janssens, G.O.A., *Ms. Thesis*, K.U. Leuven, Faculty of Agronomy (1993).

31. Baekelandt, B.G., De Tavernier, S., Schoonheydt, R.A.; *Preprints - Proceedings of the ACS Spring Meeting* (San Fransisco), Division Petroleum Chemistry, **37(2):** 596 (1992).

32. Van Dun, J.J.; *Ph. D. Thesis* #**159**, K.U.Leuven, Faculty of Agronomy (nov 1987).

33. Duffy, J.A.; *J. Solid State Chem.*, **61:** 3426 (1986).

34. Packet, D., *Ph.D. Thesis* #**151**, K.U. Leuven, Faculty of Agronomy (feb. 1987).

35. Packet, D., Schoonheydt, R.A., *ACS Symp. Ser.*, **368:** 203 (1988).

36. De Tavernier, S., Schoonheydt, R.A.; *Zeolites*, **11:** 155 (1991).

37. Schoonheydt, R.A.; *Catal. Rev.-Sci. Eng*, **35:** 129 (1993).

38. Falicov, L.M., Somorjai, G.A.; *Proc. Natl. Acad. Sci. USA*, **82:** 2207 (1985).

The Acid Function in Zeolites

Gregory J. Gajda and Jule A. Rabo
UOP
Des Plaines, IL 60017

1. INTRODUCTION

The growing importance of zeolites in industrial adsorption and catalysis applications has resulted in a broad investigation of these materials that uses a wide variety of techniques. Consequently, the chemistry of microporous crystals has become an important chapter in the story of inorganic chemistry. In catalysis, the most important characteristic of zeolites is the acidity and its interplay with the intercrystalline chemical environment. Zeolite characteristics have been extensively studied both in the absence and in the presence of probe molecules and reactants, and correlations have been developed relating performance in acid catalysis to chemical composition and structural characteristics.

Studies of zeolites have resulted in correlations that are useful in predicting such chemical changes as variations in acidity with the Si/Al ratio of a given zeolite crystal type. Generalized relationships applicable to all zeolites also have been developed to relate aluminum density to catalytic properties. In addition, theoretical modeling of zeolites greatly enhanced the knowledge of chemical bonding, interaction with adsorbed molecules, diffusion, and other important characteristics.

However, in spite of the growing general knowledge in zeolite science, some of the most basic problems in zeolite chemistry and catalysis have remained unresolved. For example, no experimental data base is available to explain the specific causes of strong zeolite acidity, namely, the difference in protic acidity between zeolites and silica-alumina gels of similar compositions. Furthermore, no basis currently exists for predicting acid strength based on zeolite structure.

J. Fraissard and L. Petrakis (eds.), Acidity and Basicity of Solids, 127–179.

The purpose of this paper is to review the chemical and structural characteristics of zeolites that are deemed relevant to acidity and to acid catalysis. Several conclusions are also drawn to interpret some of the important chemical phenomena. The catalytic data have been the subject of several detailed reviews and are not reviewed here. This review is based, in part, on an earlier paper by Rabo and Gajda [1].

2. ACID-BASE CHEMISTRY

A Bronsted acid is defined as a species that can donate a proton, and a base is defined as a species that can accept it. The dissociation of an acid (HA) in solvents is described as an acid-base equilibrium: HA + S <----> A⁻ + SH⁺. In solution, the ionization of acid HA leads to a new acid with solvent S, forming SH⁺. This acid-base concept is applied widely from neutral to positively and negatively charged systems. The extent to which an acid ionizes depends on the acid itself and on the basicity of the solvent applied. Therefore, the solvent plays a major role in the ionization process [2]. Hammett introduced a method to determine acid strength by measuring the degree of protonation of basic indicators in acid solutions. The Hammett acidity function, H_o is derived from B + SH⁺ <----> BH⁺ + S (SH⁺ is solvated proton, B is the base, and S is solvent):

$$H_o = pk_{BH+} - \log \frac{[BH^+]}{[B]} \quad (1)$$

The Hammett acidity concept postulates that the H_o function is unique for any particular series of acid solutions of varying acid concentration. This postulate has broad applications, but it also has limitations with bases that do not give a straight plot between $\log [BH^+]/[B]$ and H_o. In a solution system, the solvent may have a major effect on both the proton donor and acceptor, depending on characteristics such as polarity and polarizability, that are best represented by the dielectric constant. Thus, the acid function in solution can be adequately characterized, at least for systems for which the properties of constituent molecules are known [2].

Recent progress in organometallic chemistry has produced a large variety of complex molecules with acidic or basic character. As a result, many new observations were made that required refinements of the rules governing acid-base interaction. The refinements advanced by Chatt and Pearson categorize systems as hard and soft acids and bases [3-6]. Even though the concept was developed mainly to explain the interaction between Lewis bases and acids, it is also relevant to protic systems. In nucleophilic displacement reactions, the rate of displacement was in most cases influenced by the proton affinity of the base. However, in other cases, the rate depended mainly on the polarizability of the base.

To explain these rate phenomena, acids and bases were grouped into "hard" vs. "soft" species on the basis of low vs. high polarizability, respectively. In equilibrium measurements, a similar characterization, which was found useful for acids and bases, also allows for a base to be both soft and strongly proton bonding. In general, hard acids prefer to bind with hard bases, and soft acids with soft bases. Among the hard acids, H^+ is found to be the most typical [4,5]. In the formation of the covalent bond, the two-center coulomb integral is maximized using bonding orbitals of similar size and similar electronegativity. This maximization explains the preference of hard acids for hard bases. For soft acids and bases, a significant or even a major part of the interaction in the acid-base system is derived from various aspects of electron correlation. This interaction may rest on the polarizability interaction (London), but it may also involve electron correlation between high-energy, exited-state orbitals of the acid and base molecules.

Pearson [5] defined the terms *absolute electronegativity* and *absolute hardness* based on molecular orbital considerations. Accordingly, the absolute hardness is twice the energy gap between the highest occupied valence orbital and the lowest unoccupied orbital. With this definition in mind, he suggests the following characteristics in the chemistry and interaction of soft and hard acids and bases:

- Hard acids prefer bonding with hard bases, and soft acids with soft bases.
- Hard molecules have a large energy gap, and soft molecules have a small energy gap.
- High polarizability is characteristic of soft acids and bases.
- Although the electron chemical potential is constant in a molecule, the hardness varies from atom to atom.
- Soft molecules undergo unimolecular rearrangements more readily than hard molecules.

These generalized considerations are of particular importance because they are consistent with the results of recent theoretical work on the chemical bonding in large electron-deficient (cationic) molecules, such as carbocations, carboranes, and their metal complexes. The theoretical molecular modeling, with extensive use of supercomputers, resulted in calculated nuclear magnetic resonance (NMR) shifts and bond energies closely matching experimentally determined values. Theoretical modeling of the polarizability of carbocations shows that the molecular bonds in these species use a variety of orbitals, including excited-state orbitals, which tend to increase the polarizability [7,8]. Thus, generally, softness and high polarizability are typical properties of large carbocations.

In contemplating acid-base interaction, the important role of the solvent must always be kept in mind. Strongly polar solvents strongly enhance the stability of ionized molecules. Most solvents tend to reduce hard character, and they enhance soft character.

Solution chemistry is fortunate to deal with molecules of usually well-defined characteristics. Therefore, the interaction between the acid, base, and solvent can be readily calculated from the existing data base using well-proven chemical principles. In contrast, the understanding of acid-base interaction between solid acids and proton acceptor reactants greatly suffers from the ill-defined nature of the surface of the acidic solid. Because the acidic solid surface is ill-defined, no adequate data base is available for the calculation of the properties of complexes formed between protonated reactants and the solids. In addition, standard experimental techniques (spectrophotometry or NMR) used to measure protic acidity in solutions are severely restricted by limitations caused by molecular sieve phenomena and by other experimental difficulties. The NMR technique is particularly useful for measuring equilibrium by estimating the first-order rate constant from line-shape analysis.

In spite of obvious differences between solution acid chemistry and zeolite acids, the basic aspects of the chemistry between the solid proton donor and the acceptor molecules are the same for both systems. Therefore, the acid catalytic activity of H-zeolites depends on the following factors:

- Acid strength and concentration of zeolite hydroxyl groups
- Media (solvent) and diffusion effects in zeolites
- Base strength of reactant molecules
- Chemistry of the carbocation-like reaction intermediates

In zeolites, the interaction between acid sites and reactant molecules involves not only the zeolite's protic site but also its media-solvent effect: the contribution from the surrounding zeolite crystal [9] . The following sections discuss zeolite characteristics relevant to the acid function.

3. CHEMISTRY AND CHARACTERIZATION OF ACID SITES IN ZEOLITES

3.1. Formation of Acidic O-H Groups

After crystallization, protonation is usually carried out by thermolyzing NH_4^+ or other proton precursor cations [10-12]. Some loss of zeolite crystallinity is observed following removal of the base as is a tendency toward framework aluminum hydrolysis, especially in the presence of steam. This reaction is particularly true for aluminum-rich zeolites (A, X), where conversion of the NH_4^+ to the H^+ form by thermal treatment results in total loss in crystallinity [11]. Thus, with zeolites, the proton is not an ordinary cation, like alkali or other metal cations. Clearly, the interaction between proton and zeolite framework oxygen can be fairly called a "proton attack" because it causes substantial changes in the bonds surrounding the newly formed hydroxyl groups [13]. This change, of course, should be expected considering the high electron affinity of the proton (13.5 eV) relative to even the smallest alkali cation (Li^+ = 5.3 eV). Thus, the formation of acidic O-H groups is expected to cause substantial changes in adjacent T-O-T bonds in the crystal framework.

One of the most important questions is whether the change brought about by the "proton attack" on zeolite oxygens remains a local phenomenon limited to atoms adjacent to the O-H group or whether the zeolite crystal responds by a concerted readjustment of the whole crystal lattice to minimize loss of lattice vibrational resonance and crystallinity. The resolution of this problem is discussed in the following sections.

3.2. Theoretical Studies and Modeling

Theoretical calculations and modeling studies of zeolites fall into two broad categories: semiquantitative (which use quantitative measures such as the Si/Al ratio to give rank orderings of some property such as acidity) and quantitative (studies such as ab initio calculations that attempt to predict quantitative results of experimental zeolite properties). The first category encompasses many semiquantitative theories that attempt to identify the key parameters influencing zeolite properties such as acidity, that are often based on direct experimental data. These semiquantitative theories can encompass greater complexity than the fully quantitative theories but may suffer from a lack of direct relationship of experimental data to key zeolite properties (such as polarizability). As experimental techniques improve, these theories are ever more stringently tested and steadily improve.

The second category encompasses both ab initio calculations, which use model clusters to mimic zeolite structures, and statistical mechanics calculations, which attempt to predict dynamic distributions (such as cation site populations during "titration"). These theoretical models provide basic information on the nature of idealized zeolite properties and allow for determination of the electronic structure of the framework and the electronic potentials in the channels. They also allow the determination of the relative importance of various parameters (structure, Si/Al ratio, bond angles) in the prediction of properties. The size of the cluster and level of detail obtainable are limited by the size and speed of the available computers. Thus, considerable detail is available for clusters such as $(HO)_3Si(OH)Al(OH)_3$, but progressively less is available for clusters approximating the actual channel or cage structure of real zeolites. Also, models of zeolites are typically formed by applying symmetry relations to generate the full structure from a limited subset. This symmetry relation is important because atomic relaxations are not independent of the chosen symmetry generation. The area of theoretical calculations and modeling is making the most rapid progress because new, larger computers and improved algorithms allow the use of increasingly realistic models with fewer constraints.

Recent reviews of theoretical studies of zeolites and zeolite acidity are given by Dwyer [14-16]; Barthomeuf and Corma [17]; Barthomeuf [18], who also discusses zeolite basicity; Grant and Abrahams [19]; van Santen et al. [20]; Sauer, Schroeder, and Hill [21], and Derouane et al. [22]. Particularly comprehensive reviews are given by van Santen, van Beest, and de Man, [23] and by Sauer [24,25].

Theoretical studies of zeolites and amorphous materials have advanced significantly in recent years. Improved models, more powerful computers, and better algorithms have combined to yield significant, testable theoretical predictions. The success of these models has moved theoretical analysis and explanations of experimental data to the forefront in the understanding of acidity in zeolites and related materials.

Although quantitative ab initio techniques are still too limited to address the most general case, detailed studies of zeolite substructures have yielded insights into properties measured by various physical characterization techniques, such as infrared (IR) and NMR spectroscopy. Although present limitations on these predictions should be borne in mind, they do suggest a significant reappraisal of data provided by physical characterization of acidity. Continuing increases in computer power, algorithm development, and theoretical insight strongly suggest that theory is moving from the role of explaining experimental data to the role of predicting key properties and providing fundamental insight into the structure-function relationships of a major family of materials.

3.2.1. Semiquantitative Methods

The semiquantitative theories primarily consider Al site distributions as the primary determinant of acid strength. In explaining variations in acid activity in faujasite-based cracking catalysts, Pine et al. [26] essentially argue that the structure of faujasite requires a given Al atom to have four Si atoms in the first surrounding layer (nearest neighbors) and nine Al or Si atoms in the second layer (next nearest neighbors, or NNN). In the case of X, all of these nine atoms are Al, by Lowenstein's rule [27], but with increasing Si/Al ratio, some Al will be replaced by Si. At a sufficiently large ratio, all nine are Si, and the original Al site is isolated. The strength of the original Al site depends on the number of Al NNN; the maximum possible is 0 NNN. This argument was extended by Wachter [28,29], who provided detailed statistical calculations of Al NNN population at various Si/Al ratios and determined that the Si/Al limit when essentially all Al sites were isolated is ~5 (not including Lowenstein's rule constraints) or ~7 (including Lowenstein's rule constraints). This level of the theory predicts that all zeolites with the same number of Al NNN (e.g., faujasite, zeolite A, chabazite) show a maximum acid strength at similar Si/Al ratios. Barthomeuf [30] has extended this idea by using topological densities to include the effects of layers one through five surrounding the Al atom. For a discussion of topological density, see Barthomeuf [30] or Meier and Moeck [31]. Topological densities for several zeolites are included in Table 1. From these numbers and the experimental data regarding the isolated Al limit for faujasite activity, Barthomeuf predicts the limiting Si/Al ratios for isolated Al in these zeolites. Different predictions for the limiting Si/Al value are given by the two theories. The Al NNN theory predicts the same limit for faujasite and zeolite A of ~7, but topological density theory predicts 6.8 for faujasite and 7.6 for zeolite A. Tests of the theories are possible only by preparing faujasite and zeolite A with varying Si/Al ratios.

Table 1
Zeolite Topological Densities

Zeolite	Topological Density	Limiting Si/Al Ratio
FER	0.287	9.9
MFI	0.278	9.5
MOR	0.275	9.4
MAZ	0.248	9.3
LTL-OFF-ERI	0.222	8.3
RHO-LTA	0.200	7.6
FAU	0.181	6.8

Source: Data from Barthomeuf [30].

Table 2
Effect of Geometric Boundary Conditions (GBC) for [HO$_3$SiOAl(OH)$_3$]

Optimization	Angles, degrees			Bond Distances, pm			
	T_1-O-T_2	O-T_1-O	O-T_2-O	(T_1-O$_{br}$)	(T_1-O$_{nbr}$)	(T_2-O$_{br}$)	(T_2-O$_{nbr}$)
Free with fixed-bond distances	139	112	108	156.9	(167.1)[a]	169.5	(171.9)[a]
Free with fixed-bond angles	139	(109.5)[a]	(109.5)[a]	157.5	167.1	168.9	171.9
GBC	142	(109.5)[a]	(109.5)[a]	157.0	(158.0)[b]	168.1	(168.8)[b]

Source: Data from Sauer (25)
Note: STO-3G basis set
[a] Parameter fixed in optimization
[b] Parameter constrained by geometric boundary conditions

These theories predict the effect of changing Si/Al ratio on relative acidity with any zeolite crystal type but cannot predict relative acidities between different zeolites or amorphous aluminosilicates. Several other factors influencing acidity (media effects, diffusivity effects, bond length, and angle effects) are discussed in other sections of this paper.

Recent work by Stach et al. [32,33] shows good agreement between experimental data and the Barthomeuf prediction of the Si/Al ratio for maximum acidity in mordenite.

3.2.2. Quantitative Methods

The use of ab initio theoretical methods requires providing a definition of the range of interactions being considered for the purposes of this discussion. Three ranges may be considered: short (up to NNN T-atoms), intermediate (between NNN T-atoms and 1 to 2 unit cells of the zeolite), and long (greater than 1 to 2 unit cells, preferably much greater). Figure 1 uses faujasite as an example to illustrate these ranges. The reasons for choosing these limits are as follows:

- *Short range*: The current limits on the number of atoms that may be modeled in detail using ab initio methods argue for a small cluster size. Modeling a zeolite from an Al atom out to its NNN T-atoms (with all their O atoms terminated as needed with H) allows the smallest unit that conveys structural information about the zeolite (possibly 9 to 12 NNN atoms, depending on the structure). Although beyond current capabilities, advances in computer power and algorithmic efficiency suggest that this capability may be attained in a few years. At this level, differences in acidity resulting from crystal structural effects as well as quantitative analysis of the effects of Si/Al ratio would become accessible without requiring fixed atomic positions, determined by x-ray crystallography, for the atoms in the cluster.

- *Intermediate range*: By modeling a few unit cells, the interaction effect of protons or cations present in channels or cages with protons or cations on the opposite wall of the channel or cage is accessible for study. In addition, the full range of Si/Al ratio becomes available for modeling.

- *Long range*: Modeling many unit cells allows the calculation of long-range crystal symmetry forces, e.g., Madelung energy, to be incorporated. By studying the effects

136

Figure 1

Size Ranges for Theoretical Calculations

Short Range

Intermediate Range

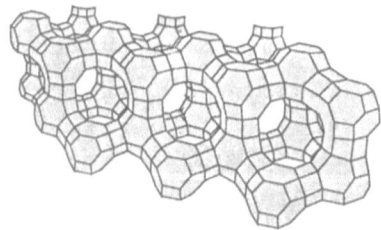

Long Range

UOP 2086-01

Figure 2

Calculated Potential Energy for Si-O⁻-Al and Si-OH-Al

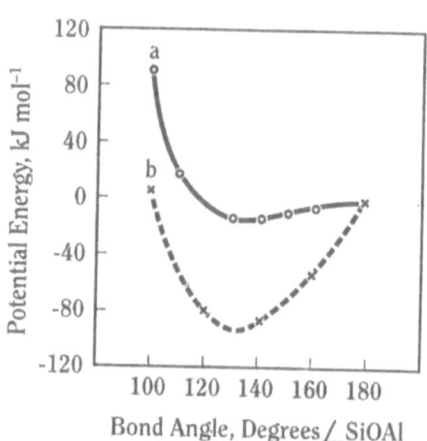

Bond Angle, Degrees ∠ SiOAl

Potential Energy Curves for (a) SiOAl⁻ and (b) SiOHAl (STO 3G Basis)
Reference ∠TOT = 180°
From Carlson and others, *Stud. Surf. Sci. Catal.*, **46**, 39-48 (1989)

UOP 2086-2

of local change, e.g., deprotonation, on the global crystal environment, tests of models differentiating zeolite from amorphous materials are possible.

In addition, these definitions help differentiate terms used by various workers. Because of the computational complexity of even moderate-sized clusters, theoreticians may refer to any effects beyond NNN as long range. Although true from the cluster's viewpoint, a further subdivision aids the discussion by minimizing the confusion of terms.

3.2.2.1. Techniques

A general review of the techniques used in quantitative calculations is given by Sauer [25]. For long-range (global) calculations, crystal orbit methods are preferred [25]. However, these methods are extremely demanding computationally and can be carried out only on materials with few, typically 10 or less, atoms in a unit cell. In addition, determining the proper approximation for the crystal field energy (or Madelung potential) further complicates the problem [34-39].

Thus, most ab initio calculations for zeolites are based on model clusters of atoms. Most examples use a simple T-O-T cluster and various terminations of the cluster. The method of terminating the cluster (e.g., the use of -H or -OH) influences the accuracy of the approximation. Similarly, geometric constraints on the bond angles and lengths result in different optima. Examples for the deprotonated, hyd' oxylated Si-O-Al cluster are given in Table 2. The effects of increasing the cluster size to improve the realism of the model are shown in Sauer et al. [21] and summarized in Table 3. In general, the larger clusters favor lower Al-O(H)-Si bond angles for maximum stability. An important point is that these small clusters do not necessarily mimic bulk materials. In particular, they are still not fully populated at the nearest-neighbor T-atom level, much less at the NNN level, where local structure first appears. These clusters are thus well within the short-range interaction zone of the zeolite model. Therefore, these clusters more closely model the amorphous silica-alumina gel, especially when fully relaxed.

3.2.2.2. Bond angle and partial atomic charge effects

Quantitative calculations attempt to provide direct predictions of magnitude differences in the physical properties of zeolites. They are generally less successful in dealing with amorphous materials because of the lack of structural models. Early calculations provided estimates of

Table 3
Bond Length and Angle for Cluster Models of Acid Sites

Model	Bond Length, pm			Bond Angles, degrees	
	r_{OH}	r_{Si-O}	r_{Al-O}	SiOH	SiOAl
$(HO)_3Si(OH)Al(OH)_3$	95.32	166	198	114.8	134
$(HO)_3Si(OH)Al(OH)_2OSiH_3$	95.29	167	198	120.5	137
$(HO)_3Si(OH)Al(OH)_2OSi(OH_3)$	95.27	167	194	115.0	133
$H_3Si(OH)Al(OH)_2OSiH_3$	95.24	170	194	120.0	132
$H_3Si(OH)Al(OSiH_3)_2(OH)$	95.02	172	193	117.4	123
Source: Data from Sauer (21)					
Note: Double zeta plus polarization basis set					

partial charge densities of framework elements and cations. Examples include Mortier [40], who calculated the partial charge on H in a faujasite hydroxyl with values of 0.12 (Si/Al = 1), 0.14 (Si/Al = 2.5), and 0.18 (Si/Al = infinity). The increase in ionic character for hydrogen can be explained by the higher electronegativity of Si vs. Al, causing an increased electron transfer from oxygen to Si and increased electron transfer from H to oxygen. Thus, the OH bond becomes more ionic. A similar effect is seen in data for Na by Beran and Dubsky [41].

Theoretical considerations of the effect of T-O-T bond angles on the bond structure and specifically on the electronegativity of the bridging oxygen is of particular importance in view of the fact that zeolites of higher protic acidity have a range of T-O-T bond angles higher than those of lower protic acidity. Compare the bond angle range of 137 to 177° for ZSM-5 [42a] or 143 to 180° for mordenite [42b] with the range of 138 to 147° for H-Y [43] as listed in Table 4. Quantitative calculations yield estimates of the effect of T-(OH)-T bond length and angle on the acidity of the bridging hydroxyl. Dwyer and O'Malley [44] calculated an optimized angle of ~144° for both protonated and unprotonated forms. Carson et al. [45] have calculated the energy for protonated and unprotonated forms (Figure 2) and noted that the unprotonated forms show little change from ~130° to 175° and the protonated forms become markedly less stable for larger angles. This energy difference suggests that the deprotonization energy decreases with increasing T-O-T angle, and thus the proton becomes more acidic. However, Mortier and Geerlings [46] calculated the hydroxyl-stretching frequency for a $(HO)_3Si(OH)Al(OH)_3$ cluster for Si-O-Al angles of 140 to 160° and find an increase of only 13 cm^{-1}, suggesting little change in bond strength. Kassab et al. [47] have used a similar cluster to calculate minimum energies at T_1-O-T_2 angles of 134.2° (protonated) and 179.3° (deprotonated). They also calculated a deprotonation energy of ~1320 kJ/mole with a H^+ charge of ~0.05 for the hydroxyl proton. In addition, they calculated deprotonization energies of ~1320 (Si/Al), ~1380 (Si/B), and ~1300 kJ/mol (Si/Ge) for the respective T_1-T_2 pairs.

Derouane and Fripiat [48] have calculated that the stability of the T-O-T bridge increases with the increasing electropositive character of the cation. Thus, the protic form of a zeolite is the least stable. They calculated a partial charge of 0.276 on the hydrogen. They also found that the substitution of Si by Al occurs with local deformation of the Si-O-Al bridge. This deformation maximizes the ionicity of the O-H bond. Derouane and Fripiat do not indicate any possible effect of this deformation on the crystal stabilization energy. Goursot et al. [49] used extended Huckel calculations for the acidity in offretite, which has the two nonequivalent sites.

They find that T_2 is more acidic than T_1 and that less acidic Al is less stable (more easily removed) in accordance with experiment (see, e.g., Fernandez et al. [50]). Finally, Derouane and Fripiat [51] have calculated preferential Al siting at T_2 and T_{12} in ZSM-5. They note that this siting implies possible Al in NNN sites, although the relatively high Si/Al in ZSM-5 makes this situation unlikely (Barthomeuf [30]) and that the framework is highly covalent with the negative charge extensively delocalized on proton abstraction.

A comparison of the T-O-T bond angles in H-Y and ZSM-5 is informative. The data are given in Table 4, which lists the angles as derived from x-ray diffraction studies. The four T-O-T angles in H-Y range in value from 139 to 147°, which is quite close to the optimized angle of 144° as predicted by Dwyer and O'Malley [44]. For ZSM-5, the 26 bond angles range from 137 to 177° (excluding the abnormally small T-O_5-T angle of 68°), suggesting a wide range of acid values, including high ones. Applying the results of Derouane and Fripiat [51] that Al preferentially sites at T_2 and T_{12}, only the angles involving O_2, O_3, O_{12}, O_{21}, O_{22}, O_{25}, and O_{26}, which range in value from 145 to 168°, are important. These angles are considerably larger than those in H-Y, which leads to a prediction of higher acid strength that is consistent with available data. The largest T-O-T angles (such as O_7 or O_{13}) are not occupied, indicating the decreasing stability as predicted from the Si-(OH)-Al bond energy at large T-O-T angles.

Relaxation effects are important. Work by van Santen and others [23,52,53] demonstrates the flexibility of the bond angles in zeolites. In particular, he notes [52] the importance of the extent of lattice relaxation on the acidity of zeolites. Kramer et al. [53] notes that neglect of relaxation may lead to spurious data. They note that relaxation is mostly local and that bond-angle distortions accommodate bond-length discrepancies. Finally, van Santen et al. [23] note that relaxation of both T-O distances and bond angles on protonation is important. The effects are mostly local, generally only over the distance of a few lattice atoms. They also note [23] that channel and cross-channel effects may be important. Sauer et al. [21] point out that examining the results for mordenite and ZSM-5 leads to the surprising conclusion that energy differences between acidic sites at the various lattice sites are small. This prediction contradicts the results of Derouane [54-56] and O'Malley and Dwyer [57]. Sauer et al. conclude that these results emphasize the importance of including lattice relaxation in the analysis of zeolite acidities. Sauer [24] warns about the possibility of artifacts resulting from cluster terminations in model relaxations not subject to geometric constraints. These effects must be evaluated to avoid erroneous conclusions.

Table 4
Zeolite T-O-T Bond Angles

H-faujasite	
T-O_1-T	138.56
T-O_2-T	147.38
T-O_3-T	139.66
T-O_4-T	145.25
Data from Olson and Dempsey [43]	

H-mordenite	
T_1-O_1-T_3	149.7
T_2-O_2-T_4	146.3
T_2-O_3-T_1	159.3
T_4-O_4-T_3	168.4
T_2-O_5-T_2	146.6
T_1-O_6-T_1	149.4
T_1-O_7-T_1	143.2
T_2-O_8-T_2	180.
T_3-O_9-T_3	147.8
T_4-O_{10}-T_4	148.2
Data from Schlenker et al. [42b]	

TPA-ZSM-5			
T_1-O_1-T_8	146.25	T_6-O_{14}-T_9	150.98
T_1-O_4-T_3	160.16	T_5-O_{15}-T_{10}	148.89
T_1-O_5-T_3	67.93	T_3-O_{16}-T_5	174.26
T_3-O_6-T_7	142.21	T_4-O_{17}-T_6	154.69
T_3-O_7-T_8	172.11	T_4-O_{18}-T_4	137.46
T_4-O_8-T_8	145.13	T_6-O_{19}-T_6	152.41
T_7-O_9-T_9	154.90	T_5-O_{20}-T_{11}	149.63
T_8-O_{10}-T_{10}	151.31	T_7-O_{23}-T_7	150.48
T_9-O_{11}-T_{11}	165.08	T_{11}-O_{24}-T_{11}	149.75
T_9-O_{13}-T_{10}	177.17		
T_2-O_2-T_7	152.76	T_6-O_{22}-T_{12}	146.92
T_1-O_3-T_2	168.16	T_2-O_{25}-T_{11}	158.19
T_{10}-O_{12}-T_{12}	162.33	T_1-O_{26}-T_{12}	144.93
T_5-O_{21}-T_{12}	157.30		
Data from Chou et al. [42a]			

3.2.2.3. Zeolite vs. amorphous aluminosilicates

Models for the acid site in amorphous aluminosilicates have been proposed by Hansford [58] and Peri [59]. Hansford proposed a surface silanol rendered more acidic by the presence of a neighboring aluminum ion. Peri has proposed a series of small (2 to 4 T-atom) ring structures and especially favors Al-O-Al bridged structures. The lack of direct structural probes greatly hinders the testing of these models, although most available data favor framework Al Lewis acid-base interactions with silanols rather than Al-O-Al bond structures.

The proper description of the primary electronic units is an important part of model construction. Pelmenshchikov et al. [60] have used $Si(OH)_4$, $Al(OH)_3$, $NaOH$, and H_2O as primary building units and have constructed simple models to determine the stability of various small linear or cyclic structures. They emphasize the short-range aspects of structure, as discussed by Zhidomirov and Kazansky [61]. Their results demonstrate the strong tendency to form Al-O-Al bonds in small clusters. However, their small structures lack any stabilization energies as a result of long-range order (crystallinity) and may be applicable primarily to amorphous materials. Based on their x-ray photoelectron spectroscopy (XPS) data, Barr et al. [62] emphasize the global nature of the electronic structure in zeolites.

A general theory regarding differences in acid strength between zeolites and aluminosilicates has been discussed by Mortier [63]. Using bond valence, Gutmann's rules for interatomic interactions, and principles of electronegativity equalization, Mortier proposed an enhanced electron donor-acceptor interaction in zeolites as shown in Figure 3. This proposal emphasizes the reactive aspect of acidity (a molecule is not an acid until a base is present for it to react with) and the electron delocalization into both Al-O and Si-O (oxygen bridge bonds). As Mortier notes, the interaction of O with Al weakens the O-H bond and increases acidity. A resonance depiction of this idea is shown in Figure 4. Structure I is a fully bridged oxygen with a weakly bonded proton, and structure II is a silanol with a weak Lewis acid interaction of the hydroxyl oxygen with the Al. The actual situation is a resonance hybrid.

These arguments lead to the proposition that zeolite crystals respond to "proton attack" by a global readjustment of the bond structure of the whole zeolite matrix to prevent major local distortions in bonding, and this reaction leads to loss of long-range symmetry and thus crystallinity. In addition, energetics favor equalization of topologically equivalent electronic

Figure 3

Model of the Interaction
of a Donor Molecule with Hydroxyls

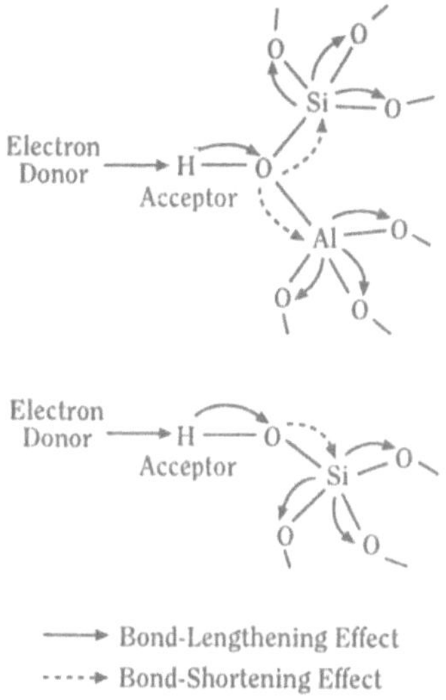

——→ Bond-Lengthening Effect
·····→ Bond-Shortening Effect

From W. J. Mortier, Proc. 6ᵗʰ IZC, 734-46

UOP 2086-3

Figure 4

Resonance Model of T-(OH)-T
Bond Structure

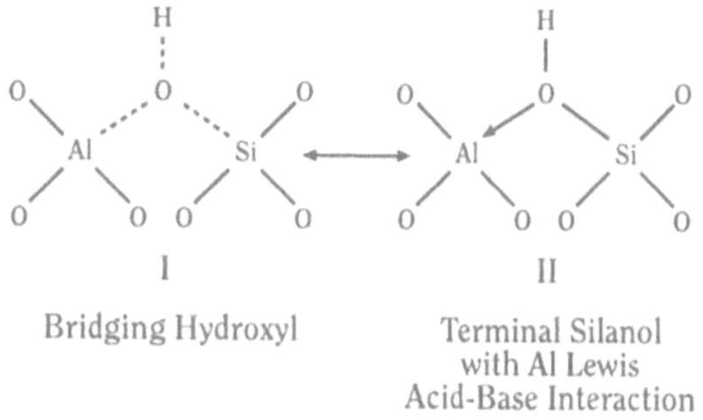

I
Bridging Hydroxyl

II
Terminal Silanol
with Al Lewis
Acid-Base Interaction

UOP 2086-4

UOP 2086-4

interactions. Thus, a significant stabilization results from the electronic structure of the Al-O and Si-O bonds becoming more equivalent. This equivalence implies a dominance of structure I in the resonance. An amorphous material has no long-range stabilization for structure I, and so structure II can dominate. The difference in long-range effects may represent the fundamental reason for the difference in acidity between crystalline zeolites and amorphous aluminosilicates. This interpretation is consistent with the available NMR evidence, which shows a large proportion of H in terminal hydroxyls in amorphous aluminosilicates. Recent work by Doremieux-Morin et al. [64] is in agreement with the predictions of this model.

Calculations by Mortier et al. [65] have examined the difference between bridging and terminal hydroxyl groups. They suggest that amorphous materials primarily consist of terminal hydroxyl groups, with few thermally unstable bridging groups to account for the acidity characteristics.

3.2.2.4. T-atom effects

Work has been undertaken on the more complex alumino-phosphate (AlPO) family of materials. The effects of bond angle on potential energy have been calculated by Carson et al. [45]. The calculation predicts a minimum at $\sim 124°$, much smaller than typical alumino-silicate minimum at $\sim 144°$. They also predict an acidity for SAPO between that of H-X and H-Y. Work by Sauer et al. [66] on AlPOs predicted that AlPOs will be hydrophobic, much like high-silica zeolites. This prediction contradicts observations and may be due to the rigid geometry and charge distribution used in the calculation. Later calculations by Sauer and Schirmer [67] predict that SAPO will have an acidity similar to a high-silica zeolite. Again, this prediction, which is contrary to experimental evidence, may be an artifact of the model used in the calculations.

Recent work by van Santen et al. [52] has calculated that the AlPO structures deform more than their silica analogues, and so generally smaller channel dimensions are the result. This prediction further increases the importance of accurately modeling the relaxed forms for reliable predictions of molecular sieve properties. Modeling of AlPO-5 and its silica analog (SSZ-24) by de Man et al. [68] indicates some persistent problems with choosing force constants. They have shown that the influence of structure on the spectra is weak. In fact, they claim that the major differences are primarily due to the interatomic force constants. Parameters for the calculation are obtained by fitting the experimental data for α-berlinite (the AlPO analog of α-

quartz). The use of larger model clusters that permit the direct calculation of force constants for a given structure (rather than using parametized data) would be a major test of their conclusions.

Work by Chamot [69] summarizes the results for B-containing zeolites. In particular, B-containing zeolites are sensitive to geometric constraints because of the short B-O bond. Using a 70-atom cluster and comparing Al or B substitution, Chamot finds that the extent of relaxation of the cluster is critical. Calculations using only the T-atoms and the bridging hydroxyl relaxations predict that the B-containing zeolite is the stronger acid. By extending the relaxation to the nearest neighbors of the Al or B (and the attached oxygens), the B zeolite is found to be a weaker acid than the Al zeolite (a difference in proton affinity of 8.7 kcal/mole). Chamot [69] also notes that the charge on deprotonation is extensively delocalized and the metal-oxygen bond (Al-O or B-O) is strengthened.

3.2.2.5. Lewis acids and defect sites

Sauer [70] has proposed a model using a threefold-coordinated Al-cation Lewis acid complex with a bridging hydroxyl as a model of defect sites. His calculations indicate that such sites should bind water about three times as strongly as ordinary bridging hydroxyls.

Sauer and Schirmer [67] have proposed a model for the enhancement of acidity by nonframework Al. They propose a Lewis acid activation of the hydroxyl group—in particular a bridging hydroxyl—that further delocalizes the charge on deprotonation and increases the acidity of the proton. Sauer and Schirmer [67] and Sauer and Zahradnik [71] also note that vicinal silanols can interact to produce sites of intermediate acidity. Their calculations suggest a decrease in deprotonation energy from ~1450 KJ/mol for an isolated silanol to ~1330 KJ/mol for vicinal silanols. For comparison [67,72], they give the deprotonization energy for a bridging hydroxyl as ~1200 KJ/mol and the calculated deprotonization energy for a Lewis acid [Al (OH)$_3$] coordinated bridging hydroxyl as ~1070 KJ/mole. Beran [73] has completed calculations for similar models and gives similar results. Lunsford [74] proposes a cyclic aluminum-oxygen dimer complex located in the beta cage to account for the acidity of steamed zeolites.

These proposals can provide models for studies of nonframework Al (or other cations, such as rare earths) enhancement of acid activity and a model for defect sites (vicinal silanols) arising from the dealumination of zeolites.

3.2.3. Theoretical Predictions of Measured Parameters

One of the major areas of progress in theoretical chemistry is the increasing number of testable predictions being made. One recent success, as noted previously, was the observed maximum in acidity of mordenite as predicted by zeolite topological densities [32,33].

A series of statistical mechanics calculations have been carried out by Mortier and co-workers [75,76] on the cation site preferences in Y. Mortier et al. [75] calculated site preferences of K in hydrated KY, and Van Dun and Mortier [76] calculated site preferences for Na in Y. They noted the predominant influence of short-range cation framework interactions and the influence of even small quantities of water. In a welcome development, Van Dun et al. [76] tested their model with experimental data on NaH-Y, CaH-Y, and SrY.

An article by van Santen et al. [23] summarizes the quantum mechanical calculations of optical frequencies in pure silica systems (α-quartz, sodalite, faujasite); these calculations yield the effective charges on atoms. Extending their results, they note certain implications for the acidity of zeolite systems. In particular, they mention the potential stabilization of the carbocation by the zeolite and the strong localized interaction (complex-like) in the pure silica system. They also note, from their work and others, the effects, particularly the weakening of the O-H bond, of the Si/Al ratio and Si-O-Al bond angle on acidity. Finally, they indicate the effects of cluster-terminating groups on the effective charges and bond strengths of the various atoms. Their article summarizes many of the current theoretical results relating to the nature of zeolite acidity.

Calculations of vibrational spectra are an interesting area. Hill and Sauer [77] have calculated ab initio force constants for aluminosilicates. Calculations of vibrational spectra for faujasite by Schroeder et al. [78] allow them to assign the observed high frequency and low frequency νOH bands to O_1H and O_3H (the supercage and sodalite cage, respectively). They can also calculate that no correlation exists between the deprotonization energy and the vibrational frequency. This result, if supported by more detailed calculations, would eliminate vibrational spectroscopy of hydroxyl-stretching frequencies as a measurement of acidity.

Calculations of vibrational spectra by de Man et al. [68] and de Man and van Santen [79] indicate little correlation of vibrational spectra with substructure. This result implies that group vibrational frequencies are poorly correlated with specific zeolite substructures and may, in fact, be more strongly influenced by structural or electronic effects on interatomic force constants. To the extent that specific substructures influence the interatomic force constants, the traditional assignment of vibrational bonds to specific substructures is valid. The work of de Man et al. suggests that the extent of validity may need to be reviewed.

Calculations of other properties, such as NMR chemical shifts for protons [80], adsorption energies [81], adsorbate interactions [20], and proton jump energies [82] have been completed. Sauer [80] notes that proton acidities are strongly correlated with partial charge on the proton, poorly correlated with the NMR chemical shift of the proton, and not correlated with the hydroxyl-stretching frequency. This result suggests that thermochemical measurements are the only direct physical measurements that provide quantitative data on acid strengths.

3.2.4. Summary

Theoretical studies of zeolite acidity and related topics have progressed rapidly. Calculated structures assist in the interpretation of experimental data and provide warnings of inappropriate analogies. In many respects, theoretical calculations now lead experiment in providing new hypothesis and tests of existing hypotheses concerning zeolite acidity.

The recent advances and continuing progress of model calculations suggest the need for a consistent definition of distance scales for interactions. The authors have suggested short-, intermediate-, and long-range scales that adequately represent the levels at which different interactions become prominent. The short-range scale (T-atoms out to the NNNs of the reference T-atom) is somewhat beyond the maximum complexity of models currently in use. However, the authors believe that increases in computer power and algorithm efficiency will render clusters of this size range (~14 to 20 T-atoms) tractable within a few years. These clusters have the advantage in that they represent the smallest size at which structure (the number of NNN T-atoms) is naturally introduced into the cluster. The cluster size allows some relaxation of fixed-symmetry requirements to simulate specific zeolites. However, the effect of local disorder on the infinite lattice must be considered to balance the effects of relaxation and crystal field energies. Moving to intermediate-range (channel and cage effects) and long-range (crystal field effects) scales with model clusters will require a longer time but provides a good reference for discussion purposes.

Semiquantitative models appear to have reached the practical limit of predictive power. Recent successful tests of the topological density model support the underlying theoretical principles, but these models cannot be improved in a straightforward manner. Their predictive power is confined to relationships within a single family of materials (e.g., faujasites).

The importance of global effects (relaxation as well as crystal-field energies) is recognized. Limitations on calculational power do not fully allow general methods of incorporating these effects into the cluster calculations. However, improved algorithms and better approximations are improving the accuracy of calculations involving global effects.

Relaxation of atoms is important even on local scales. Zeolites are surprisingly flexible frameworks. Calculations have indicated the ability of bond angles to deform to accommodate changes in local structure or electronic effects. These effects, however, must be considered on larger scales. As long-range forces are considered, the effects of crystal symmetry, which acts as a constraint against relaxation, may well limit the range of accommodation possible in a given zeolite structure. The lack of symmetry energies should allow amorphous materials to relax more completely and result in different, generally weaker, acidities. *This balance of forces (relaxation vs. crystal field energy) may be critical in differentiating the acidity of zeolites from amorphous aluminosilicates as well as the acidity of different zeolite structures.*

Also, overrelaxation of zeolite structures may result in misleading conclusions. Calculations using fully relaxed clusters mimicking ZSM-5 or mordenite predict that the acidity of most bridging oxygen sites is roughly equivalent and that little deviation occurs from a random distribution of bridging hydroxyls and Al over the possible framework sites. This prediction differs from earlier predictions of preferential Al siting using clusters mimicking the rigid geometries of specific zeolites. Experimental evidence for or against preferential Al siting in ZSM-5 or mordenite may indicate the limits to current theoretical predictions.

Calculations have been extended to systems beyond zeolites or aluminosilicates. Substituting different T-atoms (B, Ge, P) for Al or Si in these calculations predicts effects that may be seen in the newer families of microporous materials. Extending calculations to nonframework atoms (Lewis acid Al complexes) or defect framework structures, such as vicinal silanols, provides predictions of acidity that may be tested experimentally and perhaps be useful industrially. The presence of the substituent or nonframework atoms may alter the constraints on local relaxations.

Model calculations have also provided interpretations of data from physical measurements of zeolite properties. In particular, the model calculations indicate that acidity is not correlated with hydroxyl-stretching frequencies and poorly correlated with the proton NMR chemical shift. Predicted values for several measurable properties are becoming available and will increasingly provide stringent tests of the accuracy of the current theoretical models.

Statistical mechanics has proven reasonably reliable in predicting cation distributions in zeolites. Future applications using potentials derived from ab initio model calculations can be expected to expand and improve.

3.3. Thermodynamic Measurements of Acid Strength and Distribution

The measurement of the number, type, and strength of the acid sites provides key experimental data regarding zeolite acidity. This discussion focuses on Bronsted acidity, although methods to distinguish Lewis acidity also have been developed, for example, the discussion by Aboul--Gheit et al. [83]. Thermodynamic characterization techniques, including titration and thermal methods, are reviewed by Karge and Nagy [84], Karge [85], and van Hooff and Roelofsen [86]. Dimitov et al. [87] discuss the thermal characterization of zeolites. The traditional methods for measuring the strength of strong acids involve the use of calorimetric indicators, such as the Hammett indicators, which are titrated against the acid. An example using amorphous aluminosilicate is given by Hashimoto et al. [88], who find a distribution of both Bronsted and Lewis sites with maximum Hammett acidities of about -13. Unfortunately, these indicators are too large to fit into the pore system of most zeolites. Work by Umansky et al. [89] using a modified spectrophotometric approach determined the relative acidity of amorphous silica-alumina (70 to 80% H_2SO_4), H-Y (90 to 98% H_2SO_4), and H-mordenite (> 100% H_2SO_4). Other methods, such as thermometric titration, differential scanning calorimetry (DSC), temperature-programmed desorption (TPD), and catalytic rate measurements (e.g., cumene cracking and olefin oligomerization), have been proposed to measure some functions of acid strength and number. Several problems, such as diffusion effects (discussed in Section IV.B. of this paper) and lack of direct measurement of acid-strength distributions, limit the practical methods to those based on calorimetry or thermal desorption. Catalytic rate methods do not separately determine acid number and strength. Because they indicate only some function of the two combined, they are of less value for measuring acid strength distributions, although they are, of course, quite valuable for providing data from model reactions for use in catalyst development work.

3.3.1. Calorimetric Methods: Thermometric Titration and DSC

The use of calorimetric methods of measuring acid-strength distributions is well established. These methods are based on the assumption that the acid strength is directly related to the heat of adsorption (or desorption) of a standard base. The measured points must be taken at equilibrium for accurate data. Examples include the work of Tsutsumi et al. [90] and Auroux et al. [91]. Two specific methods are discussed: thermometric titration and DSC. Thermometric titration using n-butylamine in benzene was used by Bezman [92] to measure the acid-strength distribution in H-Y and by Pellet et al. [13] for various treated H-Y zeolites and amorphous aluminosilicates. This method works well for large-pore materials with three-dimensional channel systems. However, as discussed by Bezman and Pellet et al., this method is unsuccessful even with mordenite and also with smaller-pore zeolites. They propose that the problem is the blockage of the channels in mordenite by strongly adsorbed n-butylamine. Thus, although this method is of limited utility, its use could possibly be extended by more competitive solvents or higher reaction temperatures to provide greater mobility of the titrant. Extension of the experimental technique to permit the use of thermometric titration with a wider variety of zeolites would be quite valuable.

An alternative microcalorimetric measurement is discussed by Macedo et al. [93], who used ammonia to study the acid strength in H-Y zeolites after various dealumination treatments. The thermometric titration with n-butylamine shows that the initial heat of sorption is much greater for H-Y than for amorphous aluminosilicate ($\sim$35 kcal/mol and $\sim$20 kcal/mol, respectively). This result implies that H-Y is a stronger acid. The same thermometric titration also shows that the amorphous silica-alumina debris phase formed in H-Y zeolite on extensive steaming closely resembles the acidity of the amorphous silica-alumina gel.

An alternative to thermometric titration is DSC, which measures the size of the endotherm associated with desorption (usually of ammonia in tests of zeolite acidity) from a solid material. This technique thus provides the temperature and heat of desorption of standard bases. Aboul-Gheit et al. [94] applied DSC to the study of aluminosilicates and H-mordenite. The techniques developed by Aboul-Gheit et al. [95] eliminate endothermic peaks resulting from water desorption from the zeolite and thus simplify the spectrum. This technique combines features of temperature-programmed desorption (TPD) and thermometric titration. Care must be taken to ensure that the measurements are not limited by heat or mass transfer and that desorption from weak sites is not blocked by adsorbates at strong acid sites. The former can be checked by varying the rate of heating the sample, the sample size, or the flow rate, and the latter can be controlled by preloading the sample with adsorbate at various temperatures and comparing the distributions. These procedures are similar to those discussed by Kapustin et al. [96] for controlling similar effects in TPD spectra. In addition, Vannice et al. [97] discuss errors that may result from changes in the thermal conductivity of the gases surrounding the sample.

3.3.2. Temperature-Programmed Desorption

Recent reviews of TPD are given by several authors [98]. Forni and Magni [99] discuss TPD as applied to acid-strength distributions in zeolites. Temperature-programmed desorption consists of heating a sample at a constant rate and measuring the quantity of material desorbed at each temperature. As discussed by Hashimoto et al. [100], these data can be converted into a desorption energy distribution, which is related to the acid-strength distribution of the material. The assumption is that the desorption energy is primarily the reprotonization energy of the acid. Thus, a stronger acid requires a higher energy to reprotonate, but this direct relationship may be modified by the type of acid site. This technique can be extended, by suitable choice of adsorbates, to differentiate intrapore acidity from surface acidity, as demonstrated by Take et al. [101].

Typically, researchers assumed that the calculated energy is primarily the result of the desorption of the ammonia or other adsorbate. As discussed by Forni and Magni [99] and others [98], the apparent energy may be dominated by diffusional (or heat transfer) effects. Forni et al. [102] have tested the effects and propose that in the case of ammonia, the intracrystalline diffusion is controlling in zeolites. The measured activation energy for diffusion is so large (E_a ~30 kcal/mol) that they suggest a major interaction between the desorbing ammonia and available acid sites (e.g., multiple adsorption-desorption steps). These effects, and others, are discussed. The effects of heat and mass transfer can be quantified by varying the catalyst quantity and flow rate during TPD measurements. When these effects are properly taken into account, good agreement between TPD and calorimetric data is obtained, although the TPD data provide a partially averaged value for acid strength rather than a full distribution.

3.3.3. Summary

Quantitative and qualitative thermodynamic measurements in the literature show that crystalline H-zeolites have substantially larger heats of adsorption (or desorption) for nitrogen bases relative to amorphous aluminosilicates. This result implies that the H-zeolites are stronger acids.

Thermometric titration has the advantage of being an essentially isothermal calorimetric technique, thus avoiding problems of changing heat capacity with temperature. The experimental technique requires further refinement to extend the range of zeolites that can be evaluated. Where applicable, thermometric titration is the most accurate of the available techniques, if equilibrium is attained at each data point. With respect to the applicability of measurement techniques, TPD and DSC provide similar data. Differential scanning colorimetry has the advantage of directly providing the energy of desorption, which is assumed to be directly related to the acid strength, but TPD requires a mathematical transform to convert the desorption vs. temperature data into an acid-strength distribution, as discussed by Hashimoto et al. [100]. Differential scanning colorimetry is useful for materials that are not suitable for thermometric titration, but the effects of possible heat and mass transfer limitations and changes in thermal conductivity and heat capacity with changing temperature must be taken into account. Finally, TPD is a useful, simple, and versatile technique. However, its significant heat and mass transfer limitations must be taken into account if a high degree of accuracy is to be obtained. The conversion of TPD data into acid-strength distributions also causes a greater averaging of the calculated acid strength than with the other methods, but this averaging can be reduced by using smaller temperature-increase rates during data collection.

3.4. X-ray Photoelectron Spectroscopy and X-ray Emission Spectroscopy

X-ray emission spectroscopy uses a monochromatic x-ray source to excite a secondary x-ray emission with frequencies characteristic of the various elements in the sample. This technique probes the full bulk of the crystal. Dowell et al. [103] provides details of the experiments. X-ray photoelectron spectroscopy is a surface-sensitive technique that uses a monochromatic x-ray source to excite photoelectrons from the sample. The energy of these electrons is characteristic of elements in the sample for core-level electrons but may also represent hybridized molecular orbitals for valence-level electrons. A description of an XPS system is given by Barr [104]. As discussed by Okamoto et al. [105], the surface sensitivity of the XPS techniques requires appropriate precautions to ensure that the surface of the material is representative of the bulk.

Both spectroscopic techniques show changes in the binding energy (BE) of electrons on Si, Al, and O with changes in the zeolite framework Si/Al ratio. In addition, a large shift in BE occurs on changing from tetrahedrally coordinated to octahedrally coordinated Al [106]. This technique provides a method of detecting nonframework Al in steamed zeolites.

Greater detail is achieved in the XPS data, and more detailed analyses have been undertaken. Taniguchi and Takaishi [107] have correlated the shifts in the Si(1s) and Si(KLL) lines with the oxygen polarizability. West and Castle [108] determined a correlation of Auger parameter [$BE(Si_{1s}) - BE(Si_{KLL})$-h$\nu$ (incident x-ray energy)] and silicate polarizability. Taniguchi and Takaishi argue that in zeolites, the major polarizability is due to the oxygens. By extending the correlation to the measured values for Na-mordenite and NaA-zeolites, they derived oxygen polarizabilities [$\mu(O)$] of 2.2 Å^3 and 2.5 Å^3, respectively. This result compares to a calculated value for cesium mordenite of 2.45 +/- 0.15 Å^3 by Takaishi and Hosoi [109]. They also calculate the $\mu(O)$ value for zeolite A for a range of ionicities [q is the total negative charge on the four oxygens attached to a T-atom; thus q = 8 is fully ionic]: $\mu(O)$ = 3.88 Å^3 for q = 8 and $\mu(O)$ = 2.0 Å^3 for q = 2 in NaA. The q value is related to the ionicity of the M-O bond and gives a partial charge on Na of ~0.38 e$^-$. The trend toward higher polarizabilities with decreasing Si/Al ratio is in accordance with theoretical predictions, such as Derouane and Fripiat [51], based on the assumption that polarizability of oxygen is directly related to the effective negative charge on oxygen. They calculate an effective charge on oxygen of -0.695 for SiO$_4$ and -0.700 for HAlO$_4$.

Okamoto et al. [105] performed a series of experiments to determine the effects of Si/Al ratio and cation type. They determined that all XPS lines [O(1s), Si(2s), Si(2p), and Al(2p)] show increases in BE with increases in the Si/Al ratio. Because these results were relatively independent of crystal structure for Na-zeolites, they suggest that the primary cause of the shift is common to all zeolites examined (A, faujasite, and mordenite). They also investigated the effect of changing cation with the result that the O(1s) BEs decrease in the order $H^+ > Li^+ > Na^+ > K^+ > Cs^+$. The effect is expected from electronegativity considerations (see, for example, Pauling [110]), and the larger BEs imply increased electron density in the oxygen-cation bond. Thus, the ionicity of the oxygen-cation bond is $H^+ < Li^+ < Na^+ < K^+ < Cs^+$, as expected. Beran and Dubsky [41] have calculated an approximate effective charge on Na^+ in faujasite with a range of 0.23 to 0.29 e^-. Interestingly, their results indicate that the effective charge declines with decreasing Si/Al ratio (from 6.0 to 1.0). This result suggests that within the limits of their calculation, the dipolar repulsion outweighs the increase in polarizability in determining the ionicity of the bond. Mortier [40] calculates the effective charge on H^+ in faujasite and computes 0.12 e^- in zeolite X (Si/Al = 1.25) and 0.14 in zeolite Y (Si/Al = 2.5). His results are consistent with Beran et al. [41] and also reproduce the experimental XPS data that H^+ is less ionic than Na^+ in faujasite.

Barr and Lishka [111] have used high-resolution valence-band XPS to study the surfaces of several zeolites. Because they found evidence of a number of impurity phases, such as metal aluminates, on the surface of several zeolite crystal samples, they emphasize the need for careful sample preparation and analysis. Their data, however, indicate the presence of "group" rather than "elemental" shifts in the BEs. This proposal is extended by Barr et al. [62] to imply the existence of oxide units in zeolites that can be resolved into broad structural groups:

$$(SiO_2)_x \cdot (M_{y/p}^{+p} AlO_2^-)_y - zH_2O \tag{2}$$

For zeolites with an Si/Al ratio of < 5, they propose an effective structure of SiO_2-doped $NaAlO_2$, but for an Si/Al ratio of > 5, the effective structure is $NaAlO_2$- doped SiO_2. These results differ from those expected from simple solid solutions of SiO_2 and $NaAlO_2$ (a possible model for amorphous aluminosilicates), especially in the low Si/Al region. However, amorphous aluminosilicates were not examined. These data suggest the existence of global electronic effects in the crystalline zeolites. Barr et al. further distinguish between these two groups as cagelike (low Si/Al) and chainlike (high Si/Al), according to the types of zeolites they examined in each category (A, faujasite, L, and mordenite, ZSM-5, silicalite, respectively) but have no examples of high-silica cage structures or low-silica chain structures to establish this division of structure type vs. Si/Al ratio effects.

The XPS data base has two major remaining gaps. First, high-resolution core- and valence-level spectra of carefully dealuminated zeolites of individual structure types (e.g., faujasite with an Si/Al ratio of 1, 3, 5, 10, > 50) are needed to investigate the separation of structural and Si/Al effects in the BEs. Second, similar data on amorphous aluminosilicates of varying Si/Al ratio are needed to study the differences between the amorphous and crystalline materials.

3.5. Solid-State NMR Spectroscopy

Solid-state, magic-angle spinning NMR spectroscopy uses small samples of powdered materials spun at high (few kHz) frequencies at an angle of 54°44' to average out chemical-shift anisotropy. Nuclei studied include ^{29}Si, ^{27}Al, ^{13}C, ^{23}Na, ^{7}Li, ^{17}O, ^{15}N, ^{31}P, and ^{1}H. The NMR signal provides data on the electronic environment of the resonating nucleus for the following parameters: chemical shift, coupling constant, full width at half maximum, splitting pattern, and quadrupole coupling (for quadrupolar nuclei). Reviews of the technique and recent experiments are found in Nagy and Derouane [112]; Thomas and Klinowski [113]; Pfeifer, Freude, and Hunger [114-116]; Ernst [117]; Fyfe et al. [118,119]; Engelhardt [120]; and Klinowski [121]. Pines et al. [122,123] discuss techniques for removal of second-order quadrupolar effects.

As discussed in the review by Nagy and Derouane [112], solid-state magic-angle spinning NMR of several nuclei has been applied to the study of zeolites. The ^{29}Si spectrum provides a series of peaks corresponding to silicon atoms with 0 to 4 Al nearest neighbors. From this NMR data, the Si/Al framework ratio may be calculated. The exact frequencies of a given peak (e.g., Si with 1 Al nearest neighbor) vary slightly with zeolite type (see, for example, Reference 124 comparing Y and ZSM-20); these differing frequencies possibly reflect differing framework electronic polarizabilities. Also,^{29} Si NMR has demonstrated single sharp peaks for zeolite systems of varying Si/Al ratio [125]. This result is consistent with a global electronic effect that results from the electronegativity of the entire framework being averaged. Individual Si atoms still adopt geometries consistent with the crystal structure and the number of Al nearest neighbors, which produce the typical five-peak pattern of ^{29}Si NMR with Si/Al ratios greater than one. In addition, ^{29}Si NMR has shown single sharp peaks with partial cation exchange [125]. This result may be due to hydrated cations having mobility and giving a rapid-exchange-limit spectrum. The ^{29}Si NMR of carefully dehydrated partial-exchange systems may provide useful additional information on cation effects. Aluminum NMR has been applied primarily to observe the tetrahedral (framework) or octahedral (nonframework) form of Al in zeolites. A nonframework Al species has also been identified at about 30 ppm (see, e.g., Reference 125c).

Solid-state NMR of amines (using ^{15}N), phosphines (using ^{31}P), and CO (using ^{13}C) has been used as a probe of the Bronsted and Lewis acid sites in zeolites [126]. Chemical shift differences indicate the formation of protonated (Bronsted) or nonprotonated (Lewis) complexes with the acid site. In addition, chemical shift differences between protonated species indicate differences in the strength of complexation: tightly held or weakly bound. Finally, different molecules sample different subsets of acid sites, depending on the accessibility, type, or acid strength of the site.

The most direct evidence for the nature of zeolite acidity is expected to come from studies involving ^{1}H. Early work by Stevenson [127a] used broadline ^{1}H NMR to locate the protons in dehydrated H-Y and established an approximate geometry for the bridging hydroxyls. His results indicated that the protons are attached to O$_1$ and O$_3$ with an O-H bond length of ~1.00 to 1.03 Å. Work by Freude et al. [127b] has confirmed the Al-H distance of 2.38 Å in H-Y and determined an Al-H distance of 2.48 Å in H-ZSM-5. More recently, direct measurement of the chemical shift of ^{1}H in zeolites and amorphous aluminosilicates has provided further data. Pfeifer et al. [114,128] obtained ^{1}H MAS-NMR spectra of H-Y and amorphous aluminosilicate. The H-Y spectrum shows four peaks: one at ~2 ppm (a), two partially overlapping peaks at ~4.5 ppm (b and c), and one peak at ~7 ppm (d). The amorphous aluminosilicate shows one peak at ~2 ppm and a broad peak at ~4.5 ppm. They assign line a to nonacidic structural hydrogens because adsorption of pyridine shifts this line to a lower field, suggesting hydrogen-bonded molecules. They argue that protonation of pyridine leads to considerable line broadening, which they have subsequently observed [128b]. They assign line d to the residual NH$^+_4$ ions left after activation. This assignment agrees with both the reduction in intensity at increased activation temperature and the assignment of NH$^+_4$ compounds [129]. They assign the two remaining lines to bridging hydroxyl groups. A ^{1}H MAS-NMR spectrum of H-mordenite produces only one relatively broad line in the bridging hydroxyl region. Pfeifer [130] assigns peaks b and c to provide a direct relationship of the larger NMR chemical shift and the lower IR hydroxyl-stretching frequency.

Briefly discussed by Pfeifer et al. [114] is the significance of the absolute chemical shift magnitude and the effects of solvation on proton shifts. They note, for example, that liquid water gives a proton signal at 4.8 ppm [131a] and gas-phase water gives a signal at 0.31 ppm [131b]. Even in the gas phase, some hydrogen bonding may occur as may a consequent down-field shift. Pfeifer [130] provides examples of ^{1}H chemical shift vs. gas-phase acidity for several hydroxylic compounds (such as MeOH and phenol). Pfeifer also discusses the problems of interpreting the NMR experiments because of complications resulting from observing only the residence time of a proton on a given oxygen atom. He notes that the mean lifetime can be greatly reduced by the presence of even small quantities of bases.

Finally, Freude et al. [132] report that the chemical shift of the bridging hydroxyl line increases as the Si/Al ratio in faujasite increases in the range of 1.4 to 7 but remains constant for Si/Al > 10. This limit probably represents the isolated aluminum site with maximum O-H polarity, discussed in Section 3.6. of this paper.

Work by Haw et al. [133] using solid-state ^{13}C NMR has demonstrated the existence of "complex"-type reaction intermediates for propene adsorbed on H-Y at low temperatures ($\sim$200 K). They identify the complex as an isopropoxy species by its chemical shift. They also note that the species spontaneously oligomerizes as the system is warmed to room temperature and forms a new species, tentatively identified as an alkyl-substituted cyclopentenyl carbocation. This new species has much greater cation stability and is representative of a "free" carbocation as discussed later. Haw et al. [134] have also studied the products of adsorption of isobutylene on H-Y. They find no evidence of alkoxy species in contrast to the results with propylene. This lack of alkoxy species suggests the formation of a "free" carbocation. They confirm the assignment of Aronson et al. [135] for the alkoxy species from 2-methyl-2-propanol. This species would correspond to the "complex"-type carbocation. Haw et al. have also reported the formation of trimethyloxonium from dimethyl ether [136]; the species formed from cracking of ethylene oligomers [137], propylene oligomers [138] and butadiene oligomerization [139]. They also report resolving peaks for free and complexed Bronsted sites at low temperature. These peaks provide information on the strength of hydrogen bonds and suggest geometries for the H-bonded complexes [140]. Lechert et al. [141] report on NMR studies of molecular motions in faujasites.

The areas of future interest in solid-state NMR applied to acidity in zeolites are high-resolution, variable-temperature ^{1}H NMR studies of both zeolites (with varying structure and varying Si/Al ratio within the same structure) and amorphous aluminosilicates and studies with several relatively little utilized nuclei such as ^{17}O and ^2D. The results of these studies will add to the available data base and provide sufficient experimental justification for separating the effects that result from zeolite structure from those relating to Si/Al ratio. They will also address the question of acidity differences between zeolites and amorphous aluminosilicates. The observed high ratios of nonacidic to acidic protons in amorphous aluminosilicates compared with zeolites of similar Si/Al ratio may be related to the nature of the local electronic structure. The work of Doremieux-Morin et al. [64] provides data in support of this hypothesis.

3.6. Infrared Spectroscopy

As one of the earliest techniques used to study zeolite hydroxyls, IR spectroscopy provides primarily qualitative information on zeolite acidity. It is quite useful in separating Bronsted from Lewis type acidity with probe molecules such as pyridine. In addition, it provides a method of studying extra framework cations.

Several articles summarize various aspects of the study of zeolites with IR spectroscopy: Flanigan [142] and Ward [143], who discuss transmission IR studies in the framework and hydroxyl regions; Maroni et al. [144] and Janin et al. [145], who discuss the use of probe molecules to study the acidity of zeolites by diffuse reflectance and transmission techniques; and Baker et al. [146], who discuss the study of extra framework cations with far IR spectroscopy. Each of these articles summarizes the general state of the respective arts as of the date of the article.

More recent reviews are given by Barthomeuf [147], Kazansky [148], and Kazansky et al. [149]. Kazansky [148] notes the need to consider the acid-base pair (i.e., Al and neighboring O) in Lewis acid-base reactions. He notes that Lewis acid reactions can be characterized as a heterolytic cleavage (an oxidative addition rather than a coordination).

Jacobs and Mortier [150] summarize studies of zeolite acidity by hydroxyl IR spectroscopy. They rationalize the observed hydroxyl-stretching frequencies with two general principles: Sanderson intermediate electronegativity (which depends on chemical composition) and the presence of small (6- or 8-member) rings in certain zeolites. They thus propose that the primary effect is electronic because of the zeolite composition (e.g., Si/Al ratio) and cite a linear correlation between hydroxyl-stretching frequency and Sanderson intermediate electronegativity as evidence. This explanation accounts for the high-frequency hydroxyl band in a considerable number of zeolites. The second effect, the presence of restricted rings, is proposed to explain the low-frequency hydroxyl band by suggesting that hydroxyl groups located in such rings undergo a bathochromic (low-frequency) shift as a result of electrostatic interactions with the nearest oxygen atoms that are inversely proportional to the average squared distance. Thus, the effect is observed only with the smallest rings.

These two principles can rationalize the observed IR spectra but imply that all zeolites with similar Si/Al ratios would be approximately equally acidic if the hydroxyl frequency is a measure of acidity. This implication suggests that hydroxyl-stretching frequency is only an approximate indication of acid strength.

More recent work has focused on using IR to probe the types of acid sites and the nature of the local structure in zeolites. Datka et al. [151] have investigated the hydroxyl spectrum of H-Na-ZMS-5 and concluded that there is evidence for heterogeneity of hydroxyl groups in the zeolite. They propose either an effect resulting from differing numbers of Al NNN in the immediate vicinity of the hydroxyl or (their preferred explanation) a dependence on the local Al-O-Si bond lengths and angles. They calculate proton affinities for five distinct subbands ranging from ~ 1320 to ~ 1200 kJ/mol. These differences represent a range roughly comparable to the difference between acetic and trifluoroacetic acids, yet the five subbands all have the same IR frequency (~ 3609 cm^{-1}) in the absence of probe molecules.

Multitechnique studies, including IR, have been used to answer the questions raised by various treatments to dealuminate zeolites. The range of Si/Al ratios obtained by direct synthesis is generally rather limited, and various methods of altering these ratios after synthesis must be applied to obtain a series of zeolites of similar Si/Al ratio for comparison. Zi and Yi [152] have compared several techniques including hydrothermal treatment, ethylenediaminetetraacetic acid (EDTA) extraction, and ammonium hexafluorosilicate (AFS) treatment. They find that the framework asymmetric-stretching frequency is a linear function of the framework Al/(Al + Si) molar ratio.

4. MEDIA (SOLVENT) AND DIFFUSION EFFECTS IN ZEOLITES

4.1. Zeolites as Media (Solvent)

In the case of solutions, the effect of a solvent can be readily calculated by taking into account the properties, such as polarity, polarizability, or dielectric constant, of the molecules involved. The characterization of the zeolite media is much less tractable in a quantitative sense. The previous sections gave detailed information on the chemical bonding and especially on the chemical changes caused by introducing a H^+ cation. They also showed important differences between the characteristics of Al-rich and Si-rich zeolites. In this respect, laboratory experience also displays an important distinguishing characteristic, namely the strong hydrophilic character of Al-rich vs. the hydrophobic character of Si-rich zeolites. Thus, this discussion can appropriately be divided between Al-rich and Si-rich zeolites. The latter category includes pure-silica molecular sieves.

4.1.1. Aluminum-Rich Zeolites

Adsorption measurements, theoretical modeling, and extensive reaction chemistry using both NaY-zeolites and H-Y-zeolites indicate that the intracrystalline void space displays high polarity. Very large electrostatic field gradients of the order of 1 V/Å may exist in the α-cages of Y-zeolite at a distance of 2.5 Å from the center of site-2 bivalent cations. Similarly, oxygen rings unoccupied by cations provide large areas of negative electrostatic fields [153].

The large electrostatic fields derived by ab initio calculations are consistent with reports on the reaction chemistry of probe molecules adsorbed on zeolites. The results of these experiments show that Na-Y zeolite at an SiO_2/Al_2O_3 ratio of 5 can polarize and even ionize a variety of atoms and molecules to form remarkably stable occluded ions or ionic clusters in the zeolite crystal. These phenomena include the irreversible occlusion of salt molecules [154], the ionization of alkali metals forming a variety of subvalent alkali clusters (such as Na_4^{3+} or K_6^{5+}) in the cavities [153], and the ionization of $NO + NO_2$, radicals to form an $[NO^+ + NO_2^-]$ ionic complex in the zeolite cavity [155]. These reactions are endothermic by up to ~ 140 kcal/mol. Thus, the energy required for these processes is provided by the zeolite. Such interaction between occluded molecules and zeolite crystals displays the characteristics of a strong electrolyte or solvent [9,153].

In addition to theoretical modeling and reactions with probe molecules, the media (solvent) strength in zeolites can be estimated by xenon NMR spectroscopy [156]. This nondestructive method gives information on the electrostatic field in the vicinity of accessible cations. In addition, it gives information on the short- and long-range geometry of the intracrystalline pores and on the nature of occluded metal clusters.

One of the important practical results of the solvent effect in catalysis with Y zeolite is the concentration enhancement of hydrocarbon reactants within the zeolite crystal relative to the vapor phase. The result of this phenomenon is a great enhancement of the rates of bimolecular reaction steps, such as the hydride transfer reactions in catalytic cracking, relative to the rates of monomolecular reaction steps. This effect is significant even at very low Na and framework Al concentration [157]. Significantly, this selective, reaction-rate-enhancing phenomenon has been demonstrated for both nonprotic (K-exchanged) and for H-Y zeolites [9, 158]. This phenomenon shows that, directionally, the media effect is similar in both H-Y and in alkali-cation Y.

The media effect in zeolites is centered on the polarity of O-Na or O-H groups and on the polarity and polarizability of Al-O linkages. Therefore, high Al and cation concentrations contribute to increased media effect. At very high Al concentrations, however, the media effect is reversed because of dipole relaxation between adjacent polar groups and competitive, counteracting effects of adjacent polar sites with a given adsorbed molecule. The deleterious result of very high Al concentrations on media effect is similar to its impact on O-H acid strength because both are based on dipole relaxation processes. The precise value for the optimum Si/Al for maximum media effect depends both on the zeolite structure and on the size and chemistry of the reactant molecule. In general, the decline in media effect is likely to become large with Si/Al ratios below 2.5.

On the basis of model calculations and chemical probing of the intracrystalline reaction environment discussed earlier, high concentrations of framework aluminum and surface metal cations (including protic sites) result in very strong media effects. Directionally, this result is similar to the effect of polar solvents in solutions. Specific features of the media effect result from the microenvironment in the intracrystalline zeolite pores and cavities.

4.1.2. High-Silica Zeolites

The most dramatic difference in media effect between Al-rich and Si-rich zeolites is displayed by their interaction with water. Aluminum-rich zeolites are all strongly hydrophilic, and this property is similar for both metal-cation-exchanged and H-zeolites. At low pressure, they adsorb water, which fills the zeolite cavities. In contrast, Si-rich molecular sieves, particularly at high Si/Al ratios, are distinctly hydrophobic. When tested with water containing a few percent butanol, these sieves selectively adsorb the butanol. Such strong hydrophobic character shows that polarity-based interactions are minimal. The interaction between the microporous crystal and adsorbed molecules rests on London-type interactions, which are based on the relatively smaller polarizability of the Si-rich zeolite framework. These findings are fully consistent with similar conclusions deduced from mechanistic studies in the cracking of n-hexane over aprotic silicalite. This study showed strong n-hexane concentration enhancement for KY zeolite but found no concentration enhancement for silicalite [9].

4.2. **Diffusivity in Zeolites**

Diffusivity is a major factor in the kinetics of reactions on molecular sieves. Although classical diffusivity (Knudsen) is well understood, the nature of molecular diffusion in molecule-sized pores is the subject of much recent work. Diffusion is generally considered to retard the rate of reaction by restricting the rate of reactant transport to and product transport from the active site. However, with multiple active sites and narrow pores, diffusion of molecules in zeolites may increase the extent of reaction by increasing the residence time in the active zone. Thus, for all kinetic measurements of zeolite acidity, the effects of diffusion on the measurement must be taken into account.

The effects of diffusion on the interpretation of TPD data are discussed by Forni and Magni [99], who find that diffusion effects may dominate the TPD data. Forni et al. [102] analyzed experimental TPD data and found that intracrystalline diffusion controlled the desorption rate. From the large apparent activation energy ($D_a \sim 30$ kcal/mol framework Al), they suggest that interactions of the desorbing species with available acid sites (multiple desorption-adsorption steps) were occurring. Eic and Ruthven [159-161] used zero-length chromatography (ZLC) to measure diffusion rates for desorption of material from zeolites. Studies of xylene and benzene in NaX gave results much lower than NMR self-diffusivities, but they were in line with gravimetric measurements. Eic and Ruthven [159] also obtained similar results for n-butane in NaX but achieved agreement with NMR data for CaA. The NMR self-diffusion measurements determine intracrystalline diffusion coefficients as discussed by Ruthven [162]. Karger and Ruthven [163], Karger and Pfeifer [164,165], and Karger [166] provide experimental data. Unless a significant surface-to-pore barrier exists, ZLC primarily measures similar effects. The data of Eic and Ruthven suggest that a barrier does exist, as discussed later in this section.

Diffusion effects can also be used to advantage, as in the study by Caro et al. [167]. They used diffusion measurements and models to determine the location of N-bases in H-ZSM-5 by the effect on methane self-diffusion. Their results confirm previous suggestions that Bronsted acid sites are sited in the channel intersections.

Derouane et al. [168] have used sorption kinetics, heats of sorption, NMR relaxation times, and molecular graphics simulations to study the diffusion of alkanes in zeolites. They find that confinement effects regulate the diffusion, which is consistent with a "segmental diffusion" mode for the alkanes as proposed by Barrer and Davies [169]. They also considered the "floating" and "creeping" molecule concepts, as discussed later. For the zeolites considered, n-hexane has a "creeping" diffusion in large-pore zeolites but may change to "floating" in the case of AlPO-11.

Efforts have been made to apply theoretical models and related calculations to study diffusivity in zeolites. Aust et al. [170] briefly reviews the Monte Carlo models for diffusion in zeolites. They assess the influence at different model assumptions on the concentration dependence of transport and self-diffusion in a two-dimensional zeolite network.

Derouane et al. [171] discuss the effects of surface curvature on diffusion of molecules in zeolites. This model is an extension of the "nest effect," proposed by Barrer [172a] and modified by Derouane [172b], in which the adsorbed molecule and zeolite reciprocally optimize their van der Waals interactions. In zeolites, this interaction can lead to electronic structure changes, which may be manifested as changes in global crystal symmetry, when molecules such as benzene are adsorbed in ZSM-5. This model also leads to the concept of supermolecular interactions, in which the total system (zeolite plus adsorbate) must be considered. Observations of large intracrystalline diffusion rates for alkanes has led to the concept of configurational diffusion, as proposed by Weisz and by Haag et al. [173]. Derouane et al. [171] extend this concept to discuss two potential energy situations: "floating" molecules, when the channel diameter is approximately equal to the molecular diameter, and "creeping" molecules, when the channel diameter is large (approximately 2 or more times the molecular diameter). Using these concepts, Derouane et al. can semiquantitatively rationalize a number of experimental observations, including the role of zeolites as molecular traps, the origin of the surface diffusional barrier, and rapid diffusion of molecules in tight-fitting pores. Derouane et al. provide an excellent conceptual framework for describing the physical determinants of diffusional effects in zeolites.

Two points from the work of Derouane et al. [171] that discuss the difference in the diffusional characteristics of zeolites and amorphous materials need to be emphasized. Derouane et al. show that the approximate transition from "floating" to "creeping" diffusion occurs over the range of pore diameters from ~1 to ~2 times the molecular diameter for spheres (see the plot of attraction potential vs. reduced distance in Figure 5). The tendency to maximize attractive potentials for nonspherical molecules will make this transition occur at ~2 times the mean molecular diameter, although the total attractive energy will be maximized by smaller pore diameters. For typical molecules of interest, the mean molecular diameter is < ~10 Å; thus "floating" diffusion is largely restricted to pores < 20 Å in size. The effects also occur only if the pore is of relatively constant curvature; thus, long, crystallographically regular pores maximize the effect, and constantly varying pore diameters minimize the effect.

The surface diffusional barrier is a result of the sudden change in curvature on entering a small, molecule-sized pore (Figure 6). For a gradual curvature change, such as entering a large pore in an amorphous material, no barrier is expected. Derouane et al. [171] use this barrier to explain the much smaller uptake relative to NMR self-diffusion coefficients and note that rate differences of ~10^5 at 273 K are possible because of the magnitude of the potential barrier. As discussed earlier, Eic and Ruthven [159] see differences in rate even for the desorption process. This rate difference can be explained by noting that the reciprocal effect of the transition from pore to surface also has the same change in curvature, thus inducing potential energy barriers to molecular desorption from the pores. The effect should be greatest for "floating" molecules, which already have enhanced intracrystalline diffusion, and should decrease with the transition to "creeping" diffusion and with more gradual pore size changes toward the surface (e.g., funnel-shaped pores). Thus, zeolites possess potential energy barriers to desorption from the pore system that are not present in amorphous materials. This barrier can greatly increase the residence time of molecules in zeolites as a result of physical adsorption alone. Chemisorption on acid sites has additional effects, as seen in the TPD analysis by Forni et al. [102].

Figure 5
Representation of Potential in Molecule-Sized Pores

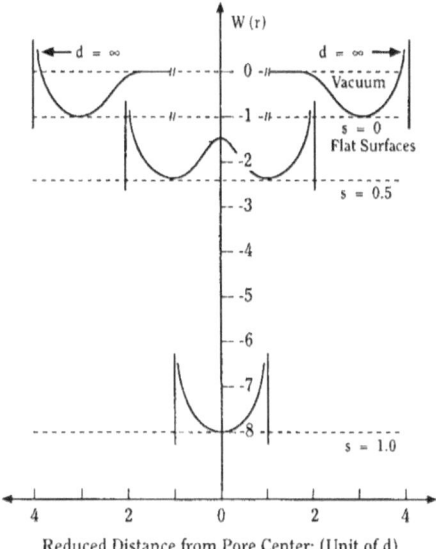

Reduced Distance from Pore Center: (Unit of d)

From Derouane and others, *J. Catal*, **110**, 58-73 (1988)

UOP 2086-5

Figure 6
Representation of Surface Energy Barrier

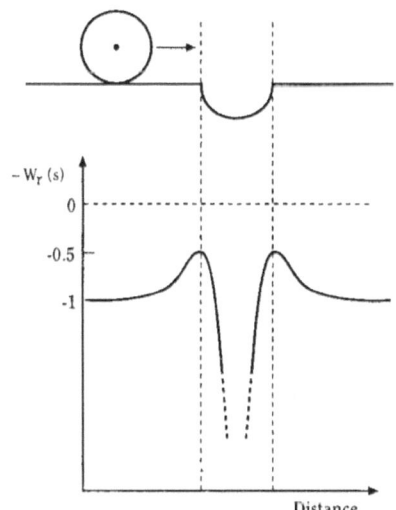

Distance

From Derouane and others, *J. Catal*, **110**, 58-73 (1988)

UOP 2086-5

5. FORMATION OF CARBOCATIONS

The direct characterization of carbocations formed in H-zeolites has been attempted by NMR spectroscopy in several studies (by Lombardo et al. [174]). The only hydrocarbon cation unequivocally identified so far is the triphenylcarbenium ion, formed on both H-Y and on amorphous silica-alumina catalysts. The chemical shift observed with this stable carbocation is almost identical to the one measured for this carbocation in a superacid. Therefore, the conclusion is that this species is a "free" cation. Unfortunately, formation of this cation is prevented in other zeolites by its large size. Other smaller carbocations have also been tentatively identified in zeolites [175]. However, the high reactivity of these carbocations to olefins, which form oligomeric species, cast doubt on the interpretation of most of the NMR assignments.

Of particular interest, considering the important role of carbocations in acid catalysis, are those zeolite acid characteristics that influence the formation, stability, and reaction chemistry of these hydrocarbon reaction intermediates. In strong acid solutions, particularly in superacids, this reaction intermediate is represented as a free solvated carbocation [2,6]. It is formed either by protonation of olefins or aromatics, by hydride abstraction from saturated hydrocarbons, or by forming five-coordinated C^+ centers [2,6,176]. In strong acid solutions, a large variety of stable carbocations have been prepared and characterized by NMR and other techniques. Extensive theoretical modeling of small as well as large and complex hydrocarbon and carborane cations has also been successfully accomplished recently using supercomputers. These calculations resulted in predicted NMR chemical shifts and bond energies closely matching the experimentally derived values.

Theoretical modeling of the polarizability of carbocations shows that the molecular bonds in these species use a variety of orbitals, including excited-state orbitals, that tend to increase the polarizability [7,8]. This characteristic of carbocations is important in acid catalysis because of the relationship between softness (polarizability) and reactivity of these species for intramolecular rearrangements.

In some cases, such as in catalytic cracking, the acid catalysis proceeds by a chain reaction involving hydride transfer from reactant molecules to chemically transformed carbocations. Thus, catalysis is initiated when the reaction conditions permit rapid hydride transfer. The initial cations are generally produced by protonated olefins. In bifunctional catalysis (using a combination of a dehydrogenation catalyst and a solid acid), the olefin produced over the metal is isomerized over the acid sites, and the isoolefins are presumably hydrogenated to isoparaffins over the metal catalyst. Therefore, a chain reaction mechanism is not needed. Whatever the sequence of steps in the H-zeolite-catalyzed reaction mechanism, the implication of the effect of both O-H acid strength and media effect on carbocation formation and structure must be considered.

The review of the chemistry of zeolite acid sites demonstrated that in the absence of strong added solvents such as water, the zeolite O-H groups carry hydrogen atoms with only a fractional positive electronic charge (0.12 to 0.25 e⁻). Also, cation- and Al-rich zeolites readily ionize a variety of adsorbed molecules [9,153,154]. Therefore, these zeolites are expected to contribute to the polarization of the hydrocarbon reactants and to the stabilization of carbocations as well. Silica-rich zeolites demonstrated no media effect, and therefore, they are not expected to contribute by media effect to the formation of carbocations. As previously discussed, the bond distortions induced by the proton are resisted by the zeolite crystal in order to maintain resonance energy and crystallinity. Therefore, upon transfer of the H^+ to the reactant molecule, the zeolite lattice should release the fraction of the deprotonation energy corresponding to the reversible structural strain introduced by H^+. This effect should also contribute to the formation of carbocations.

On the basis of this information, the zeolite media in strong electrolyte-type zeolites can be expected to contribute to the formation of weakly bound carbocations to the extent that they may be linked mainly by ionic interaction to zeolite framework oxygens. Such "free" carbocations should be relatively mobile, depending on the degree of covalent linkage with the zeolite lattice, and free to assume the thermodynamically most stable configuration in the zeolite.

In contrast to strong electrolyte-type zeolites, the media effect is small in high-Si zeolites, vide infra. However, the lack of solvation effect in these molecular sieves may be counterbalanced by a stronger O-H polarity. Because of a lack of media contribution to the stabilization of a "free" carbocation, the positively charged hydrocarbon-reaction intermediate formed by proton transfer from the H-zeolite to an olefin may be an alkoxy species, as demonstrated by Haw et al. [133], complexed to the zeolite framework.

The formation of a complex strongly linked to a zeolite framework oxygen should influence the effective charge, the lifetime (stability), and the steric hindrance imposed on this species. These characteristics should also affect the reaction rates and the reaction mechanism. Therefore, substantial differences in the reaction chemistry between positively charged hydrocarbon "complexes" and "free" carbocations may be expected. Because the formation of "free" carbocations requires higher activation energy, involving the deprotonization energy of the zeolite, "complex"-type intermediates may be formed at relatively low temperatures, probably near room temperature; "free" carbocations are generated at relatively higher temperatures. The reaction temperature used in catalytic processes such as FCC is high enough ($\sim$500°C) to provide sufficient energy to break even stable C-C or C-H bonds. Thus, the formation of "free" carbocations under these conditions is likely, particularly when the positively charged carbon has strong electron donor ligands.

The mechanistic aspects of carbocation chemistry over protic zeolites have been extensively studied by many investigators, and the work has been reviewed by Poutsma [158], who provides an enlightening discussion on the greatly diverse chemical and physical effects of H-zeolites on hydrocarbon reaction mechanisms. Some of the effects are derived from geometric (shape-selectivity) considerations, and others are derived from solvation and protic strength. These effects suggest mechanisms involving proton donors, hydride abstractions, or pentacoordinated C^+ species [158]. More recently, the formation of unusual product patterns obtained over strong acid H-zeolites has been discussed on the basis of pentacoordinated C^+ by Haag and Dessau [176a] and radical cation reaction intermediates by Rabo [176b]. In recent discussions, the superacid reaction pattern has been ascribed to very strong acid sites.

6. CONCLUSIONS

Reviewing the literature on the acid function in zeolites provides an excellent demonstration of the need for interdisciplinary research in surface chemistry and catalysis. The results show that in this complex field, a single experimental technique or theoretical calculation uncovers only a segment of the relevant chemistry. Past studies on zeolite acidity and acid catalysis cover a vast scientific effort and apply an extraordinarily wide array of techniques and theories. In spite of this effort, some of the key questions, including the cause of differences in acid strength between H-zeolites and amorphous silica-alumina and between zeolites of different structures, remained unanswered until recently. The following conclusions are based on careful consideration of a wide array of experimental data and theoretical considerations on zeolite acidity:

- The strong acidity of H-zeolites vs. amorphous aluminosilicates of similar compositions seems to be based on a global readjustment of the zeolite-lattice-bond structure to "proton attack" to minimize lattice deformations leading to loss of resonance energy and crystallinity (long-range symmetry). Without the influence of outside factors, the protonated Al-OH-Si groups prefer to form strongly covalent O-H bonds by forming relatively low (sp³-type) Al-O-Si bond angles. In zeolites, and particularly in Si-rich zeolites, the Al-O-Si bond angles range between 135 and 170°; thus, they are higher than sp³-type bond angles. Lowering these high bond angles for the Al-OH-Si units without affecting the Si-O-Si linkages also present in zeolite crystals is prevented or at least minimized by the zeolite crystal to maintain long-range order in the lattice. Thus, if maintenance of long-range order in the crystal prevails over the short-range bonding preference of Al-OH-Si groups, then the original large bond angles will be retained. In this case, the zeolite retains its crystallinity and displays strong acidity. When the local bonding preference adjacent to the OH group prevails, the crystal may collapse, and the OH groups will be of much weaker acidity. Indeed, H-zeolites with high aluminum and consequently high H-content (Si/Al < 2) all lose crystallinity. Crystalline H-zeolites require the presence of stable Si-O-Si linkages in the zeolite crystal. The observed structural loss is probably the result of the reduction in the bond angle in the Al-OH-Si groups. This hypothesis suggests that the maintenance of crystallinity and the consequent strong acidity found in Si-rich zeolites (an Si/Al ratio of > 2) depends on the balance between the favored bond formation of a single Al-OH-Si group and the long-range atomic order favored by the crystal. Here, the issue is the balance and reconciliation between the short- and long-range effects. In contrast, amorphous silica-alumina gel has no structural constraint comparable to crystalline materials to limit Al-OH-Si bond angle reductions adjacent to protic sites. These amorphous materials form chemically stable, covalent O-H linkages. Hence, the resultant protic acidity is weak.
- Differences in O-H acid strength between different zeolites rest on differences in crystal structure and related bond structure, T-O-T bond angles and lengths, and crystal resonance energy. Large T-O-T bond angles, such as those found in ZSM-5, enhance the *s* character of the T-O bonds, and consequently, the electronegativity of the oxygen. The result is lower *s* character in the oxygen lone-pair orbital bonded to the proton.

Lower *s* character means weaker O-H bonds and thus lower deprotonization energy and stronger acidity. Because part of the crystal stability is derived from lattice resonance, zeolites with cage structures (with subunits of high symmetry) may be able to relax upon "proton attack" more than channel-type structures. This additional relaxation can lead to higher covalency and to lower O-H acid strength for the former crystals.

- In general, the strength of the zeolite media (solvent) effect is expected to grow with Si/Al increasing ratios, from 1 to ~2.5, and to reach a maximum at an Si/Al ratio of 2.5 to ~5. Beyond an Si/Al ratio of ~5, the media effect should decline with lower Al content.

- In contrast to the media effect, the intrinsic O-H acid strength of individual protic sites increases with increasing Si/Al ratios to reach a near maximum value at a ratio of ~7 (depending on the zeolite structure type).

- All H-zeolites, even the strongest acids, carry O-H hydrogen with only a fractional positive charge (~0.2 e⁻). Therefore, the strength of the zeolite media is expected to strongly influence the formation of "free" carbocations. At low temperatures, where H⁻ abstraction is not dominant in the mechanism, the reaction intermediate in hydrocarbon catalysis may be an alkoxy complex. Such "complex"-like reaction intermediates should have different reactivity and reaction chemistry from "free" carbocations. The tendency to form such "complex"-like intermediates should be inversely related to the reaction temperature and to the strength of the intracrystalline-zeolite media (solvation). Carbocation mechanisms involving five-coordinated C^+ centers observed with ZSM-5 and other strong-acid zeolites result from superacid mechanisms. Furthermore, low temperatures may favor "complex"-like species, and high temperatures should favor "free" carbocations. In addition, more stable carbocations (e.g., tertiary, benzilic) should favor formation of "free" carbocations.

- The effect of hard vs. soft character of the carbocation and the zeolite lattice in acid catalysis is not well understood. However, larger polarizability for the reaction intermediate and for the lattice oxygen should affect the reaction chemistry of "complex"-type hydrocarbon reaction intermediates. Substitution of zeolite framework cations by transition metal ions is expected to increase the softness of the zeolite bond structure by providing polarizable *d* orbitals. Such phenomena may be the cause of new catalytic phenomena that are not strictly acid strength related being reported for transition-metal-substituted molecular sieves [177,178].

7. REFERENCES

1. J. A. Rabo and G. J. Gajda, Catal. Rev.-Sci. Tech., 31, 385-430 (1989-90).

2. G. A. Olah, G. K. S. Prakash, and J. Sommer, Superacids, Wiley, New York, 1985.

3. (a) R. G. Pearson, J. Am. Chem. Soc., 85, 3533-3539 (1963).

   (b) R. G. Pearson, J. Am. Chem. Soc., 107, 6801-6806 (1985).

4. J. Chatt, J. Inorg. Nucl. Chem., 8, 515 (1958).

5. R. G. Pearson, Proc. Natl. Acad. Sci., 83, 8440-8441 (1986).

6. G. A. Olah, G. K. S. Prakash, R. E. Williams, J. D. Field, and D. Wade, Hypercarbon Chemistry, J. Wiley, New York, 1987.

7. (a) J. F. Stanton, W. N. Lipscomb, R. J. Bartlett, and M. L. McKee, Inorg. Chem., 28(1), 109-11 (1989).

   (b) J. F. Stanton, P. J. Bartlett, and W. N. Lipscomb, Chem. Phys. Lett., 138(6),525-530 (1987).

   (c) K. L. Krause, K. W. Volz, and W. N. William, J. Mol. Biol. 193(3), 527-553 (1987).

8. (a) M. Bremer, P. Schleyer, and U. Fleischer, J. Am. Chem. Soc., 111(3), 1147-1148 (1989).

   (b) A. E. Reed, C. Schade, P. Schleyer, P. V. Kamath, and J. Chandrasekhar, J. Chem. Soc. Chem. Commun., (1), 67-9, 1988.

   (c) M. Saunders, K. E. Laidig, K. B. Wiberg, and P. Schleyer, J. Am. Chem. Soc., 110(23), 7652-7659 (1988).

   (d) W. Bauer, M. Feigel, G. Mueller, and P. Schleyer, J. Am. Chem. Soc. 110(18), 6033-6046 (1988).

   (e) P. Schleyer, Pure Appl. Chem., 59(2), 1647-1660 (1987).

   (f) P. Schleyer, T. W. Williams, W. Koch, A. J. Kos, and H. Schwarz, J. Am. Chem. Soc. 109(23), 6953-6957 (1987).

   (g) P. Schleyer, B. T. Luke, and J. A. Pople, Organometallics, 6(9), 1997-2000, (1987).

9. J. A. Rabo, Catal. Rev.-Sci. Eng., 23(1&2), 293-313 (1981).

10. D. W. Breck, Zeolite Molecular Sieves, Wiley-Interscience, New York, 1974.

11. J. A. Rabo, P. E. Pickert, D. N. Stamires, and J. E. Boyle, Actes Du Deuxieme Congress International de Catalyse, Paris, 1960, pp. 2055-2074.

12. J. A. Rabo, in Zeolite Chemistry and Catalysis (ACS Monograph 171), Am. Chem. Soc., 1976, Chap. 5.

13. R. J. Pellet, C. S. Blackwell, and J. A. Rabo, J. Catal., 114, 71-89 (1988).

14. J. Dwyer, Stud. Surf. Sci. Catal., 37, 333-54 (1988).

15. J. Dwyer, NATO ASI Ser., Ser. C, 352, 303-19 (1992).

16. J. Dwyer, NATO ASI Ser., Ser. C, 352, 321-45 (1992).

17. D. Barthomeuf and A. Corma, NATO ASI Ser., Ser. B, 221, 403-7 (1990).

18. D. Barthomeuf, Stud. Surf. Sci. Catal., 65, 157-69 (1991).

19. G. H. Grant and R. J. Abrahams, Catalysis, 8, 68-106 (1989).

20. R. A. van Santen, D. P. de Bruyn, C. J. J. den Ouden and B. Smit, Stud. Surf. Sci. Catal., 58, 317-58 (1991).

21. J. Sauer, K. P. Schroeder and J. R. Hill, Am. Chem. Soc. Preprints, 37(2), 666-72 (1992).

22. E. G. Derouane, J. M. Andre, L. Leherte, P. Galet, D. Vanderveken, D. P. Vercauteren and J. G. Fripiat, in "Theoretical Aspects of Heterogeneous Catalysis," J. Moffet, ed., Van Nostrand Reinhold, New York (1990) pp 1-51.

23. R. A. van Santen, B. W. H. van Beest, and A. J. M. de Man, NATO ASI Ser., Ser. B, 221, 201-24 (1990).

24. J. Sauer, NATO ASI Ser., Ser. B, 283, 533-50 (1992).

25. J. Sauer, Chem. Rev., 89, 199-255 (1989).

26. L. A. Pine, P.J. Maher, and W. A. Wachter, J. Catal., 85, 466-476 (1984).

27. W. Lowenstein, Ann. Mineral, 39, 92 (1954).

28. W. A. Wachter, Proc. 6th Int'l Zeolite Conf., 1984, pp. 141-150.

29. W. A. Wachter, in "Theoretical Aspects of Heterogeneous Catalysis," J. Moffet, ed., Van Nostrand Reinhold, New York (1990) pp 110-34.

30. D. Barthomeuf, Mat. Chem. Phys., 17, 49-71 (1987).

31. W. M. Meier and H. J. Moeck, J. Solid State Chem., 27, 349 (1979).

32. H. Stach, J. Janchen, H. G. Jerschkewitz, U. Lohse, B. Parlitz, B. Zibrowius and M. Hunger, J. Phys. Chem., 96, 8473-9 (1992).

33. H. Stach, J. Janchen, H. G. Jerschkewitz, U. Lohse, B. Parlitz, B. Zibrowius and M. Hunger, J. Phys. Chem., 96, 8480-5 (1992).

34. P. P. Ewald, Ann. Phys. (Paris), 64, 253 (1921).

35. J. Tennyson and J. N. Murrell, Mol. Phys., 42, 297 (1981).

36. J. N. Murrell, J. Tennyson and M. A. Kamel, Mol. Phys., 42, 747 (1981).

37. N. W. Winter, R. M. Pitzer and D. K. Temple, J. Chem. Phys., 86, 3549 (1987).

38. N. W. Winter, R. M. Pitzer and D. K. Temple, J. Chem. Phys., 87, 2945 (1987).

39. A. Y. S. Kung, A. B. Kunz and J. M. Vail, Phys. Rev. B, 26, 3352 (1982).

40. W. J. Mortier, Doctoral Thesis, Catholic University of Leuven, 1978.

41. S. Beran and J. Dubsky, J. Phys. Chem., 83, 2538-2544 (1979).

42. (a) K. J. Chou, J. C. Lin, Y. Wang, and G. H. Lee, Zeolites, 6, 35-38 (1986).

    (b) J. L. Schlenker, J. J. Pluth, and J. V. Smith, Mat. Res. Bull., **14,** 849-856 (1979).

43. D. H. Olson and E. Dempsey, J. Catal., **13,** 221-231 (1969).

44. J. Dwyer and P. J. O'Malley, Stud. Surf. Sci. Catal., 35, 5 (1988).

45. R. Carson, E. M. Cooke, J. Dwyer, A. Hinchliffe, and P. J. O'Malley, Stud. Surf. Sci. Catal., 46, 39-48 (1989).

46. W. J. Mortier and P. Geerlings, J. Phys. Chem., 84, 1982-1986 (1980).

47. E. Kassab, K. Seiti, and M. Allavena, J. Phys. Chem., 92, 6705-6709 (1988).

48. E. G. Derouane and J. G. Fripiat, J. Phys. Chem., **91,** 145-148 (1987).

49. A. Goursot, F. Fajula, C. Daul, and J. Weber, J. Phys. Chem., 92, 4456-4461 (1988).

50. C. Fernandez, A. Auroux, J. C. Vedrine, J. Grosmangin, and G. Szabo, Proc. 7th Int'l Zeolite Conf., 1986, pp. 345-350.

51. E. G. Derouane and J. G. Fripiat, Zeolites, 5, 165-172 (1985).

52. R. A. van Santen, A. J. M. de Man, W. P. J. H. Jacobs, E. H. Teunissen and G. J. Kramer, Catal. Lett., 9, 273-86 (1991).

53. G. J. Kramer, A. J. M. de Man and R. A. van Santen, J. Am. Chem. Soc., 113, 6435-41 (1991).

54. E. G. Derouane and J. G. Fripiat, Proc. 6th Intl Zeol. Conf., 717 (1983).

55. J. G. Fripiat, P. Galet, J. Delhalle, J. M. Andre, J. B. Nagy and E. G. Derouane, J. Phys. Chem., 89, 1932 (1985).

56. E. G. Derouane and J. G. Fripiat, Zeolites, 5, 165, (1985).

57. P. J. O'Malley and J. Dwyer, Zeolites, 8, 317 (1988).

58. R. S. Hansford, Ind. Eng. Chem., 39, 849 (1947).

59. J. B. Peri, J. Catal., **41,** 227-239 (1976).

60. A. G. Pelmenshchikov, E. A. Paukshtis, V. S. Stepanov, K. G. Ione, G. M. Zhidomirov, and K. I. Zamaraev, Proc. 9th Int'l Conf. Catal., 1988, pp. 404-411.

61. G. M. Zhidomirov and V. Kazansky, Adv. Catal., 34, 131-202 (1986).

62. T. L. Barr, L. M. Chen, M. Mohsenian, and M. A. Lishka, J. Am. Chem. Soc., **110,** 7962-7975 (1988).

63. W. J. Mortier, Proc. 6th Int'l Zeolite Conf., 1984, pp. 734-746.

64. C. Doremieux-Morin, P. Batamack, C. Martin, J. M. Bregeault and J. Fraissard, Catal. Lett, 9, 403-10 (1991).

65. W. J. Mortier, J. Sauer, J. A. Lercher and H. Noller, J. Phys. Chem., 88, 905-12 (1984).

66. J. Sauer, H. Haberlandt and W. Schirmer, Stud. Surf. Sci. Catal., 18, 313-20 (1984).

67. J. Sauer and W. Schirmer, Stud. Surf. Sci. Catal., 37, 323-32 (1988).

68. A. J. M. de Man, W. P. J. H. Jacobs, J. P. Gilson and R. A. van Santen, Zeolites, 12, 826-36 (1992).

69. E. Chamot, Am. Chem. Soc. Preprints, 37(2) 608-10 (1992).

70. (a) J. Sauer, Acta Phys. Chem., 31, 19-24 (1985).

    (b) J. Sauer, J. Phys. Chem., 91, 2315-9, (1987).

71. J. Sauer and R. Zahradnik, Intl. J. Quantum Chem., 26, 793-822 (1984).

72. J. Sauer and A. Bleiber, Catal. Today, 3, 485-92 (1988).

73. S. Beran, J. Phys. Chem., 94, 335-7 (1990).

74. J. H. Lunsford, Am. Chem. Soc. Preprints, 35(4), 654-60 (1990).

75. W. J. Mortier, D. E. W. Vaughan, and J. M. Newsam, ACS Symp. Ser., 368, 194-202 (1980).

76. (a) J. J. Van Dun and W. J. Mortier, J. Phys. Chem., 92, 6740-6746 (1988).

    (b) J. J. Van Dun, K. Dhaeze, and W. J. Mortier, J. Phys. Chem., 92, 6747-6754 (1988).

77. J. R. Hill and J. Sauer, Z. Phys. Chem. (Leipzig), 270, 203-6 (1989).

78. K. P. Schroder, J. Sauer, M. Leslie, C. R. A. Catlow and J. M. Thomas, Chem. Phys. Lett., 188, 320-5 (1992).

79. A. J. M. de Man and R. A. van Santen, Zeolites, 12, 269-79 (1992).

80. J. Sauer, J. Mol. Catal., 54, 312-23 (1989).

81. J. Sauer, H. Horn, M. Haeser and R. Ahlrichs, Chem. Phys. Lett., 173, 26-32 (1990).

82. J. Sauer, C. M. Koelmel, J. R. Hill and R. Ahlrichs, Chem. Phys. Lett., 164, 193-8 (1989).

83. A. K. Aboul-Gheit, M. A. Al-Hajjaji, M. F. Menougy, and S. M. Abdel-Hamid, Anal. Lett., 19, 529-536 (1986).

84. H. G. Karge and J. B. Nagy, NATO ASI Ser., Ser. B, 221, 387-94 (1990).

85. H. G. Karge, Stud. Surf. Sci. Catal., 65, 133-56 (1991).

86. J. H. C. van Hooff and J. W. Roelofsen, Stud. Surf. Sci. Catal., 58, 241-83 (1991).

87. Kh. Dimitov, Z. Popova, S. Mladenov, K. H. Steinberg and H. Siegel, Catal. Zeolites, D. Kallo, Kh. M. Minachev, eds., Akad. Kiado, Budapest (1988) pp 135-67.

88. K. Hashimoto, T. Masuda and H. Sasaki, Ind. Eng. Chem. Res., 27,1792-1797 (1988).

89. B. Umansky, J. Engelhardt and W. K. Hall, J. Catal., 127, 128-40 (1991).

90. K. Tsutsumi, H. Q. Koh, S. Hagiwara, and H. Takahasi, Bull. Chem.Soc. Japn., 48, 3576 (1975).

91. A. Auroux, V. Bolis, P. Wierzchowski, P. Gravelle, and J. C. Vedrine, J. Chem. Soc. Faraday Trans. I, 75, 2544 (1979).

92. R. Bezman, J. Catal., 68, 242 (1981).

93. A. Macedo, A. Auroux, F. Raatz, E. Jacquinot, and R. Boulet, ACS Symp. Ser., 368, 98-116 (1968).

94. (a) A. K. Aboul-Gheit and M. A. Al-Hajjaji, Anal. Lett., 20, 553-559 (1987).

    (b) A. K. Aboul-Gheit, J. Catal., 113, 490-496 (1988).

95. A. K. Aboul-Gheit, M. A. Al-Hajjaji, and A. M. Summan, Thermochim. Acta, 118, 9-16 (1987).

96. G. I. Kapustin, T. R. Brueva, A. L. Klyachko, S. Beran, and B. Wichterlova, Appl. Catal., 42, 239-446 (1988).

97. B. Sen, P. Chou, and M. A. Vannice, J. Catal., 101, 517-521 (1986).

98. (a) J. L. Falconer and J. A. Schwarz, Cat. Rev.-Sci. Eng., 25, 141 (1983).

    (b) R. J. Gorre, J. Catal., 75, 164-174 (1982).

    (c) R. A. Demmin and R. J. Gorre, J. Catal., 90, 32-39 (1984).

    (d) Y.-J. Huang and J. A. Schwarz, J. Catal., 99, 249-251 (1986).

    (e) J. M. Criado, P. Malet, G. Munuera, Langmuir, 3, 973-975 (1987).

99. L. Forni and E. Magni, J. Catal., 112, 437-443 (1988).

100. K. Hashimoto, T. Masuda, and T. Mori, Proc. 7th Int'l Zeolite Conf., 1986, pp.503-510.

101. J. Take, H. Yoshioka, and M. Misono, Proc. 9th Int'l Conf. Catal., pp. 372-379.

102. L. Forni, E. Magni, E. Ortoleva, R. Monaci, and V. Solinas, J. Catal., 112, 444-452 (1988).

103. L. G. Dowell, J. M. Bennett, and D. E. Passoja, 25th Ann. Denver X-ray Conf., 1976.

104. (a) T. L. Barr, Am. Lab., 10, 40 (1978).

     (b) T. L. Barr, Am. Lab., 10, 65 (1978).

105. Y. Okamoto, M. Ogawa, A. Maezawa, and T. Imanaka, J. Catal., 112, 427-436 (1988).

106. R. L. Patton, E. M. Flanigan, L. G. Dowell, and D. E. Passoja, ACS Symp. Ser., 40, 64-75 (1977).

107. K. Taniguchi and T. Takaishi, Proc. 7th Int'l Zeolite Conf. , 1986, pp. 155-156.

108. R. H. West and J. E. Castle, Surf. Interf. Anal., 4, 68 (1982).

109. T. Takaishi and H. Hosoi, J. Phys. Chem., 86, 2089-2094 (1982).

110. For example, L. Pauling, The Nature of the Chemical Bond, Cornell Univ. Press, Ithaca, NY, 1960.

111. T. L. Barr and M. A. Lishka, J. Am. Chem. Soc., 108, 3178-3186 (1986).

112. J. B. Nagy and E. G. Derouane, ACS Symp. Ser., 368, 2-32 (1988).

113. J. M. Thomas and J. Klinowski, Adv. Cat., 33, 199-374 (1985).

114. H. Pfeifer, D. Freude, and M. Hunger, Zeolites, 5, 274-286 (1985).

115. M. Hunger, D. Freude and H. Pfeifer, J. Chem. Soc. Faraday Trans., 87(4), 657-62 (1991).

116. H. Pfeifer, D. Freude and J. Karger, Stud. Surf. Sci. Catal., 65, 89-115 (1991).

117. H. Ernst, Z. Phys. Chemie (Leipzig), 269, 1073-94 (1988).

118. C. A. Fyfe, H. Grondey, Y. Feng, H. Gies and G. T. Kokotailo, NATO ASI Series, Ser. C, 352, 225-45 (1992).

119. C. A. Fyfe, H. Grondey, Y. Feng, H. Gies and G. T. Kokotailo, NATO ASI Series, Ser. C, 352, 247-70 (1992).

120. G. Engelhardt, Stud. Surf. Sci. Catal., 58, 285-315 (1991).

121. J. Klinowski, Chem Rev., 91, 1459-79 (1991).

122. K. T. Mueller, B. Q. Sun, G. C. Chings, J. W. Zwanziger, T. Terao and A. Pines, J. Magn. Res., 86, 470 (1990).

123. K. T. Mueller, Y. Wu, B. F. Chmelka, J. Stebbins and A. Pines, J. Am. Chem. Soc., 113, 32, (1991).

124. E. G. Derouane, N. Dewaele, Z. Gabelica, and J. B. Nagy, Appl. Catal., 28, 285 (1986).

125. (a) G. Engelhardt and D. Michel, High Resolution Solid-State NMR of Silicates and Zeolites, Wiley, New York, 1987, pp. 228, 257, 260.

     (b) J. M. Bennett, C. S. Blackwell, and D. E. Cox, J. Phys. Chem., 87, 3783-3790 (1983).

     (c) E. Brunner, H. Ernst, D. Freude, M. Hunger, C. B. Krause, D.Preger, W. Reschetilowski, W. Schwieger and K. H. Bergk, Zeolites, 9, 282-286 (1989).

126. (a) J. H. Lunsford, W. P. Rothwell, and W. Shen, J. Am. Chem. Soc., 107, 1540-1547 (1985).

     (b) W. L. Earl, P. O. Fritz, A. A. V. Gibson, and J. H.Lunsford, J. Phys. Chem., 91, 2091-2095 (1987).

     (c) L. Baltusis, J. S. Frye, and G. E. Maciel, J. Am. Chem. Soc., 108, 7119-7120, (1986).

     (d) L. Baltusis, J. S. Frye, and G. E, Maciel, J. Am. Chem. Soc., 109, 40-46 (1987).

127. (a) R. L. Stevensen, J. Catal., 21, 113-121 (1971).

     (b) D. Freude, J. Klinowski, and H. Hamdan, Chem. Phys. Lett., 149, 355-362 (1988).

128. (a) M. Hunger, D. Freude, H. Pfeifer, H. Bremer, M. Jank, and K. P. Wendlandt, Chem. Phys. Lett., 100, 29-33 (1983).

     (b) H. Pfeifer, Coll. Surf., 36, 169-177 (1989).

129. B. B. Whipple, P. J. Green, M. Ruta, and R. L. Bujalski, J. Phys. Chem., 80, 1350 (1976).

130. H. Pfeifer, J. Chem. Soc. Faraday Trans. I., 84, 3777-3783 (1988).

131. (a) J. W. Emsley, J. Feeney, and L. M. Sutcliffe, High Resolution NMR Spectroscopy, Pergamon, Oxford, 1966.

(b) R. K. Harris and B. E. Mann, NMR and the Periodic Table, Academic Press, London, 1978.

132. D. Freude, H. Pfeifer, and M. Hunger, Proc. 7th Int'l Zeolite Conf., 1986, p. 107.

133. J. F. Haw, B. R. Richardson, I. S. Oshiro, N. D. Lazo, and J. A. Speed, J. Am. Chem. Soc., 111, 2052-2058 (1989).

134. N. D. Lazo, B. R. Richarson, P. D. Schettler, J. L. White, E. J. Munson and J. F. Haw, J. Phys. Chem., 95, 9420-5 (1991).

135. M. T. Aronson, R. J. Gorte, W. E. Farneth and D. White, J. Am. Chem. Soc., 111, 840 (1989).

136. E. J. Munson and J. F. Haw, J. Am. Chem. Soc., 113, 6303-5 (1991).

137. F. G. Oliver, E. J. Munson and J. F. Haw, J. Phys. Chem., 96, 8106-11 (1992).

138. J. L. White, N. D. Lazo, B. R. Richardson, and J. F. Haw, J. Catal., 125, 260-3 (1990).

139. B. R. Richardson, N. D. Lazo, P. D. Schettler, J. L. White and J. F. Haw, J. Am. Chem. Soc., 112, 2886-91 (1990).

140. J. L. White, L. W. Beck and J. F. Haw, J. Am. Chem. Soc., 114, 6182-9 (1992).

141. H. T. Lechert, W. D. Basler and M. Jia, NATO ASI Ser., Ser. B, 221, 183-92 (1990).

142. E. M. Flanigan, in Zeolite Chemistry and Catalysis, ACS Monograph 171 (J. A. Rabo, ed.), 1976, pp. 80-117.

143. J. W. Ward, in Zeolite Chemistry and Catalysis, ACS Monograph 171 (J.A. Rabo, ed.), 1976, pp. 118-287.

144. V. A. Maroni, K. A. Martin, and S. A. Johnson, ACS Symp. Ser., 368, 85-97 (1987).

145. A. Janin, J. C. Lavalley, A. Macedo, and F. Raatz, ACS Symp. Ser., 368, 117-135 (1988).

146. M. D. Baker, J. Godber, and G. A. Ozin, ACS Symp. Ser., 368, 136-149 (1988).

147. D. Barthomeuf, NATO ASI Ser., Ser. C, 352, 193-223 (1992).

148. V. B. Kazansky, Stud. Surf. Sci. Catal., 65, 117-31 (1991).

149. L. M. Kustov, S. A. Zubkov, V. B. Kazansky and L. A. Bondar, Stud. Surf. Sci. Catal., 69, 303-11 (1991).

150. P. A. Jacobs and W. J. Mortier, Zeolites, 2, 226-230 (1982).

151. J. Datka, M. Boczen, and P. Rymarowicz, J. Catal., 114, 368-376 (1988).

152. G. Zi and T. Yi, Zeolites, 8, 232-237 (1988).

153. J. A. Rabo, C. L. Angell, P. H. Kasai, and V. Schomaker, Discuss. Faraday Soc., 41, 328-349 (1966).

154. J. A. Rabo and P. H. Kasai, Progress in Solid State Chemistry, 9, 1-19 1975.

155. P. H. Kasai and R. J. Bishop, ACS Monograph, **171**, 350-390 (1976).

156. (a) T. Ito and J. Fraissard, J. Chem. Soc. Faraday Trans. I., 83, 451-462 (1987).

    (b) J. Fraissard, T. Ito, M. Springuel-Huet, and J. Demarquay, Proc. 7th Int'l. Zeolites Conf., 1986, pp. 393-400.

    (c) L. Petrakis, T. Ito, M. A. Springuel-Huet, T. Hughes, I. Y. Chen, and J. Fraissard, Proc. 9th Int'l. Congress on Catal., 1988, pp. 348-355.

    (d) T. Ito, M. A. Springuel-Huet, and J. Fraissard, Zeolites, 9, 68 (1989).

    (e) M. A. Springuel-Huet, J. Demarquay, T. Ito, and J. Fraissard, in Innovation in Zeolite Material Science (P. J. Grobet et al., eds.), Elsevier, 1988, pp. 183-189.

157. S. M. Brown, W. J. Reagan, and G. M. Wolterman, U.S. Patent 4,325,813 (1982).

158. M. L. Poutsma, ACS Monograph, **171**, 437-528 (1976).

159. M. Eic and D. M. Ruthven, Zeolites, 8, 472-479 (1988).

160. D. M. Ruthven and M. Eic, ACS Symp. Ser., 368, 362-375 (1988).

161. M. Eic and D. M. Ruthven, Zeolites, 8, 40 (1988).

162. D. M. Ruthven, Principles of Adsorption and Adsorption Processes, Wiley, New York, 1984.

163. J. Karger and D. M. Ruthven, J. Chem. Soc. Faraday Trans. I, 77, 1485 (1981).

164. J. Karger and H. Pfeifer, Zeolites, 7, 90 (1987).

165. J. Karger and H. Pfeifer, ACS Symp. Ser., 368, 376-396 (1988).

166. J. Karger, J. Chem. Soc. Faraday Trans. I, 76, 717 (1980).

167. J. Caro, M. Bulow, J. Karger, and H. Pfeifer, J. Catal., **114,** 186-189 (1988).

168. E. G. Derouane, J. B. Nagy, C. Fernandez, Z. Gabelica, E. Laurent, and P. Maljean, Appl. Cat., 40, L1-L10 (1988).

169. R. M. Barrer and J. A. Davies, Proc. Royal Soc. London, Ser. A, 322, 1 (1971).

170. E. Aust, K. Dahlke, and G. Emig, J. Catal., **115,** 86-97 (1989).

171. E. G. Derouane, J. M. Andre, and A. A. Lucas, J. Catal., **110,** 58-73 (1988).

172. (a) R. M. Barrer, Zeolites and Clay Minerals as Sorbants and Molecular Sieves, Academic Press, London, 1978.

    (b) E. G. Derouane, J. Catal., 100, 541 (1986).

173. (a) P. B. Weisz, Chemtech, 3, 498-505 (1973).

    (b) W. O. Haag, R. Lago, and P. B. Weisz, Farad. Disc. Chem. Soc., 72, 317 (1981).

174. E. A. Lombardo, J. M. Dereppe, G. Marcelin, and W. K. Hall, J. Catal., 114, 167-175 (1988).

175. (a) M. C. Grady and R. S. Gorte, J. Phys. Chem., 89, 1305 (1985).

(b) M. T. Aronson, R. J. Gorte, and W. E. Farneth, J. Catal., 98, 434 (1986); 105, 455 (1987).

(c) A. Webb, Actes Deuxieme Int. Congr. Catal., Vol. 1, p. 1289, Technip, 1960; W. K. Hall comment, Actes Deuxieme Int. Congr. Catal., Vol. 1, p. 1307, Technip, 1960.

(d) M. Zardkoohi, J. F. Haw, and J. H. Lunsford, J. Am. Chem. Soc., 109, 5278 (1987).

(e) J. P. van den Berg, J. P. Wolthuizen, A. D. H. Clague, G. R. Hays, R. Huis, and J. H. van Hooff, J. Catal., 80, 130 (1983).

176. (a) W. O. Haag and R. M. Dessau, Proceedings 8th International Congress on Catalysis, Berlin, 1984, Dechema, Frankfurt-am-Main, 1984, pp. 305ff.

(b) J. A. Rabo, Proc. Sixth Int. Zeolite Conf., 1985, p. 41.

177. R. J. Pellet, P. K. Coughlin, E. Shamshoum, and J. A. Rabo, ACS Symp. Ser., 368, 512-531 (1988).

178. E. M. Flanigan, R. L. Patton and S. T. Wilson, Stud. Surf. Sci. Catal., 37, 13-27 (1988).

BASICITY IN ZEOLITES

Denise BARTHOMEUF
Laboratoire de Réactivité de Surface et Structure, URA 1106 CNRS,
Université Pierre et Marie Curie, 4 Place Jussieu, 75252 Paris Cedex 05,
France

ABSTRACT. The description of basic sites in zeolites includes framework oxygen atoms, basic hydroxyl groups, clusters of oxides or hydroxydes. They are generated usually in ion exchanged or alkali metals supported materials. Framework oxide ions form acid-base pairs with cations. The zeolite basicity is evaluated by theoretical calculations and experimental measurements. The calculations show an increase in basicity with Al content and with the presence of weakly electronegative cations. The choice of a probe for experimental studies is a major problem. The use of CO_2, acids, pyrrole, benzene... is discussed. Several techniques are used to measure either directly binding energy of O_{1s} (XPS) or indirectly the reactivity of oxygen atoms through their interaction with a probe (thermal methods, infrared, XPS...). Basic zeolites become more and more important for applications in adsorption and catalysis.

Introduction

The basicity in solids is recognized for a long time as important in catalysis (1, 2). The simultaneous participation of acid and basic sites in the transformation of alcohols was shown a long time ago, the dehydrogenation being the predominant reactions in the presence of strong basic sites (3).

In zeolites the presence of basic sites was first shown in the reaction of alkylation of toluene with methanol (4). It was observed that acidic zeolites (faujasites X and Y) favoured the ring alkylation while the chain alkylation occured on zeolites, exchanged with K,Rb ou Cs cations. This was further studied in detail and related to the basicity of framework oxygen in the latter materials (5). A clear influence of cations on framework oxygen basicity is observed in the transformation of isopropanol on X and Y exchanged with alkali cations.The dehydration activity decreases from the Li to the Cs exchanged forms (decrease in acidity) while it is the opposite for the dehydrogenation (rise in basicity) (6).

J. Fraissard and L. Petrakis (eds.), Acidity and Basicity of Solids, 181–197.

1   Nature of sites

1.1   BRÖNSTED SITES

Basic hydroxyls are rarely observed in zeolites. These materials having a negatively charged framework, the hydroxyls are usually acidic. Nevertheless, basic OH groups are seen, for instance, in faujasite exchanged with Mg or Ca cations. Their infrared vibrations at 3685 (MgY) or 3675 (CaY) cm^{-1} (7,8) correspond to OH groups in $M(OH)^+$ generated according to

$$M^{2+} (H_2O) \rightleftharpoons M(OH)^+ + H^+ \qquad\qquad (I)$$

These OH$^-$ groups are not framework hydroxyls.

1.2   STRUCTURAL BASIC SITES

They are the framework oxygens which are negatively charged due to the isomorphous replacement of Si^{4+} by Al^{3+} in a theoretical silica framework. Most of the framework oxygen are accessible to probe or reactants in the very open zeolites with windows consisting of 10 or 12 rings (10 or 12 R). They are potential basic sites. Only a part of these oxygens is accessible in denser structures. In contrast to protons which are mobile and can move toward the reactant molecules in a cage, oxygen atoms are fixed in the framework and molecules have to appproach the lattice in a configuration that is favorable to formation of the reaction intermediate. This suggests a more demanding character of the basic sites with regards to formation of the activated complex than in the case of proton sites.

In order to acquire a basic character i. e. a charge high enough, the framework oxygen has to fulfill several requirements. It has first to belong to an AlO_4 tetrahedron species which is bearing the negative charge. A second feature is that in the structure of any zeolite, the TOT angles and TO distances (T for Al or Si) are well specified. They define several types of oxygen, for instance four in faujasite, six in LTL... The electronic charge on oxygen (basicity) decreases as the TOT angles are narrower and the distances TO longer (9). It follows that the charge of one oxygen type is different from that of another one. This also implies that a mean basicity expressed, for instance, for the four oxygen types of faujasite is valid for an average of bond angles and distances, i. e. for the faujasite structure only. The six different oxygen types in LTL correspond to a completely different arrangement of tetrahedra from that in faujasite and consequently to different bond angles. Thus the basicity of LTL and faujasite framework oxygens should be different. In addition for a same oxygen type in a given zeolite, for instance, the one in hexagonal prism in faujasite (O_1 oxygen) the linkages Si-O-Si or Si-O-Al lead to different angles and different basicities. No experimental technique is presently able to give for each of the 192 oxygen of a faujasite unit cell the values of the corresponding TOT angles. Only average values for each of the four types are obtained from XRD. On oxide surfaces, such as on MgO, different basic sites are created at edges, corners and kinks (10). In zeolites, even without any defects, the various basic centers with specific basic strengths are intrinsic to the structure. In adsorption or catalysis the reactant sees all the possible oxygens and choose the best one to interact with depending on its basic strength and on its access and configuration. The interaction reactant/zeolite may modify slightly the bond angles i. e. the intrinsic basic strength of the active oxygen site.

1.3 CLUSTERS OF OXIDES OR HYDROXIDES

Some oxide materials are known as basic catalysts (MgO, ZnO...) (2). They may be dispersed in zeolites where they act as bases.

Cation hydrolysis is known to occur in zeolites exchanged with polyvalent ions giving rise to $Me^{(n-1)+}OH$ species according to

$$Me^{n+} + H_2O \longrightarrow Me^{(n-1)+} OH + H^+ \tag{II}$$

Upon dehydroxylation MeO oxides may be formed (11).

MgO and CaO clusters were shown to be generated in MgY or CaY upon heating above 500° - 600°C (12). The clusters are localized in the supercages. The number of MO entities in the clusters is estimated to be between 2 to 10.

The addition of hydroxides of alkaline cations, particularly KOH and CsOH was shown to increase the selectivity for the chain alkylation upon reaction of toluene with methanol (13, 14, 15). This indicates an increase of the catalyst basicity. It is very likely that, upon activation conditions of the catalyst, oxides are formed.

In a similar way, the introduction of Cs_2O by decomposition of Cs acetate in NaY generates strong basic sites, very active in isopropanol dehydrogenation to acetone (16). More recently MgO and M_2O (M = alkaline cations) were also shown to generate strong basicity in Y and X zeolites, the size of the cluster being important. For MgO an ensemble of Mg and O atoms to form MgO lattice is necessary to create strong basicity while isolated species of alkaline metal oxides are sufficient (17).

The enhancement of basic strength is well demonstrated by the above results when alkaline cations or Mg, Ca form clusters trapped in zeolite cages (mainly faujasite).

Many other oxides are known to form clusters in zeolites. For instance, ZnO, Ga_2O_3 supported, mainly in ZSM -5, have been studied in the aromatization of light alcanes. The basic character is not evidenced by this reaction. Nevertheless such oxides may well show basic catalytic properties in selected reactions.

1.4 ALKALI METAL SUPPORTED ON ZEOLITES

Na metal supported on MgO is known for a long time to form a superbasic catalyst (18).

Upon reaction of Na metal with surface hydroxyls O^-Na^+ species are formed. They increase the surface basicity, compared to that in the initial OH group.

The formation of sodium clusters in zeolite was reported early (19). The introduction of Na° by decomposition of NaN_3(20) generates Na clusters (21,22) and enhances the basic character of faujasite type zeolites (15, 21-23).

1.5 ACID BASE PAIRS

In oxides, acid-base pairs are well known. Conjugate acid-base pairs were shown to exist in the cationic form of zeolites. The cation acts as a Lewis acid and the close framework oxygen as a base. For instance faujasites X and Y, exchanged with the series of alkaline cations, show a prevailing acid or basic character depending on the cations electronegativity (24). Table I gives a rank obtained by titration of the cationic acid sites with pyridine and measurement of basic strength using pyrrole as a probe. It is seen that at the boundary some zeolites (KY, NaX) show an amphoteric character. They will behave as acids or bases depending on the basic and acidic properties of the probe or reactant molecules in the zeolite cages.

TABLE 1. Scale of acido-basicity for X and Y exchanged with alkaline cations[a]

Acidity (b)		Basicity (c)	
Increase in strength ↑	Li Y Na Y Na X K Y	K Y Rb Y Na X K X Rb X Cs X	Increase in strength ↓

(a) from ref (24)
(b) Pyridine titration with colored indicators
(c) Shift in infrared wavenumber of pyrrole NH vibration

1.6   ELECTRON DONOR-ACCEPTOR SITES

A good correlation is usually observed in aluminosilicates based catalysts (amorphous or crystalline) between the acid-base and electron acceptor-donor sites properties. The Lewis acidity is changing in a way parallel to that of the electron acceptor character of sites. For instance, in zeolites, perylene is ionized to the radical Pe^+ visible in EPR. The reducing centers able to transform tetracyanoethylene to $TCNE^-$ radical ions may be also observed in zeolites. Their number varies like that of basic sites.

2   Basicity evaluation

2.1   THEORETICAL ESTIMATE

Many calculations were performed to estimate the charge on the framework atoms and cations (H^+ and metal cations) in zeolites. The purpose was not to study basicity. Nevertheless, values of the charges on oxygen can be taken to look at the influence of various parameters on the "theoretical" basicity. This will be helpfull for the comparison with experimental results. For instance, ab-initio quantum chemical calculations have been performed on clusters of the type $(OH)_3$ Si-OM-T $(OH)_3$ where M is the monovalent cation (H^+, Li^+ or Na^+) and T is Si, Al or B(25). Results of table 2 show that the basicity of the oxygen close to the cation increases for the three clusters from the H^+ form, to the Li^+ and then Na^+ ones. This order is parallel to that of decreasing cation electronegativity (table 3). A similar type of calculation (26) indicates that incorporation of Ga or Ge in a cluster with Si atoms increases the oxygen basicity but keeps it lower than in the Al form.

TABLE 2. Charge on the central oxygen in the cluster $(OH)_3$ Si-OM-T $(OH)_3$ with M = H, Li or Na and T = Al or B [a]

T = Al or B	Counterion		
	H	Li	Na
Si-Al	- 0.533	- 0.582	- 0.722
Si-B	- 0.459	- 0.514	- 0.614

(a) from reference (25)

Using the Sanderson principle of equalization of electronegativities (27), Mortier (28) made a first approach to a systematic correlation between the chemical properties and the electronegativity of the framework atoms and exchanged cations.The intermediate electronegativity Sint reflects in a given material the mean electronegativiy reached by all the atoms following electron transfer. The calculation of Sint allows evaluation of an average charge on all the framework atoms by taking into account only the chemical composition of the sample. It does not consider any structure influence. A systematic calculation of a mean charge on oxygen as a function of Al content and for the protonic form or the one exchanged with the five alkaline cations gives the results of figure 1 (24, 29). It uses the values of Sanderson electronegativities reproduced in table 3.

TABLE 3. Sanderson electronegativities S [a]

Atom	Al	Si	H	Li	Na	K	Rb	Cs
S	2.22	2.84	3.55	0.74	0.70	0.42	0.36	0.28

(a) from ref (27)

The figure 1 gives the trends which may be expected for a given zeolite structure when the Al content or cation identity are changed. It shows that for any cation the mean charge on oxygen rises as the Al content goes up. This is in agreement with the lowest electronegativity of Al compared to Si (table 3), attracting less electrons from the close framework oxygen. For a given Al content, the order of increasing oxygen basicity for the various cationic zeolites : $H^+ < Li^+ < Na^+ < K^+ < Rb^+ < Cs^+$ is quite the opposite to that of cation electronegativities (table 3). This is what one might expect. The less electronegative ions (Cs) in the zeolites with the highest Al content should give the more basic materials (Cs A, Cs X for instance).

The comparison with results of table 2 obtained from ab-initio calculations show the same order for oxygen basicities in the three cationic forms H^+, Li^+, Na^+.

It is worthwhile to note that each method of calculating the charge on atoms in materials, for instance zeolites, has its advantages and disadvantages. The ab-initio

186

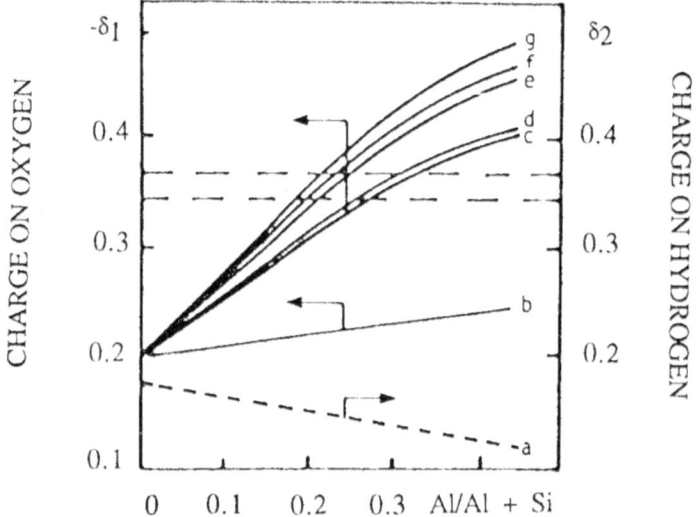

Fig. 1 Change in the calculated charge on proton[a] or oxygen[b-g] as a function of Al content and for the cations a, b : H, c : Li, d : Na, e : K, f : Rb, g : Cs (from ref. 29).

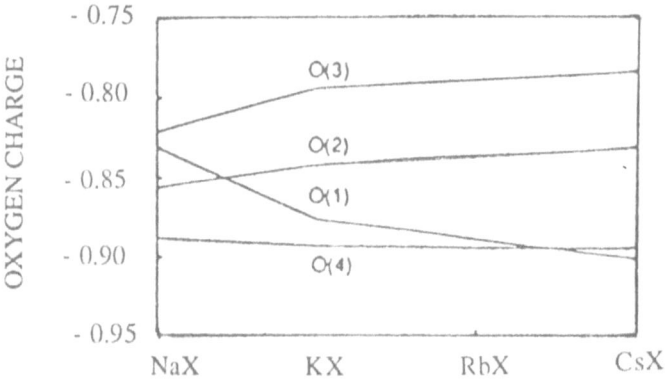

Fig. 2 Change in the charge on oxygen for X zeolites exchanged with various alkaline cations (from ref. 52).

methods consider small clusters which cannot take into account long range effects of atoms of the structure. It fixes bond angles and then does not consider a true type of oxygen of a given structure with its characteristic TOT value. Semi-empirical calculations consider larger clusters but the approach involves other approximations. The simple and easy application of Sanderson electronegativity equalization principle does not consider any effect of bond angles or bond lengths. It gives the same result for zeolites which have the same chemical composition. This may give meaningless results when extraframework phases are trapped in zeolite cages like extraframework Al species in ultrastable acidic zeolites with low Al content or oxides clusters described above in basic zeolites. In the absence of such phases a disagreement between the calculated and experimental basicities may be of major interest. It shows that in the structure considered an important parameter is involved, independent of the chemical composition, but very likely characteristic of the structure (framework ionicity for example (30)) or of the heterogeneity of Al distribution (concentration gradient, Al pairs...).

A very interesting approach considers the structure influence. In the EEM (Electronegativiy Equalization Method) calculation the long range effects can be adequately accounted for in calculating the charge distribution and average electronegativity (30). Figure 2 gives the changes in oxygen charges with cation types for alkali metal X zeolites. It shows an increase from Na to Cs forms in the calculated oxygen charge (more negative value) on O_1 and O_4 framework sites. They are the atoms belonging to the 12-R window of faujasite through which all the molecules - adsorbates or reactants - have to travel. A similar trend is obtained for the Y zeolite exchanged with the same cations. The mean oxygen charge (weighted average) increases also in the X series from NaX (- 0.849e) to KX (- 0.852e), RbX (- 0.852e) and CsX (- 0.860e) (52).

2.2   EXPERIMENTAL MEASUREMENTS

Two main types of techniques may be used to study the basicity of solids. At first a direct measurement of a representative intrinsic property of the zeolite can be carried out. Secondly, probe molecules can be used to characterize the reactivity of basic sites.

2.2.2   Direct Methods.
Evaluation of bond angles (XRD).
As said above the electronic charge on oxygen decreases as the TOT angle becomes narrower (9). XRD (or neutron diffraction) can be used to determine a mean angle for each oxygen type of a given zeolite structure. There are four values in FAU, six in LTL, ten in MOR, twenty six in MFI... Attempts were made in faujasite to correlate the bond angles to physico-chemical properties (31). The mean angles depend also on the size, volume, polarizing power, electronegativity.... of the framework atoms and of the cations. They also vary with pretreatment conditions and hydration level. It has also be shown that the s character of the bond oxygen-cation increases simultaneously with the TOT angle. This may change the reactivity of the oxygen towards adsorbates. All these effects suggest to be very careful when trying to connect experimental results of adsorption or catalysis with average bond angles obtained by structure studies. They may not reflect the actual local property when, for a given structural oxygen, the neighbouring T atom is Si or Al. In this case, the average TOT angle does not give any information on the true oxygen basicity. In fact the experimental correlation between bond angles and basic strength (or acid strength in the case of protonic zeolites) is not well established.

Bond energies (XPS)

The binding energy value of elements is well established to depend on several factors, one being the charge carried by the atom. The binding energy of $O_{(1s)}$ is then expected to decrease with the negative charge on the oxygen atom. Due to charging effects (ejection of electrons) and insulating properties of zeolites, the accuracy on shifts is sometimes rather low.

Barr conducted XPS measurements on various zeolites (32). Figure 3 reports his results on a series of zeolites in the Na or Ca forms. They are compared to values of oxygen $O_{(1s)}$ binding energies of SiO_2 and γ-Al_2O_3. The charge on oxygen is calculated as in (24, 28). From figure 1 curve d and figure 3, it is seen that the zeolites with the highest Al content (NaA, NaX) have a high mean oxygen charge i. e. a low $O_{(1s)}$ binding energy. On the opposite, the highly siliceous NaZSM-5 shows a low oxygen charge and a high binding energy close to those of SiO_2. The results clearly show that the XPS response resulting from an average binding energy of all the framework oxygen correlates rather well with the mean charge calculated from the Sanderson principle of equalization of electronegativities. The bivalent cation Ca gives higher $O_{(1s)}$ binding energies than the Na ion. The cation hydrolysis (equation II) (11) which generates protons may modify the true oxygen charge compared to the calculated one for these zeolites exchanged with bivalent cations and alter the correlation in figure 3.

A similar trend was observed more recently (33). In addition the $O_{(1s)}$ binding energy was shown to decrease as H^+ cation was exchanged by alkaline cations from Li to Cs, i.e. in the order of increasing basicity of oxygen from figure 1. Almost no change is observed in the case of ZSM-5 (33). Figure 1 shows that at the low Al level of this zeolite (Al/Al+Si < 0.1) only a small difference in oxygen charge is expected upon exchange beween the alkaline cations.

2.2.3 *Use of Probe Molecules.* The major problem in the experimental characterization of the basicity of oxides is the choice of a good probe. Specific of basic site, it should distinguish between the various basic sites and do not decompose or polymerise. CO_2 has often been used (7, 8, 12, 17, 34-36). Infrared studies showed that it interacts with several types of basic sites (7) and may give rise to ill defined carbonates. Carboxylic acids or benzoic acid have been used (37-39). Phenol (40) decomposes very easily which precludes its practical use. Pyrrole may be used if very special care is taken in order to avoid its polymerisation (fresh distillation, short contact time at room temperature) (24, 41,42). Benzene adsorbs in faujasite depending on the oxygen basicity (43, 44).

Several techniques mainly thermal or spectroscopic have been applied to the study of zeolite basicity.

Colored indicators

A qualitative approach of zeolite basicity may use the titration with benzoic acid solution. For instance CsX materials show a strong basicity (45).

Thermal methods

Thermogravimetry, calorimetry and TPD have been widely used to study the adsorption of molecules. If these adsorbates may interact with basic sites, the methods may give access to the basicity of the zeolite sites. Nevertheless, most of the probe react not only with framework oxygen atoms but also with other sites present like cations for example.

For instance table 4 gives experimental initial values of sorption energies of CO_2 on a series of X zeolites together with calculated values taking into account the various energies involved (46). The overall sorption energy has to take into account the influence of all the charges of the zeolite (positive and negative) through the field and field-gradient

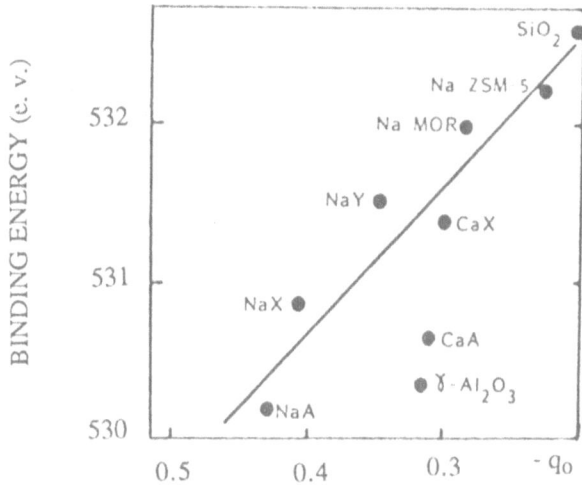

Fig. 3   XPS determination of O(1s) binding energy as a function of the average calculated charge on oxygen for various zeolites and oxides (from ref. 32).

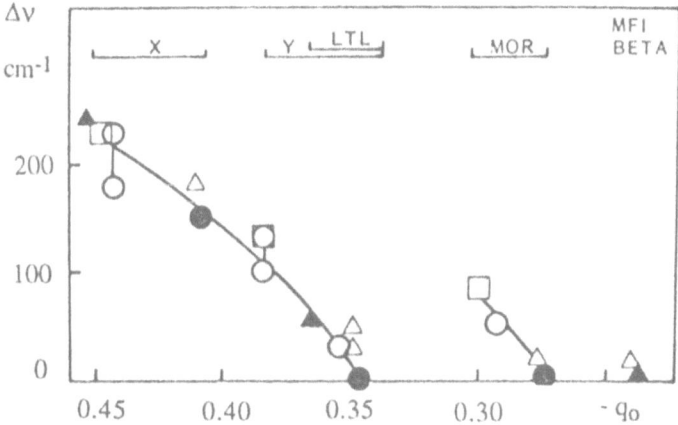

Fig.4   Decrease in the NH infrared vibration of pyrrole as the calculated charge on the oxygen decreases for zeolites exchanged with cations : ● : Li, Δ : Na, O : K, □ : Rh, ▲ : Cs (from ref. 24,51).

they create. The results show, as infrared spectroscopy did (7) that CO_2 is not specific of basic sites. This molecule may be used only to get a general view of the interaction of a small molecule, having a quadrupole moment, with sites in zeolites.

Table 4. Experimental initial values of sorption energies E and calculated values H for CO_2 adsorbed on X-type zeolite with various cations in kJ.mol^{-1} (from ref. 46).

Energy values	Cationic form				
	Li	Na	K	Rb	Cs
ΔE	51.4	45.2	43.9	42.2	36.8
ΔH	55.2	38.9	28.8	23.4	22.2
ϕ_D (oxide ions)	15.9	13.0	7.1	4.6	4.6
ϕ_D (cations)	0.4	0.8	3.3	4.6	9.2
ϕ_p	9.6	5.0	2.1	0.8	0
$\phi°_{F\text{-}Q}$	30.9	21.3	17.6	14.6	9.6
$\phi*_o$	1.7	1.3	1.3	1.3	1.3

Energies : ϕ_D : dispersion, ϕ_p : polarization, $\phi°_{F\text{-}Q}$: field gradient quadrupole,

$\phi*_o$: estimated zero point energy.

The adsorption of benzene on zeolites was studied calorimetrically and isotherms determined by chromatography (47, 48). The enthalpy and entropy were the highest for Na Y with respect to NaX and KY zeolites with a difference of 12 kJ mole $^{-1}$. Further results (43, 44) using infrared show that, like in the case of CO_2, the interaction with the cations and/or the oxygen may strongly affect these global results.

Spectroscopic methods
Infrared spectroscopy.
This is the most commonly method used to characterize the interactions of adsorbates with zeolites. For basicity measurements, several acidic (CO_2) or amphoteric (pyrrole) probes were used. The adsorption of CO_2 on cations gives rise to a band near 2 350 cm^{-1} for Ca $^{2+}$ or Mg^{2+} in Na Ca MgY zeolite (36) or Na, Ca or Co in NaA, Na Ca A and Co-A zeolites (49). The band is assigned to CO_2 physisorption. The corresponding heat may be calculated from infrared adsorption isotherms (49). If the total heat of adsorption is known, for instance by calorimetry, the substraction gives the chemisorption heat (49). The latter energy arises from interaction with oxide ions and cations giving carbonate like species adsorbing in the 1 300 - 1 600 cm^{-1} region (7, 36). The smallest polyvalent cations give the strongest chemical interaction (7, 34-36). CO_2 may also react with Brönsted basic sites (hydroxyl groups) formed in reaction II. This results in the generation of carbonated species and acidic OH groups in Ca Y and Mg Y zeolites (7).
Pyrrole was shown to interact with basic sites in zeolites (41). It is an amphoteric component. In the presence of basic oxygen in zeolites the NH vibration is shifted to low wavenumbers due to the interaction of the H atom with the oxide ions. The extent of the shift ΔvNH with regards to liquid pyrrrole evaluates the basic strength (24, 29, 41, 42,

50, 51). Figure 4 gives the shift observed on zeolites in the alkaline form as a function of the mean charge on the oxygen calculated as in (24, 29, 50, 51). It shows that FAU (X and Y) and LTL are on the same curve. This is expected from figure 1, the same calculated oxygen basicity giving the same $\Delta\nu NH$ value. MFI and BEA zeolites have a very low Al content (Al/Al + Si < 0.1) which gives a weak calculated basicity of oxygen. The Cs - and Na MFI do not give any νNH shift while Na BEA shows a 30 cm^{-1} shift. This value is the same as in KL and Na MOR. Na BEA seems to possess a higher basicity than expected. The same arises for MOR. These very interesting results demonstrate that for MOR and BEA, parameters other than chemical composition govern the basicity of oxygen atoms. This may arise in MOR, for instance, from the high asymetry of the oxygen ring, from effects of field gradient at the adsorbing site (52) or from the higher framework ionicity (30). The results of pyrrole adsorption on various zeolites together with the simultaneous change in cation Lewis acidity reported in table 1 show that cation and framework oxygen form conjugate acid base pairs (24).

A detailed infrared study of the CH out of plane bands of benzene adsorbed on zeolites allows to distinguish unambiguously the aromatic adsorbed on cations or in the 12 R window of faujasite. Bands at two different wavenumbers are obtained. The cations acting as Lewis acids interacts with the ring Π electrons while the CH of benzene interact with the oxygens of the 12 R aperture (43, 44). Figure 5 shows as an example the usual benzene adsorption isotherm decomposed in two parts (53). The one component (curve a) represents the amount of benzene adsorbed on oxygen atoms and the other one (curve b) depicts the number of aromatic molecules interacting with the Na$^+$ ions. At very low p/p$_0$ (less than 10^{-4}), infrared spectroscopy is able to measure the amount of benzene in interaction with the framework oxygen as a function of the number of molecules of aromatics adsorbed per unit cell. It is shown that this number increases with the mean oxygen charge (43, 44) whether there is a rise in Al content or there are cations with a small electronegativiy. As an example, figure 6 gives the change in absorbance of the $\nu_5 + \nu_{17}$ CH out of plane band of benzene as a function of the aromatic loading in molecules per supercage for three zeolites. The partial pressure range is 10^{-7} < p/p$_0$ < 10^{-4}. Three X zeolites in which Na cations are progressively replaced by Cs ions have an increased basicity in oxygen as the Cs content rises (44 and figure 1). The Cs 14 X sample with the weakest basicity adsorbs both on cation and oxygen already at low loading, with a preference for cations. Increasing the Cs content favours the preferential adsorption on oxygen. At low loading the Cs 57 X zeolite adsorbs the largest amounts of benzene on oxygen. The results indicate a strong dependence of benzene location on the oxygen basicity (high for Cs 57 X) and cation Lewis acidity (stronger for Cs 14 X) due to the presence of Na ions. They show that any probe molecule able to react with both sites shall also have different preferred location depending on the zeolite properties. Besides the number of sites interacting with benzene and the order of their reactivity, the infrared study also provides information on the strength of interaction. For instance, figure 7 shows that in the same Cs 57 X sample as above, the wavenumber of the C-C vibration of the aromatic which is at 1 479 cm^{-1} in liquid benzene may be shifted up to 1 486 cm^{-1} upon adsorption. The bands of the CH out of plane vibrations at 2 053 and 1 922 cm^{-1} correspond to interaction with the framework oxygen while those at 1 996 and 1 854 cm^{-1} are related to benzene adsorbed on cations (44). It follows that at low loading, benzene interacts only with framework oxygen of the 12-R window as seen also in figure 6. The corresponding C-C band is at 1 486 cm^{-1}. Bands at 1 854 and 1 996 cm^{-1} start to appear

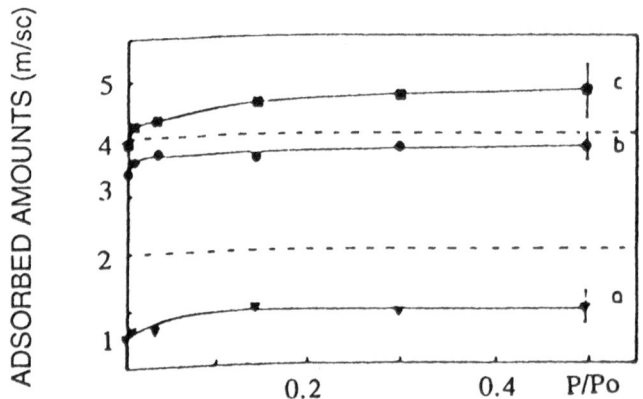

Fig. 5 Infrared isotherms of benzene adsorption on NaY at room temperature on (a) : 12-R windows, (b) : cations, (c) : (a) + (b). The adsorbed amounts are given in molecules per supercage (from ref. 53).

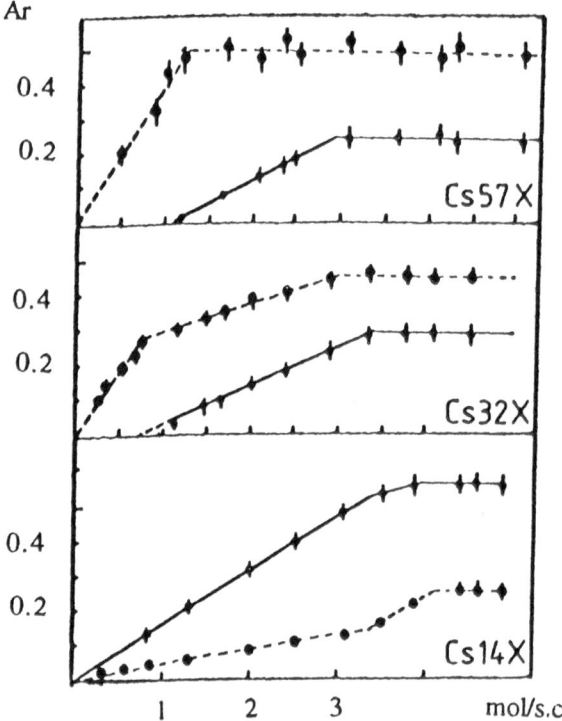

Fig.6 Changes with benzene loading of the infrared absorbance at room temperature for adsorption on cations (O) or 12-R windows (●) on X zeolites exchanged with increasing amounts of Cs (Cs per unit cell) (CH out of plane bands $\nu_5 + \nu_{17}$) (from ref. 44)

as the loading is increased showing the progressive interaction with the cations. Simultaneously the $1\ 479\ cm^{-1}$ C-C band is growing. The strong disturbance of the band when benzene is adsorbed on the oxygen ($1\ 486\ cm^{-1}$) indicate a high strength of adsorption.

All the results presented show the relevance of infrared spectroscopy for the study of the basicity of zeolites.

XPS

The adsorption of pyrrole is followed by XPS on X and Y zeolites exchanged with the series of alkaline cations (54). XPS peaks of polymerized pyrrole can be observed only when Li Y or Na Y, the more acidic zeolites, are heated with pyrrole at 65° or 250°C. This work confirms the validity of the use of pyrrole as a probe if contacted for a short time at room temperature (24, 29, 41,42). The interaction of pyrrole with the zeolite gives a decrease in $N_{(1s)}$ binding energy when moving from Li to Cs cations and from Y to X faujasites. This is explained by an increased electron transfer from an oxygen close to a cation to the N atom of pyrrole. A good correlation is obtained (54) between this $N_{(1s)}$ binding energy and the corresponding νNH infrared wavenumbers as described in ref. 24, 29, 42. The results are in line with the predicted changes in basicity in figure 1. They show in addtion that the proble molecule - pyrrole - is modified upon its interaction with the basic sites.

EPR

Few results are available on the redox properties of non-protonic zeolites. A recent study of LTL zeolites exchanged with alkaline cations was carried out by EPR using tetracyanoethylene (TCNE) as a probe (55). It measures the reducing sites forming TCNE⁻ radical anions. Table 5 gives the changes in the number of spins/g for variously exchanged LTL. This number increases in the order H < Li < K < Cs which is parallel to the theoretical oxygen basicity (figure 1) and experimental values (figure 4). The results confirm a similar trend between the basic and reducing properties. The values obtained ($\sim 10^{-17}$ spins/g) are in line with the number of oxidizing or reducing sites usually measured in zeolites.

TABLE 5. Changes in the number of reducing sites in LTL zeolites treated at 773 K[a]

Zeolite	HL	LiL	KL	CsL
Spins/g x 10^{-17} [b]	1,25	2,60	2,85	6,09

(a) from ref. 55
(b) TCNE⁻ radical ions formed

194

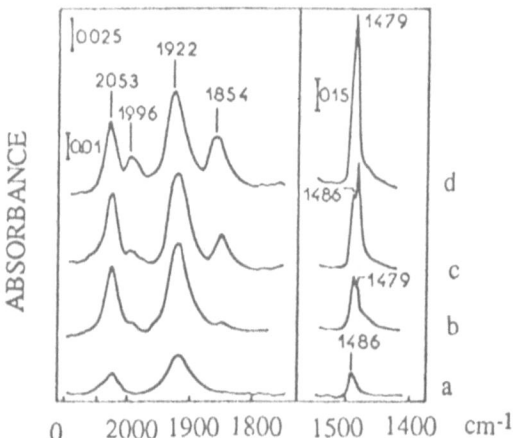

Fig. 7   Infrared spectra of benzene adsorbed on Cs_{57} X faujasite in the CH out of plane
range (left) and C-C domain (right). Benzene loading in molecules per supercage,
a : 0.5, b : 1.2, c : 2.3, d : 3.0 ($p/p_0 < 10^{-4}$) (from ref. 44).

<u>Other techniques</u>
One may expect that approaches using UV-Visible, Raman... spectroscopies will be
used to characterize specifically the basicity of zeolites.

3   Properties of basic zeolites

The implication of basic properties of zeolites in adsorption and catalysis has already been
reviewed (29, 56, 57). They will not be discussed further in this paper.
Two points may be stressed. At first, the intrinsic framework basicity may be greatly
enhanced by the presence of extra framework clusters of oxides or hydroxides. This was
clearly evidenced in the alkylation of toluene with methanol when oxides or hydroxides
like KOH, CsOH are present (13, 14). The addition of Cs_2O in Cs Na Y enhances the
dehydrogenation of isopropanol (16). The transformation of acetonylacetone to
methylcyclopentenone is favoured if Na^+ cations, forming very likely oxides, are added to
ZSM-5 zeolite (58).
A second typical characteristic of basic zeolites is the stabilization in the cages of
faujasites of carbonyl clusters of noble metals. Pd particles in Na Y form the complex Pd_{13}
$(CO)_x$ (59). It is suggested that Palladium is in the form of cubooctaedra in the supercages
of the faujasite structure. Unreduced Platinum ions in Na Y or other alkaline faujasites like
CsX give rise to bright coloured clusters (yellow, red, green...) upon CO adsorption at
room temperature. The infrared spectra show that these species are Chini complexes
stabilized by the basic properties of the zeolites. Their general formula is $[Pt_3 (CO)_3$
$(\mu_2CO)_3]^{2-}{}_n$, the value of n ranging from 3 to 5 depending on the zeolite (60).

4 Conclusion

The study of the basicity of zeolites is progressively growing as these materials become more interesting for applications. The choice of the probe molecule is, for the time being, more difficult than in the case of acidity study. Many of such molecules either are transformed or interact non specifically with various types of sites. Several of them can be used (benzene, pyrrole) if special care is taken.

For a given zeolite structure good correlations are obtained between the calculated and experimental basicity when the chemical composition is changing. Comparing various structure is more difficult, structural parameters becoming important (angles, field gradient, framework ionicity...).

The rising interest of basic zeolites leads to an increasing number of research works which will improve the knowledge of these materials.

References

1 H. Pines and J. Manassen, Adv. Catal. 16 (1966), 49.
2 K. Tanabe, "Solid Acids and Bases", Kodansha, Tokyo, 1970.
3 M. Ai, Bull. Jpn. Petrol. Inst. 18 (1976) 50.
4 Y. N. Sidorenko, P. N. Galich, V. S. Gutyrya, V. G. Il'in and I. E. Neimark, Dokl. Akad. Nauk, SSSR, 173 (1967) 132.
5 T. Yashima, K. Sato, H. T. Hayasaka and N. Hara, J. Catal. 26 (1972) 303.
6 T. Yashima, H. Suzuki and N. Hara, J. Catal. 33 (1974) 486.
7 C. Mirodatos, P. Pichat and D. Barthomeuf, J. Phys. Chim. 80 (1976) 1335.
8 C. Mirodatos, A. Abou Kais, J. C. Vedrine, P. Pichat and D. Barthomeuf, J. Phys. Chem. 80 (1976) 2366.
9 G. V. Gibbs, E. P. Meagher, J. V. Smith and J. J. Pluth, ACS Symposium Series, Washington DC, 40 (1977) 19.
10 H. Hattori, Mat. Chem. Phys. 18 (1988) 533
11 D. W. Breck , "Zeolite Molecular Sieves", Wiley and Sons, New-York, 1974, pp. 460.
12 A. Abou Kais, C. Mirodatos, J. Massardier, D. Barthomeuf and J. C. Vedrine, J. Phys. Chem. 81 (1977) 397.
13 J. Engelhardt, J. Szanyi and B. Jover, Actas Symp. Ibero-amcr. Catal. 9th, 2 (1984) 1435.
14 C. Lacroix, A. Deluzarche, A. Kiennemann and A. Boyer, J. Chim. Phys. 81 (1984) 473 and 481.
15 V. Barbarin, Thesis Paris 1987.
16 P. E. Hathaway and M. E. Davis, J. Catal. 116 (1989) 263 and 279 and J. Catal. 119 (1989) 497.
17 H. Tsuji, F. Yagi, H. Hattori and H. Kita, in L. Guczi, F. Solymosi, P. Tetenyi (Editors), Proc. 10th Intern. Cong. Catal., Budapest, Hungary, July 19-24, 1992, Akademia Kiado, Budapest, 1992, pp. 1171.
18 J. Kijenski and S. Malinowski, Bull. Acad. Pol. Sci. 25 (1977) 427 and 669.
19 P. H. Kasai and R. J. Bishop Jr, J. Phys. Chem. 77 (1973) 2308.
20 P. Fejes, I. Kiricsi, I. Hannus, T. Tihahyi and A. Kiss, in B. Imelik, C. Naccache, Y. Been Taarit, J. C. Vedrine, G. Coudurier, H. Praliaud (Editors), "Catalysis by Zeolites", Stud. Surf. Sci. Catal. 5 (1980) 135.
21 L. R. M. Martens, P. J. Grobet and P. A. Jacobs, 315 (1985) 568.

22   L. R. M. Martens, W. J. Vermeiren, D. R. Hybrechts, P. J. Grobet and P. A. Jacobs, in M. J. Phillips, M. Ternan (Editors), Proc. 9th Inter. Cong. Catal., Calgary, Canada, June-July 1988 -The Chemical Institute of Canada, 1988 pp. 420.

23   D. Barthomeuf and V. Barbarin, French patent 2.623.423

24   D. Barthomeuf, J. Phys. Chem. 88 (1984) 42.

25   E. G. Derouane and J. G. Fripiat, J. Phys. Chem. 91 (1987) 145.

26   P. J. O'Malley and J . Dwyer, Chem. Phys. Lett. 143 (1988) 97.

27   R. T. Sanderson, Chemical Bonds and Bond Energy, Academie Press, New-York, 1976.

28   W. J. Mortier, J. Catal. 55 (1978) 138.

29   D. Barthomeuf, in G. Öhlmann, H. Pfeifer, R. Fricke (Editors), "Catalysis and Adsorption by Zeolites", Stud. Surf. Sci. Catal. 65 (1991) pp. 157.

30   K. A. Van Genechten and W. J. Mortier, Zeolites 8 (1988) 273.

31   P. A. Jacobs and J. B Uytterhoeven, J. Chem. Soc. Faraday Trans. I 69 (1973) 359.

32   T. L. Barr, Zeolites, 10 (1990) 760.

33   M. Huang, A. Adnot and S. Kaliaguine, J. Am. Chem. Soc. 114 (1992) 1005.

34   L. Bertsch and H. W. Habgood, J. Phys. Chem. 67 (1963) 1621.

35   J. W. Ward and H. W. Habgood, J. Phys. Chem. 70 (1966) 1178.

36   P. A. Jacobs, F. H. Van Cauwelaert, E. F. Vansant and J. B. Uytterhoeven, J. Chem. Soc. 69 (1973) 1056.

37   K. Tanabe in J. R. Anderson and M. Boudart (Editors), "Catalysis : Science and Technology", Springer-Verlag, Berlin, 1981, vol. 2 chap. 5.

38   S. Malinowski and J. Kijenski, Catalysis, Royal Society of Chemistry, London, 4 (1981) 130.

39   P. F. Rossi, G. Busca, V. Lorenzelli, M. Lion and J. C. Lavalley, J. Catal. 109 (1988) 130.

40   O. V. Krylov and E. A. Fokina, Problemy Kinet. Katal., Akad. Nauk, Moskva 8 (1955) 246.

41   P. O. Scokart and P. G. Rouxhet, Bull. Soc. Chim. Belge 90 (1981) 983.

42   M. Huang and S. Kaliaguine, J. Chem. Soc. Faraday Trans. 88 (1992) 751.

43   A. de Mallmann and D. Barthomeuf in Y. Murakami, A. Iijima and J. W. Ward (Editors), Proceed, 7th Intern. Conf. Zeol. , Kodanska, Tokyo and Elsevier, Amsterdam, 1986, pp. 609.

44   A. de Mallmann and D. Barthomeuf, Zeolites 8 (1988) 292.

45   V. Barbarin and D. Barthomeuf, unpublished results.

46   R. M. Barrer and R. M. Gibbons, Trans. Faraday Soc. 59 (1963) 2569 and 61 (1965) 948.

47   T. R. Brueva, A. L. Klachko - Gurvich and A. M. Rubinstein, Izv. Akad. Nauk. SSSR, Ser. Khim 12 (1972) 2807.

48   D. Barthomeuf and B. H. Ha, J. Chem. Soc., Faraday Trans. I 69 (1973) 2158.

49   Y. Delaval, R. Seloudoux and E. Cohen de Lara, J. Chem. Soc. Faraday Trans. I 82 (1986) 365.

50   D. Barthomeuf and A. de Mallmann in P. Grobet, W. J. Mortier, E. F. Vansant, G. Schulz-Ekloff (Editors), "Innovation in Zeolite Materials Science", Stud. Surf. Sci. Catal. 37 (1988) 364.

51   S. Dzwigaj, A. de Mallmann and D. Barthomeuf, J. Chem. Soc. Faraday Trans. 86 (1990) 431.

52   L. Uytterhoeven, D. Dompas and W. Mortier, J. Chem. Soc. Faraday Trans. 88 (1992) 2753.
53   A. de Mallmann and D. Barthomeuf, J. Phys. Chem. 93 (1989) 5636.
54   M. Huang, A. Adnot and S. Kaliaguine, J. Catal. 137 (1992) 322.
55   Bao Lian Su, Thesis Paris 1992. Bao Lian Su and D. Barthomeuf, to be published.
56   Y. Ono, in B. Imelik, C. Naccache, Y. Ben Taarit, J. C. Vedrine, G. Coudurier and H. Praliaud (Editors), "Catalysis by Zeolites", Stud. Surf. Sci. Catal. 5 (1980) 19.
57   D. Barthomeuf, G. Coudurier and J. C. Vedrine, Mater. Chem. Phys. 18 (1988) 553.
58   R. M. Dessau, Zeolites 10 (1990) 205.
59   L. L. Sheu, H. Knözinger and W. M. H. Sachtler, Catal. Lett. 2 (1989) 129.
60   A. de Mallmann and D. Barthomeuf, Catal. Lett. 5 (1990) 293.

THE NATURE OF LEWIS ACID SITES OF ALUMINIUM OXIDE AND IN ZEOLITES AS STUDIED BY IR SPECTRA OF ADSORBED HYDROGEN.

V.B.KAZANSKY
Zelinsky Institute of
Organic Chemistry Russian
Academy of Sciences
Moscow 117071, Russia

ABSTRACT. The low temperature adsorption of molecular hydrogen with the IR diffuse reflectance control of its perturbation was used as a molecular probe for the nature of Lewis surface acid sites of aluminum oxide and in zeolites. In the case of aluminum oxide such sites were produced by high temperature dehydroxylation, while in zeolites in addition the interaction of hydrogen with monovalent and bivalent cations and extra lattice aluminum resulting from dealumination of the framework was also studied. The obtained results show that molecular hydrogen is perturbed both by low coordinated cations and by the neighboring basic oxygen atoms. Therefore it is rather a molecular test for acid-base pairs than for the Lewis sites alone. This conclusion was confirmed by the results on dissociative adsorption of hydrogen at higher temperatures. On the contrary adsorption of carbon monoxide is a probe only for Lewis sites. The importance of interaction of nonpolar molecules both with cations and basic oxygen of acid-base pairs is discussed.

I . INTRODUCTION

Lewis acid sites on the surface of oxides are often considered as the active centers of heterolytic dissociation of hydrocarbons, double bond shift in olefins, skeletal isomerization and dehydrogenation of paraffins and of some other reactions. Therefore they have been studied by number of investigators. This is, however, more difficult than the study of Broensted acid sites that can be directly obseved by Infra Red (IR) spectroscopy or NMR.

On the contrary Lewis sites as such can not be detected by these techniques. The information about their properties was mainly obtained in the indirect way with the help of molecular probes. These are the basic molecules that are perturbed when interacting with Lewis sites. The extent of such perturbation is usually followed by IR spectroscopy. The amount of Lewis sites can be also detected from the number of probe molecules specifically adsorbed by Lewis sites.

Traditionally for this purpose most often the adsorption of such strong bases as ammonia or pyridine was used. The chemical properties of these molecules have, however, very little in common with hydrocarbons that are the most interesting for catalysis. Therefore, if adsorbed ammonia is strongly held by a Lewis site, this does not yet mean that the hydrocarbons will be also strongly adsorbed. The nature of adsorption in both cases is also quite different. For instance, for hydrocarbons or hydrogen most typical is their heterolytic dissociation.

J. Fraissard and L. Petrakis (eds.), Acidity and Basicity of Solids, 199–215.

$$\begin{array}{ccc} & H^{-\delta}\ H^{+\delta} & H^-\ \ H^+ \\ \text{-O--Me--O-} + H_2 \longrightarrow & \text{O--Me--O --} \longrightarrow & \text{O--Me--O--} \qquad (1) \\ \ \ \ | & \ \ \ | & \ \ \ | \\ \ \ O & \ \ O & \ \ O \end{array}$$

while for ammonia or pyridine - their coordinative binding:

$$\begin{array}{cc} & NH_3 \\ \text{-O--Me--O-} + NH_3 \longrightarrow & \text{O--Me--O--} \qquad (2) \\ \ \ \ | & \ \ \ | \\ \ \ O & \ \ O \end{array}$$

Therefore, strong bases are the probes for Lewis sites, while molecular hydrogen or paraffins rather for the acid-base ion pairs. Thus, the conclusions on the nature, the strength or the number of Lewis sites obtained from adsorption of strong bases can be hardly transferred to adsorption of hydrocarbons.

Of course, the best molecular probes for Lewis acidity are the reactants themselves. Therefore, in [1,2] the adsorption of molecular hydrogen and light paraffins was used for this purpose. Below we will discuss some of these results in connection with the study of Lewis sites on the surface of aluminum oxide and in zeolites.

1. ALUMINUM OXIDE.

1.1. LOW TEMPERATURE HYDROGEN ADSORPTION MEASUREMENTS

Fig. 1 represents the IR diffuse reflectance spectra of hydrogen adsorbed on γ– alumina pretreated in a vacuum at different temperatures. They were recorded at 77 K without evacuation of gaseous hydrogen from the optical cell, since free hydrogen as a symmetrical molecule doesn't have any IR absorption bands.

At the lowest temperature of vacuum pretreatment (670 K), when there is still no noticeable dehydroxylation of the surface,the single band with the maximum at 4102 cm $^{-1}$ is observed (Fig. 1 a). It was ascribed to hydrogen perturbed by interaction with the surface hydroxyl groups [1,2]. The corresponding adsorption heat doesn't exceed several kcal/mol, since this band disappears after short evacuation at 77 K

The intensity of this band is also lower for the samples pretreated in a vacuum at higher temperatures, when the surface dehydroxylation becomes noticeable (Fig. 1 b, c). Instead two new bands of adsorbed hydrogen appear with the maxima at lower frequencies of 4020 and 3975 cm $^{-1}$ (2960 and 2855 cm $^{-1}$ for adsorbed deuterium).

Their positions are for 140 and 190 cm^{-1} lower than the stretching vibration of free hydrogen molecule of 4163 cm^{-1}.

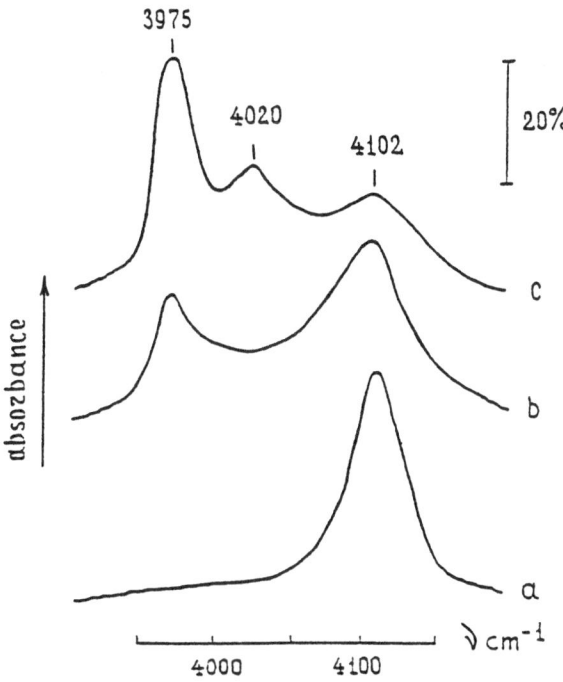

Fig. 1. The diffuse reflectance IR spectra of molecular hydrogen adsorbed at 77 K on γ – aluminum oxide pretreated in vacuum at different temperatures: a) 670 , b) 870 , c) 970 K.

These bands were attributed to hydrogen perturbed by aluminum atoms in trigonal and square pyramidal coordination created by dehydroxylation of the surface:

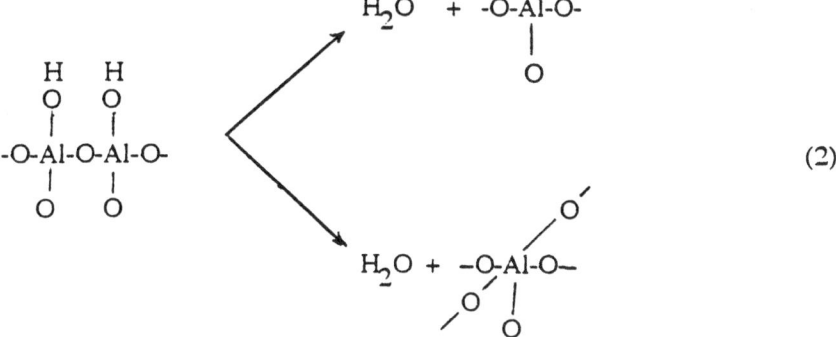

(2)

The interaction of hydrogen with such sites is stronger than with hydroxyl groups, since the corresponding bands disappear only after prolonged sample evacuation slightly above liquid nitrogen temperature.

The concentration of low coordinated aluminum atoms can be estimated from the amount of hydrogen that is desorbed at room temperature from the samples preliminary evacuated for a short time at 77 K in order to remove weakly adsorbed hydrogen. For η-alumina preheated in vacuum at 970 K it is equal to $2 \ 10^{13} \ cm^{-2}$ and for γ-alumina about three times lower. This corresponds to approximately 1 % of the surface coverage and is certainly much less than the total number of surface OH groups removed by vacuum treatment. Therefore, the desorption of the main part of surface hydroxyls results instead of low coordinated aluminum in formation of strained bridging oxygen atoms [3]:

$$
\begin{array}{cc}
H & H \\
O & O \\
| & | \\
-O-Al-O-Al-O & \longrightarrow \quad H_2O + -O-Al-O-Al-O- \\
| & | \\
O & O
\end{array}
\qquad (3)
$$

The results on hydrogen adsorption are in good agreement with the data on carbon monoxide adsorption that also indicated two different types of low coordinated aluminum atoms relsulting in C-O stretching frequencies of 2215 and 2244 cm^{-1} [4]. Their concentration is also about $10^{13} \ cm^{-2}$. In addition, CO adsorption at room temperature poisons the subsequent low - temperature hydrogen adsorption. This also confirms the identity of the sites of CO and H_2 adsorption.

On the other hand, the sites created by higher temperatures of vacuum pretreatmen perturb adsorbed carbon monoxide stronger than those formed at lower temperature, while for hydrogen this is the other way round (Fig. 1). This demonstrates the different character of adsorption of these molecular probes that will be discussed below.

2.2 .HYDROGEN ADSORPTION AT ROOM TEMPERATURE AND ABOVE.

Fig. 2(a) represents the bands of hydrides at 1895 and 1860 cm^{-1} resulting from dissociative hydrogen adsorption at room temperature:

$$
\begin{array}{cc}
H^{-\delta} & H^{+\delta} \\
\end{array}
$$
$$
H_2 \ + \ -Al-O- \ \longrightarrow \ -Al-O- \qquad (4)
$$

Simultaneously, two new OH bands that were absent in IR spectra of aluminum oxide before hydrogen adsorption appear in the region of OH bond stretching vibrations at 3510 and 3490 cm^{-1} (Fig. 2 b).

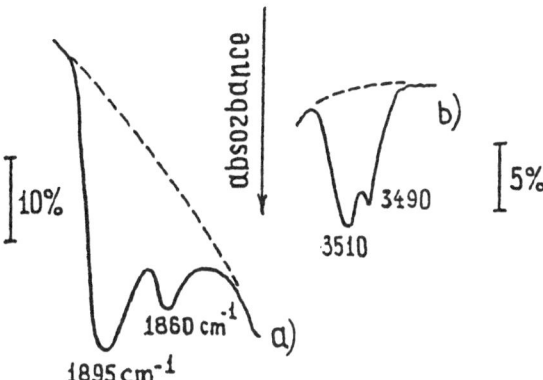

Fig. 2. Dissociative hydrogen adsorption on η-A$_2$O$_3$ at 298 K.
    a) The bands of surface hydrides. b) The bands of OH groups.

The dissociative hydrogen adsorption was also confirmed by the indirect way by the results of Fig. 3. It represents the initial spectrum of molecular hydrogen adsorbed on γ-alumina at 77 K and the spectrum of the same sample that was also recorded at 77 K, but after being preliminary kept in hydrogen atmosphere at 298 K for 24 hours. In the latter case the intensity of the low frequency band is decreasing. This can be also explained by dissociative hydrogen adsorption that blocks the surface Lewis-base pairs according to equation (4). This effect is reversible, since after evacuation above 473 K the Lewis-base pairs are regenerated and the 3975 cm^{-1} band can be again observed after low temperature hydrogen adsorption.

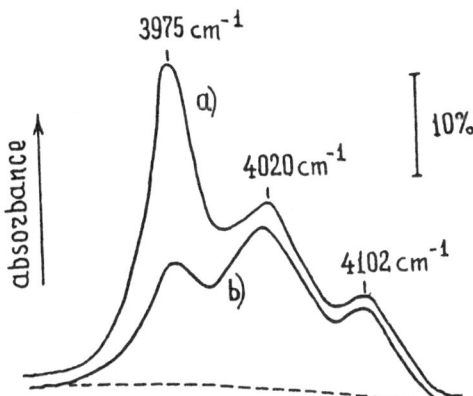

Fig. 3. H$_2$ adsorption at 77 K on η-Al$_2$O$_3$. a) Pretreatment temperature in vacuum 970 K. b) The same sample was preliminary kept in hydrogen for 24 hours at 298 K.

2.3. THE NATURE OF LEWIS SITES AND THE MECHANISM OF HYDROGEN DISSOCIATIVE ADSORPTION

The above conclusion about the different character of hydrogen and carbon monoxide adsorption on low coordinated aluminum atoms was also confirmed by quantum chemical analysis of the resulting adsorbed complexes. The semiempirical quantum chemical calculations of carbon monoxide adsorption on trigonally and square pyramidally coordinated aluminum atoms were performed in [5]. They supported the linear form of CO adsorption and indicated that the binding with trigonally coordinated aluminum is stronger than with the square pyramidally coordinated:

$$
\begin{array}{cc}
\mathrm{O} & \mathrm{O} \\
\mathrm{C} & \mathrm{C} \\
& \mathrm{O}' \\
{}_{\mathrm{O}'}\!\!\overset{\mathrm{Al}}{\underset{\mathrm{O}}{|}}\!\!{}_{\mathrm{O}} & -\mathrm{O}-\mathrm{Al}-\mathrm{O}- \\
\overset{|}{\mathrm{O}} & \overset{|}{\mathrm{O}} \\
| & |
\end{array}
\qquad (5)
$$

In both cases the adsorbed CO molecules are positively charged in agreement with the increase of their stretching frequencies. This demonstrates that carbon monoxide is a molecular probe for the aluminum Lewis sites alone.

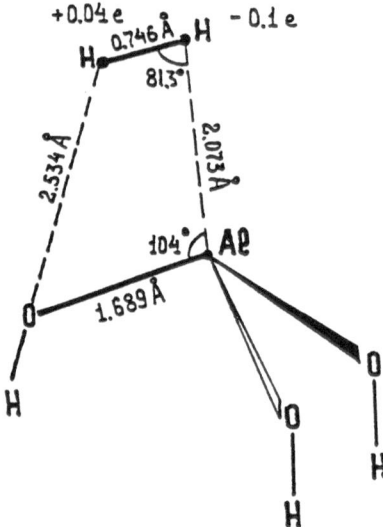

Fig. 4. Molecular hydrogen adsorption on trigonally coordinated aluminum ion as a model of Lewis acid site on the surface of aluminum oxide [6] .

For low temperature hydrogen adsorption on trigonally coordinated aluminum atoms the ab initio SCF MO quantum chemical calculations were performed in [6]. Such Lewis sites were modeled by the simplest Al(OH)$_3$ cluster with standard bond angles and bond lengths. The calculations were performed with optimization of the adsorption complex geometry (H-H distance and the position of hydrogen molecule relative to the adsorption site) with "Gaussian 80" program and 3-21 G basis set. The resulting structure of the molecular complex is represented by Fig. 4 . It clearly demonstrates the by-point nature of hydrogen adsorption that results in slight polarization of adsorbed molecule. This makes it optically active. The calculated adsorption heat was equal to 7 kcal/mol in reasonable agreement with hydrogen desorption at low temperatures. The nature of the binding is partially covalent, since the charges of both hydrogen atoms are slightly different, while the adsorbed molecule is slightly negatively charged (-0.05 e). Thus, unlike carbon monoxide, molecular hydrogen is a test for the surface acid-base pairs, where the contribution of both basic oxygen and low coordinated cation are equally important.

This conclusion was further supported by the analysis of hydrogen dissociative adsorption performed for the same trigonally coordinated aluminum cluster. The structures of the initial molecular complex and those of the transition state and the dissociatively adsorbed hydrogen are represented by Fig. 5, where the interaction of hydrogen only with one of the Al-O bonds of the cluster is shown.

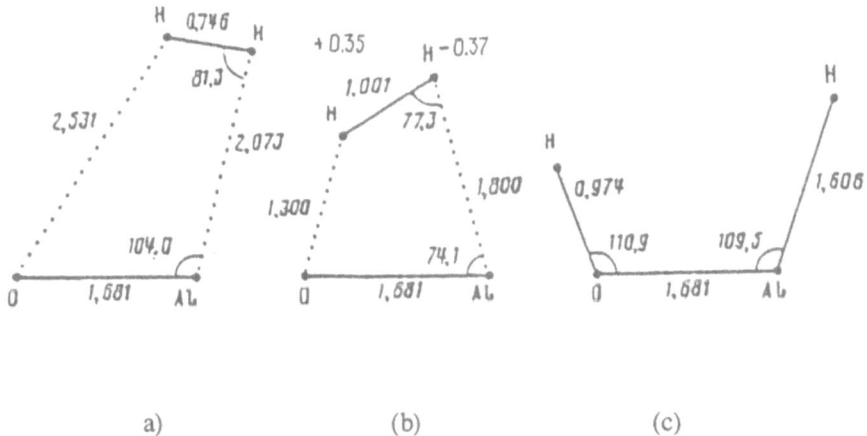

a)   (b)   (c)

Fig. 5. Reaction coordinate of dissociative hydrogen adsorption on trigonal Al atoms. Inter atomic distances in A, charges in e.

As one can see from this Figure, the reaction coordinate of the initial stage of dissociative adsorption mainly corresponds to the approaching of the left proton and the whole molecule to the basic oxygen atom of the Lewis-base pair, while the stretching of H-H bond is quite small (a) This results in electron density transfer towards the second hydrogen atom that becomes more basic in the strongly polarized transition state (b). The second part of the reaction coordinate is mainly determined by the interaction of the resulting negatively charged basic hydride with the Lewis site represented by the low coordinated aluminum atom. It mainly corresponds to the stretching of H-H bond and results in dissociative heterolytic adsorption with the activation energy of about 5 kcal/mol and the adsorption heat of about 20 kcal/mol (c).

This mechanism clearly demonstrates the equal importance of both the Lewis acid site and the basic oxygen, since the activation of hydrogen molecule is mainly connected with its polarization by the base, while the interaction of resulting hydride with the Lewis site mainly contributes the heat of adsorption.

2. LEWIS ACID SITES IN ZEOLITES

Lewis acid sites in zeolites are represented by:
i. Cations introduced by ion exchange or by any other way of modification;
ii. Extra lattice aluminum atoms created by dealumination;
iii. The sites resulting from high temperature dehydroxylation of hydrogen forms.

2.1. CATIONIC FORMS.

a) sodium forms.

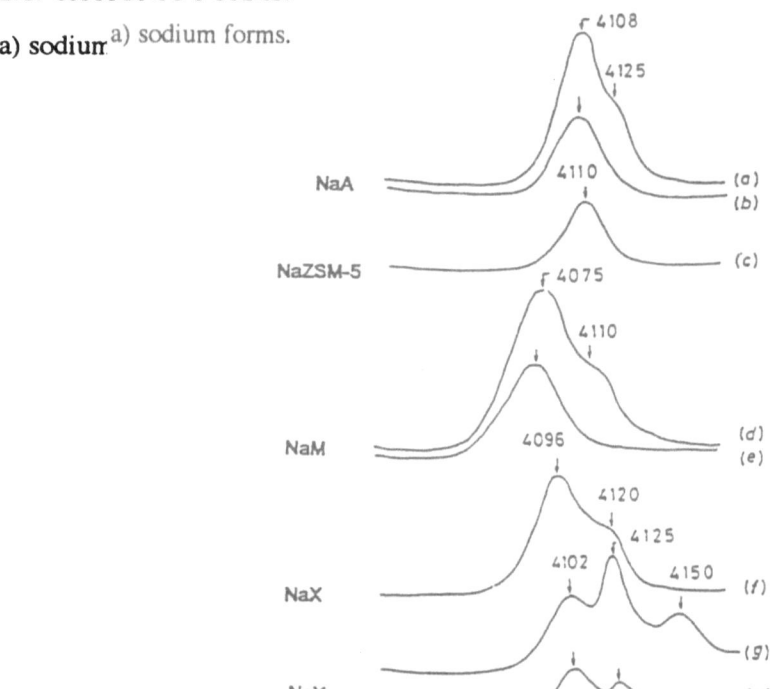

Fig. 6. Diffuse reflectance IR spectra of molecular hydrogen adsorbed at 77 K on sodium forms of different zeolites; a c,d,g - P = 100 Torr; b,e,h P = 5 Torr.

Already in the very first papers on the catalytic application of zeolites the possible role of cations as the catalytically active sites was postulated and discussed [7]. According to these ideas, they can activate adsorbed molecules due to electrostatic polarization. However, the degree of such polarization has never been experimentally studied. On the other hand, the qualitative estimation of this effect is possible from the results on low temperature molecular hydrogen adsorption.

IR spectra of molecular hydrogen adsorbed at 77 K on sodium forms of different zeolites are represented by Fig.6 [8]. Several bands in the region of H-H bond stretching vibrations are evident. They originate from the molecules perturbed by sodium ions in different positions in the zeolite frameworks.

The spectra of A zeolite and pentasils are the simplest (Fig.6,(a)-(e)). According to X-ray analysis, 8 of the overall 12 sodium ions are located in NaA near the centers of the six membered rings, which form the sodalite cages of its structure. The 3 of remaining four sodium cations are placed near the centers of 8-membered rings and one is statistically distributed among 12 fold equipoints [9]. This is consistent with the spectra (d) and (e) of Fig.6, which show that the majority of Na ions bind hydrogen more strongly and exhibit the larger low frequency shifts of H-H stretching vibrations, while the rest of sodium ions can held and perturb molecular hydrogen only at higher pressures resulting in a high frequency shoulder.

For mordenite at low hydrogen pressures also only a single band at 4108 cm^{-1} is observed with the stretching frequency by 55 cm $^{-1}$ lower than that of molecular hydrogen. This is consistent with the preferential location of sodium ions at the centers of highly distorted eight membered rings [10]. At a higher pressure (100 Torr, spectrum (a) a shoulder at 4120 cm^{-1} also appears. It may be connected with the more loosely bonded sodium cations distributed among three other available localization sites.

The data on Na cations distribution in NaZSM-5 are absent. Our results show that there is one major type of sodium in its framework (Fig.6, spectrum (c)). The corresponding stretching frequency shift of hydrogen is close to that observed for mordenite.

The spectra of hydrogen adsorbed on sodium forms of faujasites are the most complicated. They contain three different IR bands which change their intensities at different hydrogen pressures (Fig.6, spectra (f)-(h)). In [8] they were ascribed to hydrogen molecules perturbed by sodium ions located in S $_I$ (the largest shift); S $_{II}$ and in S$_I$ (the smallest shift) positions. Thus, the low - temperature molecular hydrogen adsorption allowed us to follow both the different sites of sodium ions localization and their distribution among these sites in the zeolites framework. The summary of these data is represented by Table 1 :

Table 1.

Low frequency shifts of H-H stretching vibrations resulting from molecular hydrogen interaction with sodium ions in zeolites.

Zeolithes	NaA	NaX	NaY	NaM	NaZSM-5
Shifts cm $^{-1}$	87, 58	68, 43	61, 38, 13	55	53

These results demonstrate a clear dependence of the observed low frequency H-H stretching shifts upon the aluminium content and on the type of the zeolite framework. This is in agreement with the above model of the by-point adsorption of molecular hydrogen. Indeed, for such a model the H-H stretching shifts should be dependent both on the properties of the Lewis sites and on the basic properties of neighboring oxygen. Such dependence is clearly seen from Table 1, since the low frequency shift is the largestt for the most basic NaA zeolite and the smallest for NaZSM-5. In this sense hydrogen differs from adsorbed carbon monoxide that being a test for cations exhibits very close shifts for different zeolite frameworks [8].

On the other hand , even for NaA zeolite the absolute value of H-H stretching shift is more than twice less than for Lewis sites of aluminium oxide and the dissociative adsorption of hydrogen was not observed. This shows that the rather general opinion about the exceptional polarizing ability of cations in zeolites is not supported by spectral data.

b) Alkaline earth cationic forms.

IR spectra of hydrogen, adsorbed at 77 K on alkaline earth cationic forms of mordenite preheated at 500 C in a vacuum are depicted by Fig.7. For Mg and Ca exchanged samples two bands of H_2 stretching vibrations are observed, but only the single bands for the samples exchanged with Sr and Ba ions.

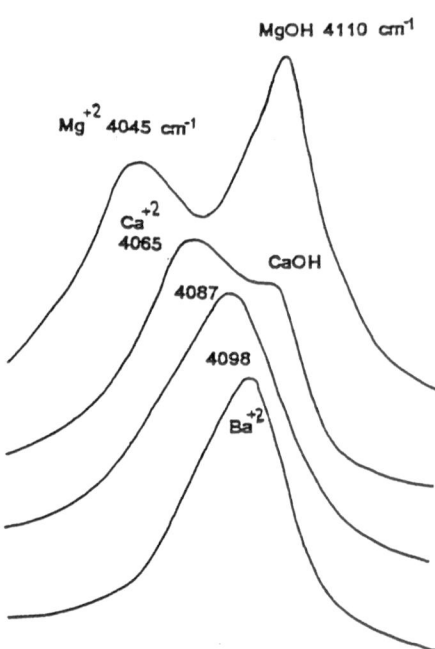

Fig. 7. Molecular hydrogen adsorption at 77 K on alkaline earth forms of mordenite. Hydrogen pressure equal to 300 Torr.

This can be easily explained by different extent of partial hydrolysis of these cations. Indeed, the second dissociation constants of (MgOH) and (CaOH) ions are rather low (3 10^{-3} and 4 10^{-4} ,respectively). Therefore these cations exist in the zeolite framework in two forms: as bivalent Me^{+2} and as hydrolyzed MeOH species. On the contrary, for $(SrOH)^+$ and $(BaOH)^+$ ions these constants are much higher (about 0.15-0.2). This results in the destruction of MeOH fragments after dehydration at elevated temperatures.

Such interpretation is supported by IR spectra of the same samples in the OH stretching region. For dehydrated Mg and Ca forms both bands of structural OH groups and those of MeOH fragments are observed. On the contrary, for the samples exchanged with Sr and Ba ions only the bands of silanol groups are evident.

The comparison of hydrogen low frequency vibrational shift for NaM of 55 cm^{-1} with those for hydrolyzed alkaline earth cations presented in Table 2 demonstrates their close values. This also supports the above interpretation, since MeOH ions hav a formally the same positive +1 charge as the univalent cations.

Table 2

Low frequency shifts of H-H stretching vibrations of hydrogen adsorbed by bivalent cations in mordenite.

Zeolites	Ions and shifts cm^{-1}	
	MeOH $^+$	Me^{+2}
BaM	-	65
SrM	-	76
CaM	50	98
MgM	53	118

More surprising are the small shifts exhibited by nonhydrolyzed bivalent cations. For Ba $^{+2}$ and Sr^{+2} they are even less than for sodium form of A zeolite, while for Ca^{+2} and for Mg^{+2} only slightly larger. Therefore the higher catalytic activity of Mg and Ca forms of mordenite in comparison with sodium forms is hardly connected with the properties of the cations themselves. More reasonable is to ascribe it to Broensted acid sites resulting from partial hydrolysis of Mg and Ca cations. Any way, these results clearly show that the formal charge of cations by no means is the major factor affecting polarization of adsorbed molecules.

C) High silica zeolites modified with Zn^{+2} ions.

The further arguments in favor of this conclusion are presented by the high silica zeolites modified with Zn^{+2} ions that are known as the catalysts for aromatization of light paffins. There are also strong indications of their bifunctional nature, while the reaction mechanism combines dehydrogenation of paraffins by modifying Zn^{+2} cations with the subsequent aromatization of resulting olefins on acid Broensted sites of the zeolite [12].

It was found in [13] that modification of HZSM-5 both by ion exchange or by impregnation with zinc nitrate results in appearance of several different Lewis sites as evident from low temperature hydrogen adsorption (Fig. 8).

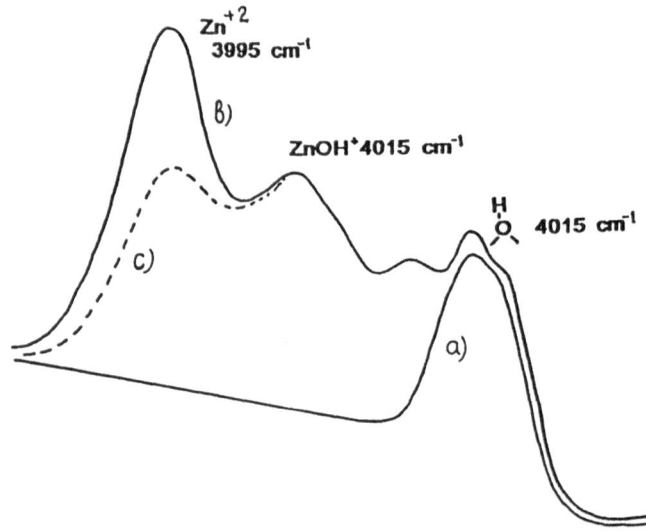

Fig. 8. Molecular hydrogen adsorption at 77 K on: a)- HZSM-5; b)- ZnZSM-5. c)- sample (b) was kept for 10 min at 300 K in hydrogen atmosphere.

The most remarkable are the large low frequency shifts of H-H stretching vibrations that considerably exceed those for alkalne earth cations. The band at 4015 cm^{-1} was ascribed to Zn OH species, while that at 3995 cm^{-1} to hydrogen molecules perturbed by Zn^{+2} cations. The low frequency shift of the latter is different from that earlier reported for hydrogen adsorption on ZnO [14]. This means, that the nature of the Lewis-base pairs on the surface of zinc oxide and in high silica zeolites is different

This was confirmed by the results on dissociative hydrogen adsorption at room temperature. Unfortunately, due to the intensive background from the zeolite crystal lattice vibrations, the direct observation of the resulting Zn-H fragments is rather difficult. Therefore, the dissociative adsorption was proved by the indirect way similar as it was done before for aluminum oxide.

Indeed, it was found that after keeping of the modified ZnZSM-5 sample at room temperature in hydrogen atmosphere the intensity of the low frequency band of adsorbed molecular hydrogen was decreasing. Simultaneously the intensty of the band from the bridging acidic hydroxyls increased (Fig. 8 c).Similar to aluminum oxide this was explained by the heterolytic dissociative adsorption of hydrogen that makes the acid-base pairs unaccessible for subsequent H_2 adsorption :

$$H^- \quad H^+$$

$$-Zn-O- \; + \; H_2 \longrightarrow \; Zn-O- \tag{6}$$

Since the stretching frequency of the resulting OH groups is the same as in HZSM-5, the oxygen atoms of these sites are those of the zeolite framework.

To understand the large low frequency shifts for Zn^{+2} ions let us compare them with the rest of bivalent cations in more detail (Table 3).

Table 3.

The comparison of characteristic features of some free bivalent cations with low frequency H-H stretching shifts.

Cation	H-H shift cm^{-1}	Ion radius A	Electronegativity of a free ion	Hardness of a free ion
Zn^{+2}	168	0.74	5.1	6.6
Mg^{+2}	118	0.66	36.2	28.4
Ca^{+2}	98	0.99	22.4	16.6
Sr^{+2}	76	1.12	22.1	16.6

It is clear from this Table that Zn^{+2} doesn't differ from Mg^{+2} or Ca^{+2} neither by its formal positive charge, nor by ion radius. On the other hand, zink ions are less electronegative and not as hard as magnesium or calcium. This has the following physical meaning: the interaction of molecular hydrogen with zink ions results from the very beginning in addition to polarization in formation of a weak covalent bond with hydrogen molecule. This creates an additional perturbation of adsorbed hydrogen and is even more important at the second step of dissociative adsorption, when the resulting hydride interacts with Lewis site. Thus, to perturb the adsorbed hydrogen most strongly the cation in the surface acid-base pair should satisfy the following requirements: it must have a high positive charge, small dimensions and must be soft to be able to form covalent bonds both with molecular hydrogen and with resulting hydride. In addition the neighboring oxygen has to be of a basic nature, since it is also involved in polarization of molecular hydrogen and binds the proton resulting from the heterolytic dissociative adsorption.

2.2 LEWIS SITES RESULTING FROM DEHYDROXYLATION OF HYDROGEN FORMS OF ZEOLITES

Historically, the first mechanism of dehydroxylation of zeolites was suggested by Uytterhoeven, Crystner and Hall for H-faujasites [15]. They postulated the formation of trigonally coordinated Al and Si pairs, when two hydroxyl groups are removed from the zeolite framework:

$$2 \underset{\overset{\displaystyle O}{\underset{\displaystyle H}{|}}}{--Si} \quad Al-- \quad ----- \quad ---Si \quad Al--- \quad + \quad ---Si \quad \overset{\displaystyle -}{\overset{\displaystyle O}{}} \quad Al \quad + \quad H_2O \qquad (7)$$

Later this mechanism was shown to be incorrect, since according to numerous solid state ^{27}Al high resolution NMR data the dehydroxylation of H-faujasites results in dealumination of their framework [16].

Unfortunately, this technique requires rehydration of the samples before NMR measurements in order to convert the extra lattice aluminum into octahedral coordination. Therefore, the resulting Lewis sites can not be directly studied by NMR as formed. However, this can be done by IR spectroscopy, as it is demonstrated below by the IR diffuse reflectance spectrum of "Deep bed" pretrated HY zeolite [17] (Fig. 9):

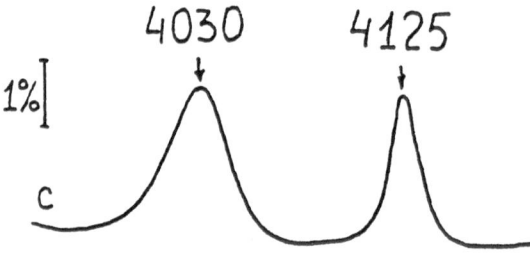

Fig. 9. Diffuse reflectance IR spectrum of molecular hydrogen adsorbed at 77 K on HY zeolite pretreated under "Deep bed conditions" at 870 K.

The spectrum contains two bands at 4125 and 4060 cm^{-1}. The former is connected with hydrogen perturbed by interaction with silanol groups, since the similar line was earlier observed after adsorption of molecular hydrogen on silica gel. The second band appears only after "Deep bed" treatment and therefore was ascribed to hydrogen interacting with the extra lattice aluminum.

According to [18], the extent of framework dealumination of Y zeolite after such pretreatment is 85 %. This is in a reasonable agreement with our results on the amount of such sites estimated from hydrogen thermodesorption and also confirms the above interpretation.

For the high silica zeolites, that are more thermally stable and resistant to dealumination than faujasites, the lattice Lewis sites resulting according to scheme (7) can be also detected by low temperature hydrogen adsorption. This is demonstrated by the results of Fig. 10.

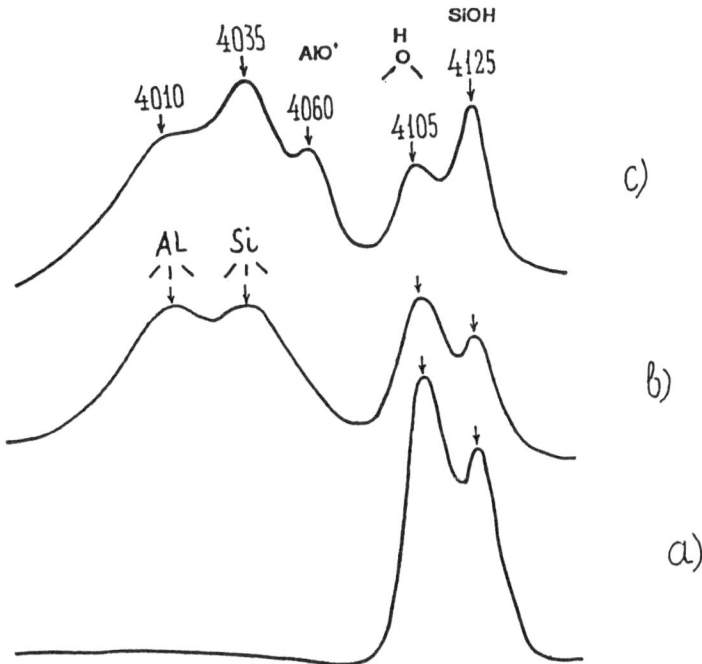

Fig. 10. Low temperature molecular hydrogen adsorption on HZSM-5 pretreated in vacuum at different temperatures. a)- 770 K; b)- 970 K; c)- 1073 K.

At moderately high temperatures, when dehydroxylation has not yet sltart (770 K), only the bands with maxima at 4125 and 4105 cm^{-1} are observed, while their intensities are decreasing with increasing pretreatment temperature (Fig. 10 a). Therefore, similar as in the above case of HY, they were attributed to hydrogen molecules perturbed by hydroxyl groups.

At the higher pretreatment temperature of 970 K (Fig.10 b) two new bands of adsorbed hydrogen with the larger low frequency shifts and the positions of the maxima at 4010 and 4035 cm^{-1} appear. They were ascribed to the molecules interacting with Lewis acid sites resulting from the removal of hydroxyl groups according to scheme (7).

At still more severe conditions of dehydroxylation the band of adsorbed hydrogen at 4060 cm^{-1} appears that was earlier attributed to hydrogen perturbed by extra lattice aluminum ions (Fig 10 c). Simlultaneously the intensity of the band with the maximum at 4010 cm^{-1} decreases. This can be easily explained if the latter line is assigned to the hydrogen perturbed by the lattice trigonally coordinated aluminum atoms, their number decreasing upon dealumination. Then the band at 4035 cm^{-1} is connected with trigonally coordinated silicon atoms.

Such a difference in dehydroxylation of H-faujasites and high silica zeolites can be explained in the following way. The removal of protons from their crystal lattices disturbs both the balance of electric charges, and changes the valence angles in resulting trigonally coordinated Si and Al atoms. Therefore, each of such sites possess a certain excess energy and is a source of strains in the zeolite structure. For faujasites at total dehydroxylation their number should be very high. For instance, according to equation (7), for Si:Al ratio equal to two one half of aluminum and a quarter of silicon atoms should be converted into trigonal coordination. Apparently, such a perturbation is too strong and results in partial decomposition of the zeolite crystal lattice. Therefore the high temperature dehydroxylation of faujasites is followed by their dealumination and partial amorphizsation.

On the contrary, in high silica zeolites the amount of lattice Lewis sites resulting from dehydroxylation does not exceed several percent. In other words, such structural defects appear in a relatively small number. Therefore, the structure of high silica zeolites remains stable up to much higher temperatures without decomposition or dealumination.

CONCLUSION

According to the well known definition, a Lewis acid is an electron pair accepter, while a base - an electron pair donor. Later Lewis developed his theory by discrimination between the "primary" and the "secondary" acids and bases [19]. The former interact with each other with zero activation energy, while the latter must be at first activated.

Hydrogen, paraffins and other nonpolar molecules certainly are the typical secondary bases. Therefore their interaction with surface Lewis sites requires the preliminary polarization of H-H or C-C bonds. According to the above discussion, this needs the by-point interaction both with acid sites and with the neighboring basic oxygen, the latter probably beeing even more important. Thus, the conception of primary Lewis sites could be hardly directly applied to activation of such molecules as hydrogen and paraffins. Instead the surface acid-base pairs should be considered. However, the most commonly used approach is still mainly focused on the properties of the Lewis acid sites alone, while their strength is correlated with the position of the lowest unoccupied molecular orbital.

As follows from the above discussion, the real requirements for the most active Lewis-base pairs are much more complicated. Some of them were already formulated

above. This are:
  i. The high basisity of the oxygen atom neighboring to the Lewis site.
  ii. The high positive charge, moderate electronegativity and the small dimensions of the low coordinated cation.
  iii.The ability of the cation to form the covalent bond both with the initial molecule and with the resulting dissociatively adsorbed fragment to avoid the excess polarization of the transition state and in this way to decrease the activation energy of dissociative aqsorption.

It is remarkable that some of these requirements well correlate with each other. For instance, the strong basic properties of oxygen suppose the moderate electronegativity of the catalyst. The highly basic oxygen in combination with the low position of LUMO means the low hardness etc.

Of course, the interaction with hydrogen is only a first step in the more realistic reformulation of the ideas about the real nature of the surface Lewis sites and the nature of their interaction with nonpolar molecules. The next step should be the discussion of these problems in connection with activation of adsorbed paraffins.

REFERENCES

1. V.B.Kazansky, V.Yu.Borovkov, L.M.Kustov. Proc of 8 th Int.Congr. on Catalysis, Berlin 1984, Dechema Verlag Chemie 1984, V. 3, P. 3.
2. V.B.Kazansky, V.Yu. Borovkov, A.V.Zaitsev. Proceedings of 9th. Int.Congr. on Catalysis, Calgary 1988, Ed. by M.J.Phillips and M.Teman, Pub. by Chem. Inst. of Canada 1988, V. 3, P.1426.
3. H.Knozinger,P.Ratnasamy. Catal.Rev.Sci.Eng. 1978, V. 17, P.31.
4. G.Gatta, B.Fubini, G.Giotti, C.Morterra. J. Catal. 1976, V.43, p. 90.
5. I.N.Senchenya, N.D.Chuvilkin, V.B.Kazansky. Kinetika i kataliz (Russ.) 1986, V. 27, P. 608.
6. I.N.Senchenya, V.B.Kazallsky. Kinetika i kataliz (Russ.) 1988, V. 29, P. 1131.
7. J.A.Rabo, P.E.Pickert, D.N.Stamires, J.E.Boyle. Proc.2nd Int. Congr. Catalysis, Ed. Tech. Paris 1960, P 2055.
8. L.M.Kustov, V.B.Kazansky, Journ. Chem. Soc. Far. Trans. 1991, V. 87, P. 2675.
9. W.J.Mortier."Compilation of extraframework sites in zeolites", Butterworth Sci. Ltd., Guildford, (1982).
10. R.Y.Yanagida, A.A.Amoro, K.Seff. Journ. Phys. Chem., 1973, V. 3, P.805.
11. C.L.Angell, P.C.Shaffer. Journ. Phys. Chem.,1966, V. 70, P. 1413.
12. D.Seddon. Catalysis Today, 1990, V. 6, P. 351.
13. V.B.Kazansky, L.M.Kustov, A.Yu.Khodakov. Proc. 8 th Int. Zeolite Conf., Amsterdam 1989, Zeolites: "Facts,Figures, Future".Ed. by P.A.Jacobs, R.A.Van Santen, Elsevier 1989, P. 1173.
14. R.J.Kokes. Proc 5th Int.Congr.Catal.,Miami Beach 1972."Catalysis". Ed.by J.W.Hightower, North Holland Pub.Co. 1973, P. 63.
15. L.B.Uytterhoeven, L.G.Crystner, W.K.Hall. J.Phys Chem., 1965, V. 69, P.2117.
16. J.M.Thomas, J.Klinowski. Adv. Catal. 1985, V. 33, Ed. by D.D.Eley, H.Pines, P.W.Weisz, P.200.
17. V.B.Kazansky. Catalysis Today, 1988, V. 3, P. 367.
18. D.Freude, T.Froelich, G.Pfeifer, G.Scheler. Zeolites 1983, V. 3, P. 171.
19. G.N.Lewis, Journ. Franklin Inst. 1938, V. 226, P. 293.

STRUCTURE AND COMPOSITION FACTORS IN THE ACIDITY AND BASICITY OF INORGANIC MOLECULAR METAL-OXYGEN CLUSTER CATALYSTS AND THEIR INFLUENCE ON THE SURFACE AND CATALYTIC PROPERTIES

J.B. MOFFAT
Department of Chemistry and Guelph-Waterloo Centre
for Graduate Work in Chemistry
University of Waterloo
Waterloo, Ontario N2L 3G1
Canada

ABSTRACT.   Metal-oxygen cluster compounds (MOCC) (sometimes known as heteropoly oxometalates) with anions of Keggin structure have acidic properties which are dependent on both the composition of the anion and the nature of the cation. MOCC with anions containing tungsten are more acidic than those which contain molybdenum, although the nature of the crystallographic structure is invariant. Replacement of the protons by larger cations may, in some cases, generate microporous structures as well as producing solids in which the cations have an inductive effect on any residual protons thus perturbing the distribution of acid strengths. The water molecules present in the hydrated forms of the MOCC influence the form of the crystallographic structure, restrict the approach of foreign species to the proton, and in combination with the proton itself, produce inductive effects both on unhydrated and hydrated protons and thus modify the distribution of acidic strengths.

1.   Introduction

Metal-oxygen cluster compounds (MOCC) (also known as heteropoly oxometalates) may be prepared in a wide variety of shapes and sizes (1, 2). Perhaps the best known of these are those with anions of Keggin structure (Fig. 1). These are large, approximately 10 Å in diameter, approximately spherical structures of T_d symmetry with a central atom (X) which may be selected from as many as 65 possibilities (2), including phosphorus and silicon. Four oxygen atoms are bonded to and arranged tetrahedrally around the central atom. Twelve octahedra with oxygen atoms at their vertices and a peripheral metal atom (M) such as tungsten or molybdenum at their approximate centers surround and share oxygen atoms with the central tetrahedron. The twelve octahedra are arranged in four groups of three edge-shared octahedra, M_3O_{13}, which share corners with each other and the central tetrahedron. Pope (2) has noted the requirements for the peripheral metal atom are considerably more stringent than those for the central atom of the anion and thus the set of those which is possible for the former position is considerably smaller than for the

J. Fraissard and L. Petrakis (eds.), Acidity and Basicity of Solids, 217–235.

latter. The octahedra are collected into four groups of three edge-shared octahedra, M_3O_{13}. There are three types of oxygen atoms, those bridging the central atom and the peripheral metal atoms, those connecting pairs of peripheral metal atoms and twelve terminal atoms, one from each of the octahedra, protruding from the anion structure. The twelve peripheral metal atoms may be of the same or two different elements. The Keggin anions are isostructural in that changes in either the central or the peripheral metal atoms leave the structure semiquantitatively unchanged. The stoichiometry of the anion may be represented as $XM_{12}O_{40}^{-n}$ or alternatively $[(XO_4)(M_{12}O_{36})]^{-n}$ where the value of n will be dependent on the oxidation states of X and M. For the present discussion X and M will be either P or Si and W or Mo, respectively.

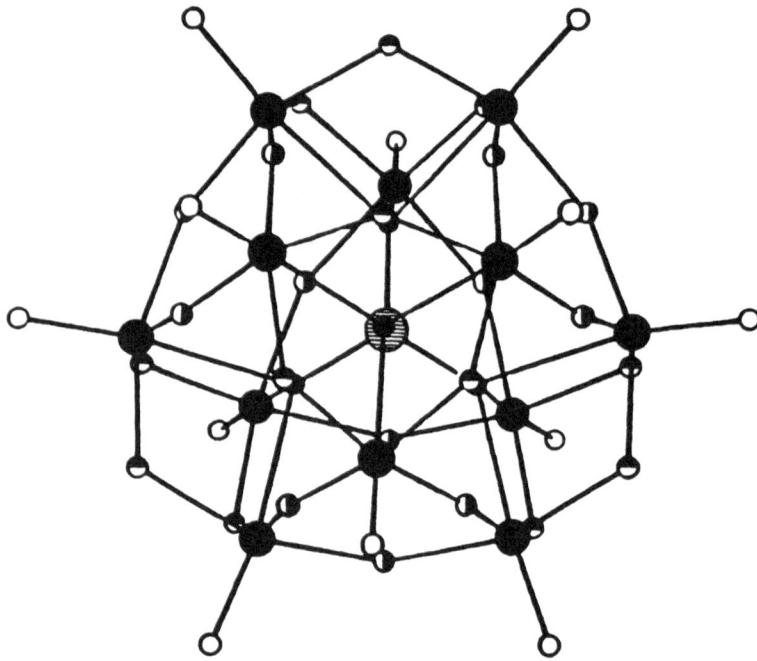

Figure 1: Anion Structure in 12-Heteropoly Acid. Central atom: ⊜; peripheral atom: ●; terminal or outer oxygen atom, O_t: ○; inner oxygen atom, O_a: ●; edge-shared, bridging oxygen atom, O_b: ◑; corner-shared, bridging oxygen atom, O_c: ◐.

2.   Crystallographic Structure

Although a variety of cations may be employed to balance the charge of the anions, the proton is the simplest. With phosphorus as the central atom and tungsten as the peripheral metal element in the hexahydrate form the proton is located in the centre of and hydrogen-bonded to two approximately coplanar water molecules (3) (Figure 2).

Although positions for four water molecules are shown only two of these are occupied in the hexahydrate and those which are occupied may vary as the result of a two-fold thermal disorder. The water molecules are, in turn hydrogen-bonded through their hydrogen atoms to the terminal oxygen atoms of the anions. The hexahydrate of 12-tungstophosphoric acid has a cubic Pn3m structure (3).

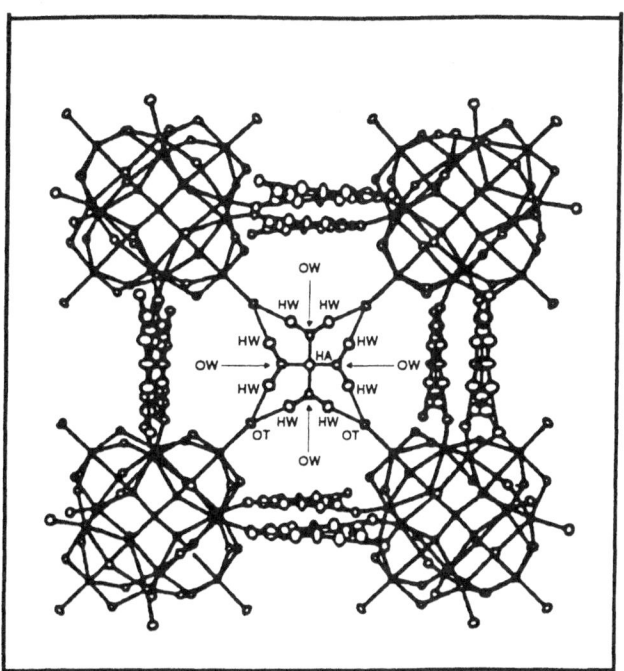

Figure 2. Crystallographic arrangement of protons, anions, and water molecules in hydrated 12-tungstophosphoric acid (3).

Recent neutron scattering studies suggest that although most of the normal coordinates of a near-planar $H_5O_2^+$ ion can be fitted to the observed spectrum, additional bands which are observed may be due to the presence of pyramidal components (4). Further studies by the same workers have identified a twofold reorientation (180° flip) of the water molecules within the centrosymmetric $H_5O_2^+$ ions (5, 6).

It is of interest to note here, somewhat parenthetically, that the higher hydrates of 12-tungstophosphoric acid ($H_3PW_{12}O_{40} \cdot nH_2O$, n = 14, 21, 29) have been shown to have the highest near-ambient temperature protonic conductivities reported for inorganic solids (7, 8). This is, of course, indicative of the protonic acidity of these materials.

The metal-oxygen cluster compounds may be considered as isostructural in the sense that the anion and crystal retain their structural forms with changes in the central atom and/or the peripheral metal atoms (Table 1). Further, for a given central atom a change in the peripheral metal atoms form tungsten to molybdenum, for example produces little or no changes in the bond lengths. It is interesting to note that, while the crystal form remains intact, a change of the central atom from phosphorus to silicon produces significant alterations in the bond distances.

Table 1. Average Bond Lengths $(\text{Å})^a$ in $[XM_{12}O_{40}]^{n-}$ (2).

X	M	$X-O_a$	$M-O_a$	$M-O_t$	M ... M		Ref
P	W	1.53	2.44	1.70	3.41^b	3.70^c	3
P	Mo	1.54	2.43	1.66	3.41	3.70	9
Si	W	1.64	2.35	1.71	3.38	3.68	10
Si	Mo	1.62	2.35	1.67	3.36	3.70	11

^{a} Averaged to T_d symmetry. ^{b} Within M_3 triplet, edge-shared MO_6 octahedra. ^{c} Between M_3 triplets, corner-shared MO_6 octahedra.

The crystallographic (secondary) structure of the metal-oxygen cluster compounds appears to depend primarily on the number of water molecules present, rather than on either the anion charge or the relative numbers of cations and anions (12, 13). In general, hydrates with five or fewer molecules of water are found to be cubic with two molecules per unit cell. The hydrates of 14 water molecules and with X as either P or Si and M as either W or Mo are apparently triclinic (13) while the same species return to their cubic structure when 29-31 water molecules are present per heteropoly anion. However, sufficiently large cations apparently distort the secondary structure of the MOCC (14). For example, the presence of cations of size greater than $MeNH_3^+$ (where Me = methyl), with 12-molybdophosphates and 12-tungstophosphates, has been shown to result in crystal structures other than the cubic (Pn 3m) type usually associated with these MOCC (14).

3.   Thermal Stability

The thermal stability of the MOCC may be evaluated by use of differential thermal and gravimetric analysis (14-20). In general the monovalent salts of the MOCC acids are found to be more thermally stable than the parent acids. Of course, the assessment of thermal stability is dependent on the method employed for the evaluation as well as the conditions under which such experiments are performed.

The acidic forms of the MOCC are known to be highly hydrated. For example, various quantities of water of crystallization, n = 30 (21), 29 (22, 23), 27.5 (24) and 24 (21) for $H_3PW_{12}O_{40} \cdot nH_2O$ at room temperature, have been reported. Freshly recrystallized 12-tungstophosphoric acid is expected to be the 29-hydrate, which has been confirmed by XRD (23) but readily converts to the 24-hydrate on standing (21).

The loss of mass during heating in air of a sample of $H_3PW_{12}O_{40} \cdot 24\ H_2O$ for 3 hours (static) and in helium for 2 hours (flow) at a given temperature can be seen in Figure 3 (19). The phosphorus content, as determined by EDTA analysis (25) decreases from 1.01_5 atom/anion in the 24-hydrate beginning at approximately 450°C to about 70% of this value at 600°C. The loss in mass was observed in three steps. In the first step a precipitous decline in the amount of water present occurs at approximately 100°C to produce the 4~5-hydrate. The anhydrous acid forms at 350-450°C as found by West and Audrieth by DTA (15). Powder X-ray data (19) show broadening and a decrease in intensity at temperatures of approximately 400°C but provides evidence for the retention of the MOCC structure. At temperatures higher than 525°C further loss in mass occurs together with a substantial change in the XRD pattern. Although DTA data (15) suggest that the MOCC structure of 12-tungstophosphoric acid decomposes at 600°C the data in Figure 3 suggest that the onset of decomposition may occur at temperatures as low as 500°C with prolonged heating. In general the tungsten-containing acids and their salts are more stable than those containing molybdenum (14).

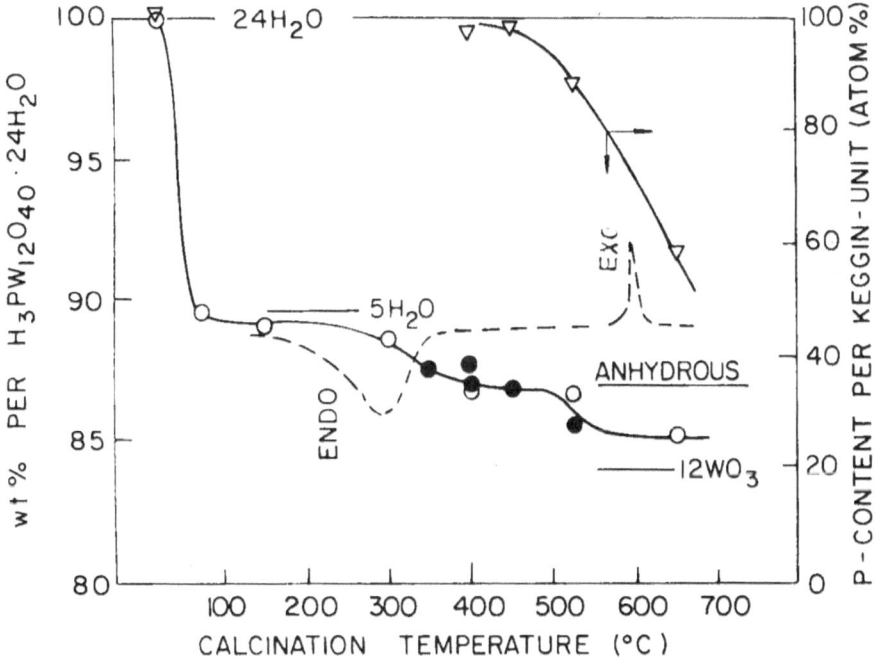

Figure 3. Decrease in weight and phosphorus content during calcination of $H_3PW_{12}O_{40} \cdot 24H_2O$. $\bigcirc$, Calcined in air for 3 hr (static); $\bullet$, calcined in helium for 2 hr (flow); X phosphorus content; dashed line, DTA by West and Audrieth (15).

The thermal stability of the anions can also be assessed through the use of photoacoustic FTIR spectroscopy (26) (Figure 4). A series of 5 or 6 bands between 800 and 1100 cm^{-1} is indicative of the anion structure. In particular two bands at approximately 1080 and 980 cm^{-1}, attributed to the triply degenerate asymmetric stretch of the central XO_4 tetrahedron and the stretch of the peripheral metal-terminal oxygen bond, respectively, may be employed as a measure of the presence of the anion. Although with 12-tungstophosphoric acid (abbreviated as HPW) at 450°C the intensity of these bands is evidently reduced, nevertheless it is evident that the anion structure remains at that temperature.

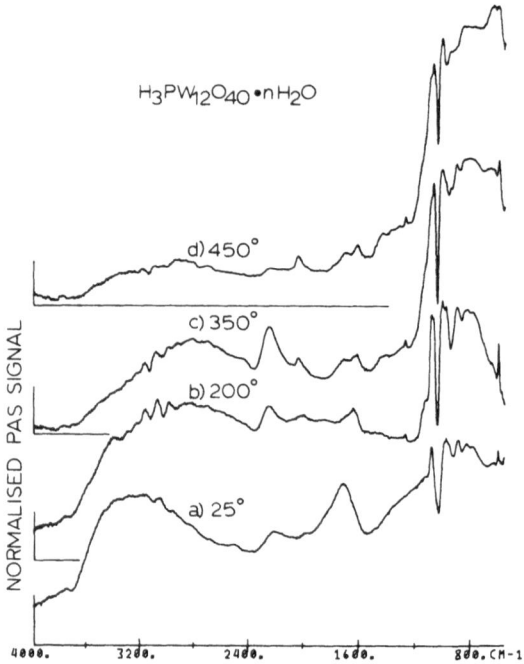

Figure 4. PAS FTIR spectra of 12-tungstophosphoric acid showing characteristic bands in 1100-800 cm^{-1} region and effect of heating in vacuo.

4.   pH Stability

The stability of the metal-oxygen cluster compounds with anions of Keggin structure has been found to be dependent on pH (27). ^{31}P and ^{183}W NMR spectra have been obtained for aqueous solutions of HPW and HPMo as well as the reactant solutions employed in their preparations at values of pH between 0 and 10. With the acidic solutions the pH was adjusted by the addition of sodium hydroxide, while that of the preparative solutions containing orthophosphate and tungstate or molybdate anions was varied by the addition of hydrochloric acid. The present discussion will, for purposes of brevity, consider only the tungsten-containing system.

The most complex NMR spectra obtained for the tungstate-phosphate system appear at $1.0 <$ pH ≤ 4.0. At pH > 8 and pH < 8, single peaks at ~ +3.5 and ~ -10.1 ppm, respectively, are observed, the former attributed to the free phosphates. This upfield shift of approximately 13 ppm is indicative of a substantial perturbation in the local phosphorus environment associated with the formation of a metal-oxygen cluster anion. This peak is attributed to $PW_{11}O_{39}^{7-}$, a known decomposition product of the $PW_{12}O_{40}^{3-}$ anion and precursor to its formation, that possesses a wide pH stability range (Fig. 5). The $PW_{11}O_{39}^{7-}$ anion exists as the major species for $2.5 <$ pH < 8.0. Although a number of peaks appear in the ^{31}P NMR spectra for $1.0 <$ pH < 2.5, that at -14.4 ppm is assigned to $PW_{12}O_{40}^{3-}$. A peak at -12.8 ppm, believed to be due to the presence of $P_2W_{21}O_{71}^{6-}$ vanishes for pH < 1. For pH < 1.5 the resonance attributed to $PW_{12}O_{40}^{3-}$ becomes the major peak, consistent with the proposed mechanism for the reversible decomposition of the $PW_{12}O_{40}^{3-}$ anion at pH 1.5-2 (2).

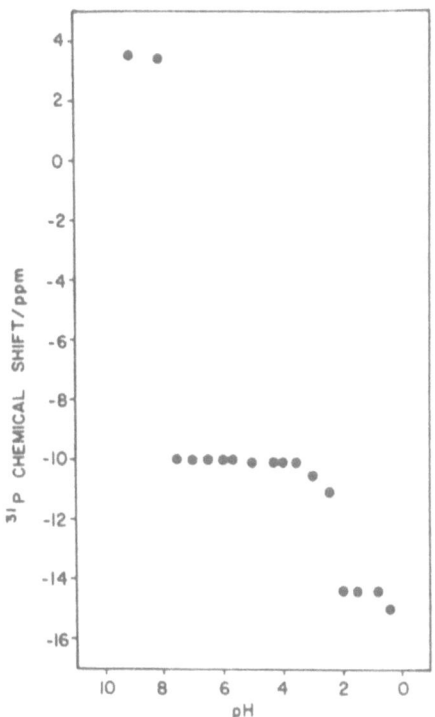

Figure 5. ^{31}P NMR chemical shifts of the major species present in tungstophosphate solutions for various values of pH.

5. MOCC as Three-Dimensional Intercalates

With the proton as cation the MOCC take on the form and function of solid acids with

224

high acidic strengths. Certain of the salts formed from these acids also have been found
to possess acidic properties. There is, however, a rather unusual feature of these materials
which is relevant to the present discussion. Although the solid acids have been found to
be nonporous and possess very small (BET N_2) surface areas (5-10 m^2/g), unexpectedly
large quantities of polar molecules are sorbed by the acids. For example, the stepwise
exposure of HPW to aliquots of gaseous ammonia at room temperature produces a band
in the PAS FTIR spectrum at approximately 1420 cm^{-1} characteristic of the ammonium
ion (27) (Figure 6). The quantity of ammonia sorbed ultimately reaches a maximum at
approximately three molecules per heteropoly anion (Keggin Unit, KU), equivalent to the
total number of protons per KU. Since the surface area is low the majority of the protons
must be found in the bulk rather than on the surface and hence the gaseous ammonia
molecules are capable of penetrating into the interior of the structure of HPW, but not the
anion itself. The PAS FTIR spectrum of the ammonium salt of HPW, as prepared by
precipitation from aqueous solutions of the parent acid and of ammonium carbonate, while

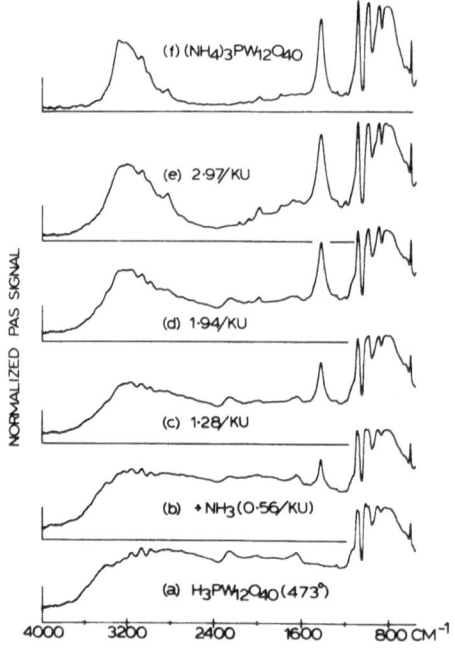

Figure 6. PAS spectra of (a) 12-tungstophosphoric acid pre-evacuated at 473 K, (b)-(e)
after stepwise dosing with ammonia at 423 K. [The ammonium salt (f) is included for
comparison.]

not quantitatively identical to that obtained for the parent acid after sorption of three
molecules of gaseous amounts per anion, is nevertheless sufficiently similar to provide
additional support for the contention that gaseous ammonia is capable of penetrating into
the bulk structure. Similar conclusions may be drawn from the PAS FTIR spectra

obtained after the sorption of gaseous pyridine (28), and methanol (29) on HPW. No bands characteristic of the presence of Lewis acid sites were found after sorption of any of the three sorbates.

With methanol sorption at room temperature the bands produced show that methanol is protonated at the oxygen atom, a process which is the precursor to the conversion of methanol to hydrocarbons on the MOCC (19, 29, 30). These observations with methanol demonstrate that, not only is HPW a Brönsted acid, but in addition the acidic strength is high, at least for a significant number of the protons which are present both on the surface and in the bulk. Adsorption measurements show that 3 molecules of methanol per anion are sorbed at room temperature, implying that methanol, as with ammonia and pyridine, penetrates into the bulk structure of the solid.

6. Acidic Properties

6.1 THEORETICAL STUDIES

A preliminary assessment of the effect of the composition of the anion on the acidity may

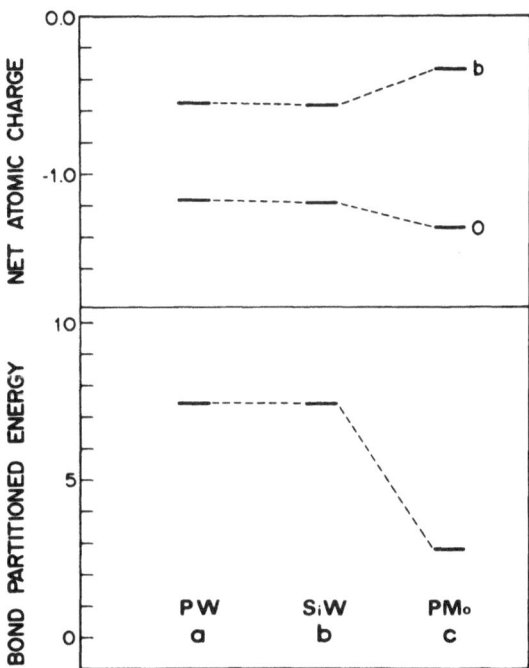

Figure 7. Magnitude of negative charges on bridging (b) and outer (o) oxygen atoms; binding energies between terminal oxygen atoms and peripheral metal atoms in the anions shown in abbreviated notation.

226

be obtained from an examination of the results of the semiempirical extended Hückel calculations on $PW_{12}O_{40}^{-3}$, $PMo_{12}O_{40}^{-3}$ and $SiW_{12}O_{40}^{-4}$ (Figure 7) (31-33). The magnitude of the negative charges on the terminal oxygen atoms in the anions containing molybdenum are higher than those in which tungsten is the peripheral metal element while the partitioned energies of the peripheral metal-terminal oxygen bonds in the former anions are smaller than those in the latter. Thus the protons in the compounds containing tungsten are expected to be more mobile and hence more acidic than those in the corresponding materials containing molybdenum, while in contrast the terminal oxygen atoms in the molybdenum-containing metal-oxygen cluster anions should be more labile than in those where tungsten is the peripheral metal element. On the basis of the results from these calculations it is expected that the tungsten-containing MOCC will be more active in reactions requiring relatively strong Brönsted acidity while those containing molybdenum may be more suitable as catalysts in oxidation processes.

6.2   MICROPOROUS SALTS

Although the acidic forms of the MOCC have been shown to be nonporous and have low surface areas, certain of the salts of monovalent cations have been found to have relatively high surface areas and possess pore structures in the microporous-mesoporous

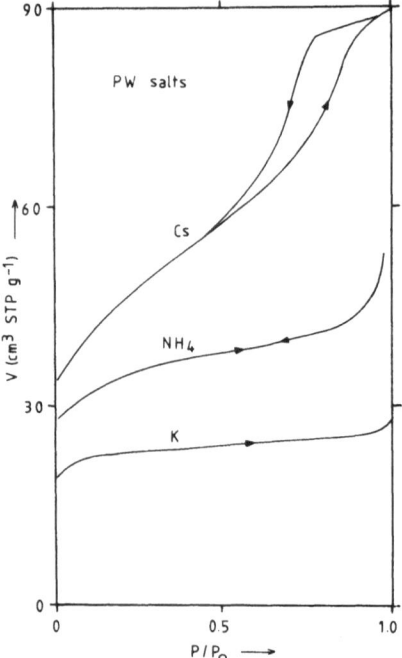

Figure 8. Nitrogen adsorption-desorption isotherms at 78 K for high surface area cesium, ammonium, and potassium salts of 12-tungstophosphoric acid.

size range (34-44) as revealed from the analyses of nitrogen adsorption-desorption isotherms (Figure 8). Although the nature of the cation is of primary importance the effect of the anion on the morphological properties cannot be dismissed (Figure 9).

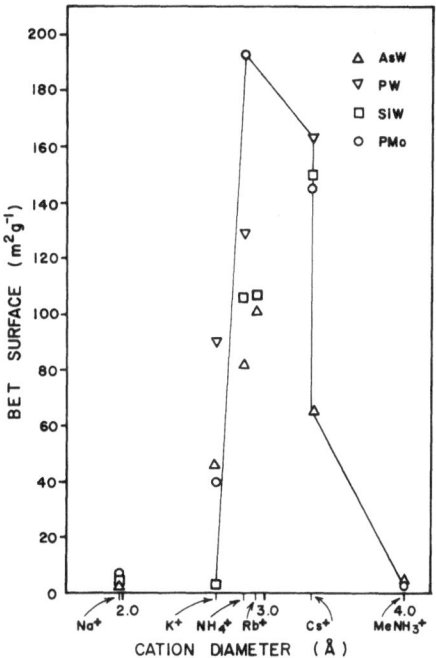

Figure 9. Surface area (BET) for salts of 12-tungstophosphoric, 12-molybdophosphoric and 12-tungstosilicic acids, with monovalent cations, as a function of cation diameter.

Earlier work in which the ammonium salt of HPW was employed as a catalyst in the conversion of methanol to hydrocarbons (45-46) provides strong evidence for the acidic characteristics of this solid. Subsequent studies of the sorption of ammonia and pyridine on NH_4PW with photoacoustic FTIR spectroscopy showed that residual protons were contained on and in the solid, although the solid had been prepared from stoichiometric quantities of the preparative reagents (27-28).

6.3   TEMPERATURE-PROGRAMMED DESORPTION, EXCHANGE AND REDUCTION

Temperature-programmed desorption (TPD) and temperature-programmed exchange (TPE) have been employed to obtain information on the effect of composition on the acidic properties of the metal-oxygen cluster anions. Temperature-programmed desorption studies of three of the acidic metal-oxygen cluster compounds show the presence of two major desorption peaks each due to the desorption of water (32, 47) (Figure 10). The peaks labelled 1 which appear at temperatures from 100 to 200°C are associated with low desorption energies and are attributed to hydrogen-bonded water in the solids. With HPMo this peak appears at approximately 100° while with HPW and HSiW the corresponding peak is seen at approximately 200°C. These results suggest that the hydrogen-bonded water in the tungsten-containing solids is more strongly bound than that in the MOCC anion containing molybdenum, presumably a reflection of the higher Brönsted acidic strengths in the former materials as predicted from the results of EXH calculations. The peaks at higher temperatures apparently result from the extraction of oxygen atoms from the anions by protons and thus their relative positions are related to the labilities of the oxygen atoms of these anions.

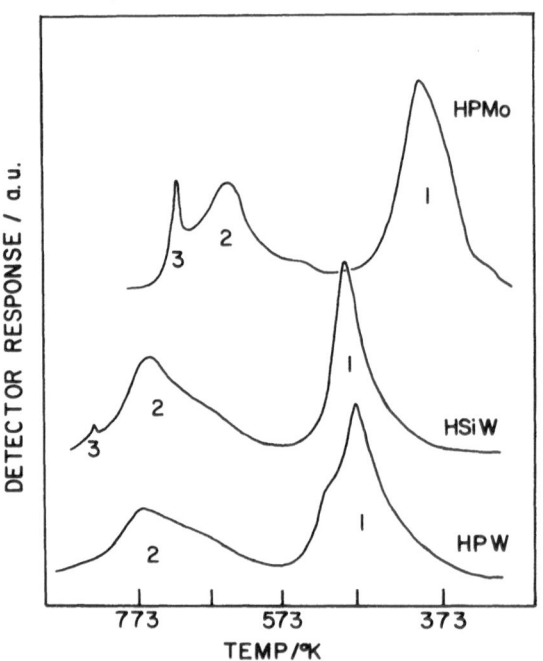

Figure 10. Temperature-programmed desorption profiles for 12-tungstophosphoric acid, 12-tungstosilicic acid, and 12-molybdophosphoric acid after pretreatment at 25°C for 16 h.

Temperature-programmed reduction (TPR) and exchange experiments have also provided information on the effect of composition on the acidic properties of the MOCC (48). The peaks observed are similar in overall shape and position to those found in the analogous TPD experiments (Figure 10). However, although D_2 exchange occurs with both HPW and HSiW no exchange was observed with HPMo (not shown). This appears to be consistent with the reduced acidic strength of the latter material in comparison with those containing tungsten.

6.4   LIQUID PHASE TITRATIONS

The Hammett indicator method as originally developed by Benesi (49) has been applied to various of the MOCC compounds (50). The effect of pretreatment temperature on the distribution of Brönsted acidic strengths can be seen in Figure 11. The most significant effect of pretreatment temperature can be observed with the most strongly acidic sites ($H_0 \le -5.6$) which increase in number with pretreatment to approximately 673K, and then decrease precipitously with further increase in temperature. The numbers of acidic sites over the entire range of strengths follow an approximately similar pattern with a decrease in number, although less abruptly than the aforementioned, at approximately 673K. It is interesting to note that the onset of the second peak observed in the TPD experiments with HPW occurs at approximately this temperature. This TPD peak is attributed to water produced by the extraction of oxygen atoms by protons which process would, of course, decrease the concentration of acid sites consistent with the results from the Hammett indicator method.

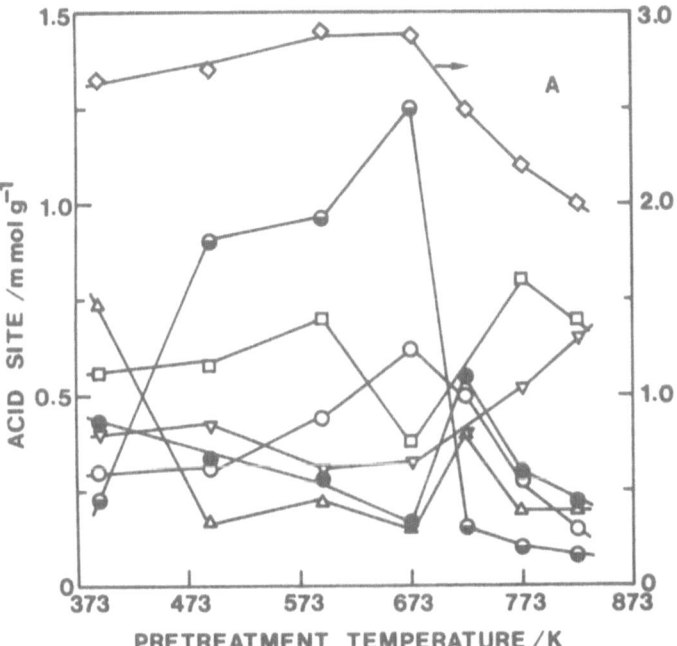

Figure 11. Effect of pretreatment temperatures on the acid strength distribution of 12-tungstophosphoric acid (A), (●) $H_0 \le -5.6$, (Δ) $-5.6 < H_0 < -3.0$, (●) $-3.0 < H_0 < 1.5$, (○) $1.5 < H_0 < 3.3$, (∇) $3.3 < H_0 < 5.0$, (□) $5.0 < H_0 < 6.8$, (◊) $H_0 \le 6.8$.

6.5   MICROCALORIMETRY

Microcalorimetry can also be advantageously employed for the study of the acidic properties of the MOCC (51). Measurements of the differential heats of the sorption of ammonia on HPW at 323 K show significant differences between the results obtained after pretreatment at 423 and 523 K (Figure 12). The differential heat, after evacuation at 423 K, for less than 0.2 molecules of NH_3/KU sorbed, is approximately 180-190 kJ mol^{-1}. For larger amounts of NH_3 taken up the differential heat remains approximately constant up to 1.25 NH_3/KU at which the value drops to approximately 75 kJ mol^{-1}. In contrast, after pretreatment at 523 K the initial values for the differential heat are approximately 150 kJ mol^{-1}. With increases in the quantity of NH_3 sorbed the differential heat decreases slightly but then remains virtually constant up to approximately 3 NH_3/KU at which point Q_{diff} decreases abruptly to approximately 75 kJ mol^{-1}.

It is interesting to note that other workers have obtained a relationship between the acid strength (H_0) and the differential heat of adsorption of ammonia on silica-alumina (52, 53). They reported that differential heats of 137.0 and 76.1 kJ mol^{-1} correspond to H_0 values of -14.5 and -5.6, respectively. Since a superacid is conventionally defined as any acid stronger than 100% H_2SO_4 ($H_0 < -12$) (54) it appears that HPW can be, at least on the basis of the microcalorimetric results, classed as a superacid.

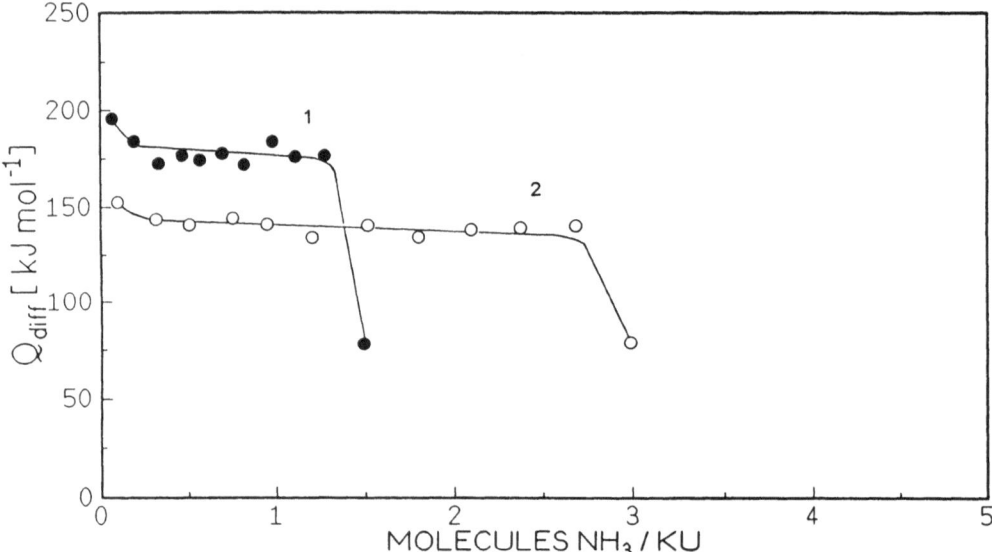

Figure 12. Differential heats of ammonia sorption on HSiW at 323 K. (1) Activation: 423 K/2 h/UHV; (2) Activation: 523 K/1 h/UHV.

The aforementioned PAS FTIR (26, 28) and TPD studies (47, 48) together with the microcalorimetric results suggest that the presence of water, presumably in the form of $H^+(H_2O)_n$ ($n \leq 4$) effectively guards the proton from interaction with sorbing species and therefore accounts for the precipitous drop in Q_{diff} at a quantity of NH_3 sorbed less than stoichiometric with the HPW pretreated at 423 K. The precipitous drop in Q_{diff} at 3 NH_3/KU with the HPW pretreated at 523 K supports the contention, derived from the PAS studies (26), that NH_3 is capable of penetrating into the bulk structure to interact with both surface and bulk protons. The higher adsorption energies found with the HPW pretreated at 423 K can be attributed to an inductive effect of the residual water on the acidic strengths of the accessible protons and/or quantitative alteration in the crystallographic structure, evidence of which is found from the decrease in the lattice parameter as the ammonium salt forms (26, 51).

Semiquantitatively similar results have been obtained in measurements of the differential heat of ammonia adsorption on 12-tungstosilicic acid (HSiW) and 12-molybdophosphoric acid (HPMo) although the value of Q_{diff} observed at the plateau for HPMo is significantly smaller than that observed for either HPW or HSiW, consistent with the expectations from both theoretical (31-33) and experimental (19, 45, 46) studies.

6.6   PYRIDINE DESORPTION

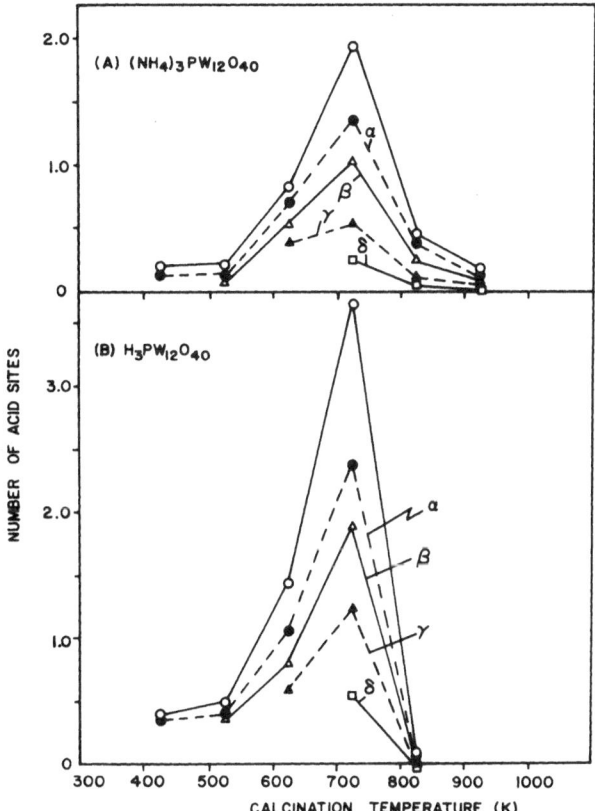

Figure 13. Dependence of acid site strength distribution of $(NH_4)_3PW_{12}O_{40}$ (A) and $H_3PW_{12}O_{40}$ (B) on calcination temperature. $\bigcirc$, $\bullet$, $\triangle$, $\blacktriangle$ and $\square$ represent number of acid sites which can adsorb pyridine irreversibly at 323, 423, 523, 623, and 723 K, respectively.

Pyridine desorption experiments can also provide valuable information on the effect of structure and composition on the acidic properties of the MOCC (55). Samples of HPW and NH_4PW pretreated at various temperatures are contacted with pyridine vapour at 298 K and 20 mbar until no further sorption occurs. The temperature of the sample is then increased in discrete steps from 298 K to 323, 423, 523, 623 and 723 K with sufficient time after each temperature to reach a steady state with respect to the pyridine desorption.

The number of acid sites can be equated, at least approximately, to the number of pyridine molecules irreversibly sorbed on and in the MOCC while the acid strength increases with the temperature at which the pyridine molecules are irreversibly sorbed. For purposes of illustration the number of pyridine molecules remaining sorbed at the various temperatures and hence the number of acidic sites are shown in Figure 13 for NH_4PW and HPW. With both of these MOCC the numbers of acid sites within each group increase exponentially with pretreatment temperature (PT) up to approximately 723 K and decrease with further increase in PT. For a PT of 823 K the numbers of acidic sites on HPW have diminished to a small fraction of the values at 423 K while those on NH_4PW remain appreciable for PT up to 923 K. This provides an explanation, however partial, of the advantageous properties of NH_4PW in acid-catalyzed processes as compared with the parent acid, HPW.

5.7 CATALYZED REACTIONS

Reactions requiring catalysts possessing acidic sites can often be employed as a means of gauging the concentration and strength of acidic sites on catalytic surfaces. As noted earlier the conversion of methanol to hydrocarbons requires the presence of Brönsted acid sites of relatively high acidity and thus can be employed for the aforementioned purpose.

The isomerization of alkenes is also an effective process for the evaluation of acidity. The isomerization of C_6-C_8 alkenes (1-hexene, 1-heptene, 1-octene) has been investigated in the liquid phase on HPW, HPMo and HSiW (56) and their microporous ammonium salts at temperatures of 303 and 343 K. Double bond and cis-trans isomerization of the olefins are strongly catalyzed by HPW, NH_4PW and HSiW and weakly by HPMo, NH_4SiW and NH_4PMo at near ambient temperatures. No skeletal isomerization of any of the olefins was observed at temperatures up to 343 K. The aforementioned results appear to be consistent, in general, with the results obtained from the various techniques discussed earlier in this report.

The catalytic alkylation of the aromatic hydrocarbons, particularly toluene, is generally recognized as acid-catalyzed (57) although basic sites may also participate in the process (58). The alkylation of toluene with methanol has been studied on two microporous MOCC, NH_4PW and NH_4SiW (59). Xylenes and trimethyl- and tetramethylbenzenes were found to be the principal products. The alkylation process appears to involve the methylation of the anions of the catalyst, similar to that which was observed in the conversion of methanol to hydrocarbons (29, 30).

The substituent-group rearrangements of alkylbenzenes is a useful process for characterizing the properties of acidic catalysts (60). Methylethylbenzene (MEB) is capable of dealkylation at several positions, ring-attachment isomerization and transalkylation. The dissimilar results obtained with MEB on NH_4PW and NH_4SiW have been attributed to differences both in the acidic natures and in the porous structures of the two catalysts (61).

The cracking of n-hexane has been frequently employed as a model reaction for the assessment of catalytic activity, and hence acidity, of a variety of heterogeneous catalysts but particulary zeolites (62). The conversion of n-hexane on NH_4PW has been investigated at 648, 673 and 698 K (63). Conversions up to 18% were obtained at 698 K while comparative values as high as 50% were obtained under similar conditions with ZSM-5.

7. Conclusions

Since the anion and crystallographic structures of the metal-oxygen cluster compounds remain essentially fixed for changes in the composition of the anion it is possible to examine the effect of anionic compositional variations while the structures remain invariant. It is evident that changes in the central atom of the anion have significantly less effect on the properties of the metal-oxygen cluster compounds than those involving the peripheral metal elements. Both theoretical and experimental studies show that the metal-oxygen cluster compounds containing tungsten have Brönsted sites of higher acidic strength than those containing molybdenum.

The effect of replacement of the protons by larger cations is less readily interpreted. Evidently, for monovalent cations the nature of the crystallographic structure remains fixed while the lattice parameters shift. The shift in the distribution of acid strengths which is observed directly and indirectly for the residual protons may be a consequence of both the production of micropores and inductive effects of the larger cations.

The changes in the apparent acidic properties of the heteropoly acids with the number of water molecules per anion may be attributed, at least in part, to the alterations in the nature of the crystallographic structures. However it appears equally likely that the water molecules restrict the approach to the proton of reactant species. The inductive effect of the hydrated protons can also not be discounted.

8. Acknowledgement

The financial support of the Natural Sciences and Engineering Research Council of Canada and of Imperial Oil Canada Limited is gratefully acknowledged.

9. References

1.   Pope, M.T. and Müller, A., Angew Chem. Intl. Ed. Eng. 30, 34 (1991).

234

2. Pope, M.T., Heteropoly and Isopoly Oxometalates, Springer-Verlag, Berlin, 1983.
3. Brown, C.M., Noe-Spirlet, M.R., Busing, W.R. and Levy, H.A., Acta. Cryst., B33, 1038 (1977).
4. Kearley, G.J., Pressman, H.A. and Slade, R.C.T., J. Chem. Soc., Chem. Commun. 1801 (1986).
5. Slade, R.C.T., Thompson, I.M., Ward, R.C., and Poinsignon, C., J. Chem. Soc., Chem. Commun. 726 (1987).
6. Pressman, H.A. and Slade, R.C.T., Chem. Phys. Letters 151, 354 (1988).
7. Nakamura, O., Kodama, T., Ogino, I., and Mikaya, Y., Chem. Letters 1, 321 (1980).
8. Hardwick, A., Dickens, P.G. and Slade, R.C.T., Solid State Ionics 13, 345 (1984).
9. D'Amour, H. and Allmann, R., Z. Krist. 143, 1 (1976).
10. Fuchs, J., Thiele, A. and Palm, R., Z. Naturforsch. 36b, 161 (1981).
11. Ichida, H., Kobayashi, A. and Sasaki, Y., Acta Cryst. B36, 1382 (1980).
12. Weakley, T.J.R., in Structure and Bonding (J.D. Dunitz, Ed.), Volume 18, p. 131, Springer-Verlag, New York, 1974.
13. Wyckoff, R.W.G., Crystal Structures, Vol. 3, p. 887, Interscience, New York, 1951.
14. McMonagle, J.B. and Moffat, J.B., J. Catal. 91, 132 (1985).
15. West, S.F. and Audrieh, L.F., J. Phys. Chem. 59, 1069 (1955).
16. Rashkin, J.A., Pierron, E.D. and Parker, D.L., ibid. 71, 1265 (1967).
17. Chumachenko, N.N., Yurchenko, E.N., et al., Kinet. Catal. 25 (3), 653 (1984).
18. Tsigdinos, G.A., Ind. Eng. Chem. Prod. R&D., 13 (4), 267 (1974).
19. Hayashi, H. and Moffat, J.B., J. Catal. 77, 473 (1982).
20. Zhang, L., M.Sc. Thesis, University of Waterloo, Waterloo, 1986.
21. Grutter, B. and Jander, G., in Handbook of Preparative Inorganic Chemistry (G. Brauer, Ed.), 2nd Ed., Vol 2, p. 1721, Academic Press, New York, 1965.
22. Tsigdinos, G.A., Topics Current Chem. 76, 1 (1978).
23. Bradley, A.J. and Illingsworth, J.W., Proc. R. Soc. (London) A157, 113 (1936).
24. Brown, D.H., J. Chem. Soc. 3189 (1962).
25. Hayashi, H. and Moffat, J.B., Talanta 29, 943 (1982).
26. Highfield, J.G. and Moffat, J.B., J. Catal. 88, 177 (1984).
27. McGarvey, G.B. and Moffat, J.B., J. Mol. Catal. 69, 137 (1991).
28. Highfield, J.G. and Moffat, J.B., J. Catal. 89, 185 (1984).
29. Highfield, J.G. and Moffat, J.B., J. Catal. 95, 108 (1985).
30. Highfield, J.G. and Moffat, J.B., J. Catal. 98, 245 (1986).
31. Moffat, J.B., J. Mol. Catal. 26, 385 (1984).
32. Highfield, J.G., Hodnett, B.K., McMonagle, J.B. and Moffat, J.B., Proceedings of the 8th International Congress on Catalysis, p. 611, Dechema, Frankfurt am Main, 1984.
33. Moffat, J.B., Proceedings of the 9th Iberoamerican Symposium on Catalysis, p. 349, Lisbon, 1984.
34. McMonagle, J.B. and Moffat, J.B., J. Coll. Interface Science 101, 479 (1984).

35. Moffat, J.B., Polyhedron 5, 261 (1986).
36. Moffat, J.B., Preparation of Catalysts IV, Studies in Surface Science and Catalysis (B. Delmon, P. Grange, P.A. Jacobs and G. Poncelet, Eds.), Vol. 30, Elsevier, Amsterdam, 1987.
37. Surface Science and Catalysis (J. Ward, Ed.), Vol. 38, Elsevier, Amsterdam, 1988.
38. Moffat, J.B., McMonagle, J.B. and Taylor, D., Solid State Ionics 26, 101 (1988).
39. Taylor, D.B., McMonagle, J.B. and Moffat, J.B., J. Coll. Interface Science 108, 278 (1985).
40. McGarvey, G.B. and Moffat, J.B., J. Coll. Interface Science 125, 51 (1988).
41. McMonagel, J.B., Nayak, V.S., Taylor, D. and Moffat, J.B., Proceedings of the 9th International Congress on Catalysis (M.J. Phillips and M. Ternan, Eds.), Calgary, Chemical Institute of Canada, 1988.
42. Moffat, J.B., J. Mol. Catal. 52, 169 (1989).
43. Moffat, J.B., McGarvey, G.B., McMonagel, J.B., Nayak, V. and Nishi, H., NATO ASI Series Plenum Press (D. Barthoment, E.G. Derouane and W. Hölderich, Eds.), N.Y., 1990.
44. Lapham, D. and Moffat, J.B., Langmuir 7, 2273 (1991).
45. Hayashi, H. and Moffat, J.B., J. Catalysis 83, 1982 (1983).
46. Moffat, J.B., Methane Conversion, A Symposium on the Production of Fuels and Chemicals from Natural Gas, Auckland, April 1987, Elsevier, Amsterdam, 1988.
47. Hodnett, B.K. and Moffat, J.B., J. Catalysis 88, 253 (1984).
48. Hodnett, B.K. and Moffat, J.B., J. Catalysis 91, 93 (1985).
49. Benesi, H.A., J. Amer. Chem. Soc. 61, 970 (1957).
50. Ghosh, A.K. and Moffat, J.B., J. Catal. 101, 238 (1986).
51. Jozefowicz, L.C., Karge, H.G., Vasilyeva, E. and Moffat, J.B., Microporous Materials, in press.
52. Taniguchi, H., Masuda, T., Tsutsumi, K. and Takahashi, B., Bull. Chem. Soc. Japan 51, 1970 (1978).
53. Taniguchi, H., Masuda, T., Tsutsumi, K. and Takahashi, B., Bull. Chem. Soc. Japan 53, 2463 (1980).
54. (a) Gillespie, R.J. and Peel, T.E., Adv. Phys. Org. Chem. 9, 1 (1971); (b) Gillespie, R.J. and Peel, T.E., J. Amer. Chem. Soc. 95, 5173 (1973); (c) Gillespie, R.J., Peel, T.E. and Robinson, E., J. Amer. Chem. Soc. 93, 5083 (1971).
55. Nayak, V.S. and Moffat, J.B., J. Mol. Catal. 80, 75 (1993).
56. Nayak, V.S. and Moffat, J.B., Appl. Catal. 36, 127 (1988).
57. Venuto, P.B. and Landis, P.S., Adv. Catal. 18, 259 (1968).
58. Itoh, H., Hattori, T., Suzuki, K. and Murakami, Y., J. Catal. 79, 21 (1983).
59. Nishi, H., Nowinska, K. and Moffat, J.B., J. Catal. 116, 480 (1989).
60. Csicsery, S., J. Catal. 110, 348 (1988).
61. Nishi, H. and Moffat, J.B., J. Mol. Catal. 51, 193 (1989).
62. Chu, C.T.W., Kuehl, G.H., Lago, R.M., and Chang, C.D., J. Catal. 90, 451 (1984).
63. Nayak, V.S., and Moffat, J.B., Appl. Catal. 47, 97 (1989).

THE ACIDITY AND BASICITY OF SOLIDS: INTRINSIC PROPERTIES OF THE SURFACE, ITS STRUCTURE AND COMPOSITION

J.B. MOFFAT
Department of Chemistry and Guelph-Waterloo Centre
for Graduate Work in Chemistry
University of Waterloo
Waterloo, Ontario N2L 3G1
Canada

ABSTRACT. The acidity and basicity of solids, as a consequence of the methods commonly employed in the measurement of these properties, are properties of the surface and its structure and composition. While surface free energy is important in characterizing a surface only recently has there been an interest in measurements of this property as a means of quantifying acid-base properties. The relevant theory is reviewed, the experimental techniques are described and recent results are summarized.

1.   Introduction

Acidity and basicity, as usually discussed in textbooks of chemistry, implicitly refers to the multicomponent liquid state, that is, to solutions. However, the qualitative definitions of Brönsted and Lewis acids as proton donors or electron acceptors, respectively, and Brönsted and Lewis bases as proton acceptors or electron donors, respectively, contain no explicit reference to the medium in which the species are found nor to its possibly heterogeneous or anisotropic nature.

Acids and bases, again on a qualitative or at best semiquantitative basis, are referred to as strong or weak where the dissociation is complete or incomplete, respectively. For more quantitative ranking of acids and bases, values of K_a or K_b, the dissociation constants of acids or bases, respectively, are frequently provided where, for a Brönsted acid (HX)

$$K_a = \frac{a_{H^+} \, a_{X^-}}{a_{HX}} \tag{1}$$

where a refers to the activity. Although not always stated, such tabulated values usually refer to liquid phase solutions with water as the solvent and, of course, assume that equilibrium exists between the undissociated acid or base and the dissociation products

237

J. Fraissard and L. Petrakis (eds.), Acidity and Basicity of Solids, 237–254.
© 1994 *Kluwer Academic Publishers.*

so that, for example,

$$\mu_{HX} = \mu_{H^+} + \mu_{X^-} \tag{2}$$

where μ_i refers to the Gibbs chemical potential of species i. Data for K_a and K_b, whether for Brönsted acids and bases or Lewis acids and bases refer to monophasic systems with properties which are homogeneous and hence isotropic. The chemical potential and hence the activity and activity coefficient are functions of a number of variables including temperature, pressure and the composition of the system. However K_a and K_b values are normally reported for the system containing only the undissociated acid and its dissociation products.

Where solids are the systems of interest a number of additional factors must be considered. The acidity and basicity of a solid are most frequently taken to be those of the surface of the solid since many of the measurements for such properties apply to the surface as opposed to the bulk structure. However it must be kept in mind that the surface properties of a solid are, in general, not identical to those of the bulk structure. The present discussion will primarily focus on the acidic and basic properties of the surface since these are observed when measurements of these properties are applied to solids.

For any liquid or solid phase work must be done in order to transfer species from the bulk to the surface. Thus the Gibbs free energy of the surface will be greater than that of the bulk. Further, species on the surface and their binding strengths will be anisotropically dependent upon nearest and next-nearest neighbours in the surface. Thus, in contrast with the observations for an aqueous solution, the acidic and basic properties of a surface will generally display a distribution of values.

For any system containing an interface the work associated with the formation of the surface may be written as

$$w = -\sigma dA \tag{3}$$

where A is the area of the interface and the surface free energy is

$$\sigma = \left(\frac{\partial G}{\partial A}\right)_{T, P, \text{composition}} \tag{4}$$

However as a result of the nature of the surface of a solid this must be modified as

$$\sigma_S = \sigma + A \frac{\partial \sigma}{\partial A} \tag{5}$$

It is obvious that the surface free energy will be dependent, in part, on the acid-base properties of the surface, and vice versa. Methods for measurement of the surface free energy of a solid and for partitioning of this energy into its components are consequently of considerable importance in the determination of the acid-base properties of the surface of a solid.

2. Methods for the Assessment of Acidity and Basicity

A number of techniques have been employed for the measurement of the acidic and/or basic properties of surfaces of solids, particularly where these solids possess catalytic properties (1-61) ranging from spectroscopic studies of adsorbed species (8-21), thermal desorption and temperature-programmed desorption (22-28), liquid phase titrations (29-33), pH at zero net surface charge (34), contact angle measurements (35-42), microcalorimetry (43-53), X-ray photoelectron spectroscopy (54) inverse gas chromatography (55-56) to theoretical methods (57-61). Several excellent reviews have appeared during the last two decades (1-7). These methods almost inevitably involve the adsorption of molecules with basic or acidic properties, respectively. These latter molecules may be adsorbed from the gas or liquid phase and the detection methods vary from indirect to direct measurements of the quantities adsorbed and/or to examinations of the surface after adsorption.

2.1 SPECTROSCOPIC METHODS

Spectroscopic detection of acidity has most frequently employed electromagnetic radiation in the infrared region although some applications of the UV-visible region have been reported (21, 31). Ordinarily, the probe molecules, either acidic or basic, have been adsorbed from the gas phase. Ammonia and pyridine are the most commonly employed of the former while carbon dioxide is probably the most common of the latter. With ammonia and pyridine adsorbed on the surface, bands in the infrared region due to ammonium and pyridinium ions reflect the presence of Brönsted acidic sites while those attributed to coordinated ammonia and pyridine provide evidence for Lewis acidity. The relative numbers of the two types of acidic sites may be estimated from the intensities of the appropriate bands. The distribution of acidic sites is difficult, if not impossible, to obtain from such experiments.

Table 1. Characteristic frequencies (cm^{-1}) for ammonia and pyridine adsorbed on acid sites (62, 63).

Base	Acid Site	
	Brönsted	Lewis
NH_3	1420	1620, 1300-1200
Pyridine	1545	1450

2.2 TITRATION FROM THE LIQUID PHASE

Another frequently employed method for the evaluation of surface acidity utilizes the titration with a weak base such as butylamine dissolved in an inert solvent (5). By the use of a series of basic Hammett indicators of a range of pKa values the distribution of acidic strengths may be estimated. It should be noted that in the usual implementation of this method Brönsted and Lewis acid sites cannot be distinguished.

The Hammett indicator method is based on the definition of acid strength as the ability to convert a base into its conjugate acid (1, 7). Thus

$$H^+ + B \rightleftharpoons BH^+ \tag{6}$$

and

$$K_a = \frac{a_{BH^+}}{a_{H^+} \cdot a_B} \tag{7}$$

and

$$K_{BH^+} = \frac{a_{H^+} \cdot a_B}{a_{BH^+}} \tag{8}$$

or

$$-\ln K_{BH^+} = \ln \left(\frac{f_{BH^+}}{a_{H^+} \cdot f_B} \right) + \ln \left(\frac{[BH^+]}{[B]} \right) \tag{9}$$

or

$$H_0 \equiv -\ln \left(\frac{a_{H^+} \cdot f_B}{f_{BH^+}} \right) = pK_{BH^+} - \ln \left(\frac{[BH^+]}{[B]} \right) \tag{10}$$

The method can be employed quantitatively or semiquantitatively. If a given Hammett indicator when adsorbed on a solid surface displays the colour of its acidic form then the value of the function H_0 of the surface is equal to or less than the pKa of the indicator. The usual set of basic Hammett indicators has pKa values ranging from -8.2 to +6.8 for anthraquinone and neutral red, respectively, the former corresponding to the greater acid strength. Thus a solid which produces the acidic colour (red) with neutral red but remains colourless (the basic colour) with anthraquinone has acid strengths falling between a value of +6.8 and -8.2 for H_0.

The Hammett indicator method may be employed in a more quantitative manner in order to determine the distribution of acid strengths. As noted earlier this involves the titration of the surface acidic sites with a weak base such as butylamine dissolved in a dry solvent such as benzene (1-7). The endpoint of the titration can be assessed by use of, for example, Hammett indicators. Thus, with neutral red as the indicator the titer which is observed when the surface of the solid changes from red to yellow corresponds to the number of acid sites with $H_0 \leq +6.8$. By a repetition of the titration with fresh aliquots of the solid and indicators of various pKa the distribution of acid sites may be estimated.

The Hammett indicator method is frequently employed for both Brönsted and Lewis acidic centres. However as is evident from the aforementioned relationships, while results can frequently be obtained with Lewis acids the validity of the comparison with those for Brönsted centres is uncertain.

The Hammett indicator method has been widely employed for the measurement of the concentrations and strengths of acidic sites. While the technique is applicable, in principle, to the determination of the corresponding properties of basic surfaces, relatively little work has been done with this method (1, 4, 6).

The problems associated with the determination of acidity by the Hammett indicator method have been studied in some detail (29-33). The adsorption of strong bases, including Hammett indicators, is a relatively slow process, particularly at room temperature. Apparently such basic molecules adsorb very strongly and thus do not readily spread over the surface of the solid (32). Further, the values obtained for the surface concentrations of acidic centres have been found to depend on the solvent employed. Decomposition of the base on the catalytic sites may also occur. Saturation of the catalyst with the base followed by its stepwise desorption before the indicator tests has been found to be advantageous (33). However there is now evidence available which suggests that the adsorption of a basic molecule on an acidic centre may induce the perturbation of neighbouring acidic sites thus altering the distribution of the strengths of the remaining acid centres. Hall and coworkers have concluded that the observation of colour changes can be misleading and have devised a spectrophotometric method for application in the measurements based on adsorbed indicators (31). The values reported by Hall and coworkers by the latter method for a number of zeolites and comparative oxides are shown in Table 2.

Table 2. H_0 values for zeolites and oxides (31)

| Catalyst | H_0 Range | |
	Most Acidic	Least Acidic
H-mordenite	-13.7	-12.4
LZ-Y82	-11.3	-8.7
LZ-210(12)		
HY (8.1)		
Beta		
SiO_2-Al_2O_3(M46)	-8.7	-3.3
SiO_2-Al_2O_3(N631L)		
Silica Gel	-3.3	

2.3   pH AT ZERO NET SURFACE CHARGE

The acidity of a solid surface has sometimes been defined as the value of the pH (pH_0) at which the immersed solid has a zero net surface charge, that is, equal surface concentrations of positive and negative charges (34). It is well known that hydroxyl groups on a solid surface can act as acids or bases. This can be represented by two processes

$$MOH \rightleftharpoons MO^- + H^+ \tag{11}$$

$$MOH + H^+ \rightleftharpoons MOH_2^+ \tag{12}$$

The equilibrium between positive and negative charges of the surface is then

$$MO^- \text{ (surf)} + 2H^+ \rightleftharpoons MOH_2^+ \text{ (surf)} \tag{13}$$

for which the equilibrium constant is

$$K = \frac{[MOH_2^+]}{[MO^-][H^+]^2} \tag{14}$$

Since $[MOH_2^+] = [MO^-]$ at pH_0 then

$$pH_0 = \frac{1}{2} \ln K \tag{15}$$

The free energy, $-\Delta G$, is assumed to be primarily determined by Z, the ionic charge of the cation and

$$R = 2r_0 + r_+ \tag{16}$$

where r_0 and r_+ are the ionic radii of oxygen and the cation, respectively. Thus

$$pH_0 = A - B\left(\frac{Z}{R}\right) \tag{17}$$

where A and B are constants. Carre and coworkers have shown that the pH_0 correlates linearly with Z/R by examining a number of oxides (34). The surface acidity was shown to increase linearly with the ionization potential (IP) of the metal of the oxide. Good agreement was obtained for the pH_0 values obtained by these two methods for a number of oxides. In addition the values of pH obtained for the surfaces by application of the aforementioned indicator method correlated linearly with the pH_0 values obtained by the two methods based on Z/R and on IP.

2.4 SURFACE FREE ENERGY, WORK OF ADHESION AND CONTACT ANGLES

As noted earlier in this report the surface free energy is a fundamental surface property. Although relatively simple techniques are available for the measurement of this function for liquid-liquid and liquid-vapour interfaces, direct evaluations for systems containing solids present a considerable challenge.

Secondary functions such as the work of adhesion and the work of cohesion are useful quantities with interfacial systems regardless of the nature of the phases. The change in Gibbs free energy when two separate phases (1 and 2) of unit surface area are brought from infinity to form an interface is given by

$$\Delta G_a = \sigma_{12} - \sigma_1 - \sigma_2 \tag{18}$$

where σ_1 and σ_2 are the interfacial tensions of the 1 and 2 phases with respect to their vapours and σ_{12} is the interfacial tension of the interface formed between the phases 1 and 2. Hence the work of adhesion is

$$W_{12} = \sigma_1 + \sigma_2 - \sigma_{12} \tag{19}$$

It is clear that the surface free energy should reflect, at least in part, the acid-base properties of the surface. Fowkes has shown that the work of adhesion can be subdivided into a dispersive or van der Waals component and an acid-base or polar component (35-39).

$$W_{12} = W_{12}^d + W_{12}^{ab} \tag{20}$$

where W_{12}^d and W_{12}^{ab} refer to the dispersive and acid-base components, respectively, of the work of adhesion. The dispersive or van der Waals component of the work of adhesion is thus related to the geometric mean of the dispersive components of the surface free energies of the two phases.

$$W_{12} = 2 (\sigma_1^d \sigma_2^d)^{1/2} + W_{12}^{ab} \tag{21}$$

Since the surface and interfacial tensions for liquid-vapour and liquid-liquid interfaces may be readily measured the aforementioned equations may be conveniently tested. The geometric mean equation for the calculation of W_{12}^d, the van der Waals contribution to the work of adhesion between liquids 1 and 2 was first tested from an examination of alkane-water and alkane-mercury systems (39). Since the alkanes (a) have only van der Waals interactions but both water (w) and mercury (m) have additional interactions the following set of equations may be written for these systems

$$W_{aw}^d = 2(\sigma_a^d \sigma_w^d)^{1/2} = 2(\sigma_a \sigma_w^d)^{1/2} \tag{22}$$

$$W_{am}{}^d = 2(\sigma_a{}^d \, \sigma_m{}^d)^{\frac{1}{2}} = 2(\sigma_a \, \sigma_m{}^d)^{\frac{1}{2}} \tag{23}$$

and
$$\sigma_{wm} = \sigma_w + \sigma_m - 2(\sigma_m{}^d \, \sigma_w{}^d)^{\frac{1}{2}} \tag{24}$$

Any alkane may be employed to determine the dispersive component of water $\sigma_w{}^d$ which is found to be 22.0 mJ m^{-2} at 20°C while that for mercury $(\sigma_m{}^d)$ is 200 mJ m^{-2}. From the surface tension of water at 20°C (72.8 mJ m^{-2}) $\sigma_w{}^{ab} = 50.8$ mJ m^{-2} while for mercury with a surface tension of 484 mJ m^{-2} at 20°C the components are $\sigma_m{}^d = 200$ mJ m^{-2} and $\sigma_m{}^{ab} = 284$ mJ m^{-2}. Finally, with the data above, σ_{wm} is calculated as 424.1 mJ m^{-2} which agrees to within approximately 1 mJ m^{-2} with experimental results.

Although only one example is cited here as a test of the relevant relationships, Fowkes and coworkers have provided further examples in support of their original hypotheses (35). The equations may be conveniently extended to systems containing a solid phase by use of the Young-Dupré equation

$$\sigma_{SV} = \sigma_{SL} + \sigma_{LV} \cos\theta \tag{25}$$

where SV, SL and LV refer to the solid-vapour, solid-liquid and liquid-vapour interfaces, respectively and θ is the contact angle between the liquid (L) and the solid(s).

The work of adhesion between the solid and liquid phase may now be written as

$$W_{SL} = \sigma_S + \sigma_L - \sigma_{SL} \tag{26}$$

or
$$W_{SL} = \sigma_{LV} (1 + \cos\theta) \tag{27}$$

$$= 2 \, (\sigma_S{}^d \, \sigma_L{}^d)^{\frac{1}{2}} + W_{SL}{}^{ab} \tag{28}$$

Although the $W_{SL}{}^{ab}$ is frequently equated to the corresponding geometric mean equation

$$W_{SL}{}^{ab} = 2 \, (\sigma_S{}^{ab} \, \sigma_L{}^{ab})^{\frac{1}{2}} \tag{29}$$

by various workers Fowkes argues that this is not valid (35). In view of the stipulations, both explicit and implicit, in the original derivation of the geometric mean relationship, its application to acid-base or polar interactions appears questionable. Unfortunately, however in its present form equation (28) is not soluble.

Van Oss, Good and Chaudhury have proposed a method for circumventing the problem (40). For an apolar liquid $W_{SL}{}^{ab}$ is zero and equation (28) becomes

$$\sigma_{LV} (1 + \cos\theta) = 2 \, (\sigma_S{}^d \, \sigma_L{}^d)^{\frac{1}{2}} \tag{30}$$

or

$$\sigma_S^d = \frac{\sigma_{LV} \, (1 + \cos \theta)^2}{4} \qquad (31)$$

For a polar liquid for which σ_L^d is known (or can be evaluated as noted earlier), W_{SL}^{ab} can then be obtained. The acid-base contributions to the surface free energy may then be obtained from

$$W_{SL}^{ab} = 2 \, [(\sigma_S^+ \, \sigma_L^-)^{\frac{1}{2}} + (\sigma_S^- \, \sigma_L^+)^{\frac{1}{2}}] \qquad (32)$$

where σ_i^- and σ_i^+ are measures of the surface basicity and acidity, respectively, of the ith phase measured by application to two polar liquids.

Van Oss, Giese and Good applied this method to polyacetylene and iodinated polyacetylene with three liquids water, glycerol and formamide (40). Their values are summarized in Table 3. It is interesting to note the relatively high value of the electron-donor parameter σ_S^- obtained for iodinated polyacetylene as compared with that for polyacetylene.

Table 3. Components of the Surface Free Energy for Polyacetylene and Iodinated Polyacetylene from Contact Angles

	σ_S	σ_S^d	σ_S^{ab}	σ_S^+	σ_S^-
Polyacetylene	44.2	39.2	5.0	0.84	7.47
Iodinated Polyacetylene	52.6	42.8	9.8	1.04	23.15

Although this method has been applied to a number of solid surfaces relatively little work has been done on catalytic surfaces.

2.5   SURFACE FREE ENERGY AND INVERSE GAS CHROMATOGRAPHY (IGC)

Although it is possible to measure contact angles directly on nonporous solids and indirectly on porous materials, it is of value to have techniques to circumvent such measurements. The direct measurement of the adsorption of polar and nonpolar materials can provide information on the specific interactions of adsorbates with solid adsorbents. Although the quantities adsorbed can be measured volumetrically or gravimetrically these methods are tedious and time-consuming. Inverse gas chromatography (55, 56) provides a rapid means for studying the interaction between gases and solids with the additional advantage that the adsorbent may be pretreated in situ.

At infinite dilution the standard free energy of adsorption is related to the retention volume by

$$\Delta G^\circ = -RT\ln \frac{BV_N}{S_m} \tag{33}$$

where B is a constant dependent upon the selection of the standard states for the gas and adsorbed phase, V_N is the retention volume and S_m is the surface area.

As proposed by Fowkes (39), a nonpolar molecule such as an n-alkane may be selected to obtain the dispersive interactions while the corresponding n-alkene provides both the dispersive and specific interactions with the surface. Hence

$$\Delta G_i = \Delta G_i^d + \Delta G_i^{ab} \tag{34}$$

may be written as

$$\Delta G_{ane} = \Delta G_{ane}^d \tag{35}$$

$$\text{and } \Delta G_{ene} = \Delta G_{ene}^d + \Delta G_{ene}^{ab} \tag{36}$$

and the specific interaction term is

$$\Delta G_{ene}^{ab} = \Delta G_{ene} - \Delta G_{ane} \tag{37}$$

where the two terms on the right of the equation are both experimentally measurable quantities since they are directly related to the retention volume

$$\Delta G_{ene}^{ab} = RT\ln \frac{V_N^{hexane}}{V_N^{hexene}} \tag{38}$$

The technique has been applied to silica and silica impregnated with alumina (55) with the results shown in Table 4. The effect of pretreatment of the silica sample D at 875 K and 1075 K is readily evident in reduced ΔG^{ab} values. The surface silanol groups which interact with the alkene are converted to siloxane groups on heating at elevated temperatures. It is interesting to note that the Benesi titration method shows no difference in the D and D-875 samples. In this regard the IGC data show a difference between the pure silica (S-875) sample and the D samples which is also not observed with the Benesi method.

Table 4. Values of ΔG^{ab} for Silica and Silica/Alumina (55)

Sample*	ΔG^{ab}	σ^d
D	3.3	41
D-875	2.8	38
D-1075	1.3	31
1 A/D	5	67
10 A/D	8.2	53
S-875	0.9	87
1 A/S	1.0	91
10 A/S	1.0	90
A-875	2.0	89
A'	3.7	72

* D = Davison 952 grade silica
D-875 = D after pretreatment at 875 K
D-1075 = D after pretreatment at 1075 K
S-875 = High-purity silica prepared from tetraethyl silicate
1 A/D = Silica D impregnated with 1% alumina
A' = commercial alumina
A-875 = Pure alumina prepared from aluminum isopropoxide

2.6   MICROCALORIMETRY

2.6.1 Acidic and basic properties of oxides. The application of microcalorimetry to catalytic systems has increased markedly in recent years as noted in a recent review (43). Kijenski and Baiker (1) consider that temperature programmed desorption and calorimetric measurements appear to be the most promising methods for the determination of acidity with both methods capable of providing the distribution of the strength of the acidic sites.

A number of recent publications have provided useful comparisons of the acidic and basic properties of a variety of common oxides (8-12, 16-18, 23-24, 28, 43, 45-52, 54). Microcalorimetric studies of the adsorption of NH_3 and CO_2, for example, are capable, in principle, of permitting the determination of both number and strength of acidic and basic surface sites, respectively.

Auroux and Gervasini (48) have itemized the possible adsorption sites on oxides and adsorption mechanisms (Figure 1). According to Auroux and Gervasini (48) ammonia can be adsorbed on the surface of an oxide by (a) hydrogen bonding, (b) formation of the ammonium ion (Brönsted acid), (c) coordination to a Lewis site, (d) dissociative adsorption, while carbon dioxide can be adsorbed by (e) adsorption of an hydroxyl group, (f) dissociative adsorption, (g) adsorption on the metal ion, (h) adsorption on the oxygen vacancy, (i, j) adsorption on oxygen.

248

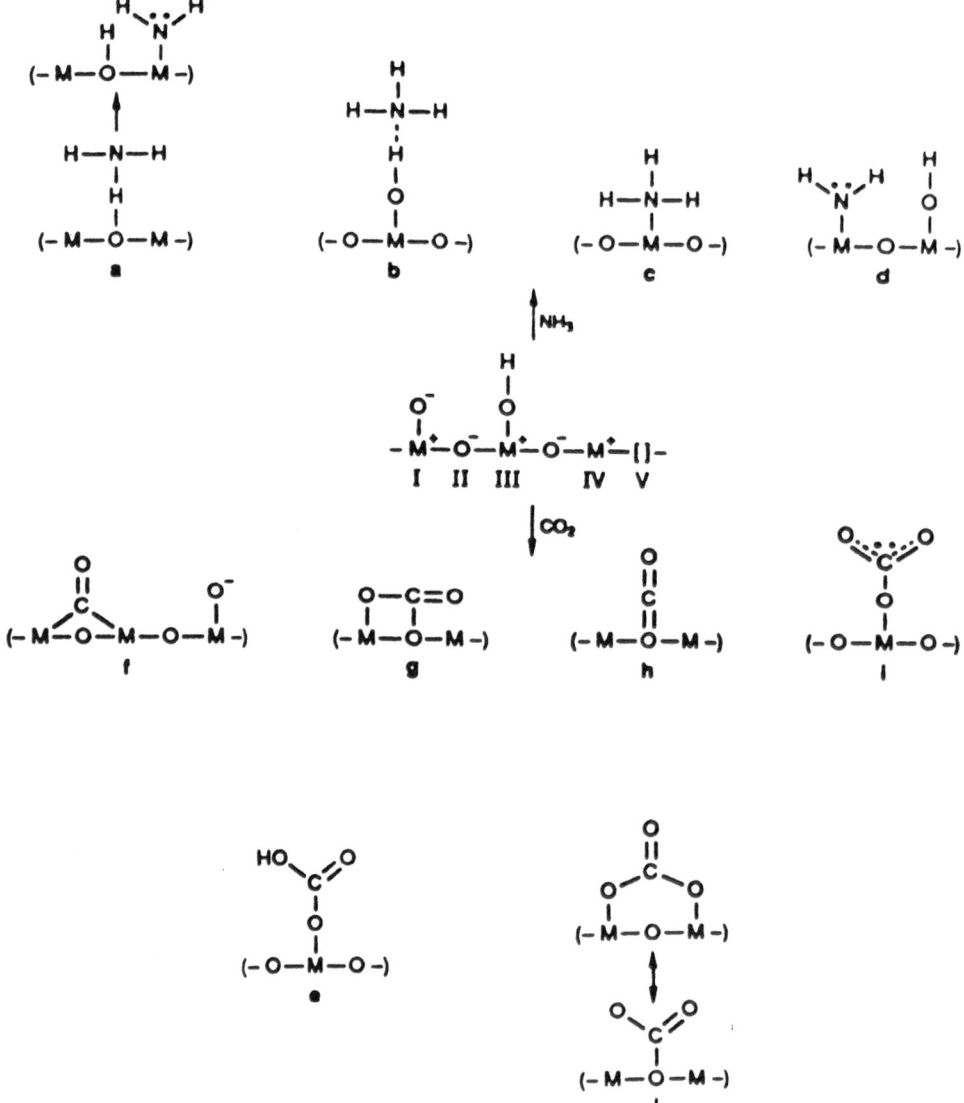

Figure 1. Adsorption sites on oxides and mechanisms for adsorption of NH_3 and CO_2 (48). Reprinted with the permission of the American Chemical Society.

Auroux and Gervasini (45-48) have measured heats of adsorption of NH_3 and CO_2 on acidic, basic and amphoteric oxides (Table 5). Particularly instructive are the representations of the data in the form of -dn/dQ versus Q where n and Q are the number of sites and differential heat of adsorption, respectively. Data for adsorption of NH_3 on the acidic oxides and the amphoteric oxides are shown as continuous and dotted lines (Figure 2). All oxides show a substantial number of acidic sites in the range of heat of adsorption from 10-20 kJ mol^{-1}. Interesting, relatively large populations of sites

Table 5. Classification of Oxides (45-48)

Acidic	Basic	Amphoteric
Cr_2O_3	ThO_2	BeO
WO_3	Pr_6O_{11}	Ga_2O_3
Nb_2O_5	Nd_2O_3	TiO_2
Ta_2O_5	MgO	Al_2O_3
V_2O_5/SiO_2	CaO	ZrO_2
MoO_3	La_2O_3	ZnO

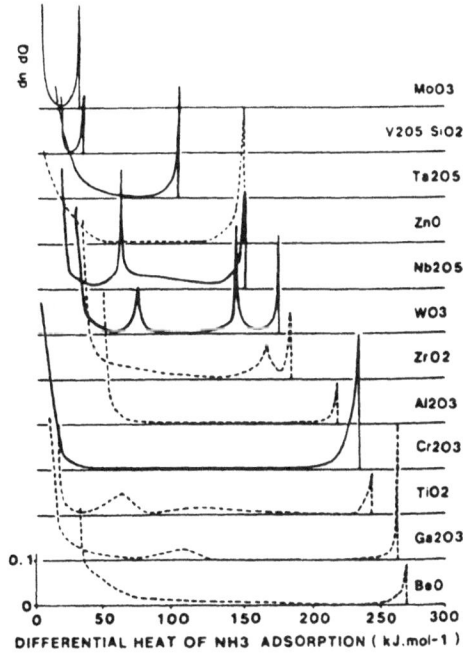

Figure 2. Acid site energy distribution for acidic and amphoteric oxides (48). Reprinted with permission. Copyright (1990) American Chemical Society.

displaying two heats of adsorption are evident for WO_3 and Nb_2O_5, both acidic oxides, and ZrO_2, an amphoteric oxide.

The corresponding data for the adsorption of CO_2 on basic and amphoteric oxides display both simple and complex basic site energy distributions (Figure 3). The oxides CaO, ZnO and MgO have single peak high energy distributions while the lanthanides and activities show several distinct strengths of basic sites.

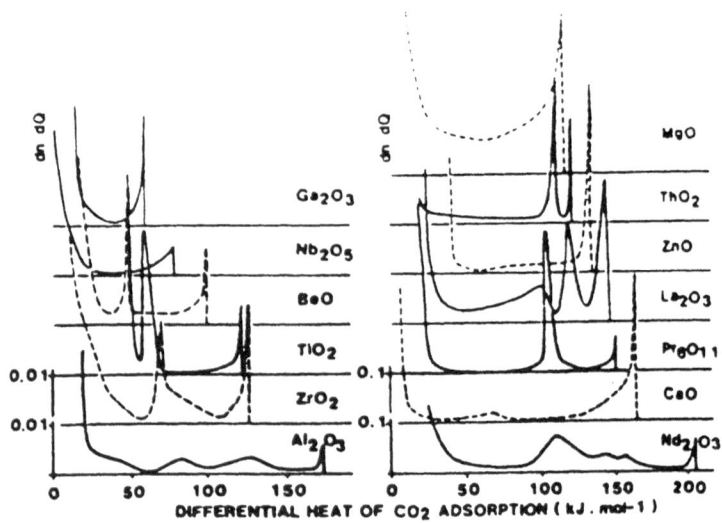

Figure 3. Basic site energy distribution for basic and amphoteric oxides (48). Reprinted with permission. Copyright (1990) American Chemical Society.

The average heats of adsorption of ammonia and of carbon dioxide were found to be dependent on the percent ionic character as calculated semiempirically (48) (Figures 4 and 5). Although the dependence on ionic character does not appear to be strong, nevertheless the observation that the average heats of adsorption of NH_3 and of CO_2 appear to decrease and increase with ionic character is consistent with the hypothesis that the acidic character is directly dependent upon, among other factors, the covalent nature of the bonding found in the solids.

The basicity of a number of oxides has been assessed from studies of the adsorption of hexafluoro-2-propanol (50). It has been shown that the acidic halogenalcohols such as fluoroalcohols adsorb selectivity on alumina cation-anion couples having a predominant basic character (51, 52). Further, unlike carboxylic acids this halogenalcohol functions only as a Brönsted acid. From a comparison of the differential heats of the fluoroalcohol on several oxides, the authors determine that the order of basicity is:

$$ThO_2 \geq MgO > Fe_2O_3 > TiO_2 > Al_2O_3 \gg SiO_2$$

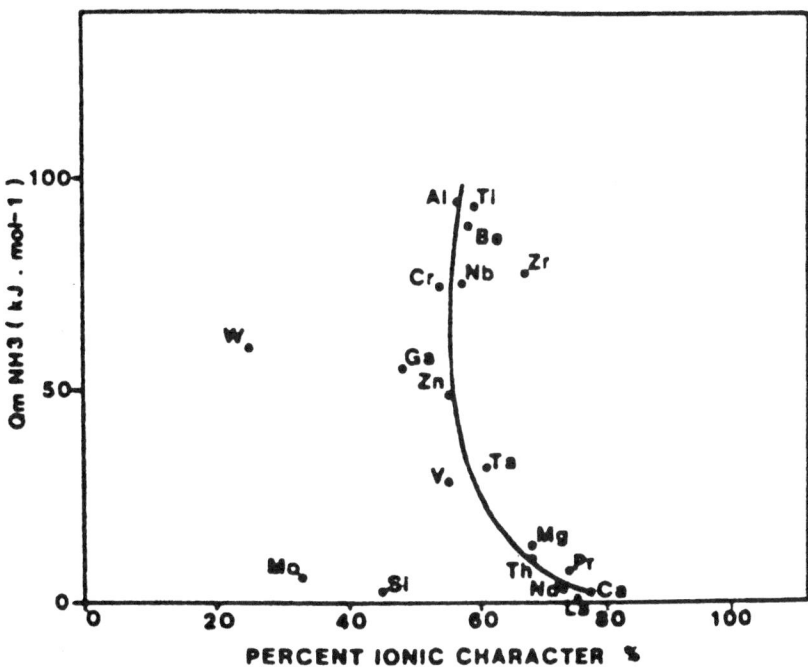

Figure 4. Average heats of NH_3 adsorption as a function of percentage of ionic character (48). Reprints with permission from the American Chemical Society.

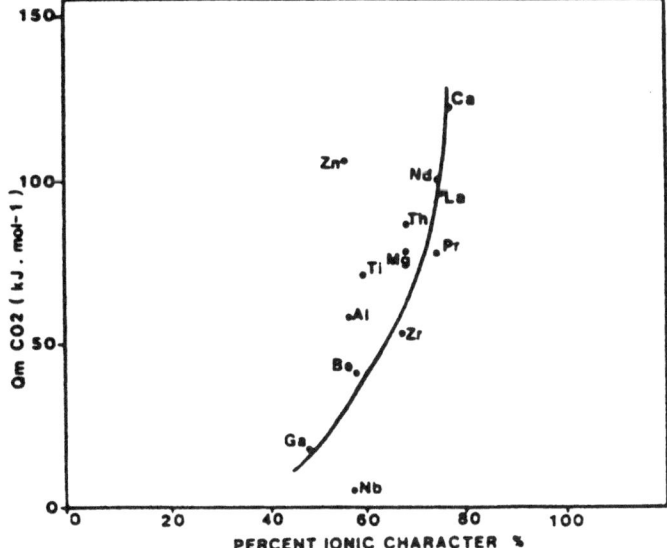

Figure 5. Average heats of CO_2 adsorption as a function of percentage of ionic character (48). Reprinted with permission from the American Chemical Society.

It is interesting to note the approximate agreement with the results obtained from the adsorption of CO_2 (48).

More recent results (49) with the aforementioned fluoroalcohol have classified ZnO as a solid of medium-high basicity, similar to TiO_2 but weaker, as a base, than ThO_2, MgO, and Fe_2O_3. Unfortunately this result appears to be in disagreement with that obtained from the CO_2 studies (48), from which it was concluded, that ZnO is a very strong base, stronger than MgO and ThO_2. The source of the differences may be due to a number of factors. The acidic strength of CO_2 and the fluoroalcohol may be dissimilar. One or both of these molecules may not adsorb selectively on basic sites. Basic sites in close proximity to one another may offer steric hindrance to larger probe molecules.

3. Conclusions

The acidity and basicity of solids, as a consequence of the methods commonly employed in the measurement of these properties, are properties of the surface and its structure. Surface free energy is one of the fundamental properties of any surface and therefore determination of its magnitude should, in principle, provide information on the acidity and/or basicity of the surface. The surface free energy can be subdivided into polar and nonpolar components, the former related to acid-base interactions. A number of techniques including contact angles, inverse gas chromatography and microcalorimetry are now available for the direct and indirect measurements of the acid-base components of the surface free energy. This approach, although not yet in common use, offers considerable advantages in the evaluation of the acidity and basicity of the surfaces of solids.

4. References

1. Kijenski, J. and Baiker, A., Catal. Today 5, 1 (1989).
2. Malinowski, S. and Marczewski, M. in Catalysis, A Specialist Periodical Report, Vol. 8, p. 107, (G.C. Bond and G. Webb, Eds.), Royal Society of Chemistry, Cambridge, 1989.
3. Jacobs, P.A., in Characterization of Heterogeneous Catalysts (F. Delannay, Ed.), p. 367, Dekker, New York, 1984.
4. Tanabe, K., Catalysis, Science and Technology (J.R. Anderson and M. Bondart, Eds.), Vol. 2, p. 231, Springer-Verlag, Berlin, 1981.
5. Benesi, H.A. and Winquist, B.H.C., in Advances in Catalysis (D.D. Eley, H. Pines and D. Weisz, Eds.), Vol. 27, p. 97, Academic Press, New York, 1978.
6. Forni, L., Catal. Rev. 8, 65 (1973).
7. Tanabe, K., Solid Acids and Bases, Academic Press, New York (1970).
8. Boehm, H.P., J. Chem. Soc. Far. Soc. Disc. 52, 264 (1971).
9. Berteau, P., and Delmon, B., Catal. Today 5, 121 (1989).

10. Kung, M.C. and Kung, H.H., Catal. Rev.-Sci. Eng. 27, 425 (1985).
11. Kiviat, F.E. and Petrakis, L., J. Phys. Chem. 77, 1232 (1973).
12. Tsyganenko, A.A., Pozdnyakov, D.V. and Filimonov, V.N., J. Mol. Struct. 29, 299 (1975).
13. Highfield, J.G. and Moffat, J.B., J. Catal. 89, 185 (1984).
14. Highfield, J.G. and Moffat, J.B., J. Catal. 88, 177 (1984).
15. Miyata, H. and Moffat, J.B., J. Catal. 62, 357 (1980).
16. Kayo, A., Yamaguchi, T., Tanabe, K., J. Catal. 83, 99 (1983).
17. Jin, T., Yamaguchi, T., and Tanabe, K., J. Phys. Chem. 90, 4794 (1986).
18. Parry, E.P., J. Catal. 2, 371 (1963).
19. Su, B.L. and Barthomeuf, D., J. Catal. 139, 81 (1993) and references therein.
20. Huang, M, Adnot, A. and Kaliaguine, S., J. Catal. 137, 322 (1992).
21. Sanders, T. and Moffat, J.B., J. Coll. Interf. Sci. 120, 451 (1987).
22. Odenbrand, C.U.I., Brandin, J.G.M., and Busca, G., J. Catal. 135, 505 (1992).
23. Morishige, K., Kittaka, S. and Katsuragi, S., J. Chem. Soc., Faraday Trans. 1 78, 2947 (1982).
24. Yori, J.C., Luy, J.C. and Parera, J.M., Appl. Catal. 41, 1 (1988).
25. Hodnett, B.K. and Moffat, J.B., J. Catal. 88, 253 (1984).
26. Hodnett, B.K. and Moffat, J.B., J. Catal. 91, 93 (1985).
27. Moffat, J.B., Chao, E. and Nott, B., J. Coll. Interf. Sci. 678, 240 (1978).
28. Schraml-Marth, M., Wokaun, A., Curry-Hyde, H.E. and Baiker, A., J. Catal. 133, 431 (1992).
29. Moffat, J.B. and Chao, E., J. Catal. 46, 151 (1977).
30. Ghosh, A.K. and Moffat, J.B., J. Catal. 101, 238 (1986).
31. Umansky, B.S. and Hall, W.K., J. Catal. 124, 97 (1990).
32. Deeba, M. and Hall, W.K., J. Catal. 60, 417 (1979).
33. Deeba, M. and Hall, W.K., Z. Phys. Chem. N.F.B. 144, 85 (1985).
34. Carre, A., Roger, F. and Varinot, C., J. Coll. Interf. Sci. 154, 174 (1992).
35. Fowkes, F.M., J. Adhesion Sci. Technol. 4, 669 (1990).
36. Fowkes, F.M., Riddle, Jr., F.L., Pastore, W.E. and Weber, A.A., Colloids and Surfaces 43, 367 (1990).
37. Fowkes, F.M., Wright, D.W. and Cole, D.A., J. Non-Crystalline Solids 120, 47 (1990).
38. Fowkes, F.M., J. Adhesion Sci. Technol. 1, 7 (1987).
39. Fowkes, F.M., Ind. Eng. Chem. 56, 40 (1964).
40. van Oss, C.J., Giese, Jr., R.F. and Good, R.J., Langmuir 6, 1711 (1990).
41. Chibowski, E., Kerkeb, M.L. and Gonzalez-Caballero, F., J. Coll. Interf. Sci. 155, 444 (1993).
42. Hüttinger, K.J., Höhmann-Wien, S. and Krekel, G., J. Adhesion Sci. Technol. 6, 317 (1992).
43. Cardona-Martínez, N. and Dumesic, J.A., in Advances in Catalysis 38, 149 (1992).
44. Chen, D., Sharma, S., Cardona-Martínez, N., Dumesic, J.A., Bell, V.A., Hoadge, G.D. and Madon, R.J., J. Catal.136, 392 (1992).

45. Gervasini, A. and Auroux, A., J. Phys. Chem. 97, 2628 (1993).
46. Gervasini, A. and Auroux, A., J. Catal. 131, 190 (1991).
47. Gervasini, A. and Auroux, A., J. Thermal. Anal. 37, 1737 (1991).
48. Auroux, A. and Gervasini, A., J. Phys. Chem. 94, 6371 (1990).
49. Rossi, P.F., Busca, G., Lorenzelli, V., Waqif, M., Saur, O. and Lavalley, J.-C., Langmuir 7, 2677 (1991).
50. Rossi, P.F., Busca, G., Lorenzelli, V., Lion, M. and Lavalley, J.C., J. Catal. 109, 378 (1988).
51. Busca, G., Rossi, P.F., Lorenzelli, V., Banaissa, M., Travert, J. and Lavelley, J.C., J. Phys. Chem. 89, 5433 (1985).
52. Benaissa, M., Saur, O. and Lavalley, J.C., Mater. Chem. 7, 699 (1982).
53. Jozefowicz, L.C., Karge, H.G., Vasilyeva, E. and Moffat, J.B., Microporous Materials, in press.
54. Casamassima, M., Darque-Cereti, E., Etcheberry, A. and Aucouturier, M., Appl. Surf. Sci. 52, 205 (1991).
55. Contescu, C.R., Jagiello, J. and Schwarz, J.A., J. Catal. 131, 433 (1991).
56. Dorris, G.M. and Gray, D.G., J. Coll. Interf. Sci. 77, 353 (1990).
57. Corma, A., Sastre, G., Viruela, R. and Zicovich-Wilson, C., J. Catal. 136, 521 (1992).
58. Mortier, W.J., in Theoretical Aspects of Heterogeneous Catalysis (J. B. Moffat, Ed.), p. 135, Van Nostrand-Reinhold, New York (1990).
59. Yoshida, S., in Theoretical Aspects of Heterogeneous Catalysis (J.B. Moffat, Ed.), p. 506, Van Nostrand-Reinhold, New York (1990).
60. Moffat, J.B., Vetrivel, R. and Viswanathan, B., J. Mol. Catal. 30, 171 (1985).
61. Moffat, J.B., J. Mol. Catal. 26, 385 (1984).
62. Basila, M.R., Kantner, T.R. and Rhee, K.H.,J. Phys. Chem. 68, 3197 (1964).
63. Peri, J.B., J. Phys. Chem. 69, 231 (1965).
64. Sanderson, R.T., Chemical Periodicity, Reinhold, New York (1960).

CHARACTERIZATION OF BRÖNSTED AND LEWIS ACIDITY BY NUCLEAR MAGNETIC RESONANCE SPECTROSCOPY

Harry Pfeifer
University of Leipzig
Linnéstr. 5
D - 04103 Leipzig
Germany

ABSTRACT. It will be shown that ^1H MAS NMR spectroscopy offers a unique possibility to characterize quantitatively the concentration, strength of acidity and accessibility of *Brönsted acid sites* of solid catalysts. The main advantage of this method with regard to infrared spectroscopy is that the intensities of the various NMR lines are directly proportional to the concentration of the respective sites which allows an absolute measurement of their concentration. In contrast, neither by a measurement of the ^{27}Al NMR spectra including the MAS and DOR technique nor by an analysis of the NMR spectra of adsorbed probe molecules it was possible up till now to characterize *Lewis acidity* quantitatively so that at present in this field NMR can not compete with infrared spectroscopy.

1. Brönsted Acidity

1.1. CONCENTRATION OF ACID SITES

With respect to a measurement of the concentration of Brönsted acid sites including non-acidic OH groups, nuclear magnetic resonance spectroscopy has an extremely important advantage compared with infrared spectroscopy: The intensity (area) of an ^1H MAS NMR signal is directly proportional to the concentration of the hydrogen nuclei contributing to this signal irrespective of their bonding state, so that any compound with a known concentration of hydrogen atoms can be used as a reference (mostly water). If the spinning rate Ω is of the order or smaller than the static line width of the NMR signal, spinning sidebands appear as artifacts of the MAS method and it becomes necessary in such a case to add their intensities to that of the central line. For hydroxyl groups of type $\equiv SiOHAl \equiv$ which are denoted as bridging OH groups and which are the typical Brönsted acid sites in zeolites and related catalysts, the magnetic dipole interaction between the hydroxyl proton and the aluminium nucleus dominates [1]. Under this supposition a general formula has been derived for the envelope of the free induction decay [2] which simplifies for room temperature or below where the mean residence time of the hydroxyl proton between two translational jumps is much larger than 0.5 ms [3] to

$$\Phi^{MAS}(t) = \exp \left\{ - \left(\gamma_p^2 M_2 / 3\Omega^2 \right) \left[2 \left(1 - \cos\Omega t \right) + \left(1 - \cos2\Omega t \right) / 4 \right] \right\}. \quad (1)$$

In this equation γ_p denotes the gyromagnetic ratio of the proton ($\gamma_p \approx 2.675 \cdot 10^8$ s^{-1} T^{-1}) and M_2 the second moment of the magnetic dipole interaction in T^2. The *average*

255

J. Fraissard and L. Petrakis (eds.), Acidity and Basicity of Solids, 255–277.
© 1994 *Kluwer Academic Publishers.*

value of M_2 has been measured [1] to be $0.4 \cdot 10^{-8}$ T^2 and $0.7 \cdot 10^{-8}$ T^2 for zeolites of type H-ZSM-5 and H-Y, respectively. In agreement with experimental results [4] a spinning frequency $\Omega/2\pi$ of ca. 10 kHz is sufficient if one wants to reduce the intensities of the the spinning sidebands to a value below the experimental error.

For the minimum number of nuclei N_{min} detectable by NMR spectroscopy the following relation can be derived [4]

$$N_{min} \propto (\gamma / \gamma_p)^{1/2} \, r^{-1} \, (T / B_0)^{3/2} \, (T_1 / T_2)^{1/2} \, T_m^{-1/2} \qquad (2)$$

with the relative signal intensity

$$r = 4 \, (\gamma/\gamma_p)^3 \, I(I + 1) / 3. \qquad (3)$$

γ denotes the gyromagnetic ratio of the resonating nucleus with spin I, T the temperature, T_m the time of the measurement, B_0 the intensity of the constant magnetic field, and T_1 and T_2 the longitudinal and transverse nuclear magnetic relaxation time, respectively. The factor of proportionality in equ. (2) depends on the quality and on the filling factor of the rf coil, on the noise figure of the electronic system and on similar other parameters.

In order to make use of equ. (2), empirical results for the number of *hydrogen* nuclei N_{min} shall be given, assuming a signal-to-noise ratio of 10, a measuring time $T_m = 600$ s, a magnetic field intensity $B_0 = 7.05$ T corresponding to a proton resonance frequency of ca. 300 MHz, and a temperature T = 300 K. In the case of ^{1}H MAS NMR of *surface OH groups*, typical values of T_1 and T_2 (MAS enhanced value) are ca. 1 s and ca. 1 ms, respectively, and one finds experimentally $N_{min} \approx 10^{18}$. For 0.2 g of a zeolite with typically $4 \cdot 10^{20}$ cavities per gram, this value corresponds to ca. 0.01 OH groups per cavity, and for 0.2 g of a catalyst with a specific surface area of S m^2/g to ca. 5/S OH groups per nm^2. Hence, one may conclude that the signal-to-noise ratio of the ^{1}H MAS NMR method is quite sufficient to study even low concentrations of Brönsted acid sites in zeolites.

With regard to infrared spectroscopy, in Fig. 1 ^{1}H MAS NMR and infrared stretching vibration spectra are compared for two differently synthesized specimens [6] of SAPO-5.

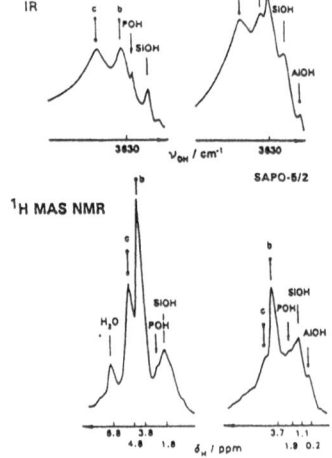

Fig. 1.
IR stretching vibration (Digilab FTS-20) and ^{1}H MAS NMR spectra (Bruker MSL 300) of two differently synthesized specimens of a zeolite SAPO-5 [6].

Line *b* corresponds to the so-called high-frequency (*HF*) band in infrared spectroscopy and is caused by bridging OH groups pointing into the large cavities of the zeolites, and line *c* corresponding to the low-frequency (*LF*) band is also due to bridging OH groups which, however, are located in the small cages (see below). While the resolution of the spectra and the positions of the various NMR and IR signals correspond to each other quite well and are in agreement with IR results published by other authors [5] there are dramatic differences in the relative intensities: From the IR spectrum of the first specimen one would erroneously conclude that the concentrations of bridging OH groups of type *b* and *c* are approximately equal, and from the IR spectrum of the second specimen that the concentration of POH groups is about three times larger than that of the bridging OH groups of type *b* while in fact the concentration of the POH groups is less than that of the OH groups giving rise to line *b*.

Therefore, in contrast to NMR, even the relative intensity of the IR signal caused by the OH stretching vibration cannot be taken as a measure for the concentration of the respective hydroxyl groups.

1.2. STRENGTH OF ACIDITY AND STRUCTURE OF ACID SITES

The strength of acidity S_a of a Brönsted acid site which is commonly described as the ability of the site to protonate an adsorbed molecule may be defined quantitatively by the inverse of the standard Gibbs free energy change ΔG^0_{DP} for the deprotonation of the OH group in vacuum [6]

$$S_a = (\Delta G^0_{DP})^{-1} \tag{4}$$

so that increasing values of S_a correspond to decreasing values of ΔG^0_{DP}. Strictly spoken, S_a should be denoted as the strength of *gas phase* acidity since ΔG^0_{DP} is independent of the particular molecule to be protonated. In order to compare the strength of the gas phase acidity with the deprotonation energy of a hydroxyl group ΔE_{DP}, a quantity which follows from quantum chemical calculations, one must take into consideration that ΔG^0_{DP} is the sum of three terms:

$$\Delta G^0_{DP} = \Delta E_{DP} + \Delta E^0_{DP} + \Delta G^t_{DP}. \tag{5}$$

ΔE^0_{DP} denotes the change of the zero point energy and ΔG^t_{DP} the change of Gibbs' free energy resulting from the conversion of the three vibrational degrees of freedom of the proton in the hydroxyl group into its three translational degrees of freedom after it has left this group. Since the change of the zero-point energy change ΔE^0_{DP} is a constant and since the contribution of ΔG^t_{DP} is only small [7], a change of ΔE_{DP} should be directly related to a change of the reciprocal value of the strength of gas phase acidity.

With respect to ^1H MAS NMR spectroscopy the question arises whether the chemical shift of the proton resonance δ_H of a hydroxyl group can be related to its strength of acidity S_a. For such a comparison however, the OH groups must be isolated which requires for gaseous samples simply a measurement at low pressure while in the case of surface OH groups three conditions must be fulfilled:

(1) The measurements have to be performed with evacuated samples since adsorbed molecules may strongly shift the proton resonance of an OH group to higher values of δ_H. This shift depends both on the strength of acidity and on the particular molecule and may be of the order of 10 ppm (see section 1.3.).

(2) The concentration of the OH groups must be not too high in order to exclude a direct interaction of neighbouring hydroxyls. This condition is fulfilled in general for surface OH groups of zeolites.

(3) Additional electrostatic interactions of the OH groups with the framework as e.g. caused by the formation of a hydrogen bond to other oxygen atoms of the surface must be excluded: If one plots the chemical shift $\delta_H(r_{OO})$ of OH groups measured in various solid samples as a function of the O-H...O distance r_{OO}, one finds a dependence which can be approximated for distances between 0.24 nm and 0.30 nm by a linear equation:

$$\delta_H(r_{OO}) = \delta_H + K \; (0.30 - r_{OO} / nm). \tag{6}$$

δ_H denotes the chemical shift of the isolated OH group ($r_{OO} \geq 0.30$ nm), and for the slope K a value of 260 ppm/nm and 380 ppm/nm follows from data published in [8] and [9], respectively. It is this dependence which leads to different values of the chemical shift for the bridging OH groups pointing into large and small cavities of zeolites (lines b and c, respectively).

With regard to a relation between δ_H and the strength of acidity S_a of isolated OH groups, experimental results shall be mentioned which have been measured [10] for seven aliphatic alcohols (methanol, ethanol, n-propanol, i-propanol, t-butanol, i-butanol, neopentanol), 2,2,2-trifluoroethanol and benzyl alcohol in the gaseous state under conditions (ca. 10^4 Pa, 148.6 °C) where hydrogen bonding has little effect on chemical shifts. A plot of δ_H versus experimental values for the standard Gibbs free energy change of deprotonation ΔG^0_{DP} [11], [12] yields a straight line with a slope derived from a linear regression [10] given by

$$d(\Delta G^0_{DP}) / d(\delta_H) = - 43 \pm 2 \; kJ \; ppm^{-1} \; mol^{-1} \tag{7}$$

with a correlation coefficient of 0.99. Qualitatively, this result which may lead to the conclusion that increasing values of the chemical shift δ_H correspond to increasing values of the strength of acidity S_a of the OH group can be explained as follows: An increasing strength of acidity which is tantamount to a decreasing value of the deprotonation energy should correspond to an increase of the net atomic charge (i.e. to a decrease of the electron density) of the hydrogen atom. On the other hand, the chemical shielding σ_H of the ^1H nucleus will decrease with an increase of the net atomic charge so that the chemical shift δ_H of the OH group which is defined by

$$\delta_H = \sigma_{TMS} - \sigma_H \tag{8}$$

with σ_{TMS} denoting the chemical shielding of the ^1H nucleus in tetramethylsilane, should increase with increasing strength of acidity.

However, this qualitative argument cannot be generalized: In a recent paper [13] ab initio SCF calculations were performed of both the chemical shielding σ_H and the deprotonation energy ΔE_{DP} for various molecules and model clusters of surface OH groups. The latter include models for terminal surface OH groups , viz. H_3SiOH and $Si(OH)_4$ for SiOH; H_2AlOH, $Al(OH)_3$ and H_3POAlH_2OH for AlOH; $OP(OH)_3$ and H_3AlOPH_2OH for POH, as well as for bridging OH groups, viz. $H_3SiOHAlH_3$ and $H_3SiOHAl(OH)_3$. Some of the results are collected in Table 1. After a comparison of all results (not shown in Table 1) including most different molecules with XOH groups, the authors of ref. [13] come to the conclusion that there is no general linear dependence between δ_H and ΔE_{DP}. The reason is that the contributions of the orbitals other than the OH bond orbital which are localized on the O atom of the OH group, namely the lone pairs and the XO bond orbital, are significant and not constant within the set of molecules studied. It is the variation of these non-OH-bond contributions which spoils the linear dependence between δ_H and ΔE_{DP} for some XOH groups.

Table 1. Results of ab initio SCF calculations of the chemical shielding σ_H and of the deprotonation energy ΔE_{DP} for various molecules and model clusters of surface OH groups [13]. [*] The chemical shielding calculated for methanol is 31.9 ppm [13] and the shift of methanol relative to gaseous TMS has been found experimentally to be 0.02 ppm [10].

	$\Delta E_{DP}/\mathrm{kJmol^{-1}}$	σ/ppm	$\delta_H^{(*)} = 31.92 - \sigma_H/\mathrm{ppm}$
H_2O	1700	30.9	1.02
H_3O^+	734	23.5	8.42
CH_3OH	1677	31.9	0.02
$CH_3OH_2^+$	804	25.5	6.22
terminal SiOH			
H_3SiOH	1561	31.0	0.92
$Si(OH)_4$	1563	29.7	2.22
terminal AlOH			
H_2AlOH	1615	29.7	2.22
$Al(OH)_3$	1636	30.7	1.22
H_3POAlH_2OH	1494	31.2	0.72
terminal POH			
H_3AlOPH_2OH	1285	28.7	3.22
$OP(OH)_3$	1438	28.5	3.42
terminal BOH			
$B(OH)_3$	1571	29.0	2.92
isolated bridging OH groups			
$H_3SiOHAlH_3$	1329	28.9	3.02
$H_3SiOHAl(OH)_3$	1353	27.8	4.12

However, for sets of hydroxyl groups in similar bonding environments and especially for surface hydroxyl groups which are responsible for the Brönsted acidity of catalysts, a linear relation has been found. This is due to the fact that these OH groups are all bonded to cations (B, Al, Si or P) whose first coordination sphere consists of oxygen atoms only. For those models studied in [13] that have the property of a complete oxygen coordination around the cations, a plot of ΔE_{DP} versus δ_H yields a straight line with a slope derived from a linear regression given by

$$d(\Delta E_{DP}) / d(\delta_H) = -84 \pm 12 \quad kJ \ ppm^{-1} \ mol^{-1} \tag{9}$$

with a correlation coefficient of 0.93.

As an experimental prove for the statement that the chemical shift δ_H can be used as a reliable and sensitive measure for the strength of acidity of isolated surface OH groups we refer to a study of the dependence of δ_H on the silicon-to-aluminium ratio n of zeolites [14]: It is apparent that an enhancement of the electronegativity of the framework of a zeolite will decrease the electron density of the OH bond of a surface OH group which should lead to an increased strength of acidity. In Table 2 values for Sanderson's intermediate electronegativity S_m are collected together with experimental values for δ_H of isolated bridging OH groups (line b, corresponding to the HF band in IR spectroscopy) in dependence on the silicon-to-aluminium ratio n. The intermediate electronegativity is defined as the geometric mean of the atomic electronegativities so that it is given for zeolites of composition $HAlO_2(SiO_2)_n$ by

$$S_m = (S_H \cdot S_{Al} \cdot S_O^{2n+2} \cdot S_{Si}^n)^{1/(3n+4)} \tag{10}$$

with the following values for the atomic electronegativities [15]: $S_H = 3.55$, $S_{Al} = 2.22$, $S_O = 5.21$, $S_{Si} = 2.84$.

Table 2. Values for the calculated intermediate electronegativity S_m (cf. equ. (10)) and the measured chemical shift δ_H of isolated bridging OH groups (line b, corresponding to the HF band in IR spectroscopy) in dependence on the silicon-to-aluminium ratio n for various zeolites [14]. The error of δ_H does not exceed $\pm$ 0.1 ppm.

n	S_m	δ_H / ppm	$\delta_H S_m^{-1}$ / ppm
1	4.00	3.8	0.95
2	4.08	4.0	0.98
5	4.16	4.2	1.01
10	4.20	4.3	1.02
20	4.23	4.3	1.02
30	4.24	4.3	1.01

The fact that the ratio of δ_H/S_m is constant within the limits of error (cf. last column in Table 2) clearly demonstrates the usefulness and sensititivy of the ^1H MAS NMR method. On the other hand, from Table 2 together with equ. (9) it can be concluded that the deprotonation energy of the Brönsted acid sites decreases by ca. 40 kJ mol^{-1} if the silicon-to-aluminium ratio increases from n = 1 to n = 10 while it remains nearly constant above that value.

Apparently, the accuracy of a measurement of the strength of acidity S_a by the ^1H MAS NMR method depends on the residual line width $\Delta \nu^{MAS}$ of the respective ^1H MAS NMR signal (central line in the case of a spinning sideband pattern). In principle, $\Delta \nu^{MAS}$ may be determined by six factors [16]:
(1) An inhomogeneity of the constant magnetic field B_0, (2) a misadjustment of the magic angle, (3) an anisotropy of the magnetic susceptibility of the zeolite crystallites,

(4) translational jumps of the hydroxyl protons, (5) the magnetic dipole-dipole interaction with quadrupole nuclei (^{27}Al), and (6) a combined influence of the homonuclear (^{1}H - ^{1}H) magnetic dipole-dipole interaction with the heteronuclear (^{1}H - ^{27}Al) magnetic dipole-dipole interaction and/or the anisotropy of the ^{1}H chemical shift.

As can be shown [16], [3], [14], the influence of (1) - (4) upon the residual line width of the ^{1}H MAS NMR signals of hydroxyl groups of zeolites and related catalysts measured at room temperature or below is negligible. The influence of (5) can be neglected if the intensity B_0 of the constant magnetic field is not smaller than 7 T corresponding to an ^{1}H NMR frequency of ca. 300 MHz. With regard to (6) two limiting cases have to be discussed: Zeolite catalysts with a low concentration of surface OH groups (catalysts of type *l*) which are characterized by a silicon-to-aluminium ratio n ≥ 15 and catalysts of type *h* defined by n < 15:

Catalysts of type l. For a zeolite H-ZSM-5 with a silicon-to-aluminium ratio n = 15, the experimentally determined residual line width $\Delta\nu$ ^{MAS} does not depend on the spinning rate Ω [16] in contrast to the $1/\Omega$ dependence which should be observed if (6) determines the residual line width. From this result and the fact that the influence of the other factors (1) - (5) can be also excluded, it follows that the residual line width of the ^{1}H MAS NMR signal of bridging OH groups is determined by the distribution width of the isotropic value of the chemical shift. Therefore, the observed value of $\Delta\nu$ ^{MAS}/300MHz ≈ 0.8 ppm for the zeolite H-ZSM-5 [16] is the natural limit for this catalyst which cannot be further reduced by an enlargement of the magnetic field B_0 or of the spinning rate Ω.

Catalysts of type h. As a typical example, a zeolite H-Y with a silicon-to-aluminium ratio n = 2.6 shall be considered. In this case the influence of (6) is not negligible at lower spinning rates. A theoretical treatment (see e.g. [14] and references cited in this review) shows that the line width $\Delta\nu$ ^{MAS} is mainly caused by the combined influence of the homonuclear with the heteronuclear magnetic dipole-dipole interaction and that it is given approximately by

$$\Delta\nu \ ^{MAS} \approx (K_2 \, M_2^{II} \, M_2^{IS} \, 2 \ln 2)^{1/2} / (\Omega \, \pi). \tag{11}$$

K_2 is a geometrical dimensionless parameter of the order of 0.1. For a system of two resonating and one non-resonating nuclei, or one resonating and two non-resonating nuclei, at the corners of an equilateral triangle, the exact value is 0.087 [17]. M_2^{II} and M_2^{IS} denote the second moments in s^{-2} for the ^{1}H - ^{1}H and the ^{1}H - ^{27}Al magnetic dipole-dipole interaction, respectively. From an analysis of the sideband pattern of a zeolite H-Y specimen where the ^{1}H - ^{1}H magnetic dipole-dipole interaction was reduced through a partial deuteration [18], the value of M_2^{IS} could be derived to be ca. $4.5 \cdot 10^8$ s^{-2}. The value of M_2^{II} has been estimated to be ca. $0.4 \cdot 10^8$ s^{-2} [1] from the static ^{1}H NMR spectrum of the same zeolite but non-deuterated.

In order to separate two lines which are ca. 1 ppm apart (e.g. lines *b* and *c*, see below), the line width must be smaller than 500 Hz for an ^{1}H NMR frequency of 500 MHz. Inserting the estimated values of M_2^{II} and M_2^{IS} together with $K_2 \approx 0.1$ into equ. (11), the spinning rate $\Omega/2\pi$ must be approximately equal to or larger than 5 kHz. Experimental results [3] are in agreement with this estimate and one may conclude that spinning rates of ca. 10 kHz are sufficient to eliminate the combined influence of the heteronuclear and homonuclear magnetic dipole-dipole interaction on the residual line width of the ^{1}H MAS NMR spectra of catalysts of type *h*. The line width achieved under these conditions is then given also by the distribution width of the isotropic value of the chemical shift (natural line width) which cannot be further reduced and which is achieved for catalysts of type *l* even at relatively small values of the spinning rate.

Summarized, the following statements can be made:

(i) The natural line width for the signals of Brönsted acid sites (bridging OH groups) in the ^{1}H MAS NMR spectra of evacuated zeolite catalysts is given by the distribution width of the isotropic value of the chemical shift (distribution width of the strength of gas phase acidity).

(ii) In the case of zeolites with a low concentration of OH groups (catalysts of type l) as e.g. H-ZSM-5, it is only necessary to apply constant magnetic fields of a sufficiently large intensity (B_0 not smaller than ca. 7 T) in order to achieve the natural line width.

(iii) For zeolites with a high concentration of OH groups (catalysts of type h), as e.g. non-dealuminated H-Y zeolites, in addition to a sufficiently strong magnetic field ($B_0 \geq 7$ T) the homonuclear magnetic dipole-dipole interaction must be reduced in order to achieve the natural line width. This can be performed either by the use of high spinning rates [3] or by a partial deuteration of the OH groups [17] or by the application of a multiple pulse sequence during the period of the free induction decay (CRAMPS technique, see e.g. [19], [14]).

In the ^{1}H MAS NMR spectra of evacuated zeolites containing only oxygen, silicon and aluminium in the framework, five lines can be separated which have been denoted [20] as lines a, b, c, d, and e:

Line a which appears in the interval between ca. 1.8 and 2.3 ppm is caused by non-acidic (silanol) groups. In contrast to an older paper [21] it seems necessary to mention that a distinction between single (SiOH) and geminal (Si(OH)$_2$) hydroxyl groups is not possible by this technique since the difference between the corresponding values of δ_H is less than 0.1 ppm [6]. This is in agreement with an ^{1}H CRAMPS study of silica gel prepared under various stages of dehydration where only three peaks at 1.7 ppm, 3.0 ppm and 3.5 ppm could be found which where identified with isolated (non-hydrogen bonded) silanols, hydrogen-bonded silanols and physisorbed water, respectively [22]. In ^{29}Si MAS NMR spectra however, well separated signals due to single and geminal silanols appear at about -100 ppm and -90 ppm, respectively [23]. Unfortunately, the possibility to ascertain these two species is limited to adsorbents built up only by silicon and oxygen since the insertion of other metal atoms as e.g. aluminium leads to overlapping signals in the ^{29}Si MAS NMR spectra [24].

Line b at 3.8 to 4.4 ppm is ascribed to isolated bridging OH groups (isolated Brönsted acid sites). The value of δ_H increases with increasing silicon-to-aluminium ratio of the zeolite (cf. Tab. 2).

Line c at 4.8 to 5.6 ppm is also ascribed to OH groups of the bridging type but under the influence of an additional electrostatic interaction as e.g. in the case of formation of hydrogen bonds to other oxygen atoms of the framework (non-isolated OH groups). The same effect has been found for the stretching vibration of OH groups giving rise to the so-called *LF* band (3540 cm^{-1}) in addition to the *HF* band (3650 cm^{-1}). In a former paper [20] line c was ascribed tentatively to OH groups located in the large cavities which are known to be responsible for the *HF* band in the IR spectrum. In a following paper [25] however, it could be shown unambiguously by the use of deuterated pyridine as a probe molecule that the correct correlation is

line b	⟷	*HF band*
line c	⟷	*LF band.*

The fact that for the bridging OH groups pointing into the large and into the small cavities separate lines appear in the ^{1}H MAS NMR spectra (lines b and c, respectively) excludes the possibility of a fast proton exchange (cf. section 2.2., equ. (24)) among the four oxygens around an aluminium atom of the zeolite framework.

Line d at 6.5 to 7.0 ppm is due to residual ammonium ions.

Line e at 2.5 to 3.6 ppm represents hydroxyl groups associated with extra-framework aluminium species. Due to the limited space available for these OH groups their chemical shift will be affected by additional electrostatic interactions. Accordingly, for isolated AlOH groups the value of δ_H is much smaller, viz. in the interval between -0.5 and +1 ppm. In a recent paper Harris et al. [26] have studied systematically H-Y zeolites containing large amounts of extra-framework aluminium. For dehydrated specimens they found a dramatic redistribution of H within the sample over the course of the aging period (24 days) with extra-framework aluminium species gaining OH groups at the expense of the other hydroxyl groups.

In Fig. 2 measured values for the ^{1}H NMR chemical shift of isolated (black) and interacting (hatched area) OH groups appearing in zeolite catalysts of various type and composition are collected [27].

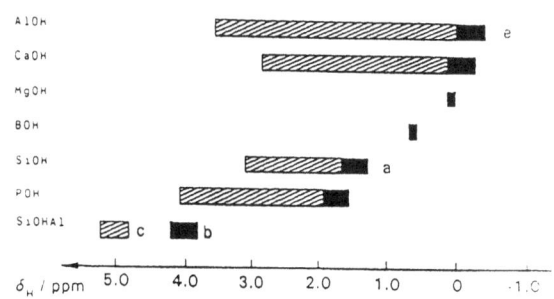

Fig. 2.
Intervals for the experimentally determined values of the ^{1}H NMR chemical shift δ_H of isolated (black) and interacting (hatched area) OH groups in zeolites [27].

If one compares the positions of the signals caused by the various sorts of OH groups in the infrared stretching vibration and ^{1}H MAS NMR spectra, a correlation between both can be found [28]. With the exception of the cationic OH groups there is a nearly linear interdependence between the wave number ν_{OH} of the IR band and the chemical shift δ_H of the ^{1}H MAS NMR signal, which can be described by the empirical relation [28]

$$\nu_{OH} / cm^{-1} = 3870 - 67.8 \cdot \delta_H / ppm. \qquad (12)$$

As has been shown above, the experimentally determined minimum line width of ca. 0.8 ppm for the ^{1}H MAS NMR signal of bridging OH groups is given by the distribution width of the isotropic value of the chemical shift. Therefore, the minimum line width set by the method itself must be smaller than that value, say 0.4 ppm or less. On the other hand, the position of a resonance line can be determined with an accuracy of ca. 1/5 of its line width so that the resolution of the ^{1}H MAS NMR method must be less than 0.08 ppm which corresponds according to equ. (12) to a resolution of less than ca. 5 cm^{-1} of the IR stretching vibration spectra. This value has to be compared with the resolution of ca. 2 cm^{-1} achieved in [29] which leads to the remarkable statement that for both methods the ultimate resolution is of the same order of magnitude (cf. also Fig. 1).

It has been mentioned already that for small spinning rates of the sample, spinning sidebands appear. For a measurement of the concentration of the OH groups they are inconvenient since their intensity must be added to that of the central line. On the other hand the spinning sideband pattern contains valuable information about the anisotropy of the chemical shift and the magnetic dipole-dipole interaction between the ^{1}H and the neighbouring ^{27}Al nuclei, or with other words, about their distance. In order to analyse these sideband patterns for each of the various signals it is necessary to separate them carefully. This can be accomplished by a simple two-dimensional NMR experiment: At first a $\pi/2$ pulse and then after a time interval t_1 a π pulse is applied to the sample. Under the condition that t_1 assumes only discrete values given by

$$t_1 \Omega = 2\pi z \qquad (13)$$

where z denotes an integer and Ω the spinning rate, the nuclear magnetic resonance signal $F(t_1,t_2)$ is measured at time t_2 after the π pulse. A twofold Fourier transformation of $F(t_1,t_2)$ then gives the two-dimensional ^{1}H MAS NMR spectrum $S(\omega_1,\omega_2)$. The important fact is that $S(\omega_1,\omega_2)$ yields for ω_1 = const. the ordinary ^{1}H MAS NMR spectrum, i.e. the central lines together with all spinning sidebands, while $S(\omega_1,\omega_2)$ gives for ω_2 = const. the ^{1}H MAS NMR spectrum without these sidebands. Therefore, the sidebands for each line can be analysed separately. In [30] a method is described which allows a determination of both the distance r_{HAl} between the ^{1}H and the neighbouring ^{27}Al nuclei and of the chemical-shift anisotropy $\Delta\sigma_H$ for the bridging OH groups (lines b and c in the ^{1}H MAS NMR spectra) of zeolites.
In Table 3 results are collected which have been found experimentally for zeolites H-Y and SAPO-5 [30]. For the zeolite H-Y the values of r_{HAl} can be compared with theoretical data found by a computer simulation [31]. The qualitative agreement

Table 3. Experimental results for the isotropic value δ_H and the anisotropy $\Delta\sigma_H$ of the chemical shift as well as for the distance r_{HAl} between the ^{1}H and ^{27}Al nuclei of the bridging OH groups (lines b and c in the ^{1}H MAS NMR spectra) in zeolites H-Y and SAPO-5 [30]. For the zeolite H-Y, the experimental values of r_{HAl} are compared with theoretical data found by a computer simulation [31].

	line b	line c
Zeolite H-Y		
δ_H	4.0±0.1 ppm	5.0±0.1 ppm
$\Delta\sigma_H$	18.3±1.5 ppm	20.2±1.5 ppm
r_{HAl}	0.248±0.004 nm	0.237±0.04 nm
r_{HAl} [31]	0.239 nm (O_1H)	0.219 nm (O_3H)
Zeolite SAPO-5		
δ_H	3.8±0.1 ppm	4.8±0.1 ppm
$\Delta\sigma_H$	14.5±1.5 ppm	19.5±1.5 ppm
r_{HAl}	0.248±0.004 nm	0.234±0.004 nm

between the values of r_{HAl} for the line b and the bridging O_1H group on the one hand side and for line c and the bridging O_3H group on the other hand side can be taken as an experimental prove for the correlation

line b ⟷ bridging O_1H groups
line c ⟷ bridging O_3H groups

where O_1 - O_4 denote the four oxygen atoms in the basic tetrahedron of the H-Y zeolite framework [31]. For the bridging OH groups which point into the small cages (line c) the H-Al distance has been found to be 0.237 ± 0.004 nm for the zeolite H-Y and 0.234 ± 0.004 nm for the zeolite SAPO-5. In contrast, for the bridging OH groups in the large cavities the corresponding distances are equal and distinctly larger, viz. 0.248 ± 0.004 nm. Within the limits of error, the values for the anisotropy of the chemical shift are equal (19 ± 2 ppm) except for line b of the zeolite SAPO-5 which exhibits a smaller value (14.5 ± 2 ppm).

In a recent paper [32] the same method has been used to study systematically a larger variety of zeolites, viz. zeolites A, X, Y, erionite, mordenite, ZSM-5, SAPO-5, SAPO-17, SAPO-34 and SAPO-37. The measurements were performed at two different magnetic field strengths (B_0 = 7 T and 9.3 T) in order to prove the consistency of the method. The values determined for r_{HAl} were compared with the mean inner diameter $<d>$ of the 6-, 8-, 10-, and 12-membered oxygen rings which are constituents of the various zeolites and to which the OH groups are attached. By a linear regression the following equation could be derived

$$r_{HAl} \, / \, nm \; = \; 0.035 \, <d> \, / \, nm \; + \; 0.277 \tag{14}$$

with a correlation coefficient of 0.94. Hence, a determination of the H-Al distance by an analysis of the sideband pattern for a bridging OH group as described in [30] allows an estimate of the mean inner diameter $<d>$ of the oxygen ring to which the hydroxyl proton is attached by the relation

$$<d> \, / \, nm \; = \; 28.6 \, r_{HAl} \, / \, nm \; - \; 6.5. \tag{15}$$

which follows immediately from equ.(14). Such a measurement may be of practical importance since $<d>$ describes quantitatively the steric accessibility of the bridging OH groups (Brönsted acid sites).

Comparing the results obtained for r_{HAl} with the values found for δ_H and $\Delta\sigma_H$ in these studies [30], [32] one observes that there is apparently no relation between these latter quantities and the H - Al distance r_{HAl} of the bridging OH groups. This surprising experimental result is a challenge for further theoretical work on the structure of Brönsted acid sites in zeolites.

By a measurement of the ^{29}Si CP MAS NMR spectra (^1H - ^{29}Si cross polarization) of unloaded zeolites it could be shown [32] that for the zeolites containing only silicon and aluminium as cations in the framework, the bridging OH groups prefer a location at those silicon atoms for which the number z of aluminium in the second coordination sphere is as large as possible ($0 \leq z \leq 4$). For zeolites of SAPO-type which contain silicon, aluminium and phosphorous as cations in the framework, the second coordination sphere of the silicon atom which bears the bridging OH group is always occupied by the same number of aluminium and phosphorous atoms if the ratio Si/(Si+Al+P) does not exceed 0.12.

Solid-state deuterium NMR has been also used [33] to study hydroxyl groups in deuterium-exchanged zeolites. Through a line-shape analysis it was possible for a zeolite of type Y to separate the signals from Brönsted acid sites and silanols, and to determine the quadrupole frequency ν_Q (cf. section 2.1.) of the former sites as 351 ± 3 kHz. However, the resolution of this method is not sufficient to separate the lines from the two sorts of Brönsted acid sites (lines b and c in the ^{1}H MAS NMR spectra), of residual ammonium ions (line d), and of OH groups at extra-framework aluminium species (line e), so that at present ^{2}H NMR can not compete with ^{1}H MAS NMR in characterizing Brönsted acidity of zeolites and related catalysts.

1.3. ACCESSIBILITY AND INTERACTION WITH ADSORBED MOLECULES

Analogously to infrared spectroscopy, the accessibility of hydroxyl groups can be determined through a study of the ^{1}H MAS NMR spectrum after loading the adsorbent with probe molecules which , however, must be fully deuterated in order to exclude an unwanted additional ^{1}H MAS NMR signal. Although this method which is widely used in infrared spectroscopy seems to be self-explanatory, the results must be interpreted with caution since they are not always unambiguous. As a typical example, the influence of piperidine upon the *LF* band of bridging OH groups shall be mentioned. Although it is well-known that piperidine is too large to penetrate into the small cavities of Y zeolites where the respective OH groups are located, the *LF* band is strongly affected after adsorption of piperidine, in contrast to the weaker base pyridine which leaves the *LF* band unchanged as it should be.

In principle, six possibilities must be taken into consideration in a comparison of OH spectra measured before and after loading the catalyst with probe molecules: An OH group may be (1) non-accessible due to steric reasons and fixed in space, (2.1) accessible, but the probe molecules are adsorbed at other (stronger) adsorption sites, (2.2) accessible with a proton exchange between the OH group and the adsorbed probe molecule, (2.3) accessible with a hydrogen bond formation to the adsorbed probe molecule, (2.4) accessible with a proton transfer to the adsorbed probe molecule (formation of carbocations), (3) indirect accessible due to a migration of the hydroxyl protons over the inner oxygen surface of the zeolite (cf. Table 4).

Table 4. Six possibilities which must be taken into consideration in a comparison of OH spectra measured before and after loading the catalyst with probe molecules.

      non-accessible OH groups (1)
          (steric barrier, no proton migration)
      accessible OH groups (2)
          no interaction due to other (stronger) adsorption sites (2.1)
          proton exchange (2.2)
          formation of a hydrogen bond (2.3)
          formation of a carbocation (2.4)
      indirect accessible OH groups (3)
          (proton migration)

This proton migration together with the stronger basicity of piperidine is the reason for the above mentioned influence of piperidine upon the OH stretching vibration band of the bridging OH groups located in the small cavities of Y zeolites (LF band).

As an example for (1) and (2.3) the interaction between non-acidic OH groups (silanols) and pyridine shall be mentioned. Using deuterated pyridine, the concentrations of accessible and non-accessible silanol groups of silica can be determined [34] since the formation of a hydrogen bond between pyridine and the silanol group shifts the ^{1}H MAS NMR signal of the latter by ca. 8 ppm to higher values of δ_H. This effect was also observed for line a of zeolites H-ZSM-5 [35].

In the case of acidic OH groups (Brönsted acid sites) however, the adsorption of pyridine leads to a protonation (cf. (2.4) in Table 4), i.e. to a formation of pyridinium ions with a larger shift of ca. 12 ppm to higher values of δ_H. This has been shown by loading a zeolite H-ZSM-5 containing ca. $5 \cdot 10^{20}$ Brönsted acid sites per g zeolite and only a negligible concentration of silanols, i.e. the ^{1}H MAS NMR spectrum of the unloaded zeolite exhibits only line b, with a concentration of $6 \cdot 10^{20}$ deuterated pyridine molecules per g zeolite [35].

Loading zeolites H-Y with deuterated pyridine, the following changes in the ^{1}H MAS NMR spectra could be observed: Small concentrations of pyridine corresponding to not more than 1 pyridine molecule per large cavity, give rise to the signal of pyridinium ions at ca. 16 ppm with a corresponding decrease of the intensity of line b. At higher concentrations, at first line b disappears and finally also line c. Hence, it must be concluded that (i) lines b and c are caused by acidic OH groups (bridging hydroxyls) located in the large and small cavities of the zeolites H-Y, respectively, and that (ii) the bridging OH groups located in the small cavities are indirect accessible (cf.(3) in Table 4) in the time scale of NMR spectroscopy (cf. section 2.2.).

The same effect of an indirect accessibility is demonstrated by Fig. 3 through proton deuteron exchange (cf. (2.2) in Table 4) between Brönsted acid sites of a zeolite SAPO-5 and deuterated cyclohexane: After adsorption of deuterated cyclohexane and

Fig.3.
^{1}H MAS NMR spectra of a SAPO-5 with $5 \cdot 10^{20}$ bridging OH groups per g zeolite: (1) Unloaded specimen; (2) loaded with $6 \cdot 10^{20}$ deuterated cyclohexane molecules per g zeolite after keeping the sample 2 hours at 320 K; (3) the same as (2) but 10 days at room temperature [27]. The asterisks denote spinning sidebands.

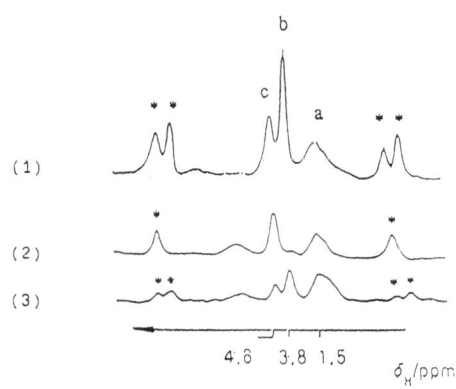

keeping the sample for 2 hours at a temperature of 320 K, only the line at about 3.8 ppm (line b) disappears due to the deuteration of the corresponding bridging hydroxyls, while the line at ca. 4.8 ppm (line c) is unaffected. After storing the sample 10 days at room temperature, the two lines reappeared but with a reduced intensity as a result of deuteron exchange among all bridging hydroxyls [27]. In contrast, the use of a stronger base like pyridine leads immediately after the loading of the SAPO-5 to the disappearance of both lines while simultaneously a signal at 16 ± 1 ppm due to pyridinium ions appears [36]: The bridging OH groups which are not accessible for pyridine due to steric reasons (line c) migrate to the main channels of SAPO-5 where they protonate, as a result of the higher proton affinity of this probe molecule, the pyridine molecules in the same way as the bridging OH groups which are located in these channels (line b).

In principle, also water may be used to study the accessibility of Brönsted acid sites and their interaction with a typical polar molecule. However, ^1H MAS NMR studies of *hydrated* zeolites will be complicated in general by a superposition of three effects:

(i) Hirschler-Plank-mechanism, i.e. the adsorption and dissociation of water molecules on extra-framework multivalent cations like Ca^{2+} with a formation of cationic and bridging OH groups [37]. The ^1H MAS NMR signals of the former hydroxyls appear at δ_H values of ca. 0 ppm and 2.8 ppm for calcium ions in the large and small cavities of zeolites Y, respectively [38], and, as has been mentioned already, the empirical relation (12) is not fulfilled for these OH groups. Up till now a simple interpretation of this experimental result cannot be given.

(ii) Formation of hydroxonium ions (chemical shift δ_{h+}) at Brönsted acid sites which, however, take part in a fast proton exchange with physically adsorbed water molecules (chemical shift δ_{ph}) and with accessible bridging OH groups (chemical shift d_{OH}). Physically adsorbed water molecules may include water molecules adsorbed on Na^+ as well as water molecules hydrogen bonded to bridging OH groups, to other water molecules and to silanol groups (adsorption energy 117 kJ/mol, 58.4 kJ/mol, 20.1 kJ/mol, and 16.4 kJ/mol, respectively [39]). Due to the fast exchange, the resulting ^1H MAS NMR shift δ is given by the following equation

$$\delta = [(c_{OH} - c_{h+}/3) \delta_{OH} + (c_w - 2c_{h+}/3) \delta_{ph} + c_{h+} \delta_{h+}] / (c_{OH} + c_w). \quad (16)$$

In this formula c_{OH}, c_{h+}, and c_w denote the concentration of the ^1H nuclei in the accessible bridging OH groups, in the hydroxonium ions, and in the physically adsorbed water molecules, respectively. With $\delta_{OH} = 4.3$ ppm, $\delta_{ph} = 3.2 - 4.8$ ppm where the upper limit is for hydrogen bonded water molecules, and $\delta_{h+} = 13$ ppm [40] it is possible to determine quantitatively the concentration of hydroxonium ions from the position δ of the ^1H MAS NMR signal. In shallow-bed (400 °C) treated zeolites H-Y the probability to find a water molecule in the state of a hydroxonium ion is ca. 25 ± 5 % for a rehydration corresponding to one water molecule per accessible bridging OH group (line b) [40]. Another method to study the formation of hydroxonium ions in hydrated zeolites was used by J. Fraissard, C. Doremieux-Morin et al. [41]. Through a measurement at very low temperatures, the fast proton exchange among the various species can be excluded. Unfortunately, the line widths become so large at these temperatures that the valuable information contained in the chemical shifts is lost. Therefore, the shape of the wide ^1H NMR line must be analysed in order to separate the contributions of the various proton-bearing species. For more details the reader is referred to the review of C. Doremieux-Morin in the present proceedings.

(iii) Adsorption of water molecules on Lewis acid sites giving rise to a narrow line at $\delta_H = 6.5$ ppm. In the case of hydrothermally pretreated zeolites H-Y (540 °C; 20 h; 4 kPa water vapour pressure) a concentration of 0.25 ± 0.06 Lewis acid sites of this type per large cavity could be found, and it is not sure at present, whether these sites are identical with extra-framework aluminium species or whether they must be ascribed to defects in the zeolite framework [40]. With increasing content of extra-framework aluminium, the situation becomes even more complicated. Harris et al. [42] have shown by a thorough investigation of the ^{1}H MAS NMR spectra of steamed zeolites H-Y that the extra-framework aluminium material changes in concentration level by reacting with water: In the dehydrated zeolite the material may be so highly condensed that it bears only few terminating OH groups while during sorption of water this 'alumina' is hydrolyzed to form material which is less concentrated and richer in hydroxyl groups.

2. Lewis Acidity

Apart from relatively weak signals in the IR spectra of hydrogen [43] and carbon monoxide [44] which were ascribed to an adsorption of these molecules at defects of the zeolite framework, it is generally accepted [45] that Lewis acidity of zeolites and and of other oxidic catalysts like silica-alumina or γ-alumina is connected with the presence of aluminium species on the surface. Therefore two possibilities exist to study Lewis acid sites by NMR methods:

(i) an analysis of highly resolved ^{27}Al NMR spectra similar to the ^{1}H MAS NMR spectroscopy described in section 1, and

(ii) the use of probe molecules as e.g. pyridine [46] or the above mentioned molecules in IR spectroscopy.

2.1. ALUMINIUM-27 NMR SPECTROSCOPY

In well-crystallized zeolites which do not exhibit Lewis acidity, aluminium is tetrahedrally coordinated with an isotropical chemical shift between 55 and 65 ppm with respect to an aqueous dilute solution of Al^{3+} [24].

After a pretreatment procedure affecting the zeolite lattice however, extra-framework aluminium species are formed (Lewis acid sites) and the ^{27}Al MAS NMR spectrum becomes more complicated: The intensity of the initial signal at ca. 60 ppm decreases due to ejection of aluminium from the framework into the intracrystalline space and on the other hand a signal appears at the same position which is caused by tetrahedrally coordinated extra-framework aluminium. This latter aluminium must exist at least partly as $Al(OH)_n$ species as could be shown by ^{1}H - ^{27}Al cross polarization (CP MAS NMR) experiments [47].

The other extra-framework aluminium species give rise to distinct resonance lines in the interval between -15 and +4 ppm, which could be assigned to polymeric aluminium species [48], and to a line near 30 ppm which is interpreted as being due to either penta-coordinated [49] or tetrahedrally coordinated [50], [51], [52] extra-framework aluminium. In addition, there appears a very broad hump below these more distinct resonances which extends from ca. -180 ppm to ca. 230 ppm for a resonance frequency of 78 MHz. As has been shown recently [47], this broad hump merges into the lines near 0, 30, and 60 ppm for ultrahigh MAS speeds ($\Omega/2\pi \geq$ 18 kHz). In *hydrated* samples a narrow line at 0 ppm is often observed which results from a few Al^{3+} cations balancing the negative charge of the zeolite framework.

The quadrupole interaction is described quantitatively by the quadrupole frequency

$$\nu_Q = 3 e^2 q Q / [2I (2I - 1) h] \tag{17}$$

with eq, eQ, I and h denoting the electric frield gradient, the electric quadrupole moment, the spin (I = 5/2 for ^{27}Al), and Planck's constant, respectively.

In recent ^{27}Al NMR experiments performed at a resonance frequency of ca. 130 MHz with a non-spinning sample [53], the following values for the ^{27}Al quadrupole frequency ν_Q have been found: 0.97 ± 0.05 MHz, 3.5 ± 0.5 MHz and 3.7 ± 0.3 MHz for the framework aluminium nuclei of zeolites Na-Y, H-Y and H-ZSM-5, respectively. Hence, it must be assumed that for extra-framework aluminium species with a lower symmetry the values of ν_Q may exceed even 3 - 4 MHz.

Large values of ν_Q give rise to a large second-order quadrupolar shift [55]

$$\nu_{CG} - \nu_L = -\nu_Q^2 \ [I(I+1) - 3/4] \ (1 + \eta^2/3) / (30\nu_L) \tag{18}$$

where ν_{CG} and ν_L denote the centre of gravity of the signal and the Larmor frequency (signal for $\nu_Q = 0$) in Hz, respectively. η is the asymmetry parameter of the quadrupole interaction $(0 \leq \eta \leq 1)$.

In addition to this effect which must be taken into consideration if one wants to determine the true chemical shift of a ^{27}Al NMR signal, a strong quadrupole interaction leads to a dramatic broadening of the lines: The widths of the spectra for a static sample $(\Delta\nu)$, for a sample under conditions of magic angle spinning $(\Delta\nu^{MAS})$ and under conditions of double rotation $(\Delta\nu^{DOR})$ are given by [54], [55], [56]:

$$\Delta\nu = \nu_Q^2 (25 + 22\eta + \eta^2) (18\nu_L)^{-1} \tag{19}$$

$$\Delta\nu^{MAS} = \nu_Q^2 (36 + 12\eta + \eta^2) (63\nu_L)^{-1} \quad \text{if} \quad \Omega \geq \Omega_{cr}^{MAS} \tag{20}$$

$$\Delta\nu^{DOR} = \Delta\nu_C \quad \text{if} \quad \Omega_i = 5\Omega_o \geq \Omega_{cr}^{DOR} \tag{21}$$

where Ω is the spinning rate of the rotor in the case of MAS, and Ω_i and Ω_o denote the corresponding rates for the inner and outer rotor of DOR, respectively. The critical values for the rates are given by

$$\Omega_{cr}^{MAS} / 2\pi = \Delta\nu^{MAS} \tag{22}$$

$$\Omega_{cr}^{DOR} / 2\pi = 5 \Delta\nu_D. \tag{23}$$

$\Delta\nu_C$ denotes the line width due to a distribution of the chemical and / or quadrupole shift and $\Delta\nu_D$ the line width caused by the *homonuclear* magnetic dipole-dipole interaction. It is the influence of these two latter quantities which determines the ultimate resolution of ^{27}Al MAS and of ^{27}Al DOR NMR spectra.

For a Larmor frequency ν_L of 130 MHz, a quadrupole frequency ν_Q of ca. 3 MHz (aluminium species of low symmetry), an axial symmetry of the electric quadrupole interaction ($\eta = 0$) and an estimated mean distance of ca. 0.3 nm between neighbouring aluminium nuclei (condensed extra-framework material) which corresponds to $\Delta\nu_D \approx 5$ kHz, the following values can be derived: $\Delta\nu = 95.9$ kHz (equ. (19)), $\Omega_{cr}^{MAS}/2\pi = 39.5$ kHz (equs. (22) and (20)) and $\Omega_{cr}^{DOR}/2\pi = 25$ kHz (equ. (23)). An inspection of these values shows that the conditions of equs. (20) and (21) cannot be fulfilled experimentally so that we must state that neither by an application of the ^{27}Al MAS NMR nor of the ^{27}Al DOR NMR spectroscopy it will be possible to detect extra-framework condensed aluminium species of low symmetry.

2.2. SPECTROSCOPY OF PROBE MOLECULES

In analogy to infrared spectroscopy, probe molecules can be used to study Lewis acid sites. In the case of NMR spectroscopy however, due to the much smaller resonance frequencies which correspond to a much longer time scale, exchange effects may lead to an average line instead of the series of separate lines caused by the molecules adsorbed on the various adsorption sites: From Heisenberg's uncertainty relation the critical value τ_{crit} of the mean residence time τ for the exchange between two states can be estimated to be

$$\tau_{crit} = (2 \pi \Delta \nu)^{-1} \tag{24}$$

so that for $\tau \gg \tau_{crit}$ the two lines appear separately with a frequency distance $\Delta \nu$ while for $\tau \ll \tau_{crit}$ they merge into a single line.
In infrared spectroscopy the typical bands of pyridine adsorbed on Brönsted and Lewis acid sites appear at ca. 1540 cm^{-1} and 1450 cm^{-1}, respectively [46], so that the value of τ_{crit} is ca. 0.1 ps. For the ^{15}N NMR of the same molecule at a resonance frequency of 30 MHz, the frequency difference between the two lines is ca. 2 kHz (cf. Table 5) corresponding to $\tau_{crit} \approx 0.1$ ms, which is nine orders of magnitude larger than the value for the IR bands. Nevertheless, also in the case of NMR spectroscopy probe molecules can be found where the spectra exhibit various lines so that at least approximately the limiting case of slow exchange ($\tau \gg \tau_{crit}$) can be assumed for an analysis of these spectra.

Table 5. Values for the ^{15}N NMR shift of pyridine [57], [58], [59] and for the ^{31}P NMR shift of trimethylphosphine [61], [62] which have been used to study Lewis acidity.

species	pyridine δ_N/ppm	trimethylphosphine δ_P/ppm
solid (-105°C)	0	
liquid	-26	0
physisorbed	-10 ± 10	0.7 ± 6
protonated (Brönsted acid sites)	88 ± 2	59 ± 2
ads. on tetr. Al^{3+} (Lewis acid sites)	22.5 ± 3.5	12 ± 10
ads. on oct. Al^{3+} (Lewis acid sites)	46.5 ± 7.7	12 ± 10

^{15}N CP MAS NMR spectra of adsorbed pyridine have been studied by Maciel et al. [57] (pyridine on silica-alumina), Ripmeester [58] (pyridine on γ-alumina, mordenite), and by Majors and Ellis [59] (pyridine on γ-alumina). Values for the resonance shifts relative to solid pyridine are collected in Table 5. The major drawbacks of these experiments are that (i) an absolute determination of concentrations is connected with large errors due to the strong dependence of the line intensities on quantities controlling the efficiency of cross-polarization, see e.g. [60], and (ii) it is not sure that the exchange rate of the molecules among the various adsorption sites is sufficiently small so that the resonance positions and intensities derived from the spectra may be only apparent quantities.

Compared with pyridine, phosphines are roughly three orders of magnitude stronger bases. ^{31}P CP MAS NMR spectra of trimethylphosphine adsorbed on zeolite H-Y and on γ-alumina were investigated by Lunsford et al. [61], and of various trialkylphosphines adsorbed on silica-alumina and on γ-alumina by Maciel et al. [62]. Values for the resonance shift relative to liquid trimethylphosphine are also shown in Table 5. For this probe molecule the same drawbacks hold as mentioned above for pyridine, although due to the larger signal-to-noise ratio of the ^{31}P NMR signal compared with ^{15}N, cross polarization must not be applied. The sensitivity of trimethylphoshine to distinguish between different sorts of Lewis acid sites seems to be less than that of pyridine since for γ-alumina only one sort of Lewis acid sites could be found (see Table 5).

Another way to study quantitatively Lewis acidity results from the fact that probe molecules could be found for which the resonance shift caused by Lewis acid sites is much larger than that caused by Brönsted acid sites or physisorption so that the system can be treated as if an exchange takes place only between molecules adsorbed on Lewis acid sites and molecules in the 'free' state (all other sites including physically adsorbed molecules). Typical examples are carbon monoxide [63], [64], [65] and dinitrogen oxide [66]. Since at room temperature and above, these molecules exchange relatively rapid between the various adsorption sites, only a single line for the ^{13}C NMR of carbon monoxide and for the ^{15}N NMR of the terminal nitrogen of dinitrogen oxide appears. The position of this single line is given by

$$\nu = (c_{LM} / c_M) \nu_{LM} + [(c_M - c_{LM}) / c_M] \nu_M. \tag{25}$$

c_M is the total concentration of the probe molecules and c_{LM} the concentration of Lewis acid sites occupied by a probe moleluce. ν_{LM} and ν_M denote the resonance frequencies of a probe moleclue if it is adsorbed on a Lewis acid site and if it is in the 'free' state, respectively. If one introduces the resonance shifts with regard to ν_M:

$$\delta = (\nu_M - \nu) / \nu_M \tag{26}$$

$$\delta_{LM} = (\nu_M - \nu_{LM}) / \nu_M \tag{27}$$

equ. (25) simplifies to

$$\delta = (c_{LM} / c_M) \delta_{LM}. \tag{28}$$

The total concentration c_L of the Lewis acid sites is connected with c_{LM} and c_M appearing in equ. (28) by the relation

$$c_{LM} = K (c_M - c_{LM}) (c_L - c_{LM}) \tag{29}$$

where K denotes the equilibrium constant of the exchange reaction. Therefore, by the use of equs. (28) and (29), a measurement of δ as a function of the concentration c_M of the probe molecules will yield results for both quantities of interest, viz. the resonance shift δ_{LM} of the probe molecule adsorbed on a Lewis acid site (measure for the strength of acidity) and the concentration c_L of these sites [14], [67].

The main disadvantage of this method results from the fact that with decreasing loading of the zeolites ($c_M \rightarrow 0$), the signal-to-noise ratio decreases so that especially in those cases where the absolute value of the slope for the plot of δ versus c_M increases strongly at the lowest loadings, the values of δ_{LM} and c_L are connected with large errors and yield too small values for δ_{LM} in general. Therefore, the numerical results for δ_{LM} which have been found in [64] and which are between 300 ppm and 400 ppm (see Table 6) should be taken as a lower limit.

Table 6. Values for the ^{13}C NMR shift of carbon monoxide [63], [64], [65] and for the ^{15}N NMR shift of dinitrogen oxide [66] which have been used to study Lewis acidity.

species	carbon monoxide δ_C / ppm	dinitrogen oxide δ_N / ppm
gas	0	0
physisorbed	-3 ± 2	4.5 ± 3
adsorbed on Brönsted acid sites	-3 ± 2	4.5 ± 3
adsorbed on Lewis acid sites	300 - 400 [64] 440 - 590 [65]	50 ± 20

In a subsequent study [65], the method of selective saturation introduced by Forsen and Hoffmann [68] has been applied to carbon monoxide adsorbed on various zeolites containing Lewis acid sites: Through a measurement of the ^{13}C NMR signal intensity as a function of the frequency of a selective ^{13}C saturation pulse, the accuracy for the value of δ_{LM} could be enhanced. As an example, in Fig. 4 the relative intensity of the ^{13}C NMR signal of carbon monoxide adsorbed on silicalite (A) and on a deep-bed pretreated zeolite H-ZSM-5 (B) is plotted as a function of the difference between the resonance frequency of the ^{13}C NMR signal (ν_{CO}) and the frequency of the saturation pulse (ν). Silicalite is an aluminium-free zeolite ZSM-5 and does not contain therefore Lewis acid sites in contrast to the deep-bed treated zeolite H-ZSM-5 for which a numerical analysis of the curve plotted in Fig. 4 yields $\delta_{LM} = 590 \pm 60$ ppm [65]. Depending on the zeolites studied and on their pretreatment, values for δ_{LM} have been found between 440 ppm and 590 ppm with an error of ca. 60 ppm (cf. Table 6).

In Table 6 results are also collected for the resonance shift of dinitrogen oxide physisorbed and adsorbed on Brönsted and Lewis acid sites [66]. The latter result has been derived by the authors using the method described above. The error given in Table 6 was added after considering the uncertainties of the experimental values and their influence upon the value of δ_{LM}.

In all cases described in the present section, the accuracy is far from the goal to take the values of the resonance shift for the molecules adsorbed on Lewis acid sites as a measure for their strength of acidity. Assuming a medium value for this resonance shift however, the use of rapidly exchanging probe molecules like carbon monoxide or dinitrogen oxide seems suitable to determine the concentration of Lewis acid sites in dependence on the pretreatment of a given zeolite catalyst.

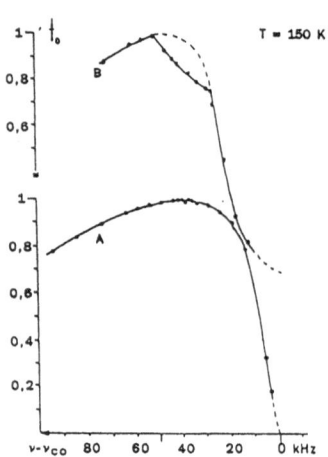

Fig. 4.
Relative intensity of the ^{13}C NMR signal of carbon monoxide adsorbed on silicalite (A) and on a deep-bed pretreated zeolite H-ZSM -5 (B) as a function of the difference between the resonance frequency of the ^{13}C NMR signal (ν_{CO}) and the frequency of the saturation pulse (ν) [65].

Summarizing the results obtained up till now with probe molecules in NMR spectroscopy, the following statements can be made:

1. At room temperature and above no separate lines for carbon monoxide and dinitrogen oxide adsorbed on Lewis acid sites can be observed due to fast molecular exchange. Hence, a measurement of coverage dependence or of selective saturation spectra becomes necessary, with the former yielding large errors for the chemical shift (strength of acidity) and for the concentration of the Lewis acid sites.

2. In those cases where due to a poor signal-to-noise ratio cross polarization has been applied (pyridine, trimethylphosphine) only a semiquantitative determination of the concentration of Lewis acid sites is possible.

3. Steric effects connected with larger probe molecules may lead to large errors. As an example the result presented in a paper of Lunsford et al. [69] shall be mentioned: With trimethylphosphine as a probe molecule, 37 protonated species per unit cell have been found for a zeolite H-Y in contrast to the real value of 54 Brönsted acid sites (framework aluminium atoms) per unit cell.

Acknowledgement

I wish to thank Dr. Horst Ernst for reading the manuscript and many helpful comments.

3. References

[1]   D. Freude, J. Klinowski, H. Hamdan: Chem. Phys. Lett. 149 (1988) 355.
[2]   D. Fenzke, B. C. Gerstein, H. Pfeifer: J. Magn. Res. 98 (1992) 469.
[3]   E. Brunner: J. Chem. Soc., Faraday Trans. 89 (1993) 165.
[4]   H. Pfeifer, H. Ernst: Annual Reports on NMR Spectroscopy, Vol. 28, Academic Press, London 1993.
[5]   S. Halik, J. A. Lercher, H. Mayer: J. Chem. Soc., Faraday Trans.I 84 (1988) 4457.
[6]   H. Pfeifer, D. Freude, J. Kärger: Studies in Surface Science and Catalysis 65 (1991) 89.
[7]   J. Sauer: Mol. Cat. 54 (1989) 312.
[8]   J. P. Yesinowski, H. Eckert: J. Am. Chem. Soc. 109 (1987) 6274.
[9]   B. Berglund, R. W. Vaughan: J. Chem. Phys. 73 (1980) 2037.
[10] J. P. Chauvel, N. S. True: Chem. Phys. 95 (1985) 435.
[11] J. E. Bartmess J. A. Scott, R. T. McIver: J. Am. Chem. Soc. 101 (1979) 6046.
[12] R. W. Taft, I. A. Koppel, R. D. Topsom, F. Anvia: J. Am. Chem. Soc. 112 (1990) 2047.
[13] U. Fleischer, W. Kutzelnigg, A. Bleiber, J. Sauer: J. Am. Chem. Soc. (submitted).
[14] H. Pfeifer: NMR, Basic Principles and Progress, Vol. 31, Springer Verlag Berlin, 1993 (in press).
[15] R. T. Sanderson: Chemical Bonds and Bond Energy, Academic Press, New York 1976.
[16] E. Brunner: J. Chem. Soc., Faraday Trans. 86 (1990) 3957.
[17] E. Brunner, D. Freude, B. C. Gerstein, H. Pfeifer: J. Magn. Res. 90 (1990) 90.
[18] D. Fenzke, M. Hunger, H. Pfeifer: J. Magn. Res. 95 (1991) 477.
[19] C. A. Fyfe: Solid State NMR for Chemists, C.F.C. Press, Guelph 1983.
[20] H. Pfeifer, D. Freude, M. Hunger: Zeolites 5 (1985) 274.
[21] E. Lippmaa, A. V. Samoson, V. V. Brey, Y. I. Gorlov: Doklady Akad. Nauk 259 (1981) 403.
[22] C. E. Bronnimann, R. C. Zeigler, G. E. Maciel: J. Am. Chem. Soc. 110 (1988) 2023.
[23] G. E. Maciel, D. W. Sindorf: J. Am. Chem. Soc. 102 (1980) 7606.
[24] G. Engelhardt, D. Michel: High-Resolution Solid-State NMR of Silicates and Zeolites, J. Wiley and Sons, Chichester 1987.
[25] D. Freude, M. Hunger, H. Pfeifer, W. Schwieger: Chem. Phys. Lett. 128 (1986) 62.
[26] P. G. Clarke, K. Gosling, R. K. Harris, E. G. Smith: preprint 1993.
[27] M. Hunger: Thesis (Dr. habil.), University of Leipzig 1991.
[28] E. Brunner, H.-G. Karge, H. Pfeifer: Z. phys. Chemie 176 (1993) 173.
[29] J. Dwyer, J. Dewing, N. E. Thompson, P. J. O'Malley, K. Karim: J. Chem .Soc., Chem. Comm. 1989, 843.
[30] D. Fenzke, M. Hunger, H. Pfeifer: J. Magn. Res. 95 (1991) 477.
[31] K.-P. Schröder, J. Sauer, M. Leslie, C. R. A. Catlow, J. M. Thomas: Chem. Phys. Lett. 188 (1992) 320.
[32] M. Hunger, M. W. Anderson, A. Ojo, H. Pfeifer: Microporous Materials 1 (1993) 17.

276

[33] T. J. Gluszak, D. T. Chen, S. B. Sharma, J. A. Dumesic, T. W. Root: Chem. Phys. Lett. 190 (1992) 36.

[34] H. Rosenberger, H. Ernst, G. Scheler, I. Jünger, R. Sonnenberger: Z. phys. Chemie, Leipzig 263 (1982) 864.

[35] M. Hunger, D. Freude, T. Fröhlich, H. Pfeifer, W. Schwieger: Zeolites 7 (1987) 108.

[36] B. Zibrowius, E. Löffler, M. Hunger: Zeolites 12 (1992) 167.

[37] A. E. Hirschler: J. Catal. 2 (1963) 428;
C. J. Plank: Proceedings of the Third International Congress on Catalysis, Amsterdam 1964, Vol. 1, p. 568.

[38] M. Hunger, D. Freude, H. Pfeifer, D. Prager, W. Reschetilowski: Chem. Phys. Lett. 163 (1989) 221.

[39] J. Sauer: Proceedings of the Intern. Symp. on Zeolite Catalysis, Siofok (Hungary) 1985, 19;
J. Sauer, P. Hobza: Theor. Chim Acta (Berlin) 65 (1984) 279 and 291.

[40] M. Hunger, D. Freude, H. Pfeifer: J. Chem. Soc., Faraday Trans. 87 (1991) 657.

[41] P. Batamack, C. Doremieux-Morin, J. Fraissard, D. Freude: J. Phys. Chem. 95 (1991) 3790.

[42] P. G. Clarke, K. Gosling, R. K. Harris, E. G. Smith: preprint 1993.

[43] L. M. Kustov, V. B. Kazanski: J. Chem. Soc., Faraday Trans. 87 (1991) 2675.

[44] L. M. Kustov, V. B. Kazanski, S. Beran, L. Kubelkova, P. Jiru: J. Phys. Chem. 91 (1987) 5247.

[45] J. Dwyer: Innovation in Zeolites Materials Science, Elsevier, Amsterdam 1987/1988, p.333.

[46] H. G. Karge: Studies in Surface Science and Catalysis, Vol. 65, Elsevier, Amsterdam 1991, p.133.

[47] L. Kellberg, M. Linnen, H. J. Jakobson: Chem. Phys. Lett. 182 (1991) 120.

[48] D. Müller, W. Gessner, H. J. Behrens, G. Scheler: Chem. Phys. Lett. 79 (1981) 59.

[49] J. P. Gilson, G. C. Edwards, A. W. Peters, K. Rajagopalm, R. F. Wormbecher, T. G. Roberie, M. P. Shatlock: J. Chem. Soc., Chem. Comm. 1987, 91.

[50] D. Freude, M. Hunger, H. Pfeifer: Z. phys. Chemie (NF) 152 (1987) 171.

[51] P. J. Grobet, H. Geerts, J. A. Martens, P. A. Jacobs: Stud. Surf. Sci. Catal. 46 (1989) 721.

[52] J. Klinowski, C. A. Fyfe, G. C. Gobbi: J. Chem Soc., Faraday Trans.I 81 (1985) 3003.

[53] I. Wolf, H. Ernst, D. Freude: 5. Deutsche Zeolith-Tagung, Leipzig 1993.

[54] A. Samoson, A. Pines: Rev. Sci. Instr. 60 (1989) 3239.

[55] D. Freude, J. Haase: NMR - Basic Principles and Progress (in press).

[56] D. E. Woessner, H. K. C. Timken: J. Magn. Res. 90 (1990) 411.

[57] G. E. Maciel, J. F. Haw, I. Chuang, B. L. Hawkins, T. A. Early, D. R. McKay, L. Petrakis: J. Am. Chem. Soc. 105 (1983) 5529.

[58] J. A. Ripmeester: J. Am. Chem. Soc. 105 (1983) 2925.

[59] P. D. Majors, P. D. Ellis: J. Am. Chem. Soc. 109 (1987) 1648.

[60] D. Schulze, H. Ernst, D. Fenzke, W. Meiler, H. Pfeifer: J. Phys. Chem. 94 (1990) 3499.

[61] J. H. Lunsford, W. P. Rothwell, V. Shen: J. Am. Chem. Soc. 107 (1985) 1540.

[62] L. Baltusis, J. S. Frye, G. E. Maciel: J. Am. Chem. Soc. 109 (1987) 40.
[63] A. Michael, W. Meiler, D. Michel, H. Pfeifer: Chem Phys. Lett.
      84 (1981) 30.
[64] A. Michael, W. Meiler, D. Michel, H. Pfeifer, D. Hoppach, J. Delmau:
      J. Chem. Soc., Faraday Trans.I 82 (1986) 3053.
[65] T. Wutscherk: Thesis, University of Leipzig, 1990.
[66] V. M. Mastikhin, J. L. Mudrakowsky, S. V. Filimonova: Chem. Phys.
      Lett. 149 (1988) 175.
[67] V. J. Borovkov, G. M. Zhidomirov, V. B. Kazanski: Zhurn. Strukt. Chim.
      16 (1975) 308.
[68] S. Forsen, R. A. Hoffmann: J. Chem. Phys. 39 (1963) 2892, 40 (1964) 1189.
[69] J. H. Lunsford, P. N. Tutunjian, P. Chu, E. B. Yeh, D. J. Zalewski:
      J. Phys. Chem. 93 (1989) 2590.

CHARACTERIZATION OF BRÖNSTED ACIDITY OF SOLIDS BY RIGID-LATTICE BROAD-LINE ^{1}H-NMR. PART 1 : METHODOLOGY

C. DOREMIEUX-MORIN
Laboratoire de Chimie des Surfaces
Associé au CNRS - URA1428
Université P. et M. Curie
Casier 196, Tour 55
4, place Jussieu
F 75252 PARIS CEDEX 05

ABSTRACT. The methodology for studying the Brönsted acid strength of solids quantitatively using simulation of ^{1}H broad-line NMR spectra recorded at 4 K is described. The oxy-protonated species formed by interaction of water molecules with the acid sites are identified and quantified.

1. Introduction

There are many methods for investigating the number of Brönsted acid sites in solids. Among them are chemical analysis and physicochemical methods, such as NMR, which is quantitative. However, it is still difficult to know the strength of such sites (1-4).

If we recall results expressed in previous lectures, it appears that O-H bonds of Brönsted acid sites in solids (denoted SOH) are often weakly ionized, even when these sites act as strong acids in chemical reactions (4). The zeolite O-H groups carry hydrogen atoms with only a fractional positive electronic charge (0.12-0.25e⁻) (4). Three examples confirm this assertion : (i) the ^{1}H chemical shift obtained for SOH in zeolites (4.3-5.5 ppm) (5-7) is small compared to that calculated for bare H^+ (about 30 ppm) ; (ii) the low ionicity of strong Brönsted acids such as amorphous polyphosphoric acids was proved long ago (8); (iii) the fact that OH group vibrations can be studied by IR spectroscopy (9-17) is not compatible with a total ionic character of the bond between these atoms. These results mean that the intra-bond in OH is mostly soft base - soft acid type. Therefore using physicochemical methods on the solid acids themselves cannot tell us much about the acid strength.

The solution to the problem is to study the interaction of the Brönsted acid sites with a base (4). The choice of this base needs some care : (i) the molecules must be small enough relative to the porosity of the solid to gain access to all the acid sites ; (ii) products of the interaction must be characterizable qualitatively and quantitatively ; (iii) the base strength and hardness must be such that the reaction is equilibrated ; (iv) the results should be related to a preexisting scale of acidity (e.g.for liquids).

Using H_2O as the base meets some of the above criteria : the molecule is small enough and the basic strength and hardness of water prove to be convenient. Moreover, water is the best base to

279

J. Fraissard and L. Petrakis (eds.), Acidity and Basicity of Solids, 279–290.
© 1994 *Kluwer Academic Publishers.*

initiate a comparison with the pH scale in aqueous solutions ; though "dilution conditions" are never achieved in such heterogeneous environments, the interacting species can be compared with those in aqueous media. The problem which remains is to characterize and quantify the oxy-protonated species formed. Compared to other non destructive spectroscopic methods, NMR methods have the advantage of being quantitative. ^{1}H, with a spin 1 / 2, is the most convenient nucleus to observe in NMR experiments, because of its 100 % natural abundance and maximum sensitivity. Therefore, also from this standpoint, water appears as an interesting base for investigating the Brönsted acid strength of solids. Another reason argues in favour of water : it contains only two H atoms per molecule. The importance of this point will appear in the "interpretation" section, where the procedure for simulating rigid-lattice broad-line ^{1}H NMR spectra is described. There is, however, a disadvantage in this choice : water molecules are numerous in our environment, interacting and exchanging with hydrogen atoms very easily. For this reason, most of the samples need to be prepared out of contact with air and maintained in sealed glass ampules for the experiments. Moreover, in spite of all precautions, NMR probeheads often contain some hydrogen atoms. Spectra of these probeheads must be recorded under conditions identical with those chosen for the samples and subtracted from them.

Our purpose is to study quantitatively the acid strength of solids since we know from ^{1}H NMR the species formed by the interaction between water molecules and the SOH groups. A priori the equilibrium would be :

$$SOH + H_2O = H_3O^+ + SO^-. \qquad [1]$$

In fact, because water molecules can form hydrogen bonds with SOH the results that we have to interpret are more complicated but also more extensive than the above equilibrium would suggest. The concentrations of the ($H_2O...HOS$) hydrogen-bonded species must also be considered and determined (18,19).

By means of the high resolution NMR method for solids, denoted "magic angle spinning" (MAS) chemical shifts and concentrations of the H atoms of the expected oxy-hydrogenated species can in principle be measured (5-7). However, at room temperature, fast chemical exchange takes place between some of these H atoms. All the exchanging H atoms resonate therefore at a weighted mean value of their individual chemical shifts. No individual chemical shifts are known ; moreover, they are strongly dependent on the hydrogen bonds formed between species (20-24). The result is that ^{1}H MAS NMR performed at room temperature is not sufficient for the study of acidity strength of solids. Lowering the temperature of the experiment would stop chemical exchange and allow measurement of the relevant chemical shifts and concentrations, if the chemical equilibrium is unchanged. For acid solids of greatest interest the temperature to which the samples need to be cooled in order to freeze the H atom motions is probably below 60 K. The width of the static signal is then enhanced. Therefore, the required sample spinning rate is higher. Techniques for fast rotation of the sample at the magic angle are not yet everyday practice at a temperature low enough.

In solids, a ^{1}H NMR effect other than the chemical shift can be used to characterize and quantify chemical groups containing hydrogen atoms. It is the direct dipolar magnetic interaction (DDMI) between protons. This effect is averaged to zero in solids using the MAS technique with high enough spinning of the samples (25) (it is also normally averaged to zero in liquids because of the fast internal motions). When two protons distance r apart are located in a magnetic field denoted B_0 the

$$DDMI = constant \cdot (1-3cos^2\theta) \, r^{-3}, \qquad [2]$$

θ being the angle between the magnetic field and the line which joins the two protons (26). The interactions for a group of H atoms must be calculated for each pair of H atoms of the group and summed. If in the group all distances between any two spins are of the same order of magnitude, these H atoms form a magnetic configuration for which the spectrum has to be calculated. Conversely, a spectrum corresponding to a particular set of : number of spins, configuration shape and r values, should characterize an oxy-protonated species ; this must be checked a posteriori from literature data. The DDMI must not be reduced ; rotations of groups and/or translations (diffusion) would narrow the spectra (27) and must be avoided. For this purpose, the experiment temperature must be low (generally 4 K). We shall come back in the second part of this contribution to the question of the equilibrium studied. The individual spectra for each configuration contribute to the total simulated spectrum in proportion to a coefficient which represents the proton weight of the corresponding chemical species. The area under the total simulated spectrum is normalized to the total number of H atoms of the sample per unit cell or per molecular group as preferred. The best parameters are assumed to be those corresponding to the best fit of the experimental spectrum. In general, the quality of the fit is appreciated visually. Minimization of the differences between the experimental and calculated spectra has been programmed, but it results in a refinement of parameters (especially distance parameters) too precise to have any physical significance. It can happen that the minimization does not converge : this means probably that the symmetry of the available configurations is too high to describe the reality. In this case it is generally the distance parameters which are poorly calculated even if the chemical form is correct.

2. Technical

We describe now a typical apparatus. The interest of this description is to show that the experiments do not need an expensive sophisticated spectrometer. The basis of the NMR apparatus is an old cw DP60 Varian (1.4092 Tesla) used with an home-made probe (Figure 1). The quality coefficient of the single coil is about 60. This coil can accommodate 5 mm sealed sample tubes. The probe can be immersed in liquid helium. Its particularity is that it can be adapted and tuned at the working temperature, using sliding cylindrical capacitors, controlled from the outside. Under these conditions, the coaxial wire, at 50 ohms, is no longer part of the active circuit. The sweep of the B_0 field is 5×10^{-3} Tesla on the resonance area. The spectra are accumulated, each scan lasting 100 s with 5 s between scans. In order to get derivative spectra relative to B_0 , this field is modulated at 20 Hz, with an amplitude of 6.8×10^{-5} Tesla peak to peak. A magnetic field spinning at the Larmor frequency for resonance is denoted B_1. Its amplitude is small (a few 10^{-7} Tesla) to avoid saturation of the resonance. The probe is related to a 90° power divider / combiner. Phase sensitive detection is performed. A sketch of the cooling installation is given in Figure 2.

3. Sample conditioning

Some samples are stable and can be studied as they are. They are introduced in convenient glass tubes and sealed (with the equipment described in the technical part and to which numerical data of the experiment will be referred, the ampule must be less than 45 mm long). Starting samples of zeolites are usually in the ammonium form. Special treatments must be performed before NMR

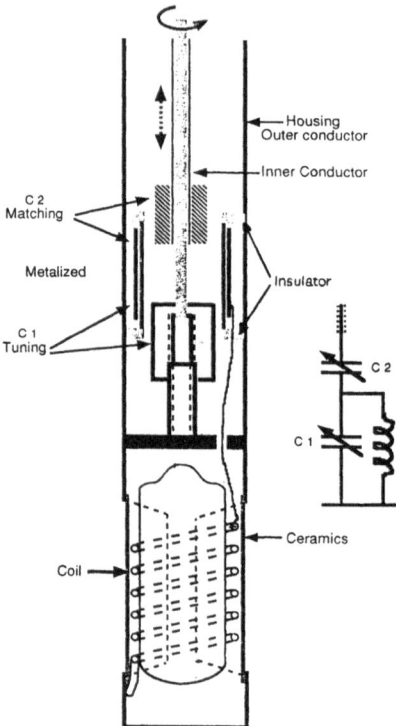

Fig.1 Probe for broad-line NMR at low temperature which can be adapted and tuned at the working temperature.

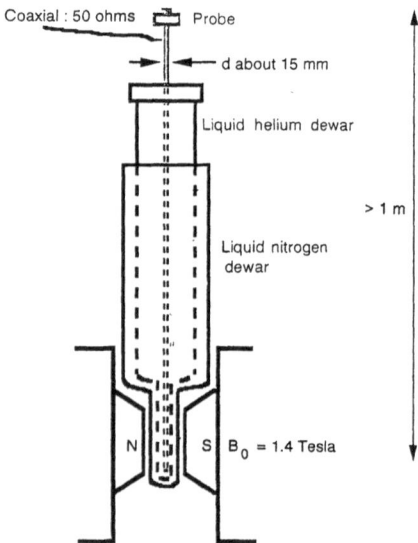

Fig.2 Cooling installation for 4 K broad-line NMR.

experiments. Their purpose is to eliminate water and ammonia from the cavities, leading to "anhydrous" samples containing SOH groups. The best treatment for zeolites is known in the specialized literature as "shallow bed". The elimination of the water vapour desorbed must be efficient enough for the sample to remain at low pressure to avoid dealumination of the framework. In practice, 0.6 g samples are degassed under vacum, first at room temperature, then at temperature increasing at $12°$ $h^{-1}$to 675 K, maintained at this temperature 15 h, the final pressure being below 10^{-2} Pascal. Water vapour is introduced at a constant temperature of 300 K in several steps, at pressures much lower than saturation; the amount of water is monitored gravimetrically. Since the adsorption of water at 300 K is generally not homogeneous, the samples are then homogenized at 395 K and sealed .

4 . Interpretation

As previously stated, both qualitative and quantitative results are obtained from the parameters used to get the best fit of the experimental spectrum. The computer program has been written in Fortran. Equations for the following magnetic configurations of spin 1 / 2 have been programmed and can be associated :

-two-spin configuration, useful for H atoms in water molecules and in some cases for OH groups ;

-three-spin at the apices of an equilateral triangle, useful for hydroxonium ions ;

-three-spin at the apices of an isosceles triangle, useful for groups formed from a water molecule hydrogen-bonded to an OH ; the symmetry of the group of H atoms is then assumed to be C_{2v}. Such an isosceles magnetic configuration can also be used for hydroxonium ions hydrogen-bonded to atoms of the negatively charged framework.

Moreover,

-a Gaussian function can be useful in some cases to describe the resonance of a collection of OH groups;

-a Lorentzian function represents a particular set of OH (28-30).

4.1.   CONFIGURATIONS OF SPIN 1 / 2 USED

4.1.1. *Two-spin 1 / 2 configuration : Pake's model*

For his calculation of two identical spin 1 / 2 DDMI Pake (26) assumed that the static magnetic field of the resonance experiment is large enough for the Zeeman effect to be much larger than the DDMI. The DDMI is then calculated as a perturbation of the Zeeman effect.

As mentioned in the "Introduction", the DDMI is direction and distance dependent (equation [2]). In order to study powder samples, Pake calculated the shape of the spectrum for a large collection of two-spin 1 / 2 configurations, isotropically oriented, characterized by a common value of r and assumed to be magnetically isolated one of the others (Figure 3A). The dependence on r is expressed as a magnetic field along the abscissa axis using the parameter $\alpha = 3 \mu r^{-3}/ 2$, μ being the magnetic moment of the proton. The interaction between neighbouring protons of distinct configurations is approximated by a Gaussian broadening of the spectrum of each configuration. The parameter of the Gaussian, denoted β, is related to the shortest distance between protons of

284

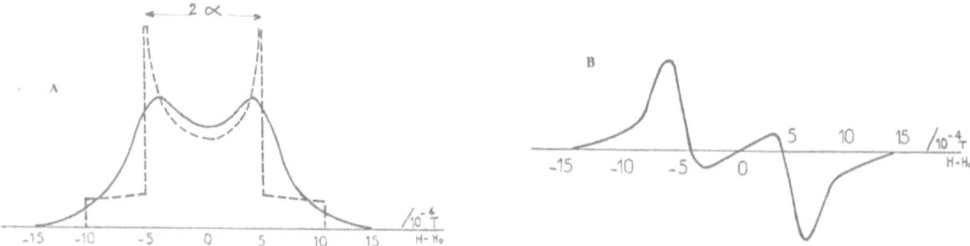

Fig.3 Calculated NMR spectra for powder samples containing identical groups of two-spin 1/2 corresponding to water molecules : A, ---- theoretical absorption spectrum for many groups isotropically oriented and assumed to be magnetically isolated one of the other ; continuous line, after assuming a moderate magnetic interaction between spins belonging to different groups ; B : derivative spectrum of the above absorption spectrum.

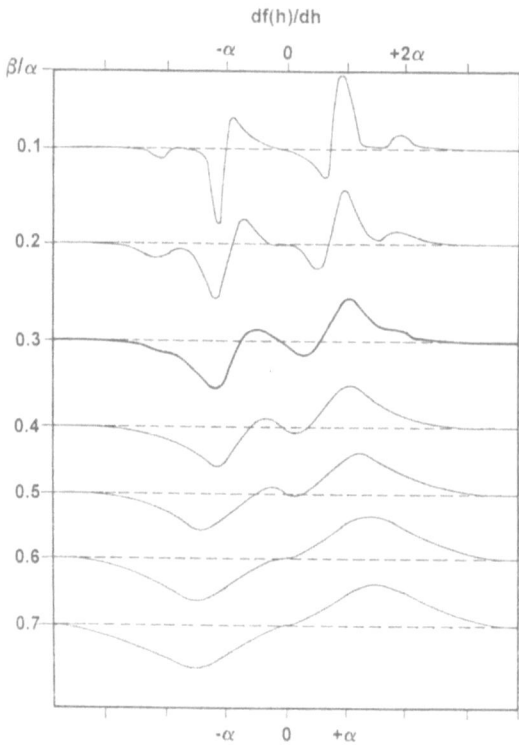

Fig.4 Derivative broad-line spectra for two-spin groups of protons for different values of the ratio β/α ; β is the coefficient of the Gaussian function which expresses the interaction between spins belonging to different groups ; α is the intra-group interaction.

solids, r values are currently found between 150 and 165 pm, which corresponds to α between 6.3 and 4.7×10^{-4} Tesla.

For hydroxonium ions, indications about the internal H H distances are less numerous. When the H atoms have C_{3v} symmetry, r values between 165 and 174 pm seem acceptable, corresponding to α between 4.7 and 4.0 10^{-4} Tesla. Sauer (37) calculated the stablest shape of H_3O^+ in zeolites and showed that it forms hydrogen bonds with O atoms of the framework. The ion has no more symmetry and the smallest distance has the same value as for a water molecule. In some cases, we use the isosceles configurations of spins (that of smallest symmetry available at the moment) to describe the resonance of hydroxonium ions: r is also typical of H_2O and r' is in the gap defined above.

For $H_2O...HOS$, we assume a C_{2v} symmetry of the H atoms, with the normal r value for H_2O ; r' is usually between 216 and 265 pm.

If the distances between H atoms of OH groups exceed 350 pm with no nearer neighbour, the resonance absorption is a Lorentzian curve.

4.3. TYPICAL SPECTRA

It must be mentioned that, for protons, the usual range of chemical shifts is 10 ppm. Working at 60 MHz means that this total scale corresponds to 1.4×10^{-5} Tesla. This is lower than 0.1 times the peak to peak width of the narrower Gaussian contributions to the experimental broad-line spectra in rigid lattice conditions that is practically found. It is therefore possible to neglect the chemical shift effect and to consider the derivative experimental signal as centro-symmetric. In practice, the two parts of the experimental spectra are averaged and the mean half curve obtained is simulated and shown in the Figures.

Values of the ratio β / α corresponding to real conditions are usually 0.2 to 0.4 for hydroxonium ions, 0.25 to 0.4 for $H_2O...HOS$ and 0.25 to 0.5 for water molecules which give clusters easily (38). In view of the shapes of the typical calculated spectra versus β / α and the values of r for the different oxy-protonated species, a maximum of the experimental derivative spectrum at 10^{-4} Tesla may be due either to free OH or to some of them hydrogen-bonded to water, or to H_3O^+ ions, depending of the other parts of the spectrum. A maximum at about 6 x 10^{-4} Tesla is to be attributed to "free" water molecules or to those hydrogen-bonded to SOH. Finally, a maximum at about 10^{-3} Tesla is typical of hydroxonium ions if the shape is appropriate.

Then, in samples containing mainly OH groups the first maximum (10^{-4} Tesla) is the most important, the two others being weak. The appearance of the spectra is significantly modified for samples with larger water content : the first maximum decreases in favour of the other two ; the second maximum (6 x 10^{-4} Tesla) becomes often the most important at high hydration levels. Typical spectra are shown in the second part of this contribution.

distinct configurations (each approximate distance is denoted X) using the equation $\beta = 3 \mu X^{-3}/2$. In fact, when the sample contains nuclei other than ^1H with non-zero spins (^{27}Al and ^{23}Na, for example, are often present in zeolites) the X value suitable for the simulation is smaller than the corresponding H H distance (31). The shape of the energy absorption spectrum (Figure 3B) depends strongly on the β / α ratio. In practice, as expressed in the "Technical" section, the recorded experimental spectra are the derivatives of the absorption spectra relative to the applied magnetic field. Figure 4 shows calculated two-spin derivative spectra for typical values of the β / α ratio.

4.1.2. *Configuration with three identical spin 1 / 2 at the apices of an equilateral triangle.*
Andrew and Bersohn (32) calculated the DDMI characteristic of a collection of configurations with three identical spins 1 / 2 at the apices of an equilateral triangle, isotropically oriented, characterized by a commun value of r and assumed to be magnetically isolated from each other. Their results have been reviewed by Richards and Smith (33) (Figure 5). As for the two-spin configuration, absorption spectra are calculated by the convolution of the equations given by a Gaussian function. The same definition of α, β and X is valid, r being the common distance between any two protons of the configuration.

4.1.3. *Configuration with three identical spin 1 / 2 at the apices of an isosceles triangle.*
Andrew and Finch (34) calculated the equations for a large collection of configurations of three identical spins 1 / 2, the spins of each configuration being at the apices of an isosceles triangles, with base r, the common value of the other sides being denoted r'. The ratio $r' / r = \lambda$ is larger than 1. The configurations are isotropically oriented and assumed to be magnetically isolated one of the others. The equations have been reviewed by Dorémieux-Morin (18). Figure 6 shows the calculated spectrum obtained for $\lambda = 1.3$, the definition of α being the same as previously. Figure 7 shows the calculated derivative spectra obtained for various values of λ and of the ratio β / α.

4.1.4. *Gaussian absorption spectrum.* The shape of a large set of spins, with numerous neighbours about equally distant is Gaussian.

4.1.5. *Lorentzian absorption spectrum.* For rigid-lattice identical spins 1 / 2, a Lorentzian absorption spectrum characterizes a set of diluted spins (28-30).

4.2. DISTANCES BETWEEN H ATOMS IN THE OXY-HYDROGENATED SPECIES

Figures 3 to 7 show some characteristic derivative spectra for different values of α. However, r values and therefore α values depend on the chemical species considered.

For water molecules, r varies quite widely depending on whether they form hydrogen bonds or not. For isolated water molecules, r can be as small as 142 pm, corresponding to a value of α equal to about 7.4 x 10^{-4} Tesla. When water is implicated in hydrogen bonds, r can attain 165 pm, which means that α equals 4.7 x 10^{-4} Tesla (35-36). Especially in non-acid solids, water molecules can show weak external interactions. Then, r can be lower than about 150 pm. In acid

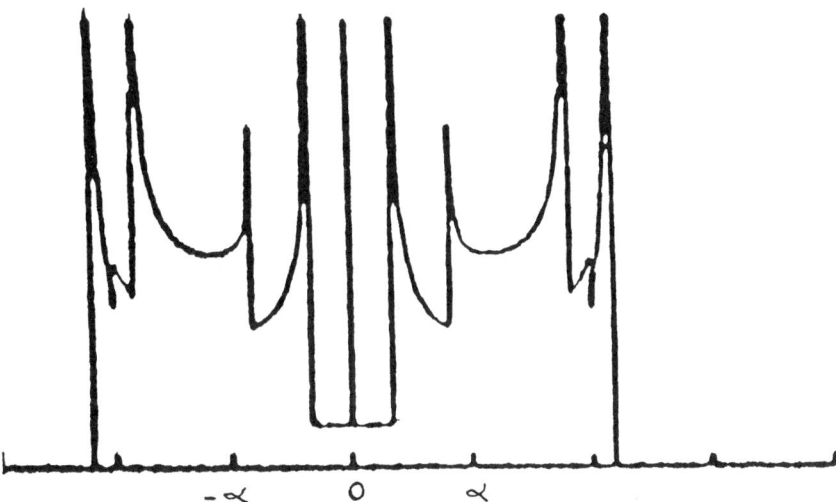

Fig.5 Calculated absorption NMR spectra for powder samples containing identical groups of three-spin 1/2 at the apices of equilateral triangles ; these groups are assumed to be isotropically oriented and magnetically isolated one of the other.

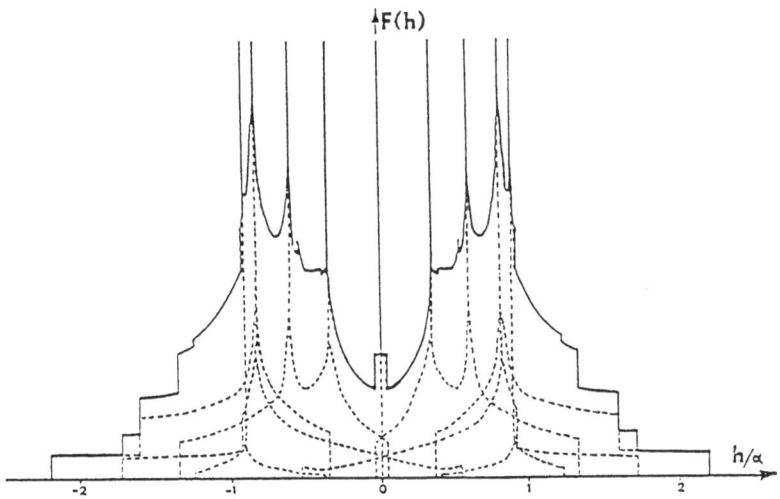

Fig.6 Calculated absorption NMR spectra for powder samples containing identical groups of three-spin 1/2 at the apices of isosceles triangles (the ratio of the equal sides to the base is equal to 1.3); these groups are assumed to be isotropically oriented and magnetically isolated one of the other.

288

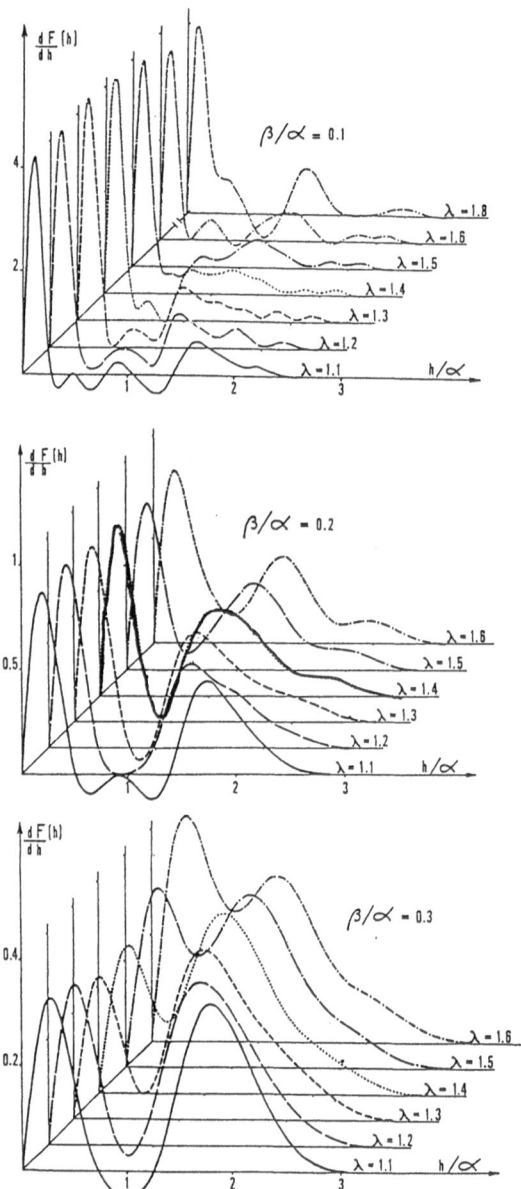

Fig.7 Calculated derivative NMR spectra for powder samples containing identical groups of three-spin 1/2 at the apices of isosceles triangles for different values of the ratio λ of the equal sides to the base and of the ratio β/α value ; β is the coefficient of the Gaussian function which expresses the interaction between spins belonging to different groups ; α is the intra-group interaction ; these groups are assumed to be isotropically oriented and magnetically isolated one of the other.

4.4. CONSTRAINTS FOR SIMULATION OF EXPERIMENTAL SPECTRA

The possible sets of shapes, abscissa and relative weights of calculated contributions of magnetic configurations to sum for the simulation of an experimental spectrum are few, even unique in many cases, especially for well resolved experimental spectra. However, the number of parameters to handle can be high (two or three distance parameters per configuration and the relative weight parameters). The larger is the number of parameters, the lower is the probability of a single fit. Distinct fits correspond to distinct chemical distributions of H atoms in oxy-protonated species and must be discussed (39). Therefore, some constraints for the results to be acceptable are useful.

Number of spins, shape and internal distance limits for a magnetic configuration to represent a real chemical species have already been discussed (Part 4.3) as well as the normalization coefficient (Part 1). Two other constraints have been chosen : (i) X values must be greater than r and r' (or equal to r'), otherwise the configuration has no physical significance ; (ii) it is assumed that no dissociation of the initial water molecule occurs nor water molecules formation from initial OH groups.

Acknowledgment

I am greatly indebted to Robert VINCENT for invaluable technical contribution over many years .

References

(1)   K. Tanabe, *Solid Acids and Bases*, Kodansho, Tokyo, Academic Press, New York and London, **1970**.
(2)   K. Tanabe, M. Misono, Y. Ono and H. Hattori, New Solid Acids and Bases, *Studies in Surface Science and Catalysis*, **1989**, *51*.
(3)   J. B. Moffat, Catalysis by Acids and Bases, *Studies in Surface Science and Catalysis*, **1985**, *20*, 157.
(4)   J. A. Rabo and G.J. Gajda, *Catal. Rev.-Sci. Eng.*, **1989-90**, *31*, 385.
(5)   H. Pfeifer, *J. Chem. Soc., Faraday Trans. 1*, **1988**, *84*, 3777.
(6)   D. Freude, *Stud. Surf. Sci. Catal.*, **1989**, *52*, 169.
(7)   H. Pfeifer, D. Freude and M. Hunger, *Zeolites*, **1985, 5**, 274.
(8)   J.R. Van Wazer, *Phosphorus and its Compounds*, Interscience, New York, **1958**, 750.
(9)   J. W. Ward, *Zeolite Chemistry and Catalysis*, ACS Monograph **1976**, *171*, 118, J. A. Rabo, Ed., ACS, Washington, D.C.
(10)   P. A. Jacobs, *Carboniogenic Activity of Zeolites*, Elsevier, Amsterdam, **1977**.
(11)   P. A. Jacobs, M. J. Mortier and J. B. Uytterhoeven, *J. Inorg. Nucl. Chem.*, **1978**, *40*, 1919.
(12)   W. J. Mortier, *J. Catal.*, **1978**, *55*, 138.
(13)   P. A. Jacobs and M. J. Mortier, *Zeolites*, **1982**, 2, 226.
(14)   P. A. Jacobs, *Catal. Rev.-Sci. Eng.*, **1982**, *24*, 415.
(15)   V. B. Kazanski, Structure and Reactivity of Modified Zeolites, *Studies in Surface Science and Catalysis*, **1984**, *18*, 61.

290

(16) L. M. Kustov, S. A. Zubkov, V. B. Kazansky and L. A. Bondar, Zeolite Chemistry and Catalysis, *Studies in Surface Science and Catalysis,* **1991**, *69*, 303.

(17) D. Dombrowski, J. Hoffmann, J. Fruwert and T. Stock, *J. Chem. Soc., Faraday Trans. 1,* **1985**, *81*, 2257.

(18) C. Dorémieux-Morin, *J. Magn. Res.,* **1976**, *21*, 419.

(19) C. Dorémieux-Morin, *J. Magn. Res.,* **1979**, *33*, 505.

(20) B. Berglund and R.W. Vaughan, *J. Chem. Phys.,* **1980**, *73*, 2037.

(21) H. Rosenberger and A.-R. Grimmer, *Z. anorg. allg. Chem.,* **1979**, *448*, 11.

(22) B. Schröter, H. Rosenberger and D. Hadzi, *J. Mol. Struct.,* **1983**, *96*, 301.

(23) D. Heidemann and A.-R. Grimmer, personal communication.

(24) R.K. Harris and P. Jackson, *J. Chem. Soc., Faraday Trans. 1,* **1988**, *84*, 3649.

(25) E.R. Andrew, A. Bradbury and R.G. Eades, *Nature (London),* **1958**, *182*, 1659.

(26) G. E. Pake, *J. Chem. Phys.,* **1948**, *16*, 327.

(27) H.S. Gutowsky, G.B. Kistiakowsky, G.E. Pake and E.M. Purcell, *J. Chem. Phys.,* **1949**, *17*, 972.

(28) A. Abragam, *The Principles of Nuclear Magnetism,* Claredon, Oxford, **1961**, Chap. IV.

(29) D. Freude, D. Muller and H. Schmiedel, *Surf. Sci.,* **1971**, *25*, 289.

(30) D. Freude and H. Schmiedel, *Surf. Phys. Status Solidi.,* **1971**, *B54*, 631.

(31) J. H. Van Vleck, *Phys. Rev.,* **1948**, *74*, 1168.

(32) E. R. Andrew and R. J. Bersohn, *J. Chem. Phys.,* **1950**, *18*, 159.

(33) R. E. Richards and J. A. S. Smith, *Trans. Faraday Soc.,* **1952**, *48*, 675.

(34) E. R. Andrew and N. D. Finch, *Proc. Phys. Soc.,* **1957**, *70B*, 980.

(35) G. Ferraris and M. Franchini-Angela, *Acta Crystallogr. Sect. B,* **1972**, *28*, 3572.

(36) G. Chiari and G. Ferraris, *Acta Crystallogr. Sect. B,* **1982**, *38*, 2331.

(37) J. Sauer, H. Horn, M. Häser and R. Ahlrichs, *Chem. Phys. Letters,* **1990**, *173*, 26.

(38) P. Hobza, J. Sauer, C. Morgeneyer, J. Hurych and R. Zahradnik, *J. Phys. Chem.,* **1981**, *85*, 4061.

(39) J. Trehoux, F. Abraham, D. Thomas, C. Dorémieux-Morin and H. Arribart, *J. Solid State Chem.,* **1988**, *73*, 80.

COMPARISON OF DIFFERENT ZEOLITES STUDIED BY TWO PROTON NMR TECHNIQUES: BROAD-LINE AT 4 K AND HIGH RESOLUTION MAS AT 300 K

L. HEERIBOUT, V. SEMMER, P. BATAMACK, C. DOREMIEUX-MORIN
AND J. FRAISSARD
Laboratoire de Chimie des Surfaces
Associé au CNRS - URA1428
Université P. et M. Curie
Casier 196, Tour 55
4, place Jussieu
F 75252 PARIS CEDEX 05

ABSTRACT. H forms of mordenites, MFI and faujasites are compared using ^{1}H broad-line NMR at 4 K and ^{1}H high-resolution MAS NMR at 300 K. The number of H_3O^+ formed versus the number of H_2O adsorbed and the nature of the oxy-hydrogenated sites are determined.

1 . Introduction

In view of the high industrial interest of the acidity of zeolites, the aim of the present paper is to compare samples of three different structures (mordenite, MFI and faujasite), using two kinds of proton NMR techniques : broad-line at 4 K and high-resolution MAS at 300 K. Both methods are carefully described in the papers by Dorémieux-Morin (1, 2) and Pfeifer (3). To reveal the acidity of zeolites, a base is necessary (4) and water has been chosen for several reasons (1). Thus the interaction between H_2O and the zeolitic framework is studied through the following equilibria :

$$H_2O + ZOH \rightleftharpoons ZOH\text{--}OH_2 \rightleftharpoons ZO^- + H_3O^+$$

Samples belonging to the following families have been studied: H-Mordenite and HZSM-5. They will be compared to results already published on HY samples (2).
The analytical characteristics of the samples are given in Tables 1 to 3.
Two commercial mordenite samples, denoted LHM and AHM, are compared. They contain 4.6 and 4.25 framework Al atoms per unit cell, respectively. Because of framework defects, the effective number of Brönsted acid sites (BAS) is lower than the number of Al atoms (Table 1).
A well crystallized HZSM-5 with 2.4 framework Al atoms per unit cell has also been studied. As for mordenite samples, the presence of framework defects diminishes the number of SiO(H)Al groups per unit cell compared to the number of framework Al atoms (Table 2).
Finally, these three samples will be compared with two HY zeolites (Table 3) : a non dealuminated sample prepared from Y64 (UOP), ND, and a partially dealuminated one, D, obtained from ND after treatment with $(NH_4)_2SiF_6$ (2).

J. Fraissard and L. Petrakis (eds.), Acidity and Basicity of Solids, 291–303.
© 1994 *Kluwer Academic Publishers.*

Table 1 : Characteristics of LHM and AHM mordenite samples.

Sample	Si/Al (NMR)	$Al(H_2O)_6^{3+}$ per unit cell of fully hydrated mordenite (NMR)	OH per unit cell of "anhydrous" zeolite		
			AlOH	SiO(H)Al	SiOH
LHM Mordenite	9.4	0.8	0.3	4.3	2.4
AHM Mordenite	10.3	0.8	0.5	3.75	3.3

Table 2 : Characteristics of the HZSM-5 sample.

Sample	Si/Al	OH per unit cell of "anhydrous" zeolite		
		AlOH	SiO(H)Al	SiOH
HZSM-5	39	0.45	1.95	2.0

Table 3 : Characteristics of ND and D HY samples.

Sample and Preparation	Si/Al (NMR)	$Al(H_2O)_6^{3+}$ (NMR)	Na+ per unit cell	OH per unit cell of "anhydrous" zeolite	
				SiO(H)Al	SiOH
HY from Y64 (UOP) :					
ND non dealuminated Na^+/NH_4^+ exchange	2.4	no	8.3	47.7	0
HY from Y64 (UOP) :					
D dealuminated $(NH_4)_2SiF_6$	4.4	no	negligible	35.8	1.5

2. Experimental Section

In order to remove the adsorbed phase, the samples were heated from ambient temperature to 673K under 10^{-2} Pa, at 12 K.h^{-1}, and maintained at this temperature for about 15 hours. Then water is adsorbed step by step at low pressure; its distribution is homogeneized at 373 K and finally the tubes are sealed.

The ^1H MAS spectra are recorded on a MSL400 MHz Bruker spectrometer with a 3 kHz spinning rate. Proton chemical shifts are given with respect to liquid TMS, as external reference, without susceptibility correction. The repetition time is between 5 and 100 s. The vertical scale on the spectra is arbitrary.

The broad-line spectra are obtained under the conditions described, for example, in the paper by Dorémieux-Morin (1) (60 MHz using a home-made probe).

3. Results

3.1. H-MORDENITES

3.1.1. ^{27}Al *MAS NMR Experiments.* The spectra, recorded with high spin rate (14500 Hz) on fully hydrated samples show, for both mordenites, a classical signal at about 57 ppm characteristic of Al atoms tetracoordinated to O atoms and an additional signal at 0.9 ppm attributed to hexacoordinated Al atoms. For both samples, this last signal corresponds to about 0.8 non framework $Al(H_2O)_6^{3+}$ per unit cell; these are not considered in the calculations.

3.1.2. ^1H *MAS NMR Experiments.* The spectrum of "anhydrous" AHM (dehydrated but still hydroxylated) presents two main Gaussian signals (Fig.1) at 2.1 ± 0.2 ppm and 4.3 ± 0.2 ppm, corresponding to silanol groups (3) and bridging SiO(H)Al groups (3), respectively. The simulation of these spectra reveals the presence of another signal, at 3.2 ppm, attributed to AlOH groups (3).

In the case of the LHM sample, an additional small signal at 6.6 ± 0.2 ppm, 1ppm wide with no spinning sideband, is detectable (Fig.1). Such a chemical shift has already been attributed to water molecules adsorbed on Lewis acid sites (5). However, no broad-line signal characteristic of water molecules could be detected. This 6.6 ppm signal has not been observed in the "anhydrous" AHM spectra.

Figure 2 shows spectra obtained for hydrated AHM samples. When water is adsorbed on "anhydrous" samples, the chemical shift of the SiOH signal does not change and the number of these groups seems to be the same for both samples. A Lorentzian signal is visible in the range from 4.3 to 6.6 ppm depending on the number of water molecules adsorbed. Its chemical shift first increases until the water concentration, $n(H_2O)$, is close to the number of initial ZOH, $n(ZOH)$, then levels off in the region where $n(ZOH) < n(H_2O) < 2n(ZOH)$ and finally decreases to about 5.5 ppm for greater hydration levels. This signal is attributed to fast chemically exchanging protons (3, 6). Its width is also greatly dependent on the water concentration; these signals are the widest when the number of water molecules adsorbed is smaller than the number of initial ZOH. For LHM, the maximum width is 8 ppm and decreases to 4 ppm when the water concentration is increased. The maximum width for AHM is even larger than for LHM (about 15 ppm) and decreases to 4.5 ppm for high water concentration. In the case of the AHM sample, the 4.3 ppm signal is still present when less than 1 H_2O per unit cell is adsorbed.

It is impossible to know if the signal attributed to AlOH groups in "anhydrous" sample spectra, and seen solely by means of simulation, still exists in the spectra of partially hydrated samples.

Moreover, for hydration levels close to or larger than 8 water molecules per unit cell, a new signal at 8.9 ± 0.2 ppm with large spinning side bands is observed for each sample. It must be attributed at least partly to $Al(H_2O)_6^{3+}$ (7). It is difficult to quantify the number of these groups because the corresponding signal contains not only the $Al(H_2O)_6^{3+}$ contribution but also the first spinning sideband of silanol groups. The presence of $Al(H_2O)_6^{3+}$ is however asserted for two reasons : (i) a comparison with the opposite spinning sideband of the silanol groups confirms the presence of hexacoordinated Al atoms; (ii) as stated above, these ions were already present in the fully hydrated samples.

The principal difference between the two mordenite samples observed by ^1H MAS NMR concerns the 6.6 ± 0.2 ppm signal (Fig. 3). Its chemical shift does not change with water concentration. This signal is present for both hydrated samples but, for the same hydration level,

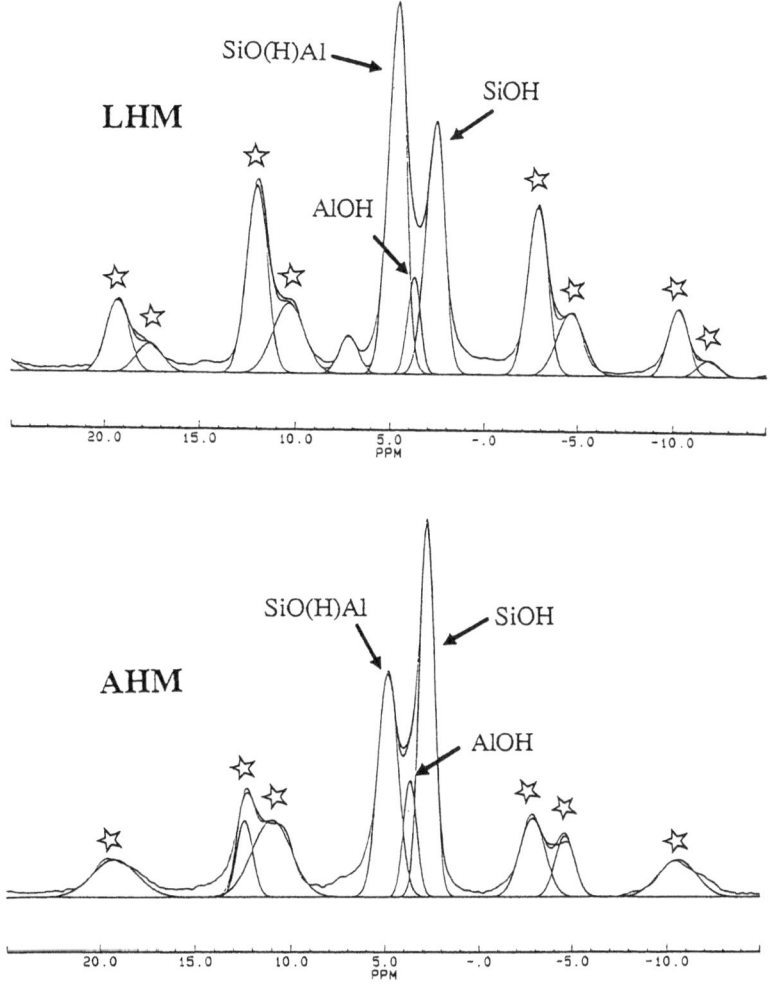

Fig. 1 : Simulation of the experimental ^{1}H MAS spectra for pretreated mordenite samples. The stars indicate spinning sidebands.

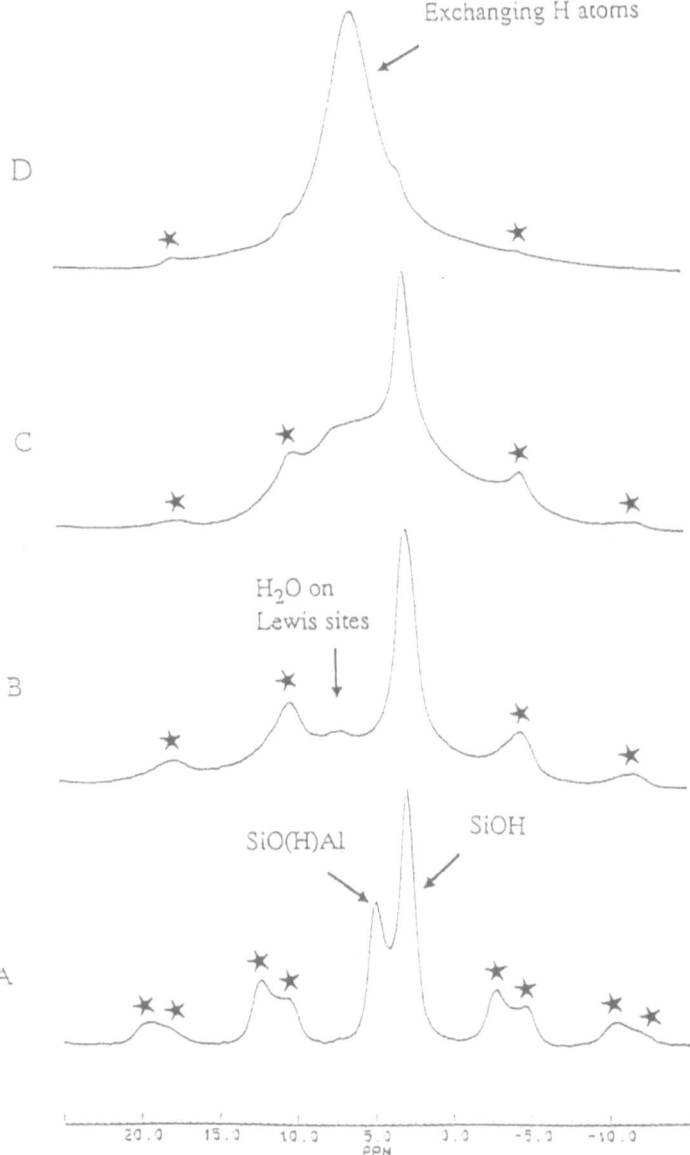

Fig. 2 : ^{1}H MAS NMR spectra for AHM after adsorption of various amounts of water. The number of water molecules adsorbed per unit cell is : A - 0 ; B - 1.9±0.1 ; C - 4.2±0.3 ; D - 14.1±0.8. The stars indicate spinning sidebands.

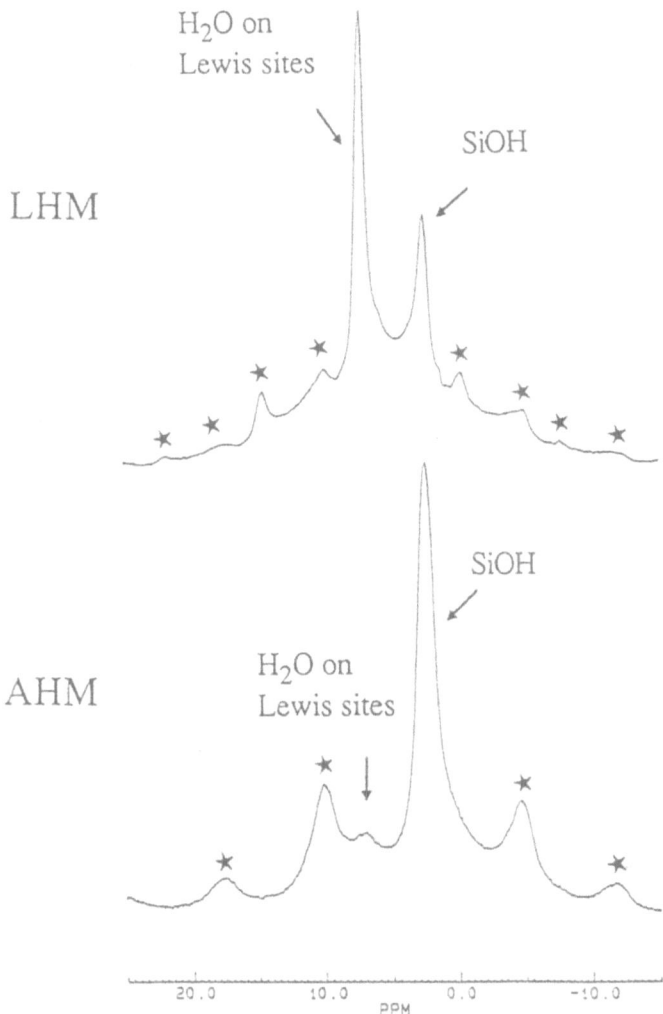

Fig. 3 : ^{1}H MAS NMR spectra for LHM and AHM samples after adsorption of the same amount of water (0.55 ± 0.05 water molecules per Brönsted acid site (BAS)). The number of silanol groups per BAS is 0.6 and 0.9 for LHM and AHM, respectively. The stars indicate spinning sidebands.

its intensity is much larger in the case of LHM. As noted above, a weak signal is detected at the same chemical shift in the pretreated LHM sample. Though the maximum number of water molecules corresponding to this 6.6 ppm signal is difficult to determine because of the presence of the wide Lorentzian signal, it can be estimated to be in the range of 1-1.5 per unit cell for LHM. This value is obtained for the smallest amount of adsorbed water tested (2.5 molecules per unit cell) and is not changed when the amount of water is increased. The 6.6 ppm signal is also observed on the AHM sample, as soon as about one water molecule per unit cell is adsorbed, whereas the exchanging H atom Lorentzian signal cannot be clearly identified. The amount of water on Lewis acid sites does not increase further when more water is adsorbed. The number of Lewis sites per BAS can be estimated to be twenty times larger in LHM than in AHM.

3.1.3. *^{1}H Broad-line NMR Experiments at 4 K.* Practically, the following equilibria are studied (1, 2) :

$$ZOH + H_2O \rightleftharpoons ZOH\text{--}OH_2 \rightleftharpoons ZO^- + H_3O^+$$

Concentrations of the different species have been determined but the final results are expressed as the number of hydroxonium ions formed with respect to the number of water molecules adsorbed. These numbers are found by simulation of the average half derivative absorption spectra using calculated spins 1/2 magnetic configurations (1). The choice of these configurations results from the properties of the direct dipolar spin-spin interactions which are : (i) proportional to r^{-3}, r being the distance between to interacting protons belonging to the same configuration, (ii) direction dependent relative to the main magnetic field. These configurations correspond to the oxy-hydrogenated species present in the above equilibrium. Figure 4 shows the dependence of the number of H_3O^+ formed per BAS on the number of H_2O adsorbed per BAS.

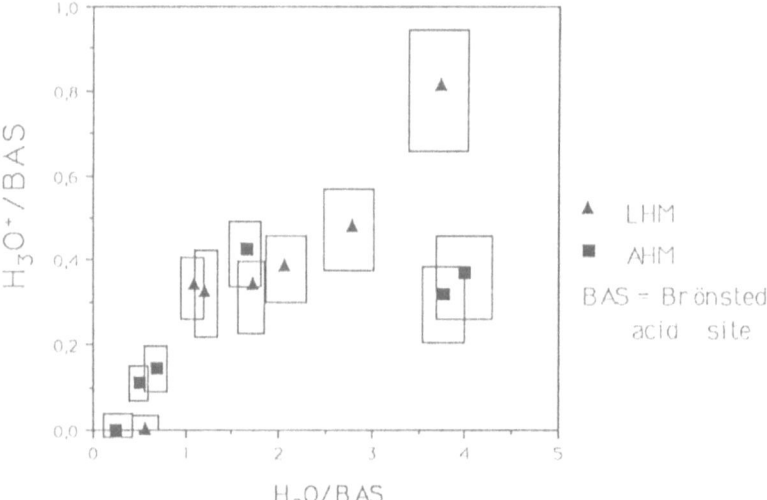

Fig. 4 : Dependence of the hydroxonium ion concentration on the amount of water adsorbed for mordenite samples

For LHM, the hydroxonium ion concentration increases when $n(H_2O)$ is lower than $n(ZOH)$; it then levels off in the region where $n(ZOH) < n(H_2O) < 2n(ZOH)$. Finally, the number of H_3O^+ again increases markedly with $n(H_2O)$ larger than $2n(ZOH)$. In the horizontal section, about 30% of the initial Brönsted acidities are found as hydroxonium ions. Experiments were run with increasing $n(H_2O)$ until 4 water molecules were adsorbed on each acid site. For such a water concentration, about 0.8 H_3O^+ per BAS are found.

For AHM, the number of H_3O^+ reaches a maximum when $n(H_2O) = n(ZOH)$ and then remains unchanged within experimental error. Insofar as all ZOH groups already interact with water, no further evolution in the number of H_3O^+ can be observed.

The horizontal section is observed for 30% of the initial Brönsted acid sites ionized to hydroxonium ions, as in LHM.

Thus, results displayed in Figure 4 show that the two mordenite samples behave rather differently when water is adsorbed.

3.2. HZSM-5

3.2.1. *^1H MAS NMR*. Figure 5 shows some spectra obtained for the HZSM-5 sample.

The spectrum of the "anhydrous" sample presents the following signals (Fig. 5A) : (i) a very narrow signal, less than 0.4 ppm wide, at 0.2 ± 0.2 ppm, with no spinning sideband; (ii) a 0.9 ± 0.2 ppm signal with no spinning sideband. When the sample is treated under oxygen pressure at 873 K, both of the 0.2 and the 0.9 ppm signals disappear : Therefore, we attribute them to a small amount of remaining organic phase, from the template used for the synthesis; (iii) a 2.4 ± 0.2 ppm signal, with spinning sidebands attributed to silanol groups; (iv) a 4.3 ± 0.2 ppm signal attributed to bridging SiO(H)Al groups (3) with large spinning side bands; (v) simulation of the "anhydrous" sample spectra reveals the presence of an additional signal, at 3.4 ppm, attributed to AlOH groups (3).

When the sample is partially hydrated (Fig. 5B, C, D) the 4.3 ppm SiO(H)Al signal is no longer visible, even when $n(H_2O)=0.5$ H_2O. A wide Lorentzian signal, highly dependent on water concentration and attributed to exchanging protons (3, 6), appears in the 4.3-6.5 ppm range. This signal is clearly visible when 1.5 H_2O are adsorbed per unit cell but for smaller concentrations its presence can not be detected because of its low intensity and its large width (more than 10 ppm).

The signals at 0.2, 0.9 ppm attributed to organic phase and the 2.4 ppm signal attributed to silanols groups are still present in the spectra of hydrated samples, with the same chemical shift. Their intensities seem to be maintained. Whatever the water concentration, no signal at 6.6 ppm is observed.

3.2.2. *^1H Broad-line NMR at 4 K*. The dependence of the number of hydroxonium ions formed on the number of water molecules adsorbed is shown in Figure 6.

When water is adsorbed, an increase in the number of H_3O^+ is observed. For one H_2O adsorbed per BAS, about 25% of the initial Brönsted acid groups are found as hydroxonium ions. Experiments were run with increasing $n(H_2O)$ until about 4 water molecules were adsorbed per BAS and, for such a hydration level, about 0.6 H_3O^+ per acid site are formed.

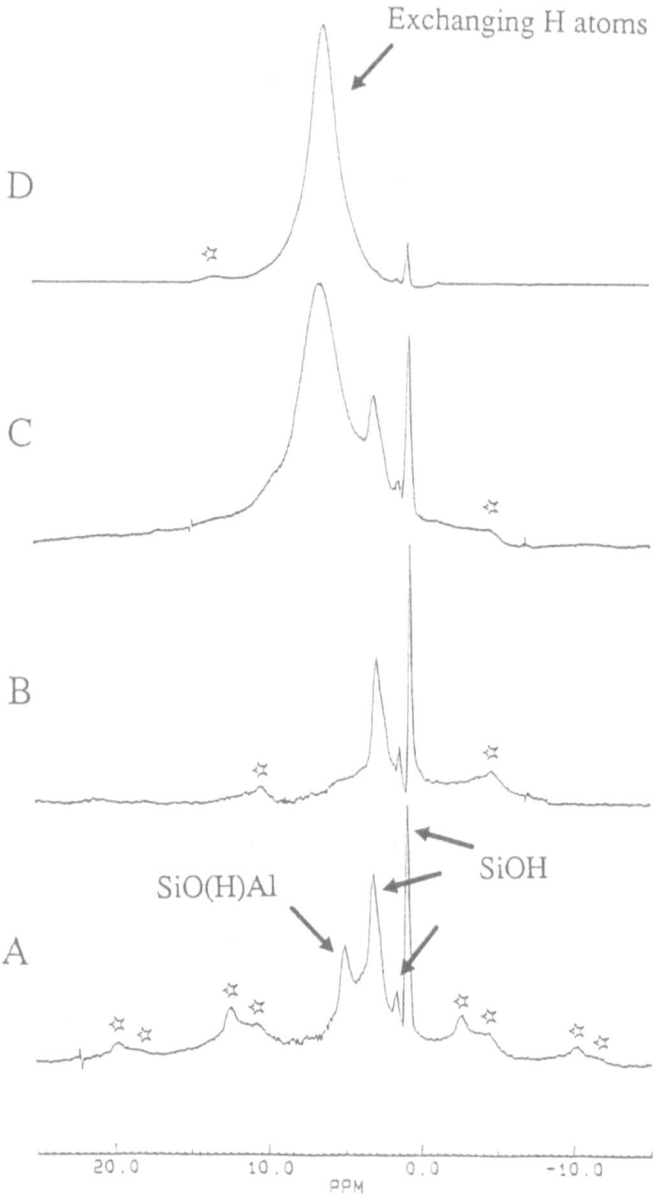

Fig. 5 : ^{1}H MAS NMR spectra for the HZSM-5 sample after adsorption of various amounts of water. The number of water molecules adsorbed per unit cell is : A - 0 ; B - 0.55±0.05 ; C - 2.7±0.3 ; D - 9.5±0.6. The stars indicate spinning sidebands.

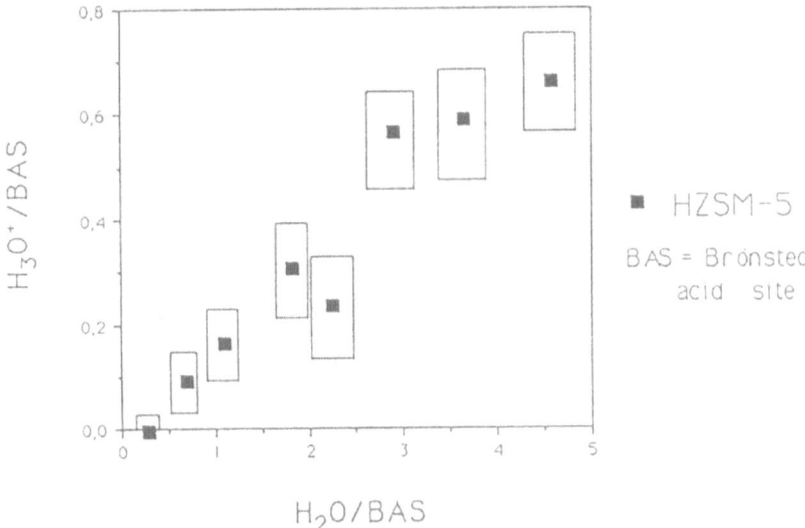

Fig. 6 : Dependence of the hydroxonium ion concentration on the amount of water adsorbed for the HZSM-5 sample.

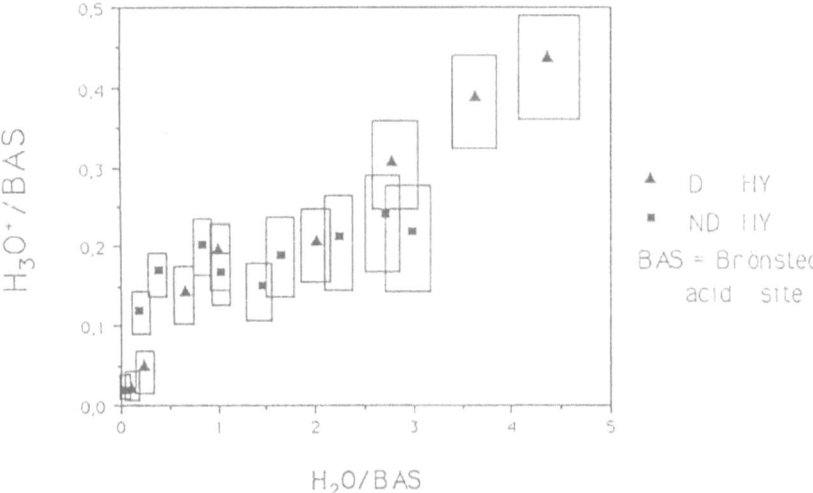

Fig. 7 : Dependence of the hydroxonium ion concentration on the amount of water adsorbed for HY samples.

4. Discussion and Conclusion

From the study of HY samples, the papers by Dorémieux-Morin (1, 2) proposed to characterize two types of Brönsted acid strength of zeolites using ^{1}H broad-line NMR after water adsorption : the Brönsted acid strength of samples without framework defects and that enhanced by the presence of framework defects. ^{1}H MAS NMR has proved to be a technique of particular interest for the detection of framework defects in zeolites (3). Using this technique, all the samples investigated here have been found to have framework defects. Moreover, several kinds of defects have been identified. For this reason, it appears premature to discuss the complete set of results now.

However, it has been found that for a given amount of water adsorbed there are important quantitative differences in the way the Brönsted acid sites of the various species ionize. Such differences can be of interest for the industrial use of zeolites as catalysts and they are summarized as follows :

- The qualitative evolution in the number of hydroxonium ions is similar for the D HY sample and LHM mordenite (Fig.4 and 7). We relate this similarity to the presence of Lewis acid sites in both "anhydrous" samples (these Lewis sites being detected after water adsorption).

- For one H_2O adsorbed per BAS, the following ranking can be given :
HY (20% ionization) < HZSM-5 (about 25%) < H-mordenites (30%).

- For four H_2O adsorbed per BAS, the following values can be given :
ND HY (20%) ; Mord AHM (30%) ; D HY (40%) ; HZSM-5 (60%) ; Mord LHM (80%).

6. Acknowledgments

We thank Bruker Analytische Messtechnik GMBH for providing us high speed ^{27}Al MAS NMR mordenite spectra from its Karlsruhe laboratory.

7. References

(1) C. Dorémieux-Morin, Part 1, *Proceedings of the NATO ASI on Acidity and Basicity : Theory, Assessment and Utility*, J. Fraissard and L. Petrakis ed., Kluwer Acad. Pub., **1993-94** (in press), and references therein.

(2) C. Dorémieux-Morin, Part 2, *Proceedings of the NATO ASI on Acidity and Basicity : Theory, Assessment and Utility*, J. Fraissard and L. Petrakis ed., Kluwer Acad. Pub., **1993-94** (in press), and references therein.

(3) H. Pfeifer, *Proceedings of the NATO ASI on Acidity and Basicity : Theory, Assessment and Utility*, J. Fraissard and L. Petrakis ed., Kluwer Acad. Pub., **1993-94** (in press), and references therein.

(4) J. A. Rabo and G. J. Gajda, *Catal. Rev.-Sci. Eng.*, **1989-90**, <u>31</u>, 385.

(5) M. Hunger, D. Freude and H. Pfeifer, *J. Chem. Soc., Faraday Trans.* , **1991**, <u>87</u>, 657.

(6) P. Batamack, C. Dorémieux-Morin, J. Fraissard and D. Freude, *J. Phys. Chem.*, **1991**, <u>95</u>, 3790.

(7) J. W. Akitt, J. M. Elders, X. L. R. Fontaine, *J. Chem. Soc., Chem. Commun.*, **1986**, 1047.

CATALYTIC CHEMISTRY OF PERFLUORINATED RESINSULFONIC ACIDS: REACTIONS, MECHANISTIC ASPECTS AND SUPERELECTROPHILIC ACTIVATION[1]

G. A. OLAH
Loker Hydrocarbon Research Institute and Department of Chemistry,
University of Southern California
Los Angeles, CA 90089-1661, USA

ABSTRACT. Solid perfluororesinsulfonic acids such as Nafion-H and its analogs are effective superacidic catalysts. Their application in author's laboratory to a variety of organic reactions such as Friedel-Crafts type electrophilic reactions, ether and ester synthesis, hydration, rearrangements, condensations, etc. is reviewed. Mechanistic aspects of the reactions are discussed, including superelectrophilic activation involving bidentate interaction with clustered acidic sites.

1. Introduction

Considerable success of liquid superacids such as FSO_3H, CF_3SO_3H and $FSO_3H:SbF_5$ (Magic Acid) and related systems in extending the scope and utility of acid catalyzed reactions, particularly hydrocarbon transformations, logically led to efforts in adopting this chemistry to solid systems allowing heterogeneous catalytic processes[2]. It is well recognized that FSO_3H ($Ho \approx -15.1$) and CF_3SO_3H ($Ho \approx -14.1$) are of comparable acid strength and appending a sulfonic acid group to either a CF_2 or a CF group in a perfluorinated back-bone chain does not diminish the acid strength of the sulfonic acid moiety significantly[3]. Such a structure besides providing high acidity sites also furnishes highly inert perfluorinated polymeric chain resistant to acid cleavage.

Acid	H_o (22°C)
CF_3SO_3H	-14.1
$C_2F_5SO_3H$	-14.0
$C_5F_{11}SO_3H$	-13.2
$C_8F_{17}SO_3H$	-12.3

Apart from oligomeric perfluoroalkanesulfonic acids (C5 to C18) as supported catalysts the system which arose our interest is the DuPont's Nafion brand membrane resins (a copolymer of perfluorinated epoxide and vinylsulfonic acid). The incentive behind the development of these polymers and related ionomers by DuPont came from the commercial application of these materials as membranes in electrochemical processes[4,5]. Their utility particularly depends on the relative inertness of these materials to corrosive environments

J. Fraissard and L. Petrakis (eds.), Acidity and Basicity of Solids, 305–334.

and permselectivity, which preferentially allows cations to diffuse faster through the membranes than anions. Typical use of this membrane includes Donnan dialysis[6-8] and ion exchange, as a separator and/or electrolyte in membrane-separated chlor-alkali cells[9,10] and other electrochemical processes[9-11], solid polymer electrolyte fuel cells[9] and batteries[9].

Nafion[11b,c,12] resins were first synthesized by DuPont chemists in the 1970s. Commercial DuPont Nafion brand ion membrane resins, such as Nafion 501 most frequently used in catalytic applications, are perfluorinated polymers $\underline{1}$ having sulfonic acid groups in the amount of about 0.01 to 5 mequi/gram catalyst.

$$-(F_2C\text{-}CF_2)X\text{-}(CF_2\text{-}\overset{|}{C}F)_y\text{-}(OCF_2\text{-}\overset{|}{C}F)_m\text{-}OCF_2\text{-}CF_2\text{-}CF_2\text{-}SO_3^-\,K^+ \quad \text{or}$$

$$(F_2C\text{-}CF_2)x\text{-}(F_2C\text{-}\overset{|}{C}F)_y\text{-}(OCF_2\text{-}\overset{|}{C}F)m\text{-}O\overset{|}{C}F\text{-}CF_3$$
$$SO_3^-\,K^+$$

$$x/y = 2\text{-}50; \quad m = 1 \qquad \qquad \mathbf{1}$$

Polymers of the above structure have been prepared in various ways. One method involves the copolymerization of the corresponding vinyl ether and sulfonyl compounds[13,14]. Related polymers $\underline{3}\text{-}\underline{5}$ have been prepared[15] by copolymerizing the perfluorinated vinyl ether $\underline{2}$ with perfluoroethylene and/or perfluoro-α-olefins.

$$F_2C=CF_2 + SO_3 \longrightarrow \begin{matrix} F_2C\text{-}CF_2 \\ |\quad| \\ O\text{-}SO_2 \end{matrix} \longrightarrow$$

$$F\text{-}O_2S\text{-}CF_2\text{-}CO\text{-}F \xrightarrow{\quad m \; F_2\overset{O}{\overset{/\backslash}{C}}\text{-}CF_2\text{-}CF_3/ \; Na_2CO_3 \; \Delta \quad}$$

$$\begin{matrix} CF_3 \\ | \\ F_2C=CF\text{-}O\text{-}(CF_2\text{-}CF\text{-}O)_m\text{-}CF_2\text{-}CF_2\text{-}SO_2\text{-}F \\ \mathbf{2} \end{matrix} \xrightarrow[\text{co-polymerization}]{\quad F_2C=CF_2 \quad}$$

$$\sim\!\!\sim\!\!\sim(CF_2\text{-}CF_2)_n\,\text{-}CF_2\text{-}\underset{|}{C}F \sim\!\!\sim\!\!\sim \; CF_3 \qquad \xrightarrow{\text{KOH}} \quad \textcircled{P}\text{—}SO_3X$$
$$O(CF_2\text{—}CF\text{—}O)_m\text{-}CF_2\text{-}CF_2\text{-}SO_2\text{-}X$$

$$\mathbf{3} \quad X = F, \text{ Nafion-F or resin}$$
$$\mathbf{4} \quad X = O^-\,K^+ , \text{ Nafion-K}$$
$$\mathbf{5} \quad X = OH, \text{ Nafion-H}$$

Typically, Nafion resins with an equivalent weight of 1200 contain tetrafluoroethylene and the perfluorovinyl ether units, in a ratio of 7:1. It is assumed[16] that the relatively high hydrophilicity of the sulfonate groups cause them to form clusters. This tendency of maximizing the interaction of similar fragments on the backbone acts as transient crosslinks in the polymer thereby imparting to it the rigidity of conventional crosslinked polymers. At higher temperatures, however, these clusters break up and the salt forms of Nafion behave as thermoplastic polymers (ionomers). Hence even though it is not crosslinked covalently, it is effectively insoluble in most solvents but can, however, be swollen by them. Relative rates of uptake of *p*-aminoazobenzene were determined[17] in three solvents for Nafion 511 resin. Water causes the greatest swelling (at 22°C, 1.0 as reference) compared to ethanol (0.3) and ethylene glycol (0.1). From a chemical point of view, besides its thermal and chemical stability, the feature that makes Nafion type polymers catalytically useful is the superacidity of the acidic form of polymers. Since the sulfonic acid group is attached to a highly electron withdrawing perfluoroalkyl backbone, a relatively high polarization of the O-H bond results. The determination of the acidity of Nafion-H and related catalysts is not easy since methods of measuring solution acidities cannot be directly applied to heterogeneous solid systems[2]. Nevertheless, studies indicate that Nafion-H resin exhibits an acidic character comparable to 100% sulfuric acid and trifluoromethanesulfonic acid in trifluoroacetic anhydride solution. The estimated Ho value (about $\geq$ -12) for Nafion-H is comparable to or stronger than that of 96 to 100% sulfuric acid. Nafion-H in contact with solvent systems affects only the surface acidity (solvent phase directly in contact with the catalytic surface). The acidity of the bulk aqueous supernatant phase of a Nafion-H/water suspension is, however, unaffected. This is easily demonstrated by dipping a piece of indicator paper into a slurry of Nafion-H in water. The acidic color of the indicator is observed only when the paper is in direct contact with the catalyst.

Over the years we have utilized Nafion-H as a versatile solid acid catalyst for a series of significant organic reactions and the presentation summarizes our work. It should be emphasized that Nafion-H has been used as a solid superacid catalyst by other researchers notably by the DuPont research group, as reviewed by Waller[11b,c], and others around the world but their discussion is outside the scope of the present review and the reader is referred to a comprehensive review[18]. Present discussion will center on the investigations of my research group and that of my colleagues associated with the Loker Hydrocarbon Research Institute.

2. Nafion-H Catalyzed Electrophilic Reactions

Unlike conventional resinsulfonic acids such as sulfonated polystyrenes (Dowex-50, Amberlite IR-112 and Permutit-Q), Nafion-H and related perfluoroalkanesulfonic acids are not only much stronger acids but are also stable in corrosive environments and at temperatures up to 210°C. These unique properties have led to an extensive development of various acid catalyzed reactions such as alkylation, transalkylation, isomerization, acylation, nitration, halogenation, etc.

2.1. ALKYLATION REACTIONS

Friedel-Crafts alkylation reactions in solution generally give complex reaction mixtures. Polyalkylation, isomerization, transalkylation, dealkylation, and polymerization all occur under the reaction conditions. Gas phase alkylations performed over Nafion-H catalyst, due to its high acidity, allows for short contact time and weak complexation of the reactants with the catalyst, thus resulting in much cleaner and less complex products.

2.1.1. *Alkylation with Olefns.* Cumene has long been produced industrially by alkylating benzene with propylene over supported acid catalyst such as phosphoric acid. On the other hand, the single largest scale industrial alkylation process, i.e., the ethylation of benzene with ethylene, is still carried out to a significant degree in the liquid phase using acid catalysts. This is mainly due to ethylene being less polar than propylene, thus, consequently requiring stronger acid catalysts for its protolytic activation.

Due to its high acidity Nafion-H shows good activity in alkylation reactions[19], including ethylation. Gas phase alkylation of benzene with ethylene and propylene, in a flow system, proceeds at temperatures as low as 110°C with short contact times[19a]. Under these conditions no s-butylbenzene is produced as a by-product.

The high acidity of the Nafion-H catalyst is further demonstrated by its ability to promote polyalkylation as well as by its high isomerizing ability. In reactions between benzene and ethylene at 190°C, 20% of the alkylated products are diethylbenzene[19a]. The isomer distribution indicates thermodynamic control.

Propene and naphthalene react over Nafion-H to give 90% 2-isopropylnaphthalene[20-22]. α-Phenylethylation of aromatics with styrene has also been reported over Nafion-H. Reaction of alkenes with electron rich aromatics like phenols gave unusual amounts of alkylation products.

2.1.2. *Alkylation with Alcohols.* When olefins are used as the alkylating agents, the catalytic activity of Nafion-H slowly decreases over a period of time, most probably due to some surface polymerization deactivating the catalytic sites. The activity decreases faster when higher alkenes are used. The use of alcohols instead of olefins as alkylating agents, on the other hand, improves the lifetime of the catalyst[19a]. With alcohols, no ready polymerization occurs since water formed as the byproduct, inhibits polymerization, but does not affect the acidic groups on the catalyst.

Reaction of alcohols with benzene over Nafion-H catalyst gave the corresponding alkylbenzenes[19a]. When propyl alcohol was the alkylating agent, no n-propylbenzene was detected, and the only product obtained was cumene[23]. This indicates the intermediacy of the isopropyl cation in the alkylation process.

Yields and conversion of methyl alcohol were much higher when the aromatic substrate was phenol, anisole or their derivatives[24] (percentage conversion: 41-63%). Unlike other metal oxide catalyzed reactions the selectivity towards *ortho* methylation when using Nafion-H as the catalyst is somewhat lower[25] probably due to the absence of basic sites on this solid catalyst capable of forming phenolates.

In the gas phase methylation reactions over Nafion-H using methyl alcohol as alkylation agent, the consumption of methyl alcohol was higher than that calculated by product analysis[24-26]. This is due to the facile formation of dimethyl ether as a by-product.

$$2 \; CH_3OH \quad \xrightarrow[\text{-}H_2O]{\text{Nafion-H, 150°C}} \quad H_3C\text{-}O\text{-}CH_3$$

Indeed, when neat methanol is passed over the catalyst at temperatures over 150°C, dimethyl ether is the only product formed with water as the by-product.

Alcohols can also undergo acid-catalyzed dehydration to give the corresponding alkenes or the corresponding ethers. The product distribution of the dehydration of alcohols over Nafion-H catalyst shows temperature dependence[19a]. Alcohols are thus efficiently dehydrated in the gas phase over Nafion-H under relatively mild conditions with no evidence for any side reactions such as dehydrogenation or decomposition. At higher temperatures, olefin formation predominates.

2.1.3. *Alkylation with Alkyl Halides.* Preparatively useful results were obtained in the Nafion-H catalyzed gasphase alkylation of aromatic hydrocarbons with alkyl halides[27].

$$C_6H_6 \; + \; R\text{-}H \quad \xrightarrow[\text{30-80\%}]{\text{Nafion-H, 140-200°C}} \quad C_6H_5\text{-}R \; + \; HX$$

Alkyl halides are reactive Friedel-Crafts alkylating agents and give high conversions in the case of benzene in the gas phase over Nafion-H catalyst. For example, in the alkylation of benzene with isopropyl chloride, conversions as high as 87% were achieved. Conversions, however, were temperature and contact time dependent[27].

The selectivity of the Nafion-H catalyst for monoalkylation has been found to be generally high. With a molar ratio of benzene:isopropyl chloride being 5:1, about 94% of the alkylate is monoalkylbenzene. This result is comparable to the highly selective monoalkylation reaction of benzene with propene (5.2:1 m/m) over phosphoric acid-quartz catalyst at ~ 200°C to give cumene in 95% yield[28].

The relatively minor differences observed in the degree of conversion in the reaction of various aromatic hydrocarbons with isopropyl chloride over Nafion-H are explained by the possibility of some dealkylation (i.e., the reverse reaction). The more nucleophilic is the alkylated aromatic product, the higher the rate of dealkylation reaction.

The sequence of various alkyl halide over Nafion-H is RF>RCl>RBr and secondary>primary alkyl halide. The reactivity of t-butyl chloride is so high that even at low temperatures (~ 130°C), complex mixture of products are observed.

Besides the advantage of their high reactivity toward alkylation reactions, primary and secondary alkyl halides show little tendency for dehydrohalogenation in Nafion-H-catalyzed gas-phase reactions[27]. Although a minor amount of olefin is reported to be formed, no polymer formation was observed on the catalyst. As a result, the catalytic activity of Nafion-H stays constant over prolonged onstream period[27].

2.1.4. *Alkylation with alkyl Esters.* The versatility of the catalytic activity of Nation-H, is also well demonstrated in the alkylation of aromatic hydrocarbons by carboxylic acid alkyl esters both in the gas-phase as well as in heterogeneous liquid-phase reactions[29]. Esters in the presence of conventional Lewis acid halide catalysts tend to give rise to acylation products along with the alkylation products[30]. Two types of alkylating agents have been studied: alkyl esters of carboxylic acids, preferentially those of oxalic acid; alkyl chloroformates (percentage conversion: 2-80%).

$$Ar\text{-}H + R^1\text{-}CO_2R^2 \xrightarrow{\text{Nafion-H}} Ar\text{-}R^2 + R^1\text{-}COOH$$

$$Ar\text{-}H + Cl\text{-}CO_2R \xrightarrow{\text{Nafion-H}} Ar\text{-}R + HCl + CO_2$$

The advantage of alkyl chloroformates lies primarily in their volatile byproducts. Diethyl oxalate shows particularly good alkylating ability even under mild conditions[29].

The gas-phase alkylation of toluene with alkyl chloroformate over Nafion-H was also reported to be an efficient one. It is interesting to compare the alkylating ability of methyl chloroformate with that of methyl alcohol on toluene under similar reaction conditions. For example a 59% conversion with methyl chloroformate was observed[29], compared with 15% conversion with methyl alcohol.

The gas-phase alkylation of toluene with dimethyl and diethyl oxalate over Nafion-H was also reported[29]. The alkylating ability of diethyl oxalate is comparable with that of ethyl chloroformate. However, the alkylating ability of dimethyl oxalate is lower than that of methyl chloroformate.

2.1.5. *Isomerization, Disproportionation and Transalkylation of Alkylbenzenes.* The isomerization and disproportionation of alkylated aromatic hydrocarbons is of commercial interest primarily as a way of converting *m*-xylene to *p*-xylene which can be further oxidized to terephthalic acid for use in the preparation of polyester fiber.

When dialkylbenzenes are passed over Nafion-H at 160°C, both isomerization and disproportionation take place (Scheme A). Monoalkylbenzenes also disproportionate under these conditions[30-31].

311

Scheme A

As expected, the aptitude for disproportionation of the aromatic compound depends upon the nature of the alkyl group, and the order of reactivity is isopropyl>ethyl>methyl. Due to their higher nucleophilicity, polyalkylbenzenes react faster than monoalkylbenzenes. This effect is pronounced in the case of methylbenzenes. Toluene itself shows little reactivity over Nafion-H at 193°C. Diethylbenzenes react much faster than dimethylbenzenes. The rate of conversion of diethylbenzenes over Nafion-H at 193°C is ~ 5.10^{-5} mol.min^{-1}.g^{-1} of catalyst[31]. This is a low rate when compared with that using aluminum-chloride/hydrogen-chloride in the liquid phase at room temperature (- 10^{-4}mol.min^{-1}.g^{-1} catalyst)[32]. However, one should bear in mind that Nafion-H is a truly insoluble heterogeneous catalyst as compared to aluminum chloride-hydrogen chloride catalyst. The equilibrium composition of the acid-catalyzed disproportionation of ethylbenzenes depends upon the nature of the catalyst.

The Nafion catalyst for the isomerization and disproportionation of m-xylene was unusually stable for a reaction involving only aromatic nuclei. After 100 h "on stream" only a 10% loss of activity was observed. This was acoompanied by a 2 to 3% gain in weight and a decrease in the titratable acidity of about 5%. These changes have been attributed to the deposition of tars on the surface of the Nafion catalyst.

Nafion-H also appears to be a very useful catalyst for transalkylation reactions. Transalkylation of benzene with diethylbenzene, as well as with diisopropylbenzene is efficiently catalyzed by Nafion-H in a flow system. The efficiency of the catalyst is, however, more limited when the transferring group is methyl[30a]. Beltrame and coworkers have also carried[30b,c] detailed mechanistic studies on the isomerization of xylenes over Nafion-H.

Transfer of a *tert-butyl* group, on the other hand, occurs very easily over Nafion-H at temperatures as low as ~ 60°C[33]. For example, 2,6-di-t-butyl-p-cresol is dealkylated to o-cresol in 0.5 h. Toluene acts as a better acceptor than benzene.

This reaction is of particular importance since *tert-butyl* group can be successively employed as an effective protecting group in achieving selective aromatic substitution[34-35]. The synthesis of an industrially important intermediate, bisphenol, has been patented[33b,c,d].

Nafion-H also very efficiently catalyzes the rearrangement of anisole, methylanisoles, and phenetole to ring-alkylated phenols and products of transalkylation when vapors of the alkylaryl ethers are passed over it at temperatures higher than 160°C[24,30a]. At these reaction temperatures, some of the starting alkyl phenyl ethers undergo cleavage of the alkyl group to give phenol.

2.1.6. *Reactions of Aliphatic Hydrocarbons.* Nafion-H has been found to be a very effective catalyst for the oligomerization of olefins at temperatures around 100°C. For example, when isobutylene is reacted with Nafion-H at 100°C for ~ 5 h, it is completely converted to its C_8 (dimer) and C_{12} (trimer) respectively. Similar reaction occurs with 1-butene and 2-butene as well as with propene, although longer reaction times (~ 24 hrs) are needed for complete conversion. Isobutane has been alkylated by 2-butene at 100°C under 500 psi of pressure over Nafion-H[37].

Nafion-H catalyst has also been reported to effect the isomerization of aliphatic hydrocarbons. Isomerization reactions are carried out in a flow system at moderate temperatures which decrease the formation of cracked compounds and further favor the isomerized product thermodynamically. A typical reaction was the isomerization of hexane to methylpentane and dimethylbutanes.

More recently, Nafion-H has been used effectively in the isomerization of *endo*-tetrahydrodicyclopentadiene to the *exo*-isomer[38]. Methane has been chlorinated with chlorine selectively to methyl chloride over Nafion-H catalyst[39].

Bromination and other superacid catalyzed electrophilic substitutions, reported previously in solution, can also be similarly effected over Nafion-H.

2.2. ACYLATION

Nafion-H is found to be an effective catalyst for heterogeneous acylation of aromatic hydrocarbons with aroyl chlorides and anhydrides[40]. These reactions were typically carried out at the boiling point of the hydrocarbon to be acylated for a period of 8-10 h. The reaction is general for aroyl halides and aromatics[40].

R = H, F, Cl, CH$_3$, NO$_2$
X = H, Cl, CH$_3$

Optimum yields were obtained when 10-30% of Nafion-H was employed relative to the aroyl halide. This procedure allows for clean reactions with no complex formation and easy work-up procedures. As Friedel-Crafts acylations in solution chemistry generally require more than molar equivalence of catalyst (due to complexation) the solid superacid catalyzed reaction represents a significant advantage.

Attempted acylation of aromatics with acetyl chloride under similar conditions led to thermal elimination of hydrogen chloride from the latter to form ketene and products thereof. In the reaction of acetyl chloride by itself with Nafion-H, diketene was observed by I.R. and ^{1}H NMR spectroscopy[40].

The reverse reaction, namely deacetylation and decarboxylation of aromatic compounds is also catalyzed by Nafion-H[41]. This reaction is, however, of synthetic value only in the case of activated arylmethyl ketones or carboxylic acids. The reaction proceeds via *ipso* protonation of the substrate followed by deacetylation or decarboxylation.

2.3. NITRATION

Introduction of a nitro group into an aromatic ring is usually carried out in solution via electrophilic aromatic substitution using nitric acid, its metal salts, mixed anhydrides and nitrate esters catalyzed by sulfuric acid or some Lewis acids[42]. The use of polystyrene sulfonic acid as a catalyst for nitration has been reported[43]. All these procedures of nitrations involve an aqueous, basic workup. Furthermore, one of the main problems with nitration using nitric-sulfuric acid mixtures is that nitration produces a molar equivalent of water, which causes dilution of the acid. The rate of nitration, consequently, slows down considerably upon dilution.

Reaction of alkylbenzenes and other aromatics with n-butyl nitrate in the presence of Nafion-H catalyst resulted in the formation of nitroarenes in excellent yields[44].

Optimum yields of products were obtained by using 10-25% Nafion-H catalysts by weight, based on n-butyl nitrate. The reaction does not proceed at room temperature thus necessitating the use of a nitrate ester of sufficiently high boiling point. Nitration with low boiling methyl nitrate gave only low yields at atmospheric pressure.

Nafion-H catalyzed nitrations of aromatics were also carried out using acetone cyanohydrin nitrate (ACN), nitric acid (azeotropic nitration) and dinitrogen tetraoxide[45]. ACN is more reactive than ordinary alkyl nitrates since increase in entropy facilitates the cleavage of the O-N bond in the intermediate O- or N-protonated ACN. This is reflected by the higher yields of nitro compounds observed in the case of deactivated aromatics[45].

Nitration of chlorobenzene, for example, showed an improved 49% yield as compared with 15% yield obtained using n-butyl nitrate. Preparative nitrations with n-butyl nitrate and ACN provide the cleanest method yet known for nitrations of aromatics. All of the by-products are volatile organic materials. Nitro compounds can, therefore, be isolated simply by filtration of the catalyst, without the need of any aqueous basic washing or work-up.

The nitration of aromatics with nitric acid over Nafion-H catalysts is particularly convenient when using excess of aromatics (such as benzene, toluene, etc.) to azeotropically remove the water formed as by-product in the nitration ("azeotropic-nitration")[45].

The nitration of aromatic compounds has also been carried out with Nafion-H in the presence of mercuric nitrate as well as Hg(II)-impregnated Nafion-H catalyst[46]. These studies indicate that Nafion-H:HNO$_3$ and Nafion-H:(HNO$_3$)Hg^{2+} operate by different mechanisms. It was proposed that the latter proceeds in part via mercurating the arene followed by nitrodemercuration of the initial product. Nafion-H catalyzed transnitration of 9-nitroanthracene in the presence of excess toluene has also been reported[47].

Nitrobenzene has also been prepared in a liquid flow system using a hollow tube of Nafion membrane. Benzene was passed over the outside of a tube containing 70% nitric acid. The yield of nitrobenzene was 19%[48]. Another patent describes the comparative yields and *p-/ o-ratio* of nitrotoluene obtained by using nitric acid and different acidic resins as follows, Nafion 501:95% and 1.41; Amberlyst 15:76% and 1.53; and Dowex 50:57% and 1.4[49].

2.4. ISOMERIZATION AND TRANSBROMINATION OF BROMOAROMATIC COMPOUNDS

The reactivity of Nafion-H catalyst towards haloaromatic compounds has also been investigated[50]. Halobenzenes are not affected by Nafion-H even at temperatures near 200°C. Even the crowded *o*-dibromobenzene did not undergo isomerization or disproportionation under these conditions. On the other hand, the more basic bromotoluenes isomerize both in the gas phase and in heterogeneous liquid phase reactions over Nafion-H[50].

The more nucleophilic bromomesitylene was Found to transbrominate aromatic hydrocarbons both in the liquid phase and in the gas phase when Nafion-H is used as the catalyst[50].

The results show that the gas phase reactions give higher yields than the liquid phase reactions although the contact time in the gas phase reactions is only between 3-10 sec[50]. This indicates the importance of the high temperatures for the reaction. Liquid phase transbromination of *m*-xylene with bromomesitylene yields ~ 77% of bromo- *m*-xylene after 5 h of reaction over Nafion-H.

In other laboratories Nafion-H has been used to carry out other electrophilic reactions such as sulfonation[51], phosphorylation[52], polymerization, etc. and the topic has been adequately reviewed[18].

2.5. ETHER AND ESTER SYNTHESIS

Nafion-H has found widespread application in the synthesis of ethers and esters as well as in the catalysis of their cleavage depending upon the appropriate conditions. Methoxy-methyl ethers are commonly used for protection of OH-groups in natural product synthesis[53]. Since the realization of chloromethyl methyl ether as an extreme carcinogen[54], alternate procedures using dimethoxymethane (DMM) with phosphorous pentoxide[55] or p-toluenesulfonic acid catalyst have been used for their preparation[56]. However, these reactions necessitate inconvenient basic work-up procedures.

Reaction of DMM and corresponding alcohols in the presence of Nafion-H gives good to excellent yields of methoxymethyl ethers. Workup procedure involves simple filteration followed by evaporation of solvent[57].

$$R\text{-OH} + H_3C\text{-O-}CH_2\text{-O-}CH_3 \xrightarrow[-CH_3OH,\ 57\text{-}96\%]{\text{Nafion-H}} R\text{-O-}CH_2\text{-O-}CH_3$$

R = alkyl, cycloalkyl

The reaction with primary alcohols is facile using DMM as a solvent and provides quantitative yields of the product. Benzene is, however, necessary as a co-solvent to raise reaction temperature and thus give convenient reaction rates with secondary alcohols. Ordinary tertiary alcohols yield only acid catalyzed olefinic dehydration products. Bridgehead tertiary alcohols are an exception, however, yielding methyl ethers instead.

Similarly, excellent yields of O-tetrahydropyranyl ethers are obtained when a solution of dihydro-4H-pyran in dichloromethane is slowly added to solution of alcohols in the same solvent in the presence of catalytic amount of Nafion-H. Reverse addition of alcohol to a solution of dihydro-4H-pyran containing Nafion-H results in a vigorous reaction which yields only a black, charry material (probably from the polymerization of dihydropyran)[58].

R = alkyl, arylalkyl, cycloalkyl

The reverse reaction i.e., the cleavage of these ethers is also facile under Nafion-H catalysis in the presence of methanol[58]. Accordingly both protection and deprotection can be effected by the Nafion-H catalyst system.

Similarly, O-trimethylsilylation of alcohols, phenols and carboxylic acids is also easily achieved by using allyltrimethylsilane and Nafion-H[58].

Aliphatic alchols also dehydrate to the corresponding ethers in the presence of Nafion-H. In the gas phase, with primary and secondary alcohols, it was found that low temperature (ca. 100°C) favored ether production (yields of nearly 100%) while at higher temperatures only alkenes were obtained with the exception of t-butyl alcohol, which gives 2-methylpropene even at low temperatures[19a]. In the liquid phase, diols are converted into cyclic ethers[59].

Nafion-H is also a convenient acid catalyst for esterification reactions[60].

$$R^1\text{-}CO_2H \ + \ R^2\text{-}OH \quad \xrightarrow[\text{18-100\%}]{\text{Nafion-H}} \quad R^1\text{-}CO_2R^2$$

R^1 = alkyl, aryl
R^2 = alkyl, aralkyl

2.6. ACETALS, THIOACETALS AND *Gem*-DIACETATES

When ketones or aldehydes are treated with trimethyl orthoformate in the presence of Nafion-H, the corresponding dimethyl acetals are formed in excellent yields[61]. The reactions do not require any pre-absorption on the catalyst, as is the case with other solid acids[62]. In the absence of water, the acetals decompose sluggishly to yield the corresponding aldehyde or ketone. In the Nafion-H catalyzed hydrolysis, formation of the corresponding carbonyl compounds is instantaneous, and these are obtained in excellent yields.

$$\begin{array}{c} R^1 \\ \diagdown \\ C{=}O \ + \ HC(OCH_3)_3 \quad \xrightarrow[\text{89-100\%}]{\text{Nafion-H/CCl}_4} \\ \diagup \\ R^2 \end{array}$$

R^1, R^2 = H, alkyl, aryl

$$\begin{array}{c} R^1 \\ \diagdown \\ C\,(OCH_3)_2 \ + \ H\text{-}CO_2CH_3 \\ \diagup \\ R^2 \end{array}$$

Ethylenedithioacetals (1,3-dithiolanes) can also be prepared in nearly quantitative yields by Nafion-H catalysis. In this case, heating under reflux a solution of the corresponding carbonyl compounds, 1,2-ethanedithiol, and Nafion-H in benzene with azeotropic removal of water from the reaction mixtures enables the isolation of pure products by simple filtration followed by crystallization or distillation. Related preparation of cyclic acetals using 1,3-butanediol or glycidiol has been patented[63].

$$\begin{array}{c} R^1 \\ \diagdown \\ C{=}O \ + \ \begin{array}{c} CH_2{-}SH \\ | \\ CH_2{-}SH \end{array} \quad \xrightarrow[\text{79-100\%}]{\text{Nafion-H/C}_6H_6} \quad \begin{array}{c} R^1 \\ \diagdown \!\diagup S{-} \\ C | \\ \diagup \diagdown S{-} \\ R^2 \end{array} \end{array}$$

The 1,1-diacetates can also be prepared by vigorously stirring the equivalent amounts of aldehyde, freshly distilled acetic arthydride, and a catalytic amount of Nafion-H[64] at ambient temperature. Most of the reactions were carried out in the absence of any solvent but dry tetrachloromethane was used in some cases without noticeable change in yields of diacetates.

$$RCH=O \ + \ (H_3CCO)_2O \ \xrightarrow[50\text{-}99\%]{\text{Nafion-H}} \ R\text{-}CH\ (OCOCH_3)_2$$

R = alkyl, aryl

2.7. HYDRATION

Nafion-H catalyzes the hydration and methanolysis of epoxides under very mild conditions generally affording high yields of the product[65]. The catalysts can be readily regenerated for further use without loss of activity. Only *trans* products are observed in the reaction with cyclohexene oxide and cyclopentene oxide indicating back-side attack. The conversion of ethylene oxide to ethylene glycol in 94% yield is patented[66].

2.8. REARRANGEMENTS

Nafion-H has been shown to catalyze the pinacol-pinacolone rearrangement[67]. The 1,2-diols rearrange to the corresponding ketones in 80-90% yield.

Another acid catalyzed rearrangement catalyzed by Nafion-H is the Fries rearrangement[68].

Refluxing a solution of the phenol ester in nitrobenzene solvent in the presence of Nafion-H (~ 5% by weight with respect to the ester) affects smooth conversion to the corresponding hydroxyphenyl carboxylic acids. Since Nafion-H also catalyzes the hydrolysis of esters, it is important to carry out the reaction under anhydrous conditions.

The Rupe rearrangement of alkynyl tertiary alcohols is one of the most feasible routes into α–β-unsaturated carbonyl compounds. The major drawback of this reaction, however, is that the unsaturated product can subsequently undergo acid-catalyzed polymerization[69] and form side products such as vinyl acetylenes and aldehydes[70].

Significant improvement of the Rupe rearrangement is observed with the use of Nafion-H as a catalyst[71] with yields better than those obtained using mercury-impregnated Dowex-50 resin[72].

The gas phase rearrangement of allyl alcohols to the corresponding aldehydes has been reported to proceed by Nafion-H catalysis at 170-190°C[73], while the isomerization of styrene oxide proceeds at room temperature in 1/2 h to give 47% yield of phenylacetaldehyde[74].

2.9. MERCURY IMPREGNATED NAFION-H CATALYST

As described earlier, Mercury(II)-promoted azeotropic nitration of aromatics over Nafion-H catalyst gave isomer ratios in the product that were quite different from those observed without the presence of mercury. The increased amounts of *meta* isomer obtained in the nitration of toluene, ethylbenzene and *tert*-butylbenzene are attributed not only to the reversible aromatic mercuration but also to the facile mercury shifts in arenemercurium ions[46]. Whereas an attempted azeotropic nitration of ethylbenzene with nitric acid over Nafion-H catalyst gives only side-chain oxidation[75], yielding acetophenone, the mercury(H) promoted nitration gave only 13% of side chain oxidation. Hydration of alkynes are usually carried out in dilute sulfuric acid solutions to afford carbonyl compounds. Except for reactive alkynes, the reaction rates in the absence of mercury(H) salts are low, even when higher acid concentrations are used. One of the serious problems arising from using mercury(H) salts is the formation of a precipitate of an inactive sludge consisting of finely divided metallic mercury mixed with insoluble mercury(H) organic compounds. Apart from difficulties in the work-up stage, this also causes loss of catalytic activity and environmental problems. Use of mercury(H) impregnated Nafion-H (replacing about 25% of the acidic protons by mercury(H)) as a catalyst alleviates all these problems affording the corresponding ketones very cleanly and easily[76].

$$R_1\!\!-\!\!\!\equiv\!\!\!-R_2 \; + \; H_2O \quad \xrightarrow[65\text{-}94\%]{\substack{\text{Hg/Nafion-H/}\\\text{EtOH or AcOH}}} \quad R_1CH_2\overset{\displaystyle O}{\underset{}{\overset{\|}{C}}}R_2$$

No loss in the catalytic activity is observed if the hydration is done at room temperature which is complete in 90 min. While refluxing in ethanol completes the reaction in 5 min, the catalyst is unsuitable for reuse.

2.10. MISCELLANEOUS REACTIONS

Although many Diels Alder reactions proceed without the necessity of catalysts, higher reaction temperatures are often required[77]. Where heat sensitive intermediates are involved, for example in complex, multistep synthyesis of natural products[78], the use of Diels-Alder reactions suffer from this limitation. While Lewis acid catalysis does help the reaction to proceed at room temperature with satisfactory yields[79-80], they are often accompanied by diene polymerization. Moreover, two molar excess amounts of the catalyst are often needed to catalyze carbonyl containing dienophiles. The use of Nafion-H as a protic acid catalyst combines the advantage of requiring only catalytic amounts since it forms reversible complexes, and also avoiding destruction and separation of the catalyst upon reaction completion[81]. Thus, while the uncatalyzed reaction of 1,3-cyclohexadiene and acrolein gives 25% of the adduct upon heating to 100°C for 3.5 h, Nafion-H catalysis gives 88% after stirring for 40 h at 25°C.

$$\text{cyclohexadiene} \;+\; \text{CHO} \quad \xrightarrow{\text{Nafion-H/CH}_2\text{Cl}_2} \quad \text{bicyclic-CHO}$$

Nafion-H has also been used to catalyze Ritter reactions between alcohols and alkyl or aryl nitriles[82].

$$R^1OH \;+\; R^2CN \quad \xrightarrow{\text{Nafion-H}} \quad R^1NHCOR^2$$

Baeyer-Villiger oxidation of ketones to lactones has been observed to proceed smoothly over Nafion-H[82] with a minimum of work-up procedure using a single equivalent of the oxidizing agent hydrogen peroxide or *m*-chloroperbenzoic acid.

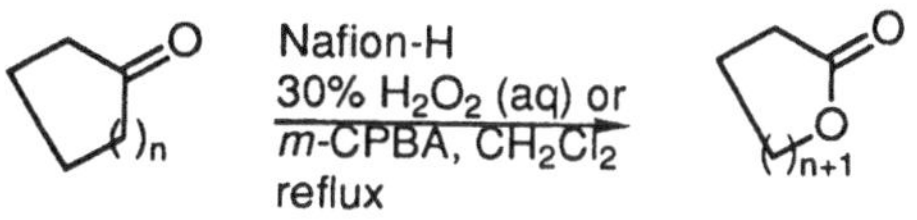

The above reaction has been further improved[83] using bistrimethylsilyl peroxide as the oxidant.

Nafion-H has also been employed[84] for the reductive cleavage of acetals and ketals to their respective ethers with triethylsilane.

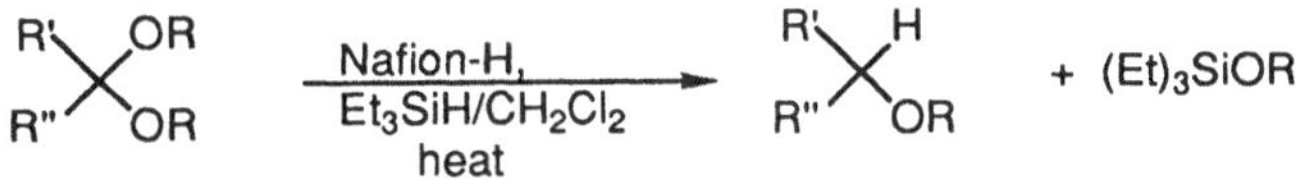

Nafion-H catalyzed allylation of acetals and ketals has also been achieved with allyltrimethylsilane[85].

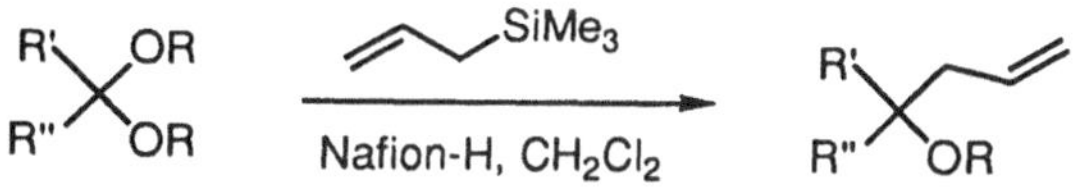

Acetone has been selectively condensed to mesitylene by Nafion-II catalysis[86].

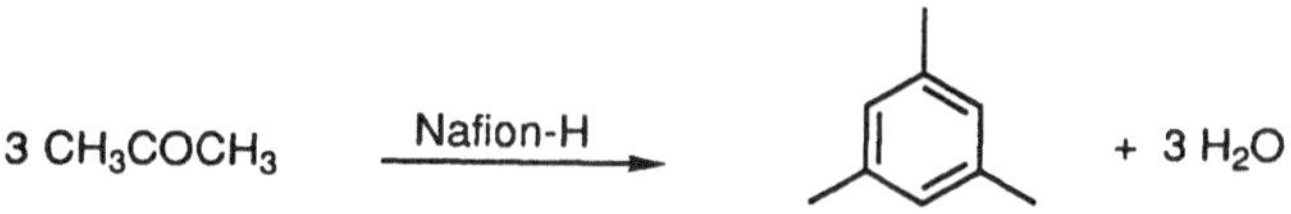

Nafion-H is also a suitable catalyst for the condensation of acetophenones to 1,3,5-triarylbenzenes[87].

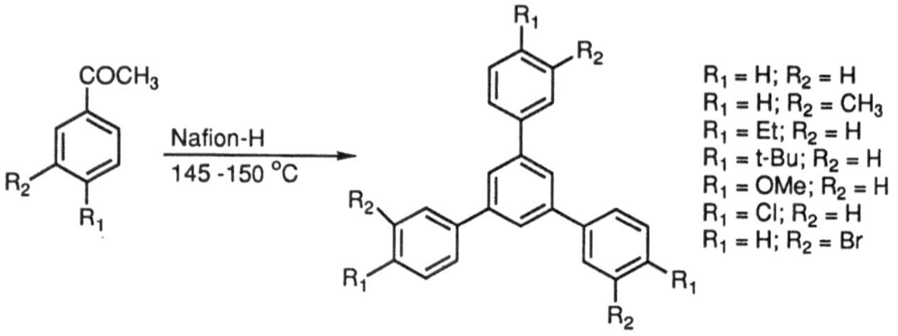

R₁ = H; R₂ = H
R₁ = H; R₂ = CH₃
R₁ = Et; R₂ = H
R₁ = t-Bu; R₂ = H
R₁ = OMe; R₂ = H
R₁ = Cl; R₂ = H
R₁ = H; R₂ = Br

Nafion-H has been used for the *tert*-butylation of aromatic compounds with 2,6-di(*tert*-butyl)p-cresol[88]. The *para:meta* isomer ratio is same as that obtained under usual Friedel-Crafts conditions using AlCl₃ as the catalyst[89].

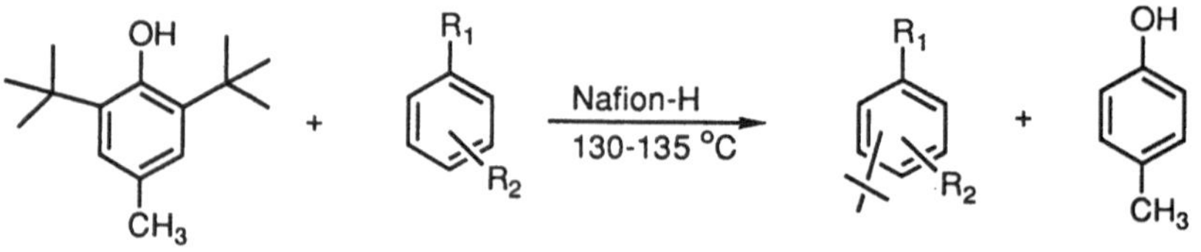

Friedel-Crafts benzylation of benzenes and substituted benzenes with benzyl alcohols has been achieved[90] under relatively mild experimental conditions with Nafion-H catalyst. The method was also found suitable for the intramolecular cycloalkylation and oligomerization of methoxybenzyl alcohols. The procedure has been used for the preparation of metacyclophanes[91].

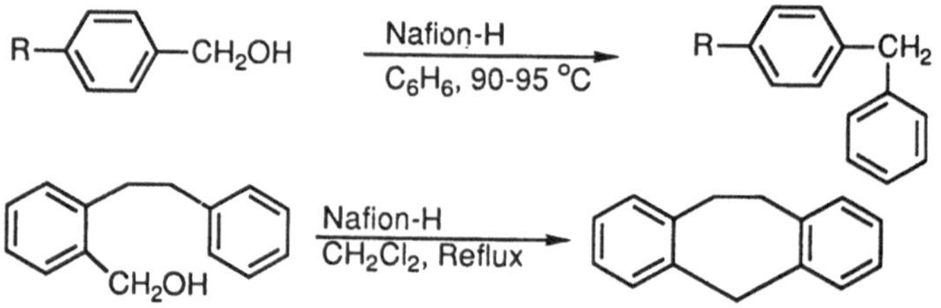

Even intramolecular cycliacylation is affected by Nafion-H[92].

An interesting ring closure of 2,2'-dihydroxybiphenyls to the corresponding dibenzofurans was also achieved over Nafion-H catalyst[93].

Similarly 2,2'-diaminobiphenyl lead to a new preparative route to carbazoles[94].

Nafion-H was also found to be a good solid acid catalyst for the Petersons methylenation of carbonyl compounds[95].

Acetals were readily converted to thioacetals over Nafion-H[96].

3.   Comparison of Nafion-H with Related Perfluorinated Resinulfonic Acid Catalysts and Supported Solid Perfluoroalkanesulfonic Acids

Perfluorinated resinsulfonic acids as discussed are efficient and versatile heterogeneous acid catalysts. Their acid strength is comparable to 100% sulfuric acid. The fluorocarbon polymer acids are chemically resistant and stable up to 200-250°C. At the same time they have the disadvantage that their surface areas compared to those of more common heterogeneous acid catalysts (metal oxides, alumino-silicates, zeolites, etc.) are very low (0.5-2 m^2/g). The acidic sites which are incorporated into the fluorocarbon matrix are not readily accessible to the reactants and thus the reaction rates are limited by diffusion. This problem can be minimized by reducing the particle size of the polymer. For example, the reactivity of a 120-mesh resin is nearly seven times greater than that of a 30-mesh resin in the alkylation of phenol with 1-decene[97]. Finely powdered catalysts, however, are usually inconvenient to use in contact catalytic processes, particularly in continuous flow systems. Another way of increasing of the surface area is to support the catalysts on suitable solid carriers. This often-used method was also applied to perfluorinated resinsulfonic acid catalysts. For example, McClure and co-workers used Nafion-H resin supported on different carriers (alumina, silica, silica-alumina, and porous glass) for the alkylation of 2-methylpropane with olefins and for the isomerization of hexane. It has been found that the activity of a 1% Nafion-H on Chromosorb-T is 2.5 times greater than that of 5% Nafion-H on silica and about 12 times greater than that of the unsupported Nafion-H resin[98]. In another example, in several aromatic alkylation reactions it was found that a supported Dow perfluorinated resinsulfonic acid catalyst (Dow XU 40036) had a much higher efficiency than unsupported 10-30 mesh solid polymer pellets.

Bucsi and Olah compared[99] three solid catalyst of the DOW Chemical Co. (XUS-40036.01, XUS-40036.02 and XU-40036.04), with DuPont's Nafion-H resin in the oligomerization of 2-methylpropene and the transformation of 2,4,4-trimethyl-2-pentene.

XU-40036.04 (for brevity '04') is an unsupported perfluorinated resinsulfonic acid which is similar in structure to Nafion-H, only the side chains carrying sulfonic acid groups are shorter:

$$-CF_2-\underset{\underset{\underset{OCF_2CF_2SO_3H}{|}}{\overset{\displaystyle [X]_n}{|}}}{CF}-CF_2-CF_2-$$

DuPont Nafion-H: n = 1,2,3, . . .
Dow resin: n = 0

$$X = CF_2CF_2$$

The equivallent weight of this polymer is 800 g/eq and was used as small pellets.

XUS-40036.01 (01) is the 04 resin supported on 1/16" α-alumina spheres. The resin content is 12 wt% and the acid capacity approximately 0.1 meq/g.

XUS-40036.02 (02) is the 04 resin supported on 1/4" hollow cyclinders of silicon carbide. It has a very low surface area, (approximately 0.02 m^2/g), the acid capacity of the dry catalyst is 0.20-0.22 meq/g, resin content is 18 wt%. The form of the 02 catalyst was not suitable for the studies, thus before use it was broken into small pieces similar in dimension to the 01 catalyst.

The Dow perfluorinated resinsulfonic acids and DuPont Nafion-H catalysts showed similar activity and selectivity during the oligomerization of 2-methylpropene.

Supporting the resinsulfonic acids on alumina or silicon carbide supports has no effect on their catalytic properties. The supported catalysts. however, show greater activity per unit of resin than the unsupported ones. Below 100°C, the investigated catalysts did not suffer noteworthy deactivation after 20 h on-stream time. Above 130°C, the lifetime of the catalysts is greatly limited.

The transformation of 2,4,4-trimethyl-2-pentene at room temperature showed significant dependence on used hydrated or dehydrated resinsulfonic acid catalysts and thus the acidity of the catalysts. In the case of hydrated ones only double-bond isomerization takes place. With dehydrated resinsulfonic acids both isomerization and oligomerization occur. From these results it is indicated that the Hammett acidity of the hydrated resinsulfonic acids is comparable to 65-70% H_2SO_4, whereas the dehydrated resinsulfonic acid is comparable to or stronger than 100% H_2SO_4.

Perfluoroalkanesulfonic acids, such as trifluoromethanesulfonic (triflic) acid or pentafluoroethanesulfonic (pentflic) acid are highly efficient superacidic catalysts in solution chemistry. Their higher C_6-C_{12} homologs, such as perfluorodecanesulfonic acid (PDSA) were studied as supported solid catalysts in heterogeneous catalytic reactions[100]. Alkylation, isomerization, polymerization and cracking reactions, *inter alia*, were found to be effectively catalyzed, particularly when the sulfonic acids were combined with TaF_5, NbF_5 and the like Lewis acid fluorides. Support generally was fluoridated alumina. Olah also found fluorinated graphite as a suitable support for superacidic catalysts based on SbF_5, TaF_5, NbF_5 and perfluoroalkanesulfonic acids[101].

4.   The Acidity of Solid Superacid Catalysts

One of the major difficulties in characterizing solid acids is to accurately determine their acidity. The most frequently used methods relate to kinetic rather than thermodynamic measurements. These can give indication of catalytic activity of the solid acids, but not necessarily of their accurate acidities. It is in my view uncertain to what extent one can extrapolate results of measurements based on assumed analogies with solution equilibria to solid acid behavior in claiming H_0 and related acidity functions. Tanabe discussed such acidities in considerable detail[102].

Bunce, Fyfe and their coworkers studied[103,104] with acid base indicators suspensions of Nafion-H in solvents capable of removing water from within the polymer beads. Proton NMR of Nafion-H samples swollen in benzene and cyclohexane showed that the ^1H chemical shift of the hydroxylic protons moved downfield (because more shielded) as the water content decreased. The chemical shift change arises from fast exchange between H_2O and Nafion-H. With an OH/SO_3^- ratio i.e. of m ≥ 3, the resin behaving as $R_FSO_3^-.H_3O^+$. The chemical shift thus can be used not only to measure the water content but as an indication of the acid strength. Nafion-H dried in vacuum to constant weight was shown to be the monohydrate, i.e. $R_FSO_3^-.H_3O^+$. This explains the great acidity of the Nafion-H catalyst, comparable at least with concentrated H_2SO_4 or CF_3SO_3H.

Spectroscopic studies, frequently based on surface acid-base interactions, calorimetric measurements and recently solid state NMR methods, including ^1H MAS NMR study of zeolites, have started to provide more direct ways for direct measurement of acidities of solid acids[106].

5. Mechanistic Considerations

By definition, electrophiles are electron acceptor reagents, i.e. electron deficient compounds[107]. The proton is the ultimate electrophile, having no electron at all. In the condensed state, however, no "naked" proton can exist and it is always solvated (associated). Trivalent carbocations are similarly highly electron deficient carbon electrophiles although they tend to fill up partially their electron deficiencies either by intra or intermolecular interactions. Carboxonium ions, acyl cations, etc. are increasingly less electron deficient carbocationic species, as participation of electron donating neighboring groups decreases the electron deficiency of the carbon center.

Over the course of 35 years my research group has studied long-lived electrophilic intermediates. These include carbocations[108], carboxonium ions and various onium ions[109] including oxonium, sulfonium, halonium[110], azonium ions, as well as other electron deficient species. Our studies initially utilized only liquid superacidic media as the essential low nucleophilicity environments[2]. It was realized that long lived highly electron deficient (i.e. acidic) species in solution can exist only in very low nucleophilicity (basicity) systems. Our studies resulted in establishing what is now generally referred to as <u>superacidic, stable ion chemistry</u> and the chemistry of stable carbocations[108,109]. Thousands of publications from around the world have and are continuing to contribute to this very active field. Subsequently studies were extended to solid superacids, primarily to Nafion-H and related perfluorinated resinsulfonic acids.

In general acid catalyzed reactions of hydrocarbons (or their derivatives) involve protolytic (or Lewis acid complexed) formation of electrophiles capable of reacting with various substrates such as arenes, alkenes, alkynes, even alkanes. Superacid catalyzed reactions fit this general pattern and their mechanistic aspects were reviewed previously[111]. The readers are referred to these reviews.

During our studies on superacidic systems we made observations which indicated that superacids besides being highly ionizing, low nucleophilicity media, in some instances were also capable of additional unexpected activation of electrophiles by further protolytic (electrophilic) interaction (coordination). In 1975 with Germain, Lin and Forsyth[112] we have first reported that the acetyl cation (CH_3CO^+) and nitronium ion (NO_2^+) in superacidic media display highly enhanced reactivity indicative of their protolytic (protosolvated) activation through dications, ($CH_3C=OH$ and NO_2H^{2+}). Later we have found similar indications for the protoformyl dication ($HCOH^{2+}$, diprotonated CO). In 1986 (with Prakash, Barzaghi, Lammerstma, Schleyer and Pople) we described studies relating to the protohydronium dication (tetrahydridooxonium dication, diprotonated water/H_4O^{2+})[113] and subsequently the tetrahydrido-sulfonium dication, H_4S^{2+}[114]. In

1989 (with Prakash and Lammerstma) we published a brief review on protonated (protosolvated) onium ions (onium dications)[115]. The overall concept and significance of these studies, however, till recently aroused relatively little response and interest in the broader chemical community, probably because they were considered of highly specialized nature involved only in still exotic superacidic systems. Not unlike most chemists reflection of mass spectrometric, gas phase ion and ion-molecule studies, superacidic studies frequently are viewed as systems in which surprising observations can be made, but this does not necessarily relate to "real" chemistry. Recently, however, the general significance of superacid activation of electrophiles started to emerge[116].

Electrophilic (proto)solvation of electrophiles such as onium, carboxonium, and related ions involves their further coordination with strong Brønsted or Lewis acids. Interaction of these ions with superacid systems in the limiting case can lead to extremely reactive *de facto* dications, although polarized donor-acceptor interactions should have similar effect. Such protolytic (electrophilic) activation of electrophiles leads to <u>superelectrophiles, i.e. electrophiles whose reactivity greatly exceeds that of their parents in aprotic or conventional acidic media</u>. The definition is arbitrary, as is that of superacids, but describes the remarkably enhanced reactivity of activated electrophiles of doubly electron deficient (dipositive) nature. Electrophilic activation can play a significant role in Friedel-Crafts and superacid catalyzed reactions, as well as in solid acid catalyzed systems and even enzymatic reactions.

When considering protosolvated carbocations (carboxonium ions) association or clustering effects may have a major influence. In particular in small dications, such effects tend to diminish the effect of charge-charge repulsion and thus could bring them into a thermodynamically more accessible region.

Protosolvation is generally associated with superacidic solution chemistry. However, solid superacids, possessing both Brønsted and Lewis acid sites, whose catalytic reaction were discussed in this review, can similarly effect superelectrophilic activation. It was established that solid acids such as Nafion-H or HZSM-5 possess superacidic nature. In order to explain how these acidic solids can frequently display remarkable activity, for example allowing the catalytic transformations of extremely low nucleophilicity alkanes (even methane), a reappraisal of the mode of electrophilic activation at the acid sites of these catalysts is worth considering.

Nafion-H is known to contain acidic -SO_3H groups in clustered areas.

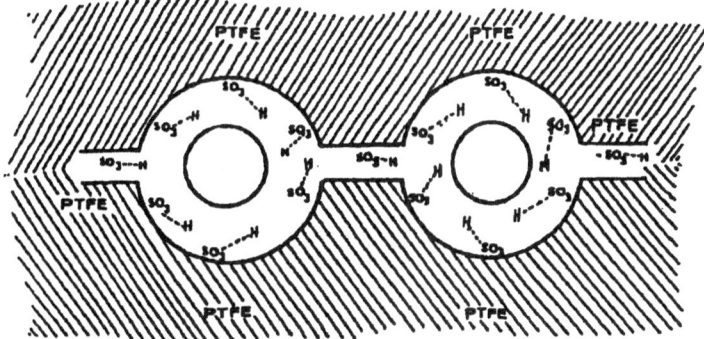

H-ZSM-5 also displays superacidic activity. , As found by Haag et al.[117] it for example readily isomerizes and catalyzes alkylation of alkanes (with H_2 formed as the protolytic byproduct in stoichiometric amounts). In this Zeolite active Brønsted and Lewis acid sites are again in close proximity (~2.5Å).

Lewis Site

Bronsted ----→
Site

$$
\begin{array}{c}
\text{OH} \qquad\qquad \text{OH} \\
| \qquad\qquad\qquad\qquad\quad | \\
-\text{Si}---\text{O}---\text{Al}---\text{O}---\text{Si}- \\
| \qquad\qquad\qquad\qquad\quad |
\end{array}
$$

It is suggested that in these (and other) solid superacid catalyst systems bi- or multi-dentate interactions at the acidic sites forming highly reactive intermediates is possible, amounting to the solid state equivalent of protosolvation discussed previously for solution systems.

Nature is able to perform its own transformations in ways which chemists have only started to understand. At enzymatic sites many significant transformations take place which are acid catalyzed. In a generalized sense electron deficient metal ion catalyzed processes can also be included. As at enzymatic sites due to their unique geometry bi- and multi-dentate interactions are again readily possible, the same principles may also be relevant to our better understanding of some enzymatic process.

In conclusion superelectrophiles i.e. electrophiles of doubly electron deficient (dipositive) nature whose reactivity significantly exceeds that of their parents observed are highly reactive, energetic, high lying intermediates. They are the *de facto* reactive intermediates of many electrophilic reactions in superacidic systems (including those involving solid superacids) and should be differentiated from low lying, much more stable intermediates, which frequently are observable and even isolable but not necessarily reactive enough without further activation.

It should be recognized, however, that some of the observed superelectrophilic reactions can also be considered to proceed involving only "electrophilic assistance" by the superacids without necessarily forming distinct dipositive intermediates

$$
H^{\oplus} + X\overset{\oplus}{-}H \longrightarrow H-\overset{(2+)}{\underset{|}{X}}-H + NU \longrightarrow \text{Product}
$$

$$
H^{\oplus} \quad \overset{\oplus}{\underset{|}{X}}-H \quad :Nu \longrightarrow \text{Product}
$$

Some representative of so far studied superelectrophiles and their parents are listed.

Parent electrophile	Superelectrophile
$R\overset{\oplus}{\underset{R}{O}}R$ R = H, alkyl or Lewis acid	$R\overset{\overset{R}{\mid}}{\underset{R}{P}}R^{(2+)}$
$\underset{R}{\overset{R}{C}}=\overset{\oplus}{O}R$	$\underset{R}{\overset{R}{\overset{\oplus}{C}}}-\overset{\oplus}{\underset{R}{O}}R$
$RC{\equiv}\overset{\oplus}{O}$	$R\overset{\oplus}{C}=\overset{\oplus}{O}R$
$R-\overset{\overset{OR}{\mid}}{\underset{OR}{\overset{\oplus}{C}}}$	$R-\overset{\overset{\oplus}{O}R}{\underset{\overset{\oplus}{O}R_2}{C}}$
$HCO^{\oplus}$	$H\overset{\oplus}{C}=\overset{\oplus}{O}H$
$HO{\cdots}\overset{\overset{OH}{\cdots}}{\underset{OH}{\overset{\oplus}{C}}}$	$HO{\cdots}\overset{\oplus}{\underset{HO}{C}}-\overset{\oplus}{O}H_2$
$R_3S^{\oplus}$	$R_4S^{(2+)}$
$R_3Se^{\oplus}$	$R_4Se^{(2+)}$
$R_3Te^{\oplus}$	$R_4Te^{(2+)}$

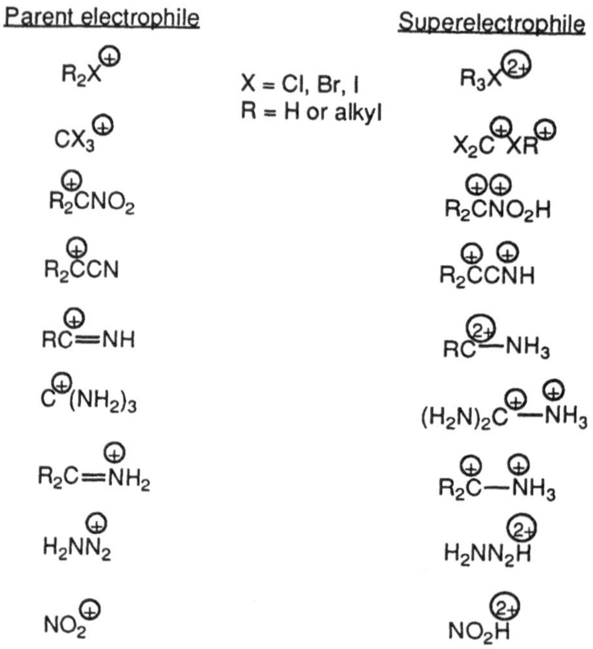

330

Parent electrophile		Superelectrophile

$R_2X^{\oplus}$

$X = Cl, Br, I$
$R = H$ or alkyl

$R_3X^{2\oplus}$

$CX_3^{\oplus}$

$X_2C^{\oplus}XR^{\oplus}$

$R_2\overset{\oplus}{C}NO_2$

$R_2\overset{\oplus\oplus}{C}NO_2H$

$R_2\overset{\oplus}{C}CN$

$R_2\overset{\oplus}{C}C\overset{\oplus}{N}H$

$R\overset{\oplus}{C}{=}NH$

$R\overset{2\oplus}{C}{-}NH_3$

$\overset{\oplus}{C}(NH_2)_3$

$(H_2N)_2\overset{\oplus}{C}{-}\overset{\oplus}{N}H_3$

$R_2C{=}\overset{\oplus}{N}H_2$

$R_2\overset{\oplus}{C}{-}\overset{\oplus}{N}H_3$

$H_2N\overset{\oplus}{N}_2$

$H_2NN_2\overset{2\oplus}{H}$

$NO_2^{\oplus}$

$NO_2\overset{2\oplus}{H}$

Scheme: Representative superelectrophiles and their parents

Acknowledgment: I thank my colleagues, whose names appear in the references, for their contributions, without which the reported work could not be achieved. Professor G. K. S. Prakash and I cooperated for many years in studies on superacid catalysis and he always played a major part in our joint efforts. The National Science Foundation, the U. S. Army Office of Research and the Loker Hydrocarbon Research Institute supported our work.

References

1. Based in part on a review by G. K. S. Prakash and G. A. Olah, *Ind. Natl.Acad.Sci.Proc.* **1988**, *100*, 143.
2. G. A. Olah, G. K. S. Prakash and J. Sommer, *Superacids*, Wiley Interscience: New York, N.Y., **1985**.
3. a) R. J. Gillespie and T. E. Peel, *J.Am.Chem.Soc.* **1973**, *95*, 5173; b) .J. Grondin, R. Sagnes and A. Commeyras, *Bull.Soc.Chim.Fr.*, **1976**, 1779.
4. S. C. Stinson, *Chem.Eng.News*, March 15, **1982**, *60*, 22.
5. A. Eisenberg and H. L. Yeager, (Eds.), *Perfluorinated Ionomer Membranes*, ACS Symposium Series, ACS, Washington, D.C. **1982**, p.180.
6. M. S. Seko, S. Ogawa and K. Kimoto, Chapter 15 in ref. 5.
7. B. Kippling, Chapter 19 in Ref. 5.
8. H. L. Yeager, Chapter 3 in Ref. 5.
9. R. Yeo, Chapter 18 in Ref. 5.
10. R. Strasser and S. Payer in M.O. Coulter (Ed.), *Modern Chlor-Alkali Technology*, Ellis Horwood, Chichester, England, **1980**, p. 51.
11. a) A. B. LaConti, paper presented at the workshop on perfluorinated ionomers of the Polymer Division of the ACS in Lake Buena-Vista, Florida, February **1982**; b) F. J. Waller, *Polymer Reagents and Catalysts*, **1986**, *308*, 42; c) F. J. Waller, *Catal.Rev.*, **1986**, *28*, 1.
12. D. J. Conolly and W. F. Gresham, U.S. Patent 3282875, 1966, DuPont.
13. British Patent 1034197, 1966, DuPont; C.A. **1967**, 66, 11326.
14. R. J. Cavanaugh and W. H. Calkins, U.S. Patent, 3882093, 1975; C.A. **1975**, 83, 165338.
15. J. D. McClure, U. S. Patent 4041090, 1977; C.A. **1977**, 87, 184185.
16. T. Gierke, *Ionic Clustering in Nafions Peruorosulfonic Acid Membranes*, Electrochemical Society Fall meeting, October 1977, Atlanta, GA, U.S.
17. B. M. Rode, A. Engelbrecht and J. Schantl, *J.Phys. Chem., (Leipzig)*, **1973**, *253*, 17.
18. For a comprehensive review see, G. A. Olah, P. S. Iyer and G. K. S. Prakash, *Synthesis*, **1986**, 513.
19. a) G. A. Olah, J. Kaspi, and J. Bukala, *J.Org. Chem.* **1977**, *42*, 4187; b) R. J. Vaughan, British Patent 2064356, 1981; C.A. **1982**, 96, 19791; c) Japanese Patent 5767525, 1982; Nippon Petrochemicals Co., C.A. **1982**, 97, 19984; d) Japanese Patent 5883640, 1983, Mitsui Petrochemicals Ind., C.A. **1983**, 99, 160306; e) R. L. Cobb, U.S. Patent 4480142, 1984; C.A. **1985**, 102, 61909.
20. T. Onada and K. Wada, Japanese Patent 7608231, 1976; C.A. **1976**, 84, 164449.
21. Belgian Patent 887202, 1981, Produits Chimiques Ugine Kuhlmann; C.A. **1981**, 95, 203634.
22. G. A. Olah, U.S. Patent 4288646, 1981; C.A. **1981**, 95, 203636; G. A. Olah, Canadian Patent 1152111, 1983; C.A. **1984**, 100, 34291.
23. G. A. Olah, U.S. Patent 4116880, 1978; C.A. **1978**, 89, 221652.
24. J. Kaspi and G. A. Olah, *J.Org.Chem*, **1978**, *43*, 3142.
25. W. FunakosM, T. Urasaki, I. Oda, and T. Shima, Japan Kokai 7397825, 1973; C.A. **1974**, 80,59675; Japan Kokai 7407235, 1974; C.A. **1974**, 80, 120534; Japan Kokai 7413128, 1974; C.A. **1974**, 80, 120532; Japan Kokai 7414432, 1974; C.A. **1974**, 80, 120527; Japan Kokai 7418834, 1974; C.A. **1974**, 81, 3586.

26. J. Kaspi, D. D. Montgomery and G. A. Olah, *J.Org. Chem.*, **1978**, *43*, 3147.

27. G. A. Olah and D. Meidar, *Nouv.J.Chim.*, **1979**, *3*, 269.

28. G. E. Langlois, U.S. Patent 2713600, 1955; C.A. **1956**, 50, 5738.

29. G. A. Olah, D. Meidar, R. Malhotra, J. A. Olah and S. C. Narang, *J.Catal.*, *1980*, *61*, 96.

30. a) G. A. Olah and J. Kaspi, *Nouv.J.Chim.*, **1978**, *2*, 581; b) P. Beltrame, P. L. Beltfame, P. Carniti, and M. Magnoni, *Gazz. Chim. Ital.*, **1978**, *108*, 651; c) P. L. Beltrame, P. Carniti and G. Nespoli, *Ind.Eng.Chem.Prod.Res.Dev.*, **1980**, *19*, 205.

31. G. A. Olah and J. Kaspi, *Nouv.J.Chim.*, **1978**, *2*, 585.

32. G. A. Olah, M. W. Meyer and N. A. Overchuk, *J.Org. Chem.*, **1964**, *29*, 2313.

33. a) G. A.Olah et al. unpublished work; b) W. M. Kruse and J. F. Stephen, U.S. Patent 4487978, 1984; C.A. **1985**, 102, 113028; c) W. M. Kruse and J. F. Stephen, U.S. Patent 4482755, 1984; C.A. **1985**, 102, 78556; d) T. Maki, T. Masuyama, T. Yokoyama and Y. Fujiyama, European Patent 45959, 1982; C.A. **1982**, 96, 218395.

34. M. Tashiro, *Synthesis*, **1979**, 921.

35. G. A. Olah, G. K. S. Prakash, P. S. lyer, M. Tashiro and T. Yamato, *J.Org. Chem.*, **1987**, *52*, 1881.

36. W. R. Cares, U.S. Patent 4065512, 1977; C.A. **1978**, 89, 6805.

37. J. D. McClure and S. G. Brandenberger, U.S. Patent, 4038213, 1977; C.A. **1977**, 87, 120384.

38. G. A. Olah et al. unpublished results.

39. G. A. Olah, B. G. B. Gupta, M. Farnia, J. D. Felberg, W. M. Ip, A. Husain,, R. Karples, K. Lammertsma, A. K. Mehrotra and N. J. Trivedi, *J.Am.Chem.Soc.*, **1985**, *107*, 7097.

40. G. A. Olah, R. Malhotra, S. C. Narang and J. A. Olah, *Synthesis*, **1978**, 672.

41. G. A. Olah, K. Laali and A. K. Mehrotra, *J.Org.Chem.*, **1983**, *48*, 3360.

42. G. A. Olah, R. Malhotra and S. C. Narang, *Nitration: Methods and Mechanisms*, VCH Publishers: New York, N.Y., 1989.

43. T. Kameo, S. Nishimura and O. Manabe, *Nippon Kagaku Kaishi*, **1974**, *1*, 122; C.A. **1974**, 80, 82273.

44. G. A. Olah and S. C. Narang, *Synthesis*, **1978**, 690.

45. G. A. Olah, R. Malhotra and S. C. Narang, *J.Org. Chem.* **1978**, *43*, 4628.

46. G. A. Olah, V. V. Krishnamurthy and S. C. Narang, *J.Org. Chem.*, **1982**, *47*, 596.

47. G. A. Olah, S. C. Narang, R. Malhotra and J. A. Olah, *J.Am. Chem. Soc.*, **1979**, *101*, 1805.

48. R. J. Vaughan, Australian Patent, 506423, 1980; C.A. **1980**, 93, 46166.

49. T. Kameo, T. Hirashima and O. Manabe, *Nippon Kagaku Kaishi*, **1983**, *3*, 414; C.A. **1983**, 99, 38110.

50. G. A. Olah, D. Meidar and J. A. Olah, *Nouv.J.Chemie*, **1979**, *3*, 275.

51. R. J. Vaughan, Australian Patent, 4308215, 1981; C.A. **1982**, 96, 87450.

52. R. J. Cozens, P. J. Hogan and M. J. Lalkham, European Patent, 24128, 1981; C.A. **1981**, 95, 150884.

53. F. W. McOmie, *Advances in Organic Chemistry*, Wiley Interscience: New York, N.Y. 1963; Vol. 3, p.191.

54. *Occupational Safety and Health Administration*, U.S. Department of Labor, Federal Register, **1974**, *39*(20), 3756.

55. K. Fuji, S. Nakano and E. Fujita, *Synthesis*, **1975**, 276.

56. J. P. Yardley and H. Fletcher, *Synthesis*, **1976**, 244.

57. G. A. Olah, A. Husain, B. G. B. Gupta and S. C. Narang, *Synthesis*, **1981**, 471.
58. G. A. Olah, A. Husain and B. P. Singh, *Synthesis*, **1983**, 892.
59. G. A. Olah, S. C. Narang and R. Malhotra, *Synthesis*, **1981**, 474.
60. G. A. Olah, T. Kcumi and D. Meidar, *Synthesis*, **1978**, 929.
61. G. A. Olah, S,. C. Narang, D. Meidar and G. Salem, *Synthesis*, **1981**, 282.
62. E. C. Taylor and C. S. Chiang, *Synthesis*, **1977**, 467.
63. a) Japanese Patent 56166185, 1981, Mitsubishi Chemical Ind.; C.A. **1982**, 96, 142832; b) O.R. Hughes, U.S. Patent 4003918, 1977; C.A. **1977**, 86, 171466.
64. G. A. Olah and A. K. Mehrotra, *Synthesis*, **1982**, 962.
65. G. A. Olah, A. P. Fung and D. Meidar, *Synthesis*, **1981**, 280.
66. L. Kim, U.S. Patent 4165440, 1979; C.A. **1979**, 91, 158329.
67. G. A. Olah and D. Meidar, *Synthesis*, **1978**, 358.
68. G. A. Olah, M. Arvanaghi and V.V. Krishnamurthy, *J.Org.Chem.*, **1983**, *48*, 3359.
69. M. F. Ansell, J.W. Hancock and W. J. Hickinbottom, *J. Chem. Soc.*, **1956**, 911.
70. R. W. Hasbrouck, A. D. Kiessling, *J.Org.Chem.*, **1973**, *38*, 2103.
71. G. A. Olah and A. P. Fung, *Synthesis*, **1981**, 473.
72. M. S. Newman, *J.Am.Chem.Soc.*, **1953**, *75*, 4740.
73. G. A. Olah, D. Meidar and G. Liang, *J.Org.Chem.*, **1978**, *43*, 3890.
74. Japanese Patent Applied 8092293, 1980, Toya Soda Mfg. Co.; C.A. **1982**, 97, 38670.
75. G. A. Olah, V. V. Krishnamurthy and S. C. Narang, unpublished results.
76. G. A. Olah and D. Meidar, *Synthesis*, **1978**, 671.
77. M. C. Kloetzel, *Org.React.* **1948**, *4*, 1; L. Butz and A. W. Rytina, *Org.React.* **1949**, *5*, 136.
78. K. Nakanishi, T. Goto, S. Ito, S. Natori and S. Nozoe, (Eds.), *Natural Products Chemistry*, Academic Press: New York, **1974**.
79. P. Yates and P. Eaton, *J.Am.Chem.Soc.*, **1960**, *82*, 4436.
80. G. 1. Fray and R. Robinson, *J.Am.Chem.Soc.*, **1961**, *83*, 249.
81. G. A. Olah, D. Meidar and A. P. Fung, *Synthesis*, **1979**, 270.
82. G. A. Olah, T. Yamato, P. S. Iyer, N. J. Trivedi, B. P. Singh and G. K. S. Prakash, *Mat.Chem.Phys.*, **1987**, *17*, 21.
83. G. A. Olah, T. Ernst, C. B. Rao, Q. Wang and G. K. S. Prakash, unpublished results.
84. G. A. Olah, T. Yamato, P. S. lyer and G. K. S. Prakash, *J.Org.Chem.*, **1985**, *51*, 2826.
85. G. A. Olah, Y. Yamato, J. Handlcy and G.K.S. Prakash, unpublished results.
86. G. A. Olah and W. M. Ip, *New J.Chem.*, **1988**, *12*, 299.
87. T. Yamato, C. Hideshima, M. Tashiro, G. K. S. Prakash, and G. A. Olah, *Catal.Lett.*, **1990**, *6*, 341.
88. T. Yamato, C. Hideshima, A. Miyazawa, M. Tashiro, G. K. S. Prakash, and G. A. Olah, *Catal.Lett.*, **1990**, *6*, 345.
89. M. Tashiro, G. Fukata and T. Yamato, *Org.Proc.Int.*, **1977**, *9*, 151.
90. T. Yamato, C. Hideshima, G. K. S. Prakash and G. A. Olah, *J.Org.Chem.*, **1991**, *56*, 2089.
91. T. Yamato, N. Sakane, T. Furusawa, M. Tashiro, G. K. S. Prakash and G. A. Olah, *J.Chem.Res(s).*, **1991**, 242.
92. T. Yamato, C. Hideshima, G. K. S. Prakash and G. A. Olah, *J.Org.Chem.*, **1991**, *56*, 3955.
93. T. Yamato, C. Hideshima, G. K. S. Prakash and G. A. Olah, *J.Org.Chem.*, **1991**, *56*, 3192.

94. T. Yamato, C. Hideshima, K. Suehiro, M. Tashiro, G. K. S. Prakash and G. A. Olah, *J.Org.Chem.*, **1991**, *56*, 6248.

95. G. A. Olah, V. P. Reddy and G. K. S. Pr;akash, *Synthesis*, **1991**, 29.

96. G. K. S. Prakash, P. Ramaiah and R. Krishnamurti, *Catal.Lett.*, **1991**, *9*, 59.

97. J. D. Weaver, E. L. Tassef and W. E. Fry, *Catalysis*, **1987**, 483.

98. J. D. McClure, S. G. Brandenberger, U.S. Patent 4,038,213 (1977), C.A. *87*, 120384 (1977).

99. I. Bucsi, G. A. Olah, *J.Catalysis*, **1992**, *137*, 12.

100. G. A. Olah, U.S. Patent 4,613,723 (1986).

101. G. A. Olah U.S. Patent 4,116,880 (1978).

102. K. Tanabe, *Solid Acids and Bases*, Academic Press: New York/London, **1970**.

103. a) S. J. Sondheimer, N. J. Bunce, M. E. Lambe and C. A. Fyfe, *Macromol.* **1986**, *19*, 339; b) N. J. Bunce, S. J. Sondheimer and C. A. Fyfe, *Macromol.* **1986**, *19*, 333.

104. L. L. Ferry, *J.Mcaromol.Sci.-Chem.* **1990**, *A27*, 1095.

105. S. J. Sondheimer, N. J. Bunce and C. A. Fyfe, *J.Macromol.Sci.-Rev.Macromol. Chem.Phys.* **1986**, *C26*, 353-413.

106. a) E. M. Arnett,, R. A. Haaksma, B. Chawla and M. H. Healy, *J.Am.Chem.Soc.* **1986**, *108*, 4888; b) P. Batamack, C. Doremieux-Morin and J. Fraissard, *Catal.Lett.*, **1991**, *11*, 119.

107. G. A. Olah, K. Wade and R. E. Williams, *Electron Deficient Boron and Carbon Clusters,* Wiley-Interscience: New York, **1991** and references therein.

108. G. A. Olah, *Carbocations and Electrophilic Reactions, Angew. Chem.* **1973**, *85*, 183; *Angew.Chem.Int.Ed.Engl.*, **1973**, *12*, 173; Verlag Chemie (Weinheim), Wiley (New York) **1974** and references therein.

109. G. A. Olah, A. M. White and D. H. O'Brien, *Chem. Rev.* **1970**, *70*, 561.

110. G. A. Olah, *Halonium Ions,* Wiley-Intersciences, New York **1975**.

111. G. A. Olah, *Carbocations and Electrophilic Reactions*, Verlag Chemie (Weinheim)-Wiley (New York), **1974**; G. A. Olah, *Angew.Chem.Int.Ed.* **1973**, *12*, 173.

112. G. A. Olah, A. Getmain, H. C. Lin and D. A. Forsyth, *J.Am.Chem.Soc.* **1975**, *97*, 2928.

113. G. A. Olah, G. K. S. Prakash, M. Barzaghi, K. Lammertsma, P. v. R. Schleyer and J. A. Pople, *J.Am.Chem.Soc.* **1986**, *108*, 1062.

114. G. A. Olah, G. K. S. Prakash, M. Marcelli and K. Lammertsma, *J.Phys.Chem.* **1988**, *92*, 878.

115. G. A. Olah, G. K. S. Prakash and K. Lammertsma, *Res. Chem. Intermediates,* **1989**, *12*, 141.

116. G. A. Olah, *Angew.Chem.*, **1993**, *105*, 000.

117. W. Haag and R. H. Dessau, Int. Catalysis Congress, Berlin, 1984, II, 105.

THE MECHANISM OF PROTON TRANSFER AND THE NATURE OF PROTONATED INTERMEDIATES IN HETEROGENEOUS ACID CATALYSIS

V.B.KAZANSKY
Zelinsky Institute of
Organic Chemistry Russian
Academy of Sciences
Moscow 117071, Russia

ABSTRACT. The diffuse reflectance IR stidy of the surface OH groups on oxides and in zeolites performed in the broad spectral range allowed to calculate their homolytic dissociation energies and demonstrated the covalent character of their ground states. In the similar way the protonation of olefins results instead of ion-molecular pairs in covalent alkoxides, while the adsorbed carbenium ions are the energetically excited transition states that can be formed from these covalent precursors. This most often results in concerted mechanisms of acidically catalyzed transformations of hydrocarbons where both the acid Broensted sites and the neighboring basic oxygen are equally important.

I. INTRODUCTION.

The practical importance of heterogeneous acid catalysis for refining chemistry, such as cracking, isomerization and alkylation of hydrocarbons is well known. On the other hand, the ideas about the nature of acid active sites and protonated active intermediates strongly influenced the general development of the catalysis theory, since acid catalysis is the simplest and most well understood example of the catalytic action.

The mechanism of conversion of hydrocarbons on solid acids was at first formulated almost 40 years ago in the way similar to homogeneous acid catalysis. Since that time it remained practically unchanged. The general scheme used in both cases is the following:

$$RH + \overset{+}{H}_{ads} \longrightarrow \overset{+}{RHH}_{ads} \longrightarrow products + \overset{+}{H}_{ads} \qquad (1)$$

According to this scheme, the only difference between homogeneous and heterogeneous systems is that protons and protonated active intermediates are considered either as solvated (homogeneous catalysis) or adsorbed on the surface (heterogeneous catalysis).

Of course, equation (1) doesn't represent a real reaction mechanism and can not explain all of its details. For instance, it is well known that migration of the double bond in alkenes requires weaker acid sites than cracking or skeletal isomerization. This difference can not be explained by (1), since the heterogeneous catalyst is even not considered explicitly. In other words, such a scheme does not give the real description

335

J. Fraissard and L. Petrakis (eds.), Acidity and Basicity of Solids, 335–352.
© 1994 *Kluwer Academic Publishers.*

of catalysis on molecular level and therefore is out of date.

The aim of this paper is to present and to discuss more adequate mechanisms of heterogeneous acid catalysis that take into account the interaction of active intermediates with the surface of the catalysts on molecular level.

1. INFRARED SPECTROSCOPY OF FREE OH GROUPS: THEIR INTRINSIC BROENSTED ACIDITY.

Originally the idea about the nature of Broensted acid sites was formulated for amorphous silica aluminas [1]. It was postulated that some of aluminum atoms in these materials can have a tetrahedral coordination and, therefore, require nonstoichiometric protons to compensate the excessive negative charges of $[AlO_4]$ units. Indeed, the strong Broensted acidity of amorphous silica alumna was well proved by Hammett indicators [2,3]. However, such protons were never directly observed neither in IR nor in high resolution NMR spectra.

The reason of such a discrepancy consists in the nonstoichiometric character of the strong acid sites in amorphous silica alumina. Therefore their amount is always much less than the total aluminum content. This is consistent with the octahedral coordination of the overvelming part of aluminum in amorphous catalysts according to NMR data and is reflected by their chemical formula x SiO_2 y Al_2O_3 that doesn't contain any protons or hydroxyl groups. In other words, acid protons at tetrahedrally coordinated aluminum atoms are a nonstoichiometric defects in the amount negligible as compared with the total aluminum content [4,5].

Recently the amorphous silica alumina was replaced by crystalline aluminosilicates or zeolites. Their chemical composition is quite different. For sodium forms it corresponds to the empirical formula:

$$O .5x\ Na_2O\ O. 5\ x\ Al_2O_3\ y\ SiO_2$$

where y:x ratio can change from one to several hundreds. Upon decationation sodium atoms are stoichiometrically replaced by protons, while at a 100 % of exchange their amount becomes equal to aluminum content. Therefore the composition of hydrogen forms can be represented as x $HAlO_2$ y SiO_2.

The crystal structure of zeolites is also quite different from amorphous silica alumna, since all aluminum and silicon atoms are tetrahedrally coordinated and form a three dimensional four-connected framework of AlO_2 and SiO_2 tetrahedra linked to each other by the sharing of oxygen ions. Thus, the Broensted acidity of decationated zeolites is of a stoichiometric character, while the number of strongly acidic protons in their structure is equal to the aluminum content.

Such Broensted acid sites are represented by the so called bridging or structural OH groups:

$$
\begin{array}{ccc}
 & \text{H} & \\
\diagdown\text{O} & \diagup\text{O}\diagdown & \text{O}\diagup \\
-\text{O}-\text{Si} & & \text{Al}-\text{O}- \\
\diagup\text{O} & & \text{O}\diagdown
\end{array}
\qquad (2)
$$

Table 1

Zeolite	Hydroxyl group	Frequency cm^{-1}		
		Fundamental	in plain	out of plain
HZSM-5	bridging OH	3610	1050	290
HM	bridging OH	3610	1050	290
HY	bridging OH	3555	1050	?
HY	bridging OH	3645	1020	280
HX	bridging OH	3660	995	?
MgNaM	terminal MgOH	3620	960	
CaNaM	terminal CaOH	3600	930	
All zeolites	terminal SiOH	3745	830	

Since the in-plain and out-of-plain frequencies are quite different, these results indicate the planar structure of Si-OH-Al fragments. They also show that the bending frequencies are more sensitive to the zeolite structure than the fundamental frequencies. For instance, the bending frequency of silanol groups in amorphous silica alumina is about 40 cm^{-1} higher than in silica gel, while their stretching frequencies practically coincide with each other. Fig 1 (b) represents another example, where in spectrum of HX zeolite three bands of bridging OH groups are seen in the combination region, while only two bands are observed in the fundamental region.

Another source of important information are the overtones of stretching vibrations. Their positions provide the unique possibility to reconstruct from the spectral data the shape of the potential curves of O-H bond of surface hydroxyls with different acid strength. This was done in our papers [10-14] within the formalism of Morse potential:

$$U(r) = D_O \left[1 - e^{-\beta(r - r_O)} \right]^2 = \frac{\omega_e}{4X_e} \left[1 - e^{-\beta(r - r_O)} \right]^2 \tag{3}$$

This expression includes two parameters: the harmonic frequency at the bottom of the potential well ω_e and the anharmonicity coefficient X_e. To find both of them, the experimental frequencies of any two vibrational transitions are required. Of course, if several transitions are known, then the results of such calculations are even more precise and reliable. From such data the shape of O-H potential curve can be easily reconstructed and the O-H dissociation energy D_O can be calculated.

Indeed, despite the Morse function was originally introduced for diatomic molecules, it also nicely describes the O-H potential shape in surface hydroxyl groups, since their vibrations are highly characteristic and don't interact with those of the oxide crystal lattices. Fig 2 represents such reconstruction for bridging hydroxyl groups in HY zeolite and Table 2 depicts the positions of the experimentally observed fundamental transitions and overtones for some other zeolites and oxides with basic (MgO) and weakly

The most common technique employed for their direct observation as well as for the study of Broensted acid sites on the surface of other oxides is IR spectroscopy. However, in most cases, only the fundamental OH stretching frequencies are observed and investigated.

In our papers [6-9] the application for this purpose of the diffuse reflectance IR spectroscopy was suggested. It allowed us to study the surface OH-groups in the NIR region. Each of their type is characterized then instead of the only fundamental stretching frequency by a number of IR bands originated from overtones or from different combinations of stretching and bending vibrational modes. This provides the information that quite often can not be obtained by any other physical or spectroscopic method.

For example, Fig. 1 represents the diffuse reflectance spectra of HY and HX zeolites recorded in a wide spectral range. In addition to the bands of fundamental OH stretching vibrations at 3550 - 3750 cm^{-1} they contain very weak bands of the combination of the out-of-plain (3940 cm^{1-}) and in-plain (4600-4700 cm^{-1}) bending vibrations and the bands of overtonels of stretching vibrations in the region of 7000-7300 cm^{-1}.

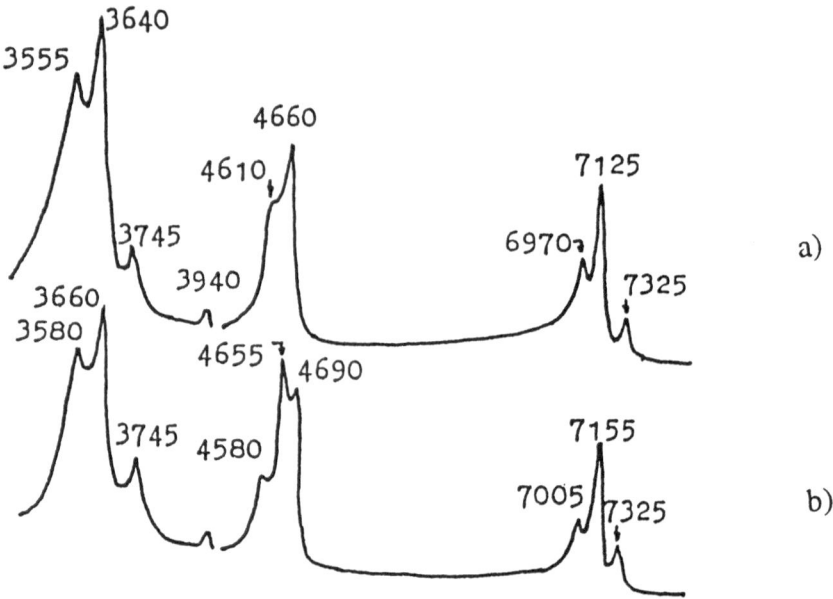

Fig 1. Diffuse reflectance IR spectra of HY (a) and HX (b) zeolites recorded in the broad spectral range.

The in-plain and out-of-plain bending frequencies can be obtained from such spectra as the difference of combination of stretching plus bending vibrations and the corresponding stretching frequencies. Such data for some zeolites are collected in Table 1.

acidic (SiO_2) properties. The OH dissociation energies calculated from the spectral data are also included.

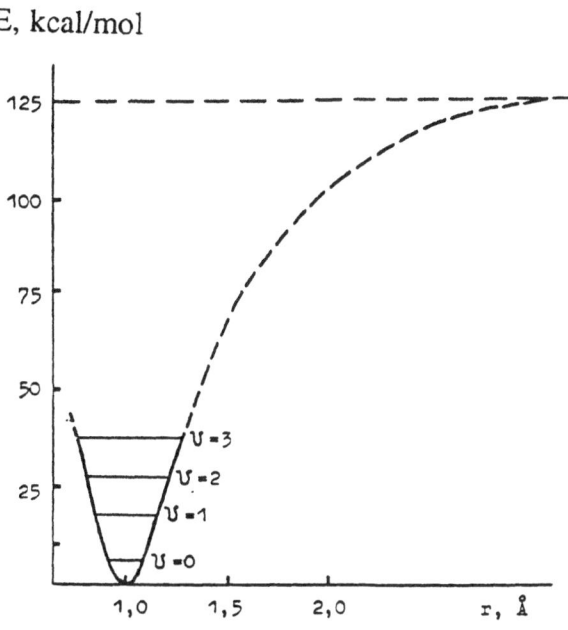

Fig. 2 The potential shape of 3645 cm^{-1} OH bond in HY zeolite reconstructed from spectral data.

Table 2.

The dissociation energies of the surface hydroxyl groups with different acid-base properties calculated from spectral data [6].

Oxide or catalyst	Frequencies of fundamental vibrations and overtones cm^{-1}			D_O kcal/mol
MgO	3750	7350	10760	140
SiO_2	3749	7326	10735	124
B/SiO_2	3705	7246	10621	124
AlPO4	3679	7147		126
HY	3645	7130	10460	125
HM	3611	7065	10360	125

The most striking in this Table are the dissociation energies D_O. All of them are close to each other and relatively low in their absolute values. This definitely indicates the homolytic character of dissociation of surface hydroxyl groups (compare for instance the calculated O-H bond strengths of about 125 kcal/mol with the energy of homolytic dissociation of water, that is equal to 118 kcal/mol or with that of hydrogen chloride of 103 kcal/mol). It is also remarkable, that hydroxyls with very different acid-base properties have almost the same homolytic dissociation energies and the shapes of corresponding potential curves. Thus, the ground states of all of them are definitely covalent.

These results show that the rather general opinion about unusual properties of bridging hydroxyl groups in zeolites, or the naive belief that the zeolites are solid electrolytes with almost free protons incorporated in their cages is certainly a very rough simplification.

On the other hand, there is no doubt that the OH heterolytic dissociation energies are quite different from those represented in Table 2. Unfortunately, for solid acids and bases there is no direct way to measure them experimentally. However, they may be easily estimated with the help of the following cycle:

Let us at first split the surface O-H bond homolytically, then ionize the resulting hydrogen atom and transfer the electron to the solid. This results in the heterolytic splitting of the surface hydroxyl with the following total energy:

$$D_{het} = D_O + U_H - \phi \qquad (4)$$

The hydrogen atom ionization potential U_H is equal to 13.6 e.V. and the photo electron work functions ϕ of oxides never exceeds 6-7 e.V. Therefore it follows from this expression that the heterolytic dissociation energies of surface OH groups are at least for 6-7 e.V. higher than those of their homolytic dissociation or are equal to 11-12 eV (250-280 kcal/ mol).

This estimation is consistent with results of semiempirical and nonempirical quantum chemical calculations [15]. It also fits the heterolytic proton abstraction energies in zeolites estimated from bathochromic shifts of OH stretching vibrations created by adsorption of different bases [16]. Thus, abstraction of protons corresponds to the electronically excited heterolytic terms, that are quite different from the covalent ground states typical of homolytic dissociation.

In other words, the ground states of free surface OH groups both in zeolites or on the surface of any other oxide are rather insensitive to their acid-base properties. This was already demonstrated by the similar parameters of the potential curves presented by Table 2. Proton NMR chemical shifts represent another example of this kind. According to [17,18] for the OH groups with different acid-base strength they are within 2-5 p.p.m. These values are more close to the gas phase chemical shifts in water or to those in paraffins than to the proton shift in hydroxonium ion.

Thus, any attempt to determine the acid strength of a given surface OH group from its stretching frequency or from any other property of its covalent ground state will fail. Probably some qualitative correlations of this kind for the narrow groups of similar compounds are still possible [19]. However, the only proper way to measure the acidity strength of surface OH groups is to follow their response to the interaction with adsorbed bases. This is actually the most common approach when the low frequency shifts of OH fundamental stretching frequencies caused by hydrogen bonding with an adsorbed molecule, the heat of ammonia adsorption, the position of thermodesorption peak or the adsorption of Hammett indicators are used.

2. THE REAL NATURE OF ADSORBED CARBENIUM IONS AND THE MECHANISM OF CATALYTIC TRANSFORMATIONS OF OLEFINS

Originally the carbenium ion mechanisms were introduced for amorphous aluminosilicates in the end of 40s [20-24]. They were based on purely chemical analogy with low temperature conversions of hydrocarbons in presence of strong acids in solutions (the higher reactivity of olefins, their isomerization, hydrogen transfer, selective formation of isoparaffins etc.). Later these ideas were generally accepted and automatically transferred to heterogeneous acid catalysis. They remained practically unchanged. For instance, the catalytic transformations of 1-butene through adsorbed (ads) carbenium ion intermediates are usually represented as follows:

$$CH_2=CH-CH_2-CH_3 \ + \ \overset{+}{H}_{ads} \ \rightarrow \ CH_3-\overset{+}{C}H-CH_2-CH_{3ads} \longrightarrow$$

$$CH_3-CH=CH-CH_3 \ + \ \overset{+}{H}_{ads} \quad \text{(double bond shift)}$$

$$(5)$$

$$\begin{array}{c} CH_3 \\ CH_3-\overset{|}{C}H-H_{3ads} \\ + \end{array} \overset{+}{} — \ i\text{-}C_4H_8 + \overset{+}{H}_{ads} \quad \text{(isomerisation)}$$

$$CH_4 + C_3H_6 + \overset{+}{H}_{ads} \quad \text{(cracking)}$$

However, adsorbed carbenium ions were never been directly observed by any physical method. Moreover, there are strong evidences that they don't exist as the real active intermediates, but rather represent the transition states or activated complexes of corresponding reactions.

At first, such conclusion was made in our papers [14,25,26] from NIR data. They allowed us to reconstruct the potential profile of proton transfer to adsorbed molecules similar as above for free hydroxyl groups. This was done from the positions of fundamental OH stretching bands and overtones perturbed by hydrogen bonding with adsorbed olefins. The obtained results showed only a moderate tendency of the direct proton transfer to adsorbed hydrocarbons. They also demonstrated the importance of relaxation effect e.g. of the interaction of adsorbed protonated species with the surface. This makes them more covalent than it was believed before.

The further evidence was provided by quantum chemical calculations performed with the full optimization of the geometry of protonated species. For example, the comparative analysis of the electronic structure of ethyl fragment in ethyl sulfuric acid and on the surface of a high silica zeolite was carried out in [27]. The results of both semiempirical and nonempirical calculations showed that the carbon-oxygen bond is covalent and that the electronic structure and geometry of ethyl fragments in both cases has little in common with the free carbenium ions. In other words, the alkyl groups resulting from the proton transfer to adsorbed olefins represent the covalent surface ester rather than ion pairs.

For ethyl sulfuric ester this was also confirmed by a small low field chemical shift in the ^{13}C NMR spectrum of the α-carbon atom in the alkyl substituent [28]. It is practically the same as in ethyl alcohol that certainly is a covalent compound. Recently, similar ^{13}C MAS NMR shifts of about 60-70 ppm were also reported for adsorbed alkyl fragments resulting from the reaction of bridging hydroxyl groups on the surface of HY zeolite with tert-butyl alcohol [29] or propylene [30]. The nature of corresponding surface esters is therefore also definitely covalent.

The semiempirical quantum chemical calculations were also carried out for methyl, ethyl and isopropyl surface alkoxy groups in a bigger cluster [31, 32]:

$$\begin{array}{c} R \\ H-O \quad O \quad O-H \\ H-O-Si \quad Al-O-H \\ H-O \qquad O-H \end{array} \tag{6}$$

The net positive charges of alkyl fragments were found rather low and practically constant, when passing from the methoxy to the isopropoxy substituent. Only the slight increase of C-O distance in this series and geometry of alkyl groups confirmed their covalent nature. On the contrary, the heterolytic C-O bond dissociation energies, D_{het}, strongly decreased in this sequence (Table 3):

Table 3.

The semiempirical quantum chemical calculations for alkyl groups adsorbed on the surface of high silica zeolite [32].

Alkyl fragmen	$\overset{+}{CH_3}$	$\overset{+}{C_2H_5}$	$\overset{+}{C_3H_7}$
qR	+0.32	+0.33	+0.33
qO	-0.42	-0.4 1	-0.41
qC	+0.18	+0.13	+0.07
Dhet eV	7.25	6.05	4.93

This is similar to the relationship between the homolytic ground states and the heterolytic dissociation of the surface OH groups, as already discussed above. Therefore it was suggested by analogy that the carbenium ion properties of the surface alkoxyls are exhibited not by their covalent ground states, but by the excited transition states with elongated C-O bonds. The importance of more ionic character of stretched chemical bonds was also discussed earlier in [33].

This idea was recently further developed in [34], where nonempirical quantum chemical calculations were performed for interaction of ethylene with acidic OH groups of a high silica zeolite and for the reverse reaction of the decomposition of surface ethoxy groups. The geometry of adsorbed ethylene or of resulting alkoxide was fully optimized with the help of a gradient procedure of a "Gaussian 80" program. Acidic groups of the high silica zeolite were modeled by the simplest HO(H)Al(OH)$_3$ cluster.

At first ethylene is adsorbed as π-complex (Fig. 3 a). The energy of its formation is equal to 6.90 kcal/mol. No considerable change in the ethylene geometry was found after adsorption. It resulted only in a very small (+0.021e) positive charging.

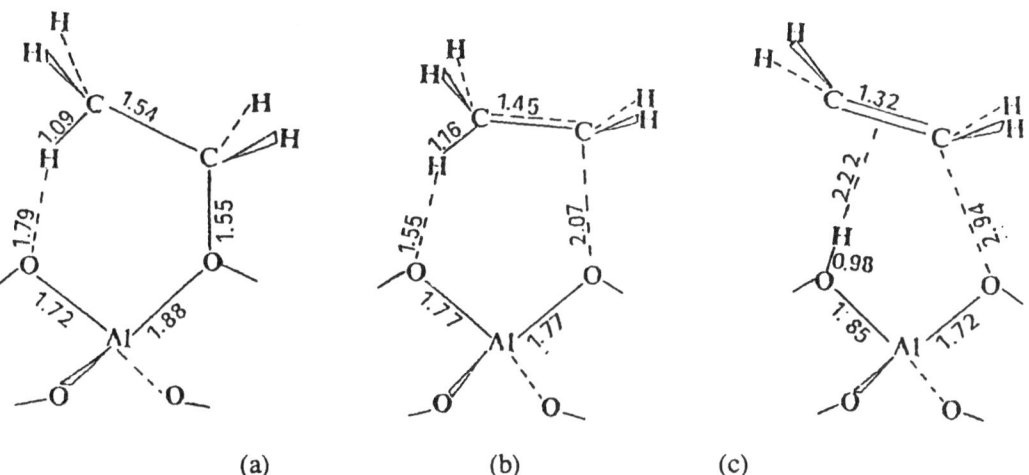

(a)   (b)   (c)

Fig. 3 The results of quantum chemical calculations of ethoxylation of bridging OH groups in zeolites.

On the other hand, in accordance with the above results, the most stable protonated structure is not an ion pair, but a covalent ethoxy group (Fig. 3 c). This follows from a relatively low positive charge of the ethyl fragment (+0.384 e) and from its geometry that is typical of covalent organic compounds. Indeed, the equilibrium C-O bond length is close to the mean length of these bonds in alcohols or esters. The obtained C-C distance of 1.553 A and the nearly tetrahedral O-C-H and O-C-C angles in the alkyl fragment also confirm its covalent character. Finally, the calculated heat of zeolite surface ethoxylation (11 kcal/mol) is also consistent with the thermochemical data on the heat of decomposition of covalent sulfuric acid ester [35].

The transformation of adsorbed ethylene into the surface ethoxide proceeds via a transition state of Fig. 3 (b), with an activation energy barrier of 15.4 kcal/ mol (Fig.4).The geometry and the electronic structure of the alkyl fragment in such an activated complex resemble those of the classical form of ethyl cation. This follows from the marked enhancement of the positive charge of the ethyl group from +0.384 up to +0.565 e and from the new length of C-C bond intermediate between those of a double and single bonds. In addition, an essential flattening of the CH_3-fragment occurs with O-C-H angles now close to 90°. Moreover, the deprotonation tendency is clearly revealed in the formation of the hydrogen bond between one of the protons of CH_3 group with the neighboring basic oxygen.

The reaction coordinate of ethoxylation of the zeolite surface is very complicated. It includes a stretching of the O-H bond in the acidic hydroxyl group and of the C-C bond in the adsorbed molecule as well as some change of the angles in methyl and methylene fragments and a shift of the adsorbed ethylene toward the surface basic center. The activation barrier of this elementary step is low enough only if all of these parameters are changing simultaneously. Thus, the possibility of proton transfer over low potential barrier can be explained by a concerted mechanism, which enables maximum energy compensation of the bonds to bebroken by those to be formed

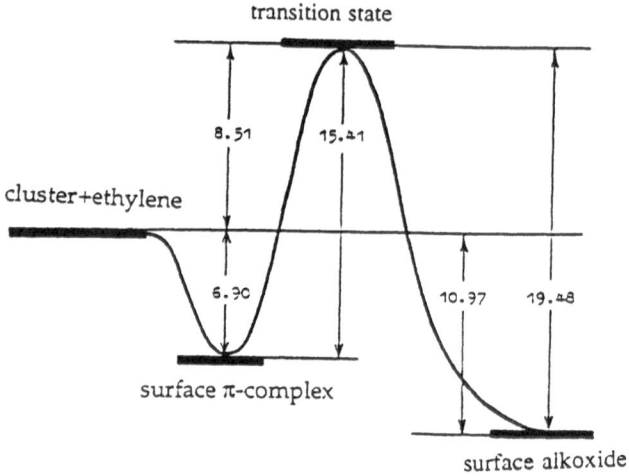

Fig. 4 The energy diagram of ethoxylation of bridging OH group in zeolites.

The decomposition of the surface alkoxy structure follows the opposite pathway. The reaction coordinate mainly includes here the stretching of C-O bond in the surface ester and the deprotonation of the resulting transition state resembling a carbenium ion. Although the homolytic abstraction of an ethyl group is considerably more favorable than its heterolytic dissociation, the initial stages of C-O bond stretching result in increase of its polarity. This effect is maximal in the transition state, which resembles a carbenium ion as just said above.

The furter heterolytic dissociation is extremely energetically unfavorable. Therefore, instead of abstraction of a carbenium ion or alkyl radical, the deprotonation of the transition state occurs. This requires to overcome an activation energy barrier of only about 20 kcal/ mol as compared with a homolytic dissociation energy of about 90 kcal/mole or a heterolytic dissociation energy of about 150 kcal/mole (Table 3, Fig. 4).

According to this ideas, the acid-catalyzed double bond shift in olefins also proceeds via a concerted mechanism. It includes a subsequent formation and decomposition of the surface alkoxy groups. At high temperature, when π-complexes are unstable, this probably occurs by interaction of gaseous olefins directly with the free surface OH-groups. The reaction limiting step is the decomposition of the resulting alkoxyls. This is consistent with experimental activation energies of double bond isomerization on zeolites of about 15-20 kcal/mole [36], that is close to the height of activation barrier estimated by us. On the other hand, since the transition state has cationic character, the double bond isomerization obeys all the rules of the carbenium ion formalism.

A similar approach can be applied to some other acid-catalyzed transformations of olefins. For instance, their oligomerization is also believed to proceed via adsorbed carbenium ions. They are easily formed from branched olefins, but not from ethylene. In our opinion, this should be understood in the sense that the formation of the carbenium-ion like transition states becomes easier with increasing branching of the corresponding covalent alkoxide intermediates. The reaction coordinate probably includes here mainly a concerted C-O bond elongation and approaching of the second olefin molecule to the surface carbenium-ion-like exited complex. Again, a low activation barrier is possible only if in addition to these main parameters some others also contributing to the reaction coordinate are changing simultaneously. Other examples of such concerted mechanisms of acid-catalyzed heterogeneous reactions are discussed in more detail elsewhere [37].

3. CATALYTIC TRANSFORMATIONS OF PARAFFINS

The adsorbed carbenium ions play also the role of active intermediates in cracking and isomerization of paraffins. The traditional mechanism represents the chain reaction, with the adsorbed classical carbenium ions as the chain carriers. The chain initiation occurs through interaction of bridging hydroxyls with olefins similar to the reaction already discussed above. They are always present in the reaction mixture at the steady state conditions as the final products. The formation of adsorbed carbenium ions is followed then by their cracking and by chain propagation through hydrogen transfer in the rate determining step:

$$\overset{+}{H}_{ads} + olefin \longrightarrow \overset{+}{R}_{ads}$$
$$\overset{+}{R}_{ads} \longrightarrow \overset{+}{R}_{1ads} + olefin \qquad (7)$$
$$\overset{+}{R}_{1ads} + RH \longrightarrow R_1H + \overset{+}{R}_{ads} \quad etc..$$

The similar chain mechanism is also considered for isomerization of paraffins with the only difference that instead of cracking the skeletal isomerization of carbenium ions takes place:

$$\overset{+}{R}_{ads} \longrightarrow i\text{-}\overset{+}{R}_{ads} \qquad (8)$$
$$i\text{-}\overset{+}{R}_{ads} + RH \longrightarrow i\text{-}RH + \overset{+}{R}_{ads} \quad etc. \qquad (8)$$

The adsorbed carbenium ions involved in both of these reactions certainly are similar to those already discussed above. Therefore they also don't really exist as adsorbed active intermediates, but rather represent the transition states or activated complexes. For instance, the mechanism of their cracking could be represented as follows:

$$
\begin{array}{ccc}
\underset{\underset{\underset{\underset{O}{Si}\underset{O}{O}}{\overset{O}{\diagup}\overset{O}{\diagdown}}}{\overset{CH_3}{\underset{CH}{\overset{CH_2}{CH_3}}}} & \longrightarrow & \underset{\underset{\underset{O}{Si}}{\overset{O}{\diagup}}}{CH_3-CH\mathrel{\ddot{=}}CH_2 \overset{\varepsilon+}{\underset{\varepsilon-}{\diagdown}}CH_3} & \longrightarrow & \underset{\underset{\underset{O}{Si}\underset{O}{O}}{\overset{O}{\diagup}\overset{O}{\diagdown}}}{\overset{C_3H_6}{\overset{CH_3}{}}} & (9)
\end{array}
$$

It starts with covalent alkoxide that is excited into the carbenium-ion-like transition state by stretching of the C-O bond. This is followed by the concerted splitting of the olefin and by stronger binding of the resulting alkyl fragment with the basic surface oxygen.

Recently it was also shown that at low partial pressure of olefins and high temperatures, when the concentration of adsorbed carbenium ions is low, the mechanism of chain initiation is different. It involves the adsorbed carbonium ions that are formed by direct protonation of paraffins [38]. Their consequent dehydrogenation or cracking with splitting of methane or hydrogen also results in adsorbed carbenium ions:

$$
\overset{+}{H}_{ads} + \underset{H}{\overset{R}{R-C-R}} \longrightarrow \underset{H\ \ H}{\overset{R+}{R-C-R}}
\begin{array}{l}
\nearrow \quad H_2 + \underset{+}{\overset{R}{R-C-R_{ads}}} \quad (a) \\
\\
\searrow \quad CH_4 + \underset{H}{\overset{R+}{R-C_{ads}}} \quad (b)
\end{array}
\quad (10)
$$

The most convincing evidences of this mechanism were obtained for cracking of light paraffins at very low conversions. For instance, at 450-500 C the isobutane conversion on HZSM-5 zeolite with a steady stay flow reactor was less than 1 %. The main products were methane and propylene formed in equimolecular amounts with the apparent activation energy of 57 kcal/mol [39]. This indicates that cracking proceeds through a carbonium ion intermediate according to the scheme (10 b)

The similar conclusion that the chain initiation occurs by direct protonation of C-H and C-C bonds by Broensted acid sites was also made for initial steps of i-butane and n-pentane cracking on faujasites. Cracking of i-butyl carbonium ions results both in splitting of molecular hydrogen and methane, while for n-pentyl carbonium ions only in splitting of methane. The resulting carbenium ions propagated chain reactions, mainly by isomerization followed by hydrogen transfer. The estimated chain length varied depending on the reaction conditions within 3-15 [40].

Similar to carbenium ions the adsorbed carbonium ions certainly also represent the activated transition states or highly excited intermediates. This follows from the high activation energy of methane formation according to mechanism (10 b) [39] and is consistent with the about 1.5 eV lower proton affinity of paraffins in comparison with olefins (Table 4):

Table 4.

The hydrocarbon	Proton affinity eV.
CH_4	5.1
C_2H_6	5.6
C_3H_8	6.1
C_2H_4	6.9
C_4H_8	7.8
C_3H_6	7.9

The lower tendency of the proton transfer to paraffins is also supported by spectral data (smaller low frequency shifts of OH stretching frequencies caused by interaction with paraffins).

The experimental data on the structure of adsorbed carbonium ions at the moment are absent, despite in the gas phase their formation and reactions have been well studied by mass spectrometry [41,42]. There were also many ab initio quantum chemical calculations performed for free ethyl carbonium ion with very different degrees of accuracy. They indicated that these species exist in two isomeric forms. The more stable isomer is represented by structure (11 a) where two methyl fragments are connected through hydrogen atom:

$$
\begin{array}{cc}
\overset{+}{\underset{CH_3\ \ CH_3}{\overset{H}{\diagup}}} & CH_3{-}\overset{+H\ H}{\underset{H\ \ H}{C}} \quad\quad (11)\\
(a) & (b)
\end{array}
$$

According to the highest MP2/6-31G level of calculations structure (a) is nonlinear with C-H-C angle of 105.8 °. In the isomer (b) that is less stable for about 8 kcal/mol only one methyl group is protonated resulting in pentacoordinated carbon atom [43].

As a very first attempt of the quantum chemical analysis of the real nature of adsorbed carbonium ions we recently considered the interaction of the more stable structure (11a) with the simplest HO(H)Al(OH)3 cluster as a model of the zeolite framework. The results of nonempirical calculations with 6-31G* basis set are represented by Fig. 5.

It was found that the interaction of the ethyl carbonium ion with the surface corresponds to the local minimum of the energy and is exothermic for about 70 kcal/mol. Despite this, the formation of such adsorbed complex from free ethane molecule and the surface hydroxyl group is still highly endothermic for about 80 kcal/mol. At the first glance this figure looks very high, however, is rather reasonable, since the corresponding experimental value for isobuthyl carbonium ion that certainly is formed easier that the ethyl cation is about 60 kcal/mol [43]. In addition our calculation should be considered only as the very preliminary.

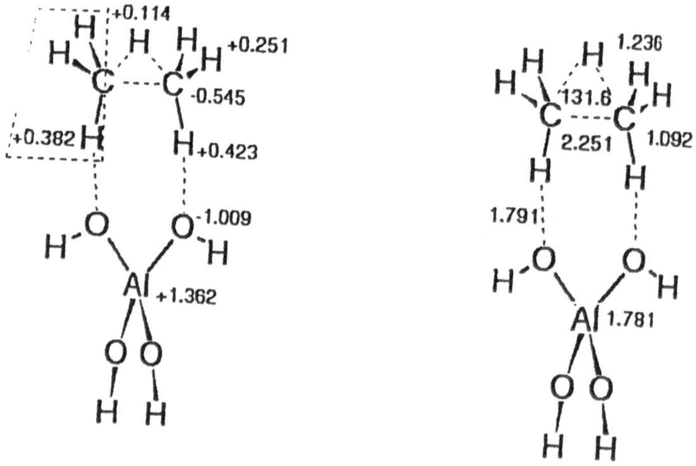

(a)   (b)

Fig. 5. The results of 6-31 G nonempirical quantum chemical calculations of ethyl carbonium ion adsorbed on zeolite.

The positive charge of the adsorbed ethyl cation is rather high (+0.878 e, Fig 5 a). It is mainly concentrated on methyl groups, while the positive charge of the central hydrogen atom is only +0.114 e. Thus, the adsorbed nonclassical ethyl carbonium ion in some sense resembles two methyl cations connected by the hydride ion.

The geometry of the cation very little changes upon adsorption (Fig. 5 b), despite the protons of C-H bonds directed to the surface have about +0.2 e higher positive charge than those directed from the surface. The adsorbed structure is very unstable, since even the small displacement of the central hydrogen atom toward one of the methyl groups results in the redistribution of the positive charge, and in formation of methane and methyl cation. The latter certainly easily forms a strong chemical bond with the basic oxygen of the surface. Therefore the activation energy of protolytic cracking seems to be close to the endothermic effect of alkane protonation.

4. THE INHOMOGENEITY AND THE BIFUNCTIONAL NATURE OF BROENSTED ACID SITES

In accordance with the above discussion, the close connection between Broensted acid sites and neighboring basic oxygen in zeolites should be always kept in mind [13,26,27,37]. Therefore, an important feature of heterogeneous acid catalysis is the bifunctional nature of bridging hydroxyl groups: their acid moiety acts as a proton donor, while the basic moiety either stabilizes intermediate protonated species or favors the proton abstraction from the transition states. This results in a broad variety of concerted reaction mechanisms as discussed above.

This idea suggests the need for a proper description of the acidity strength of such dual acid-base active sites. Indeed, any approach, that does not take into account the interaction of protonated species with the surface basic oxygen is obviously not sufficient, e.g., using only the proton abstraction energies.

In this connection, let us compare the most commonly used methods of the surface acidity measurement: determination of adsorption heat, temperature programmed desorption, and the shifts of OH stretching vibrations caused by adsorption of bases. It is clear from Fig. 4, that each of them deals with the different features of the interaction of adsorbed molecules with the acid active sites.

For instance, the adsorption heats characterize most directly the adsorption or desorption enthalpy, but not the corresponding activation energy. The latter is more closely related with thermodesorption data, that are controlled by the desorption activation barrier, while the bathochromic shifts of OH stretching vibrations can be correlated with the adsorption activation barrier.

As follows from the Fig. 4, both of these characteristics are also connected with the activation energies of adsorbed carbenium ions formation from the side of the less stable π-complexes or from the more strongly adsorbed alkoxides. Therefore, it is not surprising that acid strength measurements performed with help of these very different methods some times fail to agree with each other.

Finally, the bifunctional nature of the active sites brings forth one more new important concept: the relative geometry of Broensted acid and neighboring basic site. The influence of O-Al-O angle on the height of the activation energy barrier is qualitatively clear from Fig. 3. The role of Si-O-Al angles is not so obvious, but could be illustrated by comparing of structures I and II, that represent the π-complexes formed on dual sites with different Si-O-Al angles:

$$(12)$$

I   II

Structure II seems to be more prepared for proton transfer as it requires less relaxation of the resulting alkyl fragment. In addition the basic properties of the neighboring oxygen are also influenced by the values of Si-O-Al angles. This should also result in the dependence of the reaction rate on the geometry of the active sites. The similar is also correct for the mechanism of cracking represented by (9).

Such structural inhomogeneity of active sites is most important for high silica zeolites, where the local distortions of the crystal lattices are pronounced the most strongly. For instance, there are twelve crystalographically different tetrahedral sites in ZSM-5 zeolite with the O-Si-O (or O-Al-O) angles varying between 96 and 129°, and Si-O-Si angles between 140 and 175°. The similar broad distribution of these angles is also typical of mordenite, ZSM-11 and other high silica zeolites prepared both by the direct syntheses or by dealumination. Thus, depending on the local distortions of [AlO4] tetrahedra, the activity of such dual active sites could be different even for the same zeolite structure.

The important point in this connection is that these effects should be specific to the geometry of the reagents. Therefore they may play a decisive role in catalysis, but at the same time have only a little influence on adsorption of different probe molecules. This results in apparent uniformity of acid strength, but different activity and selectivity of corresponding sites in catalytic reactions.

Probably such structural nonuniformity may explain the recent results [44] where it was shown by the ammonia poisoning technique that only about 10 % of total aluminum content in high silica zeolites is responsible for their catalytic activity in cracking of paraffins. The similar conclusion was reached on the basis of poisoning of the acidity of dealuminated Y-type zeolite by sodium [45]. It was found that only one fifth of the framework aluminum atoms are associated with the catalytic activity. It is also notable that these effects are definitely not connected with different distribution of aluminum in the crystal framework, since in both cases the high silica zeolites with presumably isolated aluminum atoms were studied.

CONCLUSION

It follows from the above discussion, that the ground states of the surface OH groups with different acid strength are essentially covalent, while the main difference between the acidic and basic group consists in the positions of their electronically exited heterolytic terms. In the similar way the interaction of Broensted acid sites with olefins results instead of adsorbed carbenium ions in formation of surface alkoxides, that are much less ionic, than it is usually believed. Therefore the adsorbed carbenium ions should be considered rather as the transition states of the heterogeneous catalytic reactions resulting from excitation of covalent alkyl groups, than as the really existing surface active intermediates.

This conclusion is even more likely for adsorbed carbonium ions resulting from direct protonation of paraffins. Due to about 1-2 eV lower electron affinity in comparison with olefins they probably represent a highly energetically excited short living states resulting from active collisions of paraffins from the gas phase with Broensted acid sites.

The interaction of protonated active intermediates with the surface of heterogeneous catalysts puts forward an idea about the bifunctional nature of the active sites in heterogeneous acid catalysis. Their Broensted acid part protonates the adsorbed molecules, while the interaction with the neighboring basic oxygen converts the resulting transition states into more stable covalent intermediates. Such dual nature of the active sites may also result in their structural inhomogeneity connected with the relationship of the geometry of Broensted surface acids and the neighboring basic sites.

Finally, an important feature of the heterogeneous acid catalysis is the significance of concerted mechanisms with the very complicated reaction coordinates. This results in the reaction pathways with the low activation energies due to most complete compensation of the energy required for the rupture of the reacting chemical bonds in the initial compounds by those newly formed in the final products. Such concerted mechanism in some sense represents the analog of the S_{N2} mechanism that is known to be effective for many acidically catalyzed reaction in organic chemistry.

All of these ideas should be certainly regarded not as a criticism of the carbenium ion conception but rather as the attempt of its further development and modernization.

REFERENCES

1. B.C.Gates, J.R.Katzer,G.C.A.Schuit. "Chemistry of Catalytic   Processes", NY, McGraw Hill, 1979.
2. M.W.Tamele. Disc. Far. Soc. 1950, P. 270.
3. H.A.Benesi. J. Amer. Chem. Soc. 1956, V. 78, P. 5490.
4. Y. Amenomiya, K.J.Cvetanovich. Journ. Catal., 1971, V. 23, P. 331.
5. V. Yu . Borovkov, A . A . Alekseev, V . B . Kazansky . Journ . Catal. 1983, V. 80, P. 462.
6 . V.B.Kazansky. Chem.Rev. (Sov. Sci.Rev.), 1979,V. 1, P. 69 .
7 . L . M . Kustov, V . Yu . Borovkov, V . B . Kazansky . Journ . Catal .   1971, V. 72, P.149-159.
8. V.B.Kazansky, L.M.Kustov, V.Yu.Borovkov. Zeolites 1983, V. 3, P. 77.
9. V.B.Kazansky. "Structure and Reactivity of Modified Zeolites". Ed. by P. Jacobs, P.Jiru and V.Kazansky. Elsevier Sci . Pub ., Amsterdam 1984 , P . 61 .
10. B.Kazansky, A.M.Gritskov, V.M.Andreev. Dokl. Phys. Chem. lg77 ,V. 235, P . 635 .
11. B.Kazansky. Kinet. Katal. 1977, v. 18, P. 966.
12. V.B.Kazansky, A.M.Gritskov, V.M.Andreev, G.M.Zhidomirov. Journ. Molec. Catal. 1978, V. 3, P. 135.
13. V.B.Kazansky. Kinet. Katal. 1980, V. 21, P. 128.
14 . V.B . Kazansky. Account . Chem. Res . 1991, V. 24, P . 379 .
15. I.N.Senchenya, V.B.Kazansky, S.Beran. Journ.. Phys Chem.1986, V. 90, P. 4857.
16 L.Kubelkova, S.Beran, J.A.Lercher. Zeolites 1989, V. 9, P. 539.
17. J.M.Thomas, J.Klinowski. Adv. Catal. 1985, V. 33, Ed.by D.D.Eley,   H.Pines, P.W.Weisz, P. 200.
18. H.Pfeifer. Journ. Chem. Soc. Far. Trans I 1988, V. 84, P. 3777.
19. R. Beamont, D.Bartomeuf. Journ. Catal, 1973, V. 27, P.45 .
20. G.M.Bremer. Research 1948, V. 1, P. 281.
21. R.C.Hansford. Ind. Eng. Chem. 1947, V. 39, P. 849.
22. F.G.Ciapetta, S.J.Macuga, L.W.Leum. Ind. Eng.Chem 1948, V. 40, P. 2091.
23 . C . L . Thomas. Ind. Eng . Chem . 1949, V41, P . 2564 .
24 . B. S. Greensfelder, H. H. voge, G.M. Good. Ind. Eng. Chem 1949, V. 41. P. 2573.
25 . V. B . Kazansky. In Proc . 5th Int . Symp . on Relation between Homogeneous and

Heterogeneous Catalysis; VMSC/EHCL Press : Utrecht, Holland, 1986, P. 93 .

26. V.B.Kazansky. React. Kinet. Catal. Lett. 1987, V.35, P. 237 .

27 . I.N.Senchenya, N.D.Chuvilkin, V.B.Kazansky. Kinet. Katal. 1985, V. 26, P. 1073.

28. H.A.Klinovski, S.Berger, P.Braun. "l3C-NMR Spectroscopy" . Tholone Verlag: Stuttgart, 1985.

29. M.T.Aronsom, R.J.Gorte, W.E.Farneth, D.J.White. Journ. Amer. Chem. Soc. 1989, 111, P. 840.

30.J.F.Haw, B.R.Richardson, L.S. Oshiro, N.D. Lazo. Journ. Amer. Chem. Soc. 1989, V. 111, P. 2052.

31. I N Senchenya, V.B.Kazansky. Kinet. Katal. 1987, V. 28,

32.V B.Kazansky, I.N.Senchenya. Journ. Catal. 1989, V. 119,

33. V.B.Kazansky. In Proceedings of the 6th International Congress on Catalysis, Imperial College, London 1976, Royal Society of Chemistry, Lethworth, Herts 1977.

34. V B.Kazansky, I.N.Senchenya. Catal. Lett. 1991, V. 8, P.317.

35. "Thermochemistry of Organic and Metalloorganic Compounds". Ed. by J.D.Cox, G.L.Pilcher, Academic Press, N. Y. 1979, P. 390.

36 P.A.Jaqcobs. "Carbogenic Activity of Zeolites", Elsevier, Amsterdam, 1977

37. V.B.Kazansky. Sov. Chem. Rev. 1988, V.1, P.ll09.

38. W.O.Haag, R.M.Dessau. Proc. 6th Int. Congr. on Catalysis, Berlin 1984, V.2, P.305.

39. C.Stefanadis, B.C.Gates, W.O.Haag. J. Molec. Catal. 1991, V. 67, P. 363.

40 P.V.Shertugde, G.Marcelin, G.AStill, W.K.Hall. Journ. Catal, 1992, V. 136, P.446.

41. K.Hiraoka, P.Kebarle. Journ. Am. Chem. Soc. 1976, V.98, P. 6119.

42. Bohme D.K. "Interaction between Ions and Molecules" Ed by P Ausloos, Plenum Press, N.-Y., 1975.

43. W.J.Hehre, L.Radom, P.v.R.Schleyer, J.A.Pople "Ab initio molecular orbital theory", John Wiley & Sons, N.-Y., 1985.

44. E.A.Lombardo, W.K.Hall. Journ. Catal 1988, V. 112, P. 565.

45. P.O.Fritz, J.H.Lunsford, Journ. Catal. 1989, V. 118, P. 85.

ACID-BASE BIFUNCTIONAL CATALYSIS

KOZO TANABE
Central Research Laboratory
Nippon Shokubai Co., Ltd.
5-8, Nishi Otabi-cho, Suita,
Osaka 564, Japan

ABSTRACT. The examples of acid-base bifunctional catalysis by hetero-geneous catalysts such as ZrO_2, MgO, CaO, ThO_2, Al_2O_3, sulfates, phosphates, and mixed oxides of ZrO_2 are summarized. The characteristic feature of the pronounced catalytic activity, selectivity, and life of the heterogeneous acid-base bifunctional catalysts are discussed. Industrial applications are demonstrated, the importance of the hetero-geneous acid-base bifunctional catalysis being emphasized.

1. Introduction

In homogeneous acid- or base-catalyzed reactions, many examples of acid-base bifunctional catalysis have been reported and the catalytic mechanism fully discussed [1]. A remarkable example is seen in the mutarotation of tetramethyl glucose catalyzed by 2-hydroxy pyridine [2].

The acid and base strengths of 2-hydroxy pyridine are 1/100 of acid strength of phenol and 1/10000 of base strength of pyridine. Nevertheless, the catalytic activity of 2-hydroxy pyridine is 7000 times higher than that of the mixture of phenol and pyridine. Phenol alone or pyridine alone does not show any activity. The surprisingly high activity of 2-hydroxy pyridine is considered due to the concerted acid-base bifunctional catalysis and the stereospecific orientation of acidic and basic groups of the glucose to basic and acidic groups of the catalyst as shown in Fig. 1 [2]. In this case, the acidic group (-OH) and the basic group (-N=) of the catalyst act as a proton donor and a proton acceptor, respectively, and electron shift occurs in the direction indicated by arrows. The activity difference between 2-hydroxy pyridine and the mixture of phenol and pyridine is indeed $10^2 \times 10^4 \times 7 \times 10^3 = 7 \times 10^9$, if the difference in acid and base strength is taken into consideration.

Another striking example is the dissociation of hydrogen molecule catalyzed by hydrated copper ion [3]. As shown in Fig. 2, Cu^{2+} abstracts

J. Fraissard and L. Petrakis (eds.), Acidity and Basicity of Solids, 353–373.
© 1994 Kluwer Academic Publishers.

H^- to form Cu^+H, while hydrated water abstracts H^+ to form H_3O^+. In this case, Cu^{2+} acts as a Lewis acid, while H_2O acts as a Bronsted base.

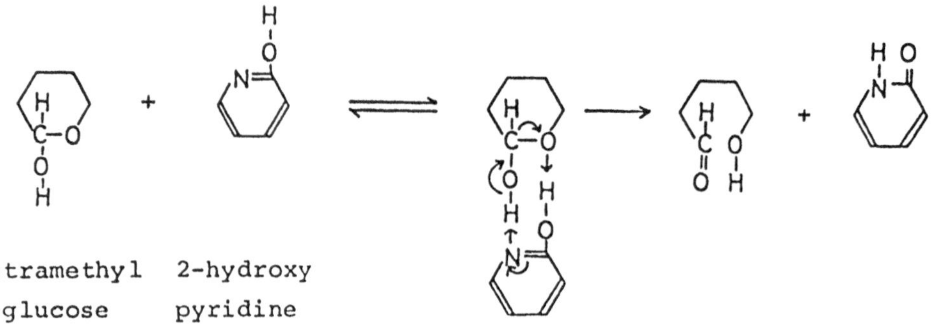

tetramethyl glucose   2-hydroxy pyridine

Fig. 1   Mutarotation of tetramethyl glucose catalyzed by 2-hydroxy pyridine.

The heterolytic splitting of hydrogen molecule takes place by the co-operation of acid with base.

$$H_2 + Cu(H_2O)_n^{2+} \longrightarrow \begin{array}{l} H^- \;----\; Cu^{2+} \\ H^+ \;----\; H_2O \cdot (H_2O)_{n-1} \end{array}$$

$$\longrightarrow \begin{array}{l} Cu^+H \\ H_3O^+ \cdot (H_2O)_{n-1} \end{array} \longrightarrow CuH(H_2O)_{n-1}^+ + H_3O^+$$

Fig. 2   Dissociation of hydrogen molecule catalyzed by hydrated copper ion.

These kinds of catalysis have been termed acid-base concerted (terna-ry, synchronous or push-pull) bifunctional catalysis [1], which is ex-pressed simply as acid-base bifunctional catalysis in this paper.

In heterogeneous reactions using solid acids and bases as catalysts, Turkevich and Smith showed more than forty years ago that the isomerization of 1-butene to 2-butene is catalyzed by metal sulfates and mounted sulfuric and phosphoric acid on which the separations between acid and base sites are 3.46 - 3.50 A [4]. However, little activity was observed for the mounted acetic acid and hydrogen chloride. The higher catalytic activity of catalysts of the former group was attributed to acid-base bifunctional catalysis as illustrated in Fig. 3. Many other examples suggesting heterogeneous acid-base bifunctional catalysis have been reported in the cases of the catalysts such as Al_2O_3, Al_2O_3-ThO_2-H_2SO_4, $NiSO_4 \cdot 0.5H_2O$, CaO, KBO_2, $Ca_3(PO_4)_2$, etc. [5], though any strong evidence was not provided.

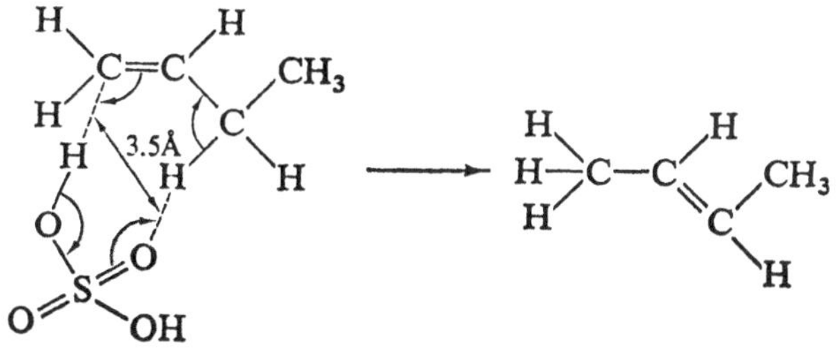

Fig. 3 Acid-base bifunctional catalysis by metal sulfate and mounted sulfuric acid for isomerization of 1-butene

It is recently that ZrO_2, ThO_2, MgO, and modified ZrO_2 and MgO have been reported to exhibit high catalytic activities and selectivities and long catalyst life for particular reactions and some reasonable evidence was given for the heterogeneous acid-base bifunctional catalysis [6 - 10].

The cooperation of a weakly acid site with a weakly base site of which the acid-base pair site is suitably oriented for a reacting molecule is surprisingly powerful for giving pronounced catalytic selectivity and long catalyst life. Those examples are summarized and the acid-base bifunctional nature is discussed. Several industrial processes which have been achieved by using acid-base bifunctional catalysts are demonstrated.

2. Acid-base Bifunctional Catalysis by Solid Acids and Bases

2.1. SYNTHESIS OF α-OLEFINS FROM ALCOHOLS

The selectivity for the formation of 1-butene in the dehydration of 2-butanol was 27% over Al_2O_3, but 90% over ZrO_2 [11]. The high catalytic selectivity of ZrO_2 was attributed to the acid-base bifunctional catalysis on the basis of poisoning experiments with n-butylamine and carbon dioxide.

Very recently, ZrO_2 calcined at 400 °C was found to exhibit fairly good conversion (42%) and selectivity (66%) in the dehydration of 1-cyclohexyl ethanol to vinylcyclohexane at 623 K [12]. The yield was better than those of TiO_2, Y_2O_3, ThO_2, Al_2O_3, SiO_2, MgO, La_2O_3 and CeO_2.

In the case of ZrO_2 treated with NaOH, the conversion and the selectivity were more than 80% and almost 90%, respectively, almost no deactivation being observed in 3000 h [12, 13]. Vinylcyclohexane is a useful compound, since the transparency of polypropylene is improved by the addition of several ppm of polyvinylcyclohexane and the dehydration process of synthesizing the -olefin (vinylcyclohexane) has been industrialized by Sumitomo Chemical Co., Ltd. [13]. The evidence for the acid-base bifunctional catalysis over ZrO_2 and ZrO_2-NaOH was given by the fitness of the distance(2.455 A) between an acid site (Zr^{4+}) and a base site (O^{2-}) of ZrO_2 calcined at 673 K with the distance(2.64 A) between a basic group (C-OH) and an acidic group (C-H) of 1-cyclohexyl ethanol and also by the values of overlap population calculated according to the theory of Paired Interacting Orbitals [12].

2.2. FORMATION OF 1-BUTENE FROM 2-BUTANAMINE

Selective formation (61%) of 1-butene in the elimination of ammonia from 2-butanamine was observed at 623 K over ZrO_2 [14]. The reaction is considered to proceed by the carbanion mechanism as shown below.

$$CH_3-CH_2-CH-CH_3 \longrightarrow CH_3-CH_2-\overset{-}{CH}-CH_2 + H^+ \longrightarrow CH_3-CH_2-CH=CH_2 + NH_3$$
$$\quad\quad\quad | \quad\quad\quad\quad\quad\quad\quad\quad\quad | $$
$$\quad\quad\quad NH_2 \quad\quad\quad\quad\quad\quad\quad NH_2$$

The carbanion formed by the abstraction of H^+ by a base site (O^{2-}) is stabilized by an acid site (Zr^{4+}) on the surface of ZrO_2.

2.3. C-H BOND CLEAVAGE

The H-D exchange reaction between methyl group of adsorbed isopropyl alcohol-d_8 and surface OH group was studied over several metal oxides by means of reflectance IR spectroscopy. ZrO_2 and ThO_2 showed significant catalytic activities for the exchange reaction, though strongly acidic $SiO_2-Al_2O_3$ and Al_2O_3 were inactive and strongly basic MgO and CaO caused the dehydrogenation of alcohol to form ketones, suggesting the acid-base bifunctional behavior of ZrO_2 and ThO_2 [15].

2.4. ACETONITRILE FROM TRIETHYLAMINE

The formation of acetonitrile from triethylamine was observed over ZrO_2, but not over strongly acidic $SiO_2-Al_2O_3$ and strongly basic MgO [16, 17]. The evidence for the acid-base bifunctional catalysis of ZrO_2 was provided by the TPD study of NH_3 and CO_2 coadsorbed on ZrO_2 [18].

2.5. ALLYL ALCOHOL FROM EPOXIDE

The rearrangement of 2-carene oxide over ZrO_2-TiO_2 (molar ratio = 1) which possesses high acidity and high basicity gave an allyl alcohol (cis-2,8(9)-p-menthadiene-1-ol) with 75% selectivity, the conversion being 77% at 353 K for 10 min [19]. The selectivity increased to 100% at 303 K for 1 h. However, the selectivities over $SiO_2-Al_2O_3$ and Al_2O_3 were 0 and 36 - 43%, respectively. The reaction over ZrO_2-TiO_2 is considered to take place according to the following scheme.

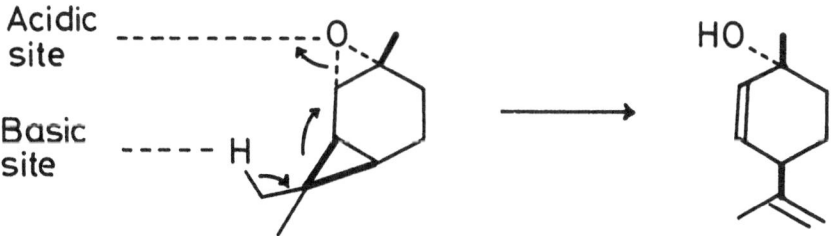

Another mixed oxide of ZrO_2 which has both acidic and basic property is ZrO_2-SnO_2 [20], of which application as a bifunctional catalyst to organic synthesis is promising.

2.6. SYNTHESIS OF KETONES

Various ketones are known to be synthesized from aldehydes, alcohols, carboxylic acids, and esters by use of ZrO_2 and its mixed oxides as catalysts [21 - 28]. Some examples of ketone synthesis are summarized in Table 1. The addition of 2 - 3% K and Li or 0.02% Mg to ZrO_2 was effective for prolonging the catalyst life. Several kinds of ketones have been commercialized by Chisso Corporation, about 1,000 t/y of diisopropyl ketone, methyl-n-propyl ketone, diethyl ketone, methyl-n-amyl ketone, etc. being manufactured since 1973.

Table 1 Synthesis of various ketones by ZrO_2 catalysts

Reactions	Catalyst	Reaction temp./K	Conv. %	Select. %
Isobutyraldehyde $\longrightarrow$	ZrO_2	743	90	88
diisopropyl ketone	ZrO_2-Li_2O	713	90	91
2-Ethylhexanal $\longrightarrow$	ZrO_2	668	35	85
di-2-ethylbenzyl ketone				
Isobutanol $\longrightarrow$	ZrO_2	733	77	72
diisopropyl ketone	ZrO_2-K_2O	693	81	84
Isobutyl isolactate $\longrightarrow$	ZrO_2	773	87	90
diisopropyl ketone	ZrO_2-K_2O	713	86	88
n-Butanol + n-Bulyraldehyde $\longrightarrow$ di-n-propyl ketone	ZrO_2-MgO	723	85 95	90
Acetic acid + Isobutyraldehyde $\longrightarrow$	ZrO_2-Na_2O	723	99 96	
acetone methyl-isopropyl ketone diisopropyl ketone				26 34 37

2.7. REDUCTION OF ALDEHYDES, KETONES, CARBOXYLIC ACIDS, AND ESTERS WITH ALCOHOLS

Hydrated ZrO_2 calcined at relatively low temperature (573 - 603 K) was found to be highly active for the reduction of various carbonyl compounds with alcoholes. When the solutions of aldehydes or ketones in 2-propanol are passed through the ZrO_2 catalyst at 423 or 473 K, the corresponding alcohols were obtained with high yields as shown in Table 2 [29]. For these reactions, which are known as Meerwein-Ponndorf-Verley reduction, aluminum isopropoxide has been used as a catalyst, but its use is disadvantageous because of the consumption of large amounts of the catalyst. Alumina is also used, but its activity is deactivated with moisture. Hydrated ZrO_2 is said to have no such disadvantages.

The catalytic activity of hydrated ZrO_2 for the reduction of carboxylic acids and esters with isopropanol was also high [30].

Table 2 Reduction of aldehydes and ketones

Reactants	Temp./K	Yield/%
∧∧∨CHO	423	100
(benzaldehyde) CHO	423	99
(pentan-2-one)	473	97
(acetophenone)	423	98

2.8. ESTERIFICATION, ESTER EXCHANGE, AMINATION, AND ACETALIZATION

Table 3 shows that esterification and ester exchange take place efficiently over hydrated ZrO_2 mentioned in 2.7, the yields being 100% at 473 K [31]. Hydrated ZrO_2 also showed high catalytic activities for the amination of carboxylic acids and esters with amines (Table 4) [32] and the acetalization of aldehydes and ketones with ethylene glycol [33].

The reactions over hydrated ZrO_2 (2.7, 2.8) are considered to be catalyzed by the pair site of weakly acidic and basic hydroxyl groups on the surface of hydrated ZrO_2.

2.9. CANNIZZARO REACTION OF BENZALDEHYDE

Benzaldehyde is known to form benzoic acid and benzyl alcohol in the presence of sodium hydroxide in aqueous solution (Cannizzaro reaction) and to form benzyl benzoate in the presence of metal benzylate (Tishchenko reaction). Both reactions are homogeneous base-catalyzed reactions. However, in a heterogeneous reaction of benzaldehyde over CaO, benzyl benzoate is mainly formed. The mechanism elucidated by kinetic and spectroscopic study is illustrated by the scheme of Fig. 4 [34].

Table 3 Esterification and ester exchange

$$\text{RCOOH} + \text{R'OH} \xrightarrow{\quad ZrO_2 \quad} \text{RCOOR'}$$

Reactants	Temperature/K	Yield/%
∿COOH + C_2H_5OH	473	100
⬡-COOH + C_2H_5OH	473	100

Table 4 Amination

$$R^1COOR^2 \xrightarrow[\quad ZrO_2 \quad]{R^3R^4NH} R^1CONR^3R^4$$

Reactants	Temperature/K	Yield/%
CH₃C(O)OH + ∿NH₂	473	100
CH₃C(O)O∿ + ∿NH₂	473	100
CH₃C(O)O∿ + ⬡-NH₂	473	100
CH₃C(O)OH + NH₃	473	97

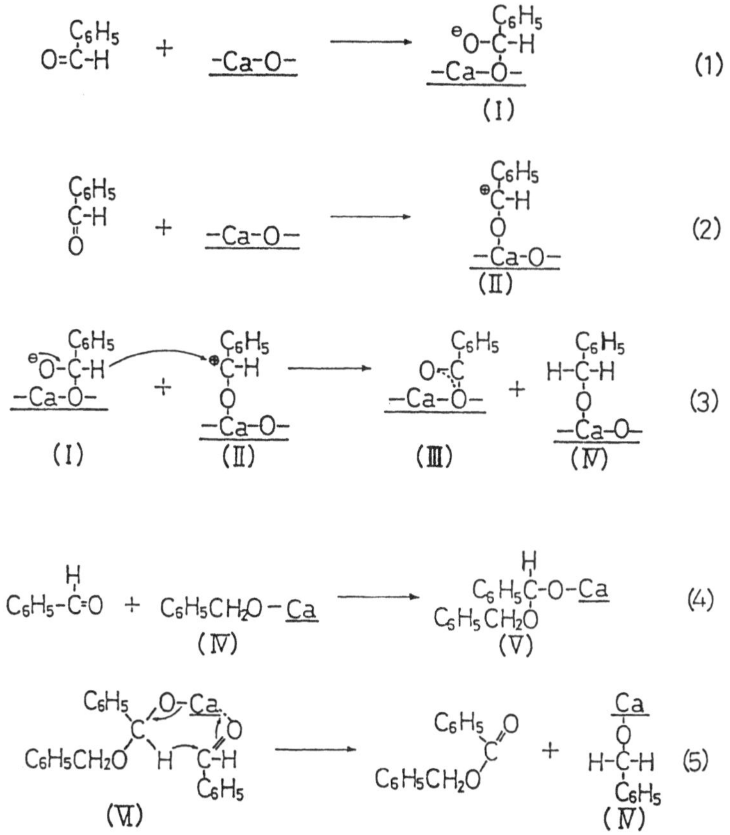

$$C_6H_5\text{-}O\!=\!C\text{-}H \; + \; -Ca\text{-}O- \; \longrightarrow \; {}^{\ominus}O\text{-}\overset{C_6H_5}{\underset{-Ca\text{-}O-}{C}}\text{-}H \qquad (1)$$
$$(\text{I})$$

$$\overset{C_6H_5}{\underset{O}{C}}\text{-}H \; + \; -Ca\text{-}O- \; \longrightarrow \; {}^{\oplus}\overset{C_6H_5}{\underset{\underset{-Ca\text{-}O-}{O}}{C}}\text{-}H \qquad (2)$$
$$(\text{II})$$

$$\overset{C_6H_5}{\underset{-Ca\text{-}O-}{{}^{\ominus}O\text{-}C\text{-}H}} \; + \; \overset{C_6H_5}{\underset{\underset{-Ca\text{-}O-}{O}}{{}^{\oplus}C\text{-}H}} \; \longrightarrow \; \overset{C_6H_5}{\underset{-Ca\text{-}O-}{O\!=\!C}} \; + \; \overset{C_6H_5}{\underset{\underset{-Ca\text{-}O-}{O}}{H\text{-}C\text{-}H}} \qquad (3)$$
$$(\text{I}) \qquad\qquad (\text{II}) \qquad\qquad (\text{III}) \qquad (\text{IV})$$

$$C_6H_5\text{-}C\!=\!O \; + \; C_6H_5CH_2O-\underline{Ca} \; \longrightarrow \; \overset{H}{\underset{C_6H_5CH_2O}{C_6H_5\overset{|}{C}\text{-}O\text{-}\underline{Ca}}} \qquad (4)$$
$$(\text{IV}) \qquad\qquad\qquad\qquad (\text{V})$$

$$\overset{C_6H_5\;\;\;O\text{-}\underline{Ca}}{\underset{(\text{VI})}{C_6H_5CH_2O\;\;\;C\;\;\;H\text{-}C\text{-}H}} \; \longrightarrow \; \overset{C_6H_5}{\underset{C_6H_5CH_2O}{C\!=\!O}} \; + \; \overset{\underline{Ca}}{\underset{\underset{C_6H_5}{H\text{-}C\text{-}H}}{O}} \qquad (5)$$
$$(\text{VI}) \qquad\qquad\qquad\qquad\qquad (\text{IV})$$

Fig. 4 Mechanism of Cannizzaro reaction (1)(2)(3) and Tishchenko reaction (4)(5) of benzaldehyde catalyzed by CaO.

The active species for the ester formation are the calcium benzylates whose formation is facilitated by both the basic sites (O^{2-}) and acidic sites (Ca^{2+}) on the surface. The mechanism of formation of benzylate is very similar to the homogeneous Cannizzaro reaction. However, the difference is that a Lewis acid as well as a basic site plays an important role as an active site in the heterogeneous reaction. In fact, the poisoning of either acid sites or basic sites with basic ammonia or acidic carbon dioxide stopped the reaction.

On the other hand, the mechanism for the formation of benzyl benzoate in heterogeneous phase is different from that in homogeneous phase, judging from the big difference in order of reaction, isotope effect, and energy and entropy of activation [34].

2.10. ALKYLATION OF PHENOL WITH METHANOL

Alkylation of phenol with methanol is industrially important as a reaction to synthesize 2,6-xylenol, a monomer of a good heat-resisting PPO resin. The reaction is known to be easily catalyzed by solid acids such as SiO_2-Al_2O_3, Al_2O_3, etc. However, SiO_2-Al_2O_3 forms various products such as o-, m-, p-cresol, o-, m-, p-xylenol, anisol derivatives, 2,4,6-trimethylphenol, etc., as shown in Fig. 5 and the selectivity for 2,6-xylenol is very low (several %). In 1965, General Electric found that MgO is highly selective (more than 90%) for the formation of 2,6-xylenol [35]. What causes such a big difference?

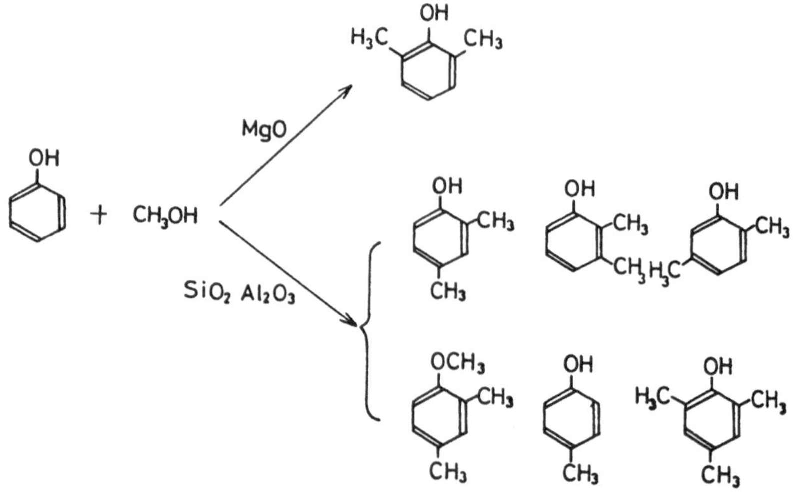

Fig. 5 Alkylation of phenol with methanol over acidic and basic catalysts.

An infrared study of phenol adsorbed on SiO_2-Al_2O_3 and MgO revealed that both of the catalysts adsorb phenol to split it heterolytically to proton and phenolate ion. Strongly acidic sites (Al^{3+}) and weakly basic sites (O^{2-}) act as active sites for the heterolytic splitting in the case of SiO_2-Al_2O_3, while strongly basic sites (O^{2-}) and weakly acidic sites (Mg^{2+}) cooperate to split phenol in the case of MgO as shown in Fig. 6 [36]. The difference is that the plane of benzene ring of phenolate is parallel to the surface of SiO_2-Al_2O_3, but almost perpendicular to the surface of MgO. This causes the difference in the ortho-selectivity.

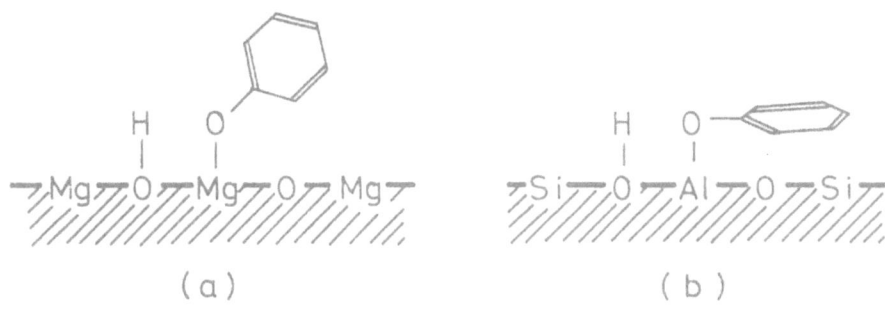

Fig. 6 Adsorbed states of phenol on MgO and SiO_2-Al_2O_3.

The difference in the adsorbed state of phenolate is considered to depend on the acid strength of the catalysts. Since the acid strength of SiO_2-Al_2O_3 is very high, the acid sites interact with the π-electrons of benzene ring of phenolate, giving the adsorbed form (b). However, such an interaction does not occur on very weakly acidic MgO and, hence, the adsorbed form (a) is produced.

Thus, it is concluded that, in both cases of SiO_2-Al_2O_3 and MgO, both acidic and basic sites are necessary for the dissociation of phenol, but the existence of strong acid sites causes low ortho-selectivity.

2.11. SIDE-CHAIN ALKYLATION OF TOLUENE WITH METHANOL

The side-chain alkylation which is important for direct synthesis of styrene or ethylbenzene has been known to be catalyzed by basic materials such as RbX zeolite [37], MgO [38], and Cs-C [39], though the activities were low. Recently, solid superbases (MgO-NaOH-Na, Al_2O_3-NaOH-Na, etc.) were attempted to be used as catalysts for the side-chain alkylation, However, unexpected results of low activity and low selectivity for the side-chain alkylation were obtained [40]. It is intriguing that the addition of boric acid to RbX enhanced the activity [41]. Suggesting an acid-base bifunctional catalysis. There have been active discussions concerning the nature of active sites. In 1988, Miyamoto et al. clearly demonstrated by using computer graphics that the acid-base cooperative catalysis proceeds in the side-chain alkylation on alkali-ion exchanged zeolites and that the geometrical factor of acid-base pair sites plays an important role [42].

2.12. SYNTHESIS OF ETHYLENIMINE

Ethylenimine derivatives are commercially important chemicals which are used as pharmaceuticals and in coatings for paper and textile. Ethylenimine has been produced by intramolecular dehydration of monoethanolamine in liquid phase using sulfuric acid and sodium hydroxide as shown in Fig. 7, but the process has some problems such as low productivity, formation of large amounts of sodium sulfate, etc. The vapor phase process which is more advantageous than liquid phase process has been attempted using various solid catalysts. However, ethylenimine was found to be very reactive on catalysts, resulting in undesirable reactions as shown in Fig. 8 [43, 44].

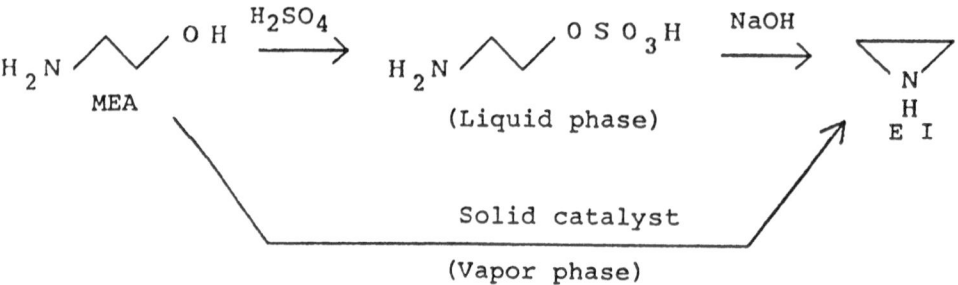

Fig. 7 Synthesis of ethylenimine from monoethanolamine

Recently, Nippon Shokubai has been successful in developing new efficient catalysts for the vapor phase process of ethylenimine production, 2,000 t/y being manufactured since 1991. With a catalyst, Si-Ba-Cs-P-O (1 : 0.1 : 0.1 : 2.4), the conversion of monoethanolamine was 86% and the selectivity for ethylenimine was 81% at 683 K and space velocity of 1,500 h^{-1} [44]. The acid and base strengths of the catalyst are weaker than H_0 = + 4.8 and weaker than H_- = 9.4, respectively, and the reaction is considered to proceed by an acid-base bifunctional mechanism [44, 45].

2.13. HYDROGENATION OF OLEFINS

Hydrogenation of olefins was found to take place over MgO evacuated at very high temperature, the optimum evacuation temperature of MgO being 1373 K, as shown in Fig. 9 [46]. The hydrogenation of butadiene took place even at room temperature over MgO evacuated at 1373 K, cis-2-

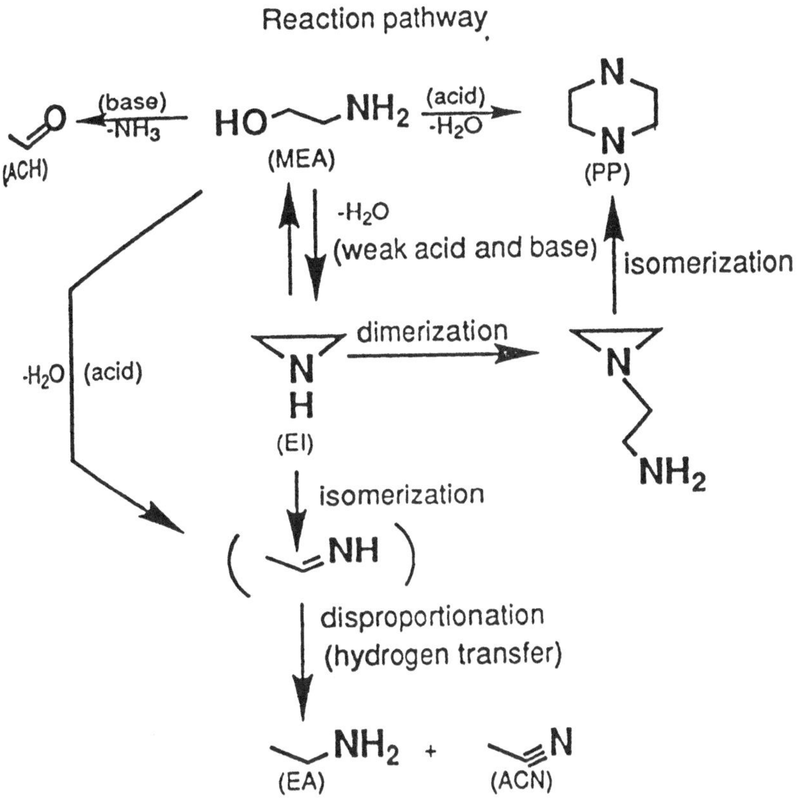

Reaction pathway

Fig. 8 Catalyst; Si-Ba-Cs-P-O (1 : 0.1 : 0.1 : 0.1 : 2.4)
Reaction temp.; 410°C, SV; 1500/h
Converson; 85.5% Selectivity; 81.3%

butene being formed selectively [47]. The hydrogenation is considered
to proceed via an anion intermediate of stable cis-form which is formed
from adsorbed butadiene and hydride ion. Characteristic nature of MgO
is to provide a pair of an acid site (Mg^{2+}) and a basic site (O^{2-}) suitably oriented so as to split a hydrogen molecule into a proton (H^+) and
a hydride ion (H^-), as shown in Fig. 10. The unique nature of MgO pretreated at 1373 K is that hydrogen ions (H^+ and H^-) attack only 1 and 4
carbons of butadiene keeping molecular identity of a hydrogen molecule
and that the exchange reaction between H_2 and D_2 does not take place.

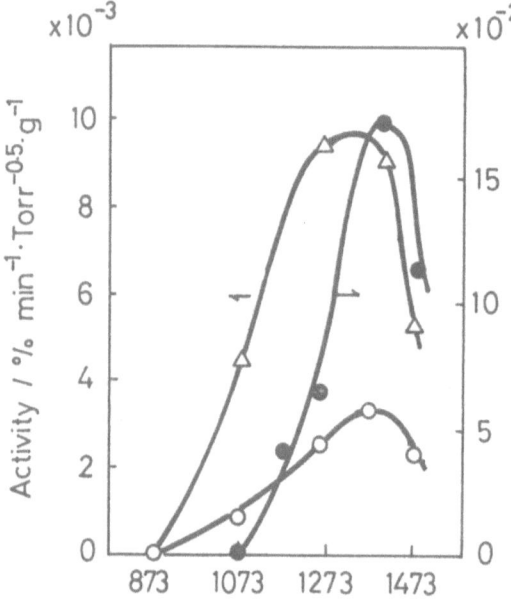

Fig. 9 Variation of hydrogenation activity of MgO with
        evacuation temperature.
        ● ; ethylene, △ ; propylene, ○ ; 1-butene

The hydrogenations of 1,3-butadiene and 2-methyl-1,3-butadiene with hydrogen or cyclohexadiene as hydrogen source take place at 323 - 343 K over ZrO$_2$ [48]. The reaction products in the hydrogenation of 1,3-buta-diene with hydrogen was mainly trans-2-butene (80%), while that with cy-clohexadiene was mainly 1-butene (64%). It is interesting to compare with the case of MgO catalyst where the main product was cis-2-butene (77%) [49].

Although both acidity and basicity of ZrO$_2$ do not change much with the change of evacuation temperature [50], ZrO$_2$ evacuated at 873 K shows a maximum activity for the hydrogenation of 1,3-butadiene with hydrogen, whereas ZrO$_2$ evacuated at 1073 K gives a maximum activity for the same hydrogenation with cyclohexadiene as seen in Fig. 11 [7, 9, 48]. Since

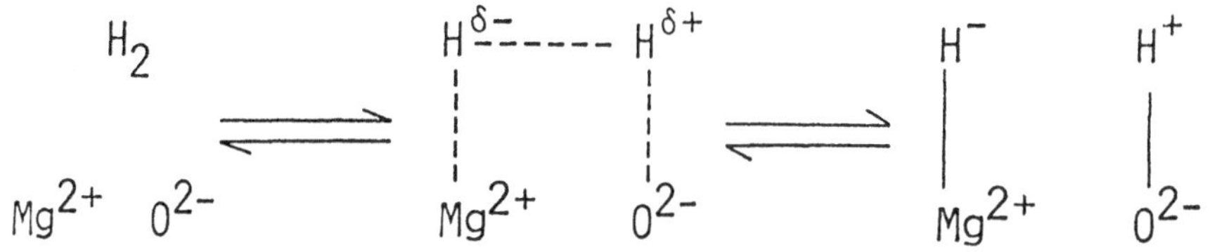

$$H_2 \qquad \rightleftharpoons \qquad \overset{\delta-}{H} \text{-----} \overset{\delta+}{H} \qquad \rightleftharpoons \qquad H^- \qquad H^+$$

$$Mg^{2+} \quad O^{2-} \qquad\qquad Mg^{2+} \quad O^{2-} \qquad\qquad Mg^{2+} \quad O^{2-}$$

Fig. 10   Heterolytic splitting of hydrogen molecule on MgO.

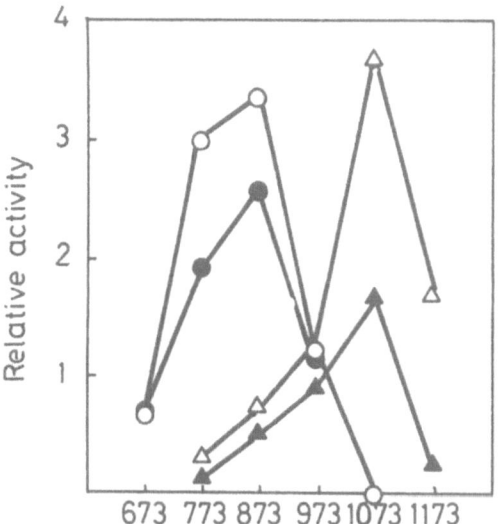

Fig. 11   Catalytic activities of ZrO_2 pretreated at different
temperatures.
O ; hydrogenation of 1,3-butadiene with H_2, ● ; H_2-D_2 exchange,
Δ ; isomerization of 1-butene, ▲ ; hydrogenation of 1,3-buta-
diene with cyclohexadiene.

it is known that the lattice constant of ZrO_2 changes considerably with the change of evacuation temperature, the appearance of two kinds of maximum activities is considered mainly due to the difference in the distance between an acid site (Zr^{4+}) and a base site (O^{2-}), suggesting the importance of orientation of the acid–base pair sites for acid–base bifunctional catalysis.

For the activation of hydrogen, an IR study [51] revealed that ZrO_2 provides the acid–base pair sites suitably oriented so as to heterolytically split H_2 into H^+ and H^-, similarly as in the case of MgO.

It was reported recently that 1-butene is hydrogenated with hydrogen to give n-butane at higher temperature (473 K) over a ZrO_2 aerogel catalyst [52].

2.14. HYDROGENATION OF CARBON MONOXIDE

On the basis of the knowledge that a hydrogen molecule dissociates to form H^+ and H^- and carbon monoxide forms $[(CO)_2]^{2-}$ on the surface of MgO pretreated at 1373 K, Wang et al. observed by TPD and IR methods that CO adsorbed on MgO reacts with H_2 to form adsorbed HCHO in the temperature range of 343 – 583 K [53]. In fact, HCHO was detected as a product of the reaction of CO (50 Torr) with H_2 (100 Torr) at 483 K over MgO and Na/MgO, while CH_3OH over ZrO_2 and La_2O_3. At higher temperature (573 K), CH_3OH was formed on MgO [54].

Isobutene is produced from CO and H_2 over ZrO_2 pretreated at 773 K under moderate conditions, the selectivity of butenes among hydrocarbons and that of isobutene among C_4 hydrocarbons are 82 and 97 mol%, respectively, at 623 K and 0.68 atm. [55]. At lower reaction temperatures, methanol is selectively formed [55, 56]. The active sites for the hydrogenation are also considered to be both acid and base sites on ZrO_2.

2.15. HYDROGENATION OF AROMATIC CARBOXYLIC ACIDS TO THE CORRESPONDING ALDEHYDES

Aromatic aldehydes are important intermediates in the production of organic fine chemicals such as pharmaceuticals, agrochemicals, and perfumes. These aldehydes have been produced mainly by a halogenation method. However, the method has some disadvantages such as poor yield and undesirable by-products formation and environmental influence.

A novel process for synthesizing aromatic aldehydes by the direct hydrogenation of the corresponding carboxylic acids over ZrO_2 has been developed recently. Mitsubishi Kasei Corporation has successfully commercialized p-t-butylbenzaldehyde, m-phenoxybenzaldehyde and p-methylbenzaldehyde by this process, 2000 t/y of the aldehydes being manufactured since 1988 [57, 58].

Yokoyama et al. found that ZrO_2 showed a high selectivity for the hydrogenation of benzoic acid as shown in Table 5 [59]. In particular, ZrO_2 modified with a small amount of Cr^{3+} (atomic ratio of Cr/Zr = 0.05) was highly active and selective (see Table 6) [59]. According to the

Table 5 Hydrogenation of benzoic acid over various metal oxide catalysts

Reaction conditions: H_2GHSV = 625 h^{-1},

benzoic acid/H_2 = 2/98 vol%, under atmospheric pressure

Catalyst	Temperature/K	Conversion of benzoic acid/%	Selectivity to benzaldehyde/%
20% $Y_2O_3/\alpha-Al_2O_3$	713	100	87
γ-Al_2O_3	713	20	53
MgO	713	62	4
TiO_2	713	15	10
ZrO_2	673	53	97
ZnO	673	55	54

Table 6 Hydrogenation of benzoic acid with modification of ZrO_2 catalyst

Reaction conditions: H_2GHSV = 625 h^{-1},

benzoic acid/H_2 = 2/98 vol%, under atmospheric pressure

Additives	Temperature/K	Conversion of benzoic acid/%	Selectivity to benzaldehyde/%
None	623	51	97
Pb	583	89	96
In	603	100	91
Cr	623	98	96
Mn	623	70	97
Ca	623	50	98

measurement of acid-base property by the indicator method and the TPD method using NH_3 and CO, ZrO_2 and Cr^{3+}-ZrO_2 which have weak acid and base strengths are concluded to be efficient catalysts. Similarly as in 2.6, the addition of a small amount of a metal oxide to ZrO_2 was effective for prolonging the catalyst life. ZrO_2 of which acid strength is

high gave poor activity and the selectivity was low due to decarboxylation of carboxylic acid. On the basis of the reaction kinetics and FTIR
spectroscopy, the reaction mechanism was proposed in which both an acid
site (Zr^{4+}) and a base site (O^{2-}) act as active sites as shown in Fig.
12 [59].

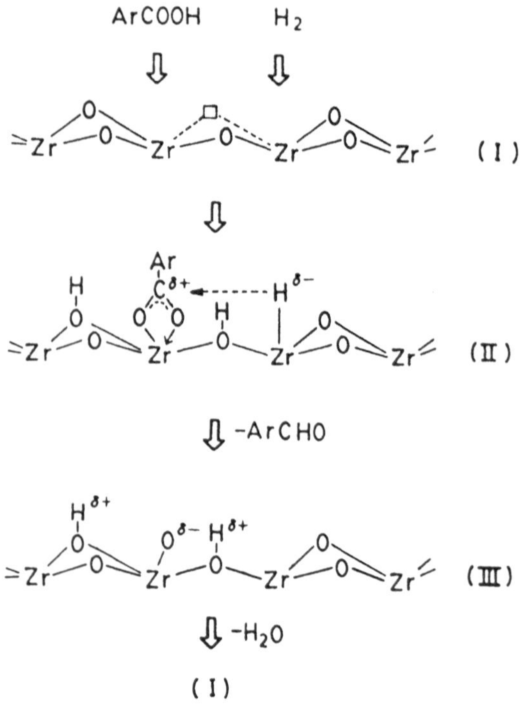

Fig. 12 Proposed mechanism of hydrogenation of
         aromatic carboxylic acid over ZrO_2

3. Conclusion

Fifteen reactions which are considered to proceed by acid-base bifunctional mechanism over solid catalysts are described, some evidence for
the bifunctional catalysis being given. It is most intriguing that ZrO_2,
hydrated ZrO_2, modified ZrO_2, high temperature pretreated MgO, Si-Ba-Cs-

P-O, etc. which are almost neutral materials exhibited high catalytic activity and selectivity and long catalyst life. The very weak acid-base strengths cause high selectivity and long life, though on strong acid or base sites undesirable side reactions and catalyst deactivation due to coking often occur. The high activities are due to the acid-base bifunctional catalysis in which the orientation of acid-base pair sites is vitally important. The orientation can be largely changed by the preparation method, the pretreatment condition, and the addition of small amounts of the other metal oxides. Thus, the four industrial processes of the syntheses of α-olefins, ketones, aromatic aldehydes, and ethylenimine have been successfully achieved as discussed in 2.1, 2.6, 2.12, and 2.15.

The future of heterogeneous acid-base bifunctional catalysis is promising and its ultimate goal would be the approach to an ideal bifunctional reaction shown in Fig. 1 and further to catalysis by enzyme.

4. References

1. R. P. Bell, The Proton in Chemistry, Cornell Univ. Press, 1959, Chapt. 7.
2. C. G. Swain and J. F. Brown, J. Am. Chem. Soc., 74, 2534, 2538 (1952).
3. J. Halpern, Advances in Catalysis, 11, 301 (1959).
4. J. Turkevich and R. K. Smith, J. Chem. Phys., 16, 446 (1948).
5. K. Tanabe, Solid Acids and Bases, Kodansha, Tokyo, Academic Press, New York, 1970, p. 145 - 153.
6. K. Tanabe, Catalysis by Acids and Bases, ed. B. Imelik et al., Elsevier, Amsterdam, 1985, p. 1 - 14.
7. K. Tanabe, Materials Chem. and Phys., 13, 347 (1985).
8. K. Tanabe, Acid-Base Catalysis, ed. K. Tanabe et al., Kodansha, Tokyo, VCH, Weinheim, 1989, p. 513 - 527.
9. K. Tanabe, Proc. 9th Intern. Congr. Catal., ed. M. J. Phillips and M. Ternan, Chem. Inst. Canada, Ottawa, 1988, Vol. 5, p. 85 - 113.
10. K. Tanabe, M. Misono, Y. Ono, and H. Hattori, New Solid Acids and Bases, Kodansha, Tokyo, Elsevier, Amsterdam, 1989, Chapt. 4.
11. T. Yamaguchi, H. Sasaki, and K. Tanabe, Chem. Lett., 1017 (1973).
12. K. Takahashi, T. Hibi, Y. Higashio, and M. Araki, Shokubai(Catalyst), 35, 12 (1993).
13. Sumitomo Chemical Co. Ltd., Japan Kokai Patent, 61-130240 (1986).
14. A. Satoh, H. Hattori, and K. Tanabe, Chem. Lett., 497 (1983).
15. T. Yamaguchi, Y. Nakano, T. Iizuka, and K. Tanabe, Chem. Lett., 677 (1976).
16. B.-Q. Xu, T. Yamaguchi, and K. Tanabe, Chem. Lett., 1053 (1987).
17. B.-Q. Xu, T. Yamaguchi, and K. Tanabe, Chem. Lett., 281 (1988).
18. B.-Q, Xu, T. Yamaguchi, and K. Tanabe, Chem. Lett., 1663 (1988).
19. K. Arata, J. O. Bledsoe, and K. Tanabe, Tetrahedron Lett., 43, 3861 (1976); J. Org. Chem., 43, 1660 (1978).
20. G.-W. Wang, H. Hattori, and K. Tanabe, Bull. Chem. Soc. Jpn., 56, 2407 (1983).

372

21. Chisso Corporation (M. Fukui et al.), Japan Kokai Patent, 48-409 (1973).
22. Chisso Corporation (M. Fukui et al.), Japan Kokai Patent, 48-26719 (1973).
23. Chisso Corporation (M. Fukui et al.), Japan Kokai Patent, 48-76806 (1973).
24. Chisso Corporation (M. Fukui et al.), Japan Kokai Patent, 48-76807 (1973).
25. Chisso Corporation (M. Fukui et al.), Japan Kokai Patent, 48-76808 (1973).
26. Chisso Corporation (M. Fukui et al.), Japan Kokai Patent, 49-61110 (1974).
27. Chisso Corporation (T. Okamoto et al.), Japan Kokai Patent, 49-109308 (1974).
28. Chisso Corporation (T. Okamoto et al.), Japan Kokai Patent, 49-48614 (1974).
29. M. Shibagaki, K. Takahashi, and H. Matsushita, Bull. Chem. Soc. Jpn., 61, 3283 (1988).
30. K. Takahashi, M. Shibagaki, and H. Matsushita, Chem. Lett., 1141 (1989).
31. K. Takahashi, M. Shibagaki, and H. Matsushita, Bull. Chem. Soc. Jpn., 62, 2353 (1989).
32. K. Takahashi, M. Shibagaki, H. Kuno, H. Kawakami, and H. Matsushita, Bull. Chem. Soc. Jpn., 62, 1333 (1989).
33. M. Shibagaki, K. Takahashi, H. Kuno, and H. Matsushita, Bull. Chem. Soc. Jpn., 63, 1258 (1990).
34. K. Tanabe and K. Saito, J. Catal., 35, 247 (1974).
35. General Electric Co., U.S. Patent, 3,446,856 (1964); Neth. Appl., 6,506,830 (1965).
36. K. Tanabe and T. Nishizaki, Proc. 6th Intern. Congr. Catalysis, 2, 863 (1977).
37. T. Yashima, K. Sato, T. Hayasaka and N. Hara, J. Catal., 26, 303 (1972).
38. K. Tanabe, O. Takahashi and H. Hattori, React. Kinet. Catal. Lett., 7, 347 (1977).
39. Mitsubishi Petrochemical Co., Japan Patent Appl., 52-133,932 (1977).
40. K. Takahashi, T. Yamaguchi and K. Tanabe, unpublished results.
41. Monsanto Co., U.S. Patent, 4,115,424 (1978).
42. A. Miyamoto, S. Iwamoto, K. Agusa, and T. Inui, Acid-Base Catalysis, ed. K. Tanabe et al., Kodansha, Tokyo, VCH, Weinheim, 1989, p. 497.
43. M. Ueshima, Y. Shimasaki, Y. Hino, and H. Tsuneki, Acid-Base Catalysis, ed. K. Tanabe et al., Kodansha, Tokyo, VCH, Weinheim, 1989, p. 41.
44. M. Ueshima and H. Tsuneki, Catalytic Science and Technology Vol. 1, ed. S. Yoshida et al., 1991, p. 357.
45. M. Ueshima, Y. Shimasaki, K. Ariyoshi, H. Yano, and H. Tsuneki, Proc. 10th Intern. Congr. Catalysis, Budapest, 1992, Part C, p. 2447.
46. H. Hattori, Y. Tanaka, and K. Tanabe, Chem. Lett., 659 (1975).
47. H. Hattori, Y. Tanaka, and K. Tanabe, J. Am. Chem. Soc., 98, 4652 (1976).
48. Y. Nakano, T. Yamaguchi, and K. Tanabe, J. Catal., 80, 307 (1983).

49. Y. Tanaka, Y. Imizu, H. Hattori, and K. Tanabe, Proc. 7th Intern. Congr. Catal., Kodansha, Tokyo, B 1254 (1981).
50. Y. Nakano, T. Iizuka, H. Hattori, and K. Tanabe, J. Catal., <u>57</u>, 1 (1979).
51. T. Onishi, H. Abe, K. Maruya, and K. Domen, J. Chem. Soc., Chem. Commun., 617 (1985).
52. G. M. Pajonk and A. El Tanany, React. Kinet. Catal. Lett., <u>47</u>, 167 (1992).
53. G. Wang, H. Hattori, and K. Tanabe, J. Chem. Soc., Chem. Commun., 1256 (1982).
54. G. Wang, H. Hattori, and K. Tanabe, Shokubai (Catalyst), <u>25</u>, 359 (1983).
55. T. Maehashi, K. Maruya, K. Domen, K. Aika, and T. Onishi, Chem. Lett., 747 (1984).
56. H. Hattori and G. Wang, Proc. 8th Intern. Congr. Catal., Verlag Chemie, Weinheim, <u>3</u>, 219 (1984).
57. Mitsubishi Kasei Corporation (T. Maki and T. Yokoyama), Japan Kokai Patent, 85-126242 (1985).
58. Mitsubishi Kasei Corporation (T. Maki and T. Yokoyama), U.S. Patent, 4,613,700 (1986).
59. T. Yokoyama, T. Setoyama, N. Fujita, M. Nakajima, and T. Maki, Appl. Catal., <u>88</u>, 149 (1992).

GENERATION OF ACIDITY (AMOUNT AND STRENGTH) IN SILICOALUMINOPHOSPHATES (SAPO ZEOLITES). EXAMPLES OF SAPO-5, -11, -34 AND - 37

Denise BARTHOMEUF
Laboratoire de Réactivité de Surface et Structure, URA 1106 CNRS,
Université Pierre et Marie Curie, 4 Place Jussieu, 75252 Paris Cedex 05,
France

ABSTRACT. The introduction of Si atoms in microporous crystalline $AlPO_4$ type materials generates, after the removal of the organic template, a negative framework, neutralized by protons. Bridging Si-OH-Al species are the acid centers. The possible formation of Si rich islands during synthesis or upon heat treatment determines the amount and strength of protonic acidity. The "isolated" Si with four Al as first neighbours give rise to a medium acid strength. Si at the border of the siliceous patches may generate very strong acidity depending on the number of Al and Si as neighbours. The thermal and hydrothermal stability of the structure is quite different from that of Si-Al zeolites. It may greatly affect the number and strength of bridging hydroxyls.

1   Introduction

Crystalline microporous aluminophosphate materials ($AlPO_4$) constitute a large family of solids with a variety of crystalline structures (1-3). The introduction in their framework of different atoms generates a very broad class of molecular sieves which may be used as adsorbents or catalysts. SAPO (S for Si, A for Al and P and O for the corresponding elements) are issued from the presence of Si in $AlPO_4$. The introduction of metal ions (Me = Fe, Mg, Mn, Co, Zn...) gives MeAPO or MeAPSO while that of elements (El = Be, Ga, Ti, Ge, Li, As...) forms ElAPO or ElAPSO (2). Some of the SAPO's have the same structure as well known Si-Al zeolites (3). Table 1 reports examples of SAPO's together with other chemical forms of $AlPO_4$ - based materials having the same structure.

The usual SAPO anhydrous composition is $(Si_x Al_y P_z)O_2$, x varying typically from 0.04 to 0.20 depending on the structure type and synthesis conditions. No evidence for the presence of Si-O-P is found and it is proposed that these bonds are not likely (3-5).

2   Origin of the protonic acidity

The acidity of these new materials depends on the atoms present in the structure and on the possible metal cations compensating the framework charge. Only the case of SAPO's will be considered here.

J. Fraissard and L. Petrakis (eds.), Acidity and Basicity of Solids, 375–390.
© 1994 *Kluwer Academic Publishers.*

TABLE 1. AlPO$_4$-based materials with the structure of Si-Al zeolites[a]

Structure	SAPO	Other possible chemicals[b]	Pore size nm
Faujasite	37		0.74
Erionite	17	Co, Fe, Mg	0.36 x 0.51
Chabasite	34	Co, Fe, Mg, Mn, Zn	0.38
A	42		0.41
Sodalite	20	Co, Fe, Mg, Mn, Zn	0.3

(a) from reference (2)
(b) in the AlPO$_4$ or SAPO structure

    With regards to the overall charge of the materials, AlPO$_4$ are formed of AlO$_4$ and PO$_4$ tetrahedra containing Al^{3+} and P^{5+} ions respectively. As a result, the network has no significant protonic acidity (6,7). The replacement of P^{5+} by Si^{4+} in some TO$_4$ tetrahedra (T for Al, P or Si) (1, 2, 4) creates a charge deficit. The negative charges of the framework may be compensated by protons. The first neighbours of SiO$_4$ are AlO$_4$ as in a Si-Al zeolite. The bridging hydroxyls in the bonds Si-OH-Al are then the protonic centers. It has to be noted that by contrast with Si-Al zeolites where the presence of Al is at the origin of protonic acidity, one proton neutralizing one AlO$_4^-$ tetrahedron, in SAPO's the Si atoms are responsible for the creation of acidity. There is a large number of AlO$_4$ connected only to PO$_4$ which are not then involved in acidity.
    In order to better understand the generation of acidity a detailed picture of the SiO$_4$ location has to be envisioned.

3 Expected number and strength of protonic sites

An easy way which is proposed to look at SAPO acidity is to consider the introduction of Si in an AlPO$_4$ framework (1, 2, 8). Some SAPO structures do not have an isostructural AlPO$_4$ counterpart. For instance the faujasite structure exists for SAPO (SAPO-37) but not for AlPO$_4$. In this instance one has to consider a hypothetical AlPO$_4$ framework where Si atoms are progressively introduced. Depending on which Si atoms (Al or P) is replaced by Si the y and z atomic fraction of these elements in the formula (Si$_x$ Al$_y$ P$_z$) O$_2$ are variously affected and the resulting charge on the framework may be very different (2). Three types of mechanisms are accepted to describe the substitution. Si can replace Al (mechanism 1), P (mechanism 2) or Al +P pairs (mechanism 3) (1). Figure 1 describes the principle of such substitutions (9). It is clear that schemes (b) and (c) give rise to Si-O-P bonds which are not likely to exist (3-5). The system tends to the replacement of P by Si giving Si rich islands (scheme d, figure 1) (5, 8). The experimental observation of such islands has been described in many cases using mainly ^{29}Si MAS NMR (5, 8-16)). The contribution of mechanisms 2 and 3 to the incorporation of Si in SAPO can be evaluated, in SAPO-37 for instance, from the ratio of the intensities of the -92 and -110 ppm ^{29}Si MAS NMR signals, the peaks corresponding respectively to mechanism 2 (Si with four Al

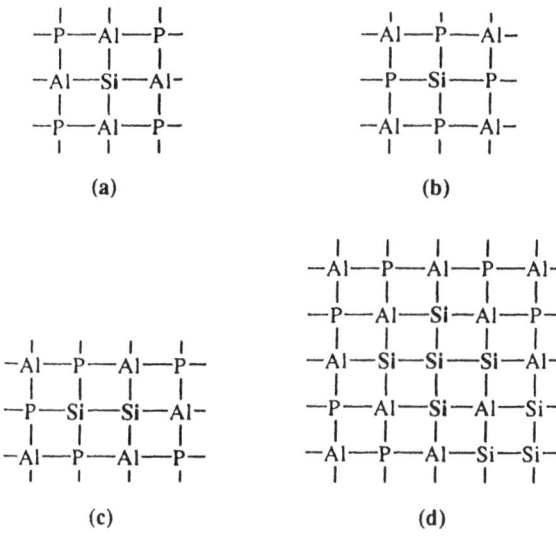

Fig. 1   Principle of the introduction of Si in AlPO₄ framework.
a :   Si replaces P : (Si (4Al))
b :   Si replaces Al : (Si (4P))
c :   Si replaces a pair P-Al : (Si (3Al, Si) and Si (3P, 1Si))
d :   Formation of Si islands (Si (3Al, 1Si) and Si (4Si))
From ref. (9).

```
        Al—P—Al—P—Al—P—Al—P—Al
        |  |  |  |  |  |  |  |  |
        P—Al—P—Al—P—Al—P—Al—P
        |  |  |  |  |  |  |  |  |
        Al—P—Al—P—Al—P—Al—P—Al
        |  |  |  |  |  |  |  |  |
  (a)   P—Al—Si—Al—P—Al—P—Al—P
        |  |  |  |  |  |  |  |  |
        Al—P—Al—P—Al—Si—Al—P—Al
        |  |  |  |  |  |  |  |  |
        P—Al—P—Al—Si—Si—Si—Al—P
        |  |  |  |  |  |  |  |  |
  (b)   Al—P—Al—P—Al—Si—Si—Si—Al
        |  |  |  |  |  |  |  |  |
        P—Al—P—Al—Si—Si—Si—Al—P
        |  |  |  |  |  |  |  |  |
        Al—P—Al—P—Al—Si—Al—P—Al
        |  |  |  |  |  |  |  |  |
        P—Al—P—Al—P—Al—P—Al—P
```

Fig. 2   Various possible environments of Si in a planar configuration.
a :   isolated Si
b :   Si island with 5 Si(3 Al, 1 Si), 2 Si (2 Al, 2 Si), 1 Si (1 Al, 3 Si),
       3 Si (4 Si)
From ref. (23).

first neighbours) and 3 (siliceous islands) (9-11). Siliceous islands with Si-Al-Si connections may also be formed (10).

As will be seen further (parts 3.2 and 5.6) the location of Si may depend not only on synthesis conditions but also on heat treatments (17, 18). One has to be very careful in correlating properties measured after the SAPO activation with NMR results given by the as synthesized materials.

In order to characterize the acidic properties both the number of protons and their strength have to be considered.

3.1   NUMBER OF CHARGES

Assuming that the replacement of P^{5+} by Si^{4+} creates one negative charge while that of Al^{3+} by Si^{4+} gives one positive charge in the framework, the average charge is estimated from the formula (Six Aly Pz) O_2 by (z-y) (1).

TABLE 2. Calculation of framework charge on SAPO's (Six Aly Pz)O_2 for some idealized characteristic compositions[a]

Six	Aly	Pz	Charge[b]	Mecha-nism	Information
0.10	0.40	0.50	+ 0.10	(1)	positively charged framework. Si-O-P unlikely : extraneous P (c)
0.15	0.50	0.35	- 0.15	(2)	Si for P only (d)
0.14	0.45	0.41	- 0.04	(2) (3)	Si for P and Si for (Al + P) Si islands (e)

(a) from reference (3)
(b) (z-y)
(c) figure 1b
(d) figure 1a
(e) principle of substitution in figure 1d. Due to the formation of Si rich islands the charge is less than z-y

The results of table 2 may be obtained for some potential characteristic compositions. It was early recognized that due to the presence of Si rich islands the number of charges to be neutralized is less than the Si content suggests (9-11, 19). Nevertheless the approach described below shed some light on the formation of acidity in SAPO's. Table 2 describes three cases which may be expected. The first line corresponding to mechanism (1) shows a positive framework charge. This has not been observed experimetally. This means that there is very likely an excess of P which is titrated but not really involved in the SAPO structure. The second line describes a case often observed. Half of the atoms of the hypothetical AlPO$_4$ are either Al or P. In mechanism (2) Si replaces only P leaving the Al level (0.5) untouched. It is represented in figure 1a and is referred to as "isolated Si" where all the first neighbours are Al and the second ones P. It gives in ^{29}Si MAS NMR a single peak in the range -89, -92 ppm (5, 8, 9, 12, 14, 15, 20) where the chemical shift is slightly depending on the SAPO structure. In the third line in table 2 both Al and P are replaced by Si. This is described in figure 1d where both mechanism (2) and (3) are involved. This gives rises to Si islands characterized by [29] Si MAS NMR peaks in the range -110, -112 ppm (Si-O-Si bonds). In addition lines corresponding to Si located at the border of the islands, i. e. with different Si or Al neighbours, are observed (8, 10, 12, 14, 16). The existence well proved of siliceous islands has a consequence on the true number of charges. The hypothesis that one negative charge is created at each SiO_4 tetrahedra is no longer valid. Figure 1d shows that the Si atom at the center of the island does not share any Si-O-Al bond. The corresponding SiO_4 tetrahedra does not bear any negative charge to be neutralized by a proton. In addition the charge at each Si at the border of the island is less than one, depending on the number of Al neighbours. A detailed calculation showed that the average number of proton per Si atom decreases from 1 to 0.44 as the size of the island increases (5 to 14 Si in the island) (14). The low value of -0.04 in table 2 line 3 is then due to the presence of Si islands. Generally speaking, an increase in the Si level in SAPO's favors the mechanism (3) and as consequence the formation of islands. The number of protons in the framework becomes then lower than the atomic Si content (11, 14, 16, 19, 21-24). An absolute decrease may even be observed (14, 24).

A practical interest of the approach described in table 2 is a rapid check of the consistency of the results of the elemental analysis and of the potential protonic acidity. For atomic fraction smaller for Al than for P (line 1 table 2) one should be suspicious as to the purity of the SAPO phase. If both Al and P values are less than 0.5 the presence of siliceous islands is likely. In this case the true charge, i. e. number of protons, can not reach the upper limit calculated assuming one proton per Si (line 3 table 2).

3.2 PROTONIC STRENGTH

The "isolated" Si atoms with four Al as first neighbours, noted Si (4Al) (figure 1) are recognized for a long time (1, 25, 26) to generate acidity of medium strength.

Theoretical CNDO/2 calculations show that in four rings units the acid strength of protons is less in SAPO's than in HY or HZSM-5 (27).

In Si-Al zeolites it is well accepted that the strength of the proton in the Si-OH-Al species increases as the number of Si, second neighbours of Al, increases. The order of increasing acid strength is then for the various environments in faujasite :
Al (4Si) (9Al) < Al (4Si) (8Al, 1 Si) <... < Al (4Si) (1Al, 8Si) < Al (4Si) (9 Si).
This is quite in agreement with the weak acidity of X or A type zeolite and the strong one of siliceous zeolites. This is also in line with theoretical calculations. As said above Si plays in SAPO's the role of Al in Si-Al zeolites with regards to acidity formation. One may then strongly expect that the acid strength should vary with the environment of Si atoms (14). A difference with Si-Al zeolites is that Si-O-Si bonds may exist while Al-O-Al ones are forbidden. It follows that the influence of the neighbours of Si, on the acid strength created, involves the first neighbours and the following sequence should exist as to the increase in acid strength (28) :
Si (4 Al) < Si (3 Al) < Si (2 Al) < Si (1 Al)
Since Si-O-P bonds are unlikely, the other close neighbours different from Al are Si. This means that at the border of islands like in figure 2 (23), Si atoms will generate different acid strengths. The scheme figure 2 shows that in the island there are Si (3 Al, 1 Si), Si (2 Al, 2 Si), Si (1 Al, 3 Si) and Si (4 Si) species. A quantitative calculation of the number of each type of environment can be carried out for each known structure (14). The exact number depends on the size of the island and on the topology of the structure. For instance in SAPO-37 the order of occurrence of Si environment for islands containing from 5 to 14 Si atoms is :

Si (3 Al) > Si (0Al) > Si (2 Al) > Si (1 Al).

This suggests a distribution of acid strengths with a small number of sites with a strong acidity (Si (1 Al)) (28). Such an influence of the first neighbours was also noted in SAPO-5 (29).

The second neighbours affect the acid strength in Si-Al zeolites. They have also an influence in SAPO's. It was shown that SAPO-37 with only Si (4Al) (9P) environments i. e. no Si island and only P as second neighbours has a stronger acidity than X zeolites with Si (4 Al) (9 Si) entities (28). Since both materials have the same structure i. e. the same topology only the chemical influence of P or Si in the second shell is important. This may be related to the higher electronegativity of P.

Any parameter which may change the Si location and environment, like synthesis conditions (8,11) or heat treatments (17, 18) will influence the acid site distribution even for a same overall framework composition.

Some examples of studies of protonic acid site number and strength are detailed further for typical SAPO's.

4   Expected Lewis acidity

In Si-Al zeolites the Lewis acidity is related to the hexacoordinated Al species (Al^{VI}) present as an extraframework aluminic phase. In SAPO's after the removal of the template and of adsorbed water only the ^{27}Al MAS NMR peak of tetrahedral Al (Al^{IV}) is observed (5, 30, 31). The adsorption of water is known to generate Al^{VI} peak (5, 30, 31). Depending on the SAPO structure the effect is reversible in some cases with no decrease in crystallinity. This may suggest an interaction of H_2O with some AlO_4 tetrahedra as in Lewis acid-base reaction. Other examples with different bases will be given for specific structures.

5   Measurement of the acidity

5.1   OBTENTION OF THE ACIDIC SAPO'S

SAPO's are always synthesized using organic templates. The organic molecules are very often trapped in small cages with apertures too small to allow their departure. It follows that despite their high cost, the templates are usually destroyed by heating between 600 to 850 K. An extraction with methanolic HCl was described to eliminate triethylamine from $AlPO_4$-5 (32).

By contrast with Si-Al zeolites, it is clear that some SAPO's are very sensitive to water at room temperature i. e. at ambient atmosphere while they resist to steaming and have very stable structures up to around 1200 K. For instance SAPO-11 changes its space group from the dehydrated to the hydrated forms (33), water at room temperature may reversibly open Si-OH-Al bonds in SAPO-34 (34) or the structure of SAPO-37 is irreversibly damaged at ambient conditions (35). It is then of major importance to check carefully the hydration level and the sensivity to water of the SAPO considered before any further study of the properties. It turns out that among the usual SAPO's studied, SAPO-5 is the most stable. This is probably the reason why the number of publications is the most important for this material, the other ones giving possibly irreproducible results if they have been contacted with water.

TABLE 3. Structure characteristics of four SAPO's[a]

SAPO	Structure type	ϕ pore (nm)	Aperture [b]	Dimensionality[c]
SAPO-5	AFI	0.73	12R	1
SAPO-11	AEL	0.39 x 0.63	10R	1
SAPO-34	CHA	0.38 x 0.38	8R	3
SAPO-37	FAU	0.74	12R	3

(a) from reference (62)
(b) number of tetrahedra in the ring R
(c) pores in one or three directions of space

5.2 SAPO'S MOST COMMONLY STUDIED

The table 3 reports the structural and textural characteristics fo SAPO-5, SAPO-11, SAPO-34 and SAPO-37 and figure 3 gives a schematic view of their structure. Two of these materials show very interesting potential applications in catalysis (25, 26). SAPO-11 is active for the skeletal isomerization of light olefins (like butenes) with little coking (25, 26). The transformation of methanol to ethylene is obtained very selectively on SAPO-34 (26) particularly with the Ni-SAPO-34 form (36, 37). SAPO-5 has a very open structure which may be of interest for adsorption or catalysis of rather large molecules. A same structure, faujasite, in X, Y and SAPO-37 allows a study to be done on the influence of the incorporation of P in a framework on its stability, acidity, catalytic properties.

5.3 SAPO-5 (STRUCTURE AlPO$_4$-5)

The template-free SAPO-5 is very stable in ambient atmosphere at room temperature or in the presence of water vapour up to around 1200 K (35).

Various techniques are used to characterize the OH groups in SAPO's. For instance figure 4A represents the infrared spectra of HSAPO-5 after calcination at 623 K to remove the template (6) and figure 4B the ^{1}H MAS NMR spectra of H-SAPO-5 (38).

The hydroxyls bands observed in infrared spectra have often been reported (3, 6-9, 19, 22, 38-42). Depending very likely on the Si content and on the experimental conditions changing from one laboratory to the other, the bands are observed at slightly different wavenumbers for silanols (3740 - 3746 cm^{-1}), P-OH (3672-3680 cm^{-1}) and bridging Si-OH-Al (3618-3630 and 3520-3528 cm^{-1}) (table 4).

The acidity of the OH groups has been studied in various ways using probe molecules like strong bases (ammonia (7, 21, 22, 24, 40, 43, 44) or pyridine (6, 19, 22)) or weak bases forming H-complexes. In this last case the shift of the hydroxyl band is followed by infrared (7-9, 39,42) or ^{1}H MAS NMR (22). The sorbates are benzene (7-9), ethylene (7, 42), CO (39) or CD$_3$ CN (42). The shift observed are reported in table 5. A general agreement is that the bridging OH vibrating at around 3625 cm^{-1} are accessible to all the above molecules while the ones giving the 3525 cm^{-1} band are accessible only to the smaller ones. A second conclusion from the results is the following order of acid strengths :

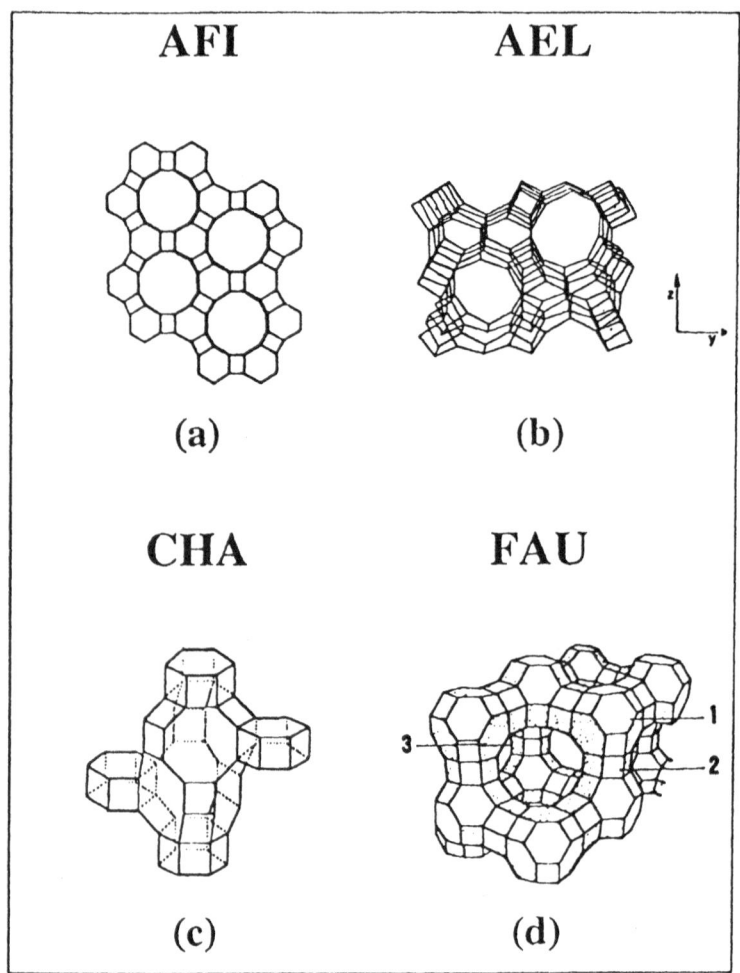

Fig. 3   Structure of a : SAPO-5, b : SAPO-11, c : SAPO-34, d : SAPO-37 (1 : sodalite cage, 2 : hexagonal prism, 3 : supercage)

HZSM-5 > H SAPO-5 > HY

The adsorption of acetonitrile suggests that in addition to interaction with bridging OH groups, some CD_3CN molecules interact with AlO_4 tetrahedra giving rise to new coordination (42). This could be considered as a Lewis acid-base interaction.

SAPO-5 contains, at high Si levels, a large amount of Si in siliceous patches, as seen by ^{29}Si MAS NMR (8). One may expect (see part 3.2) that the number of protons is lower than that of Si atoms. This is experimentally observed (21).

5.4   SAPO-11 (STRUCTURE AlPO$_4$-11)

No drastic change in stability of the template free material is reported. A modification of the crystalline space group is observed in the rehydrated form (33) and a reversible change from tetrahedral Al (Al^{IV}) to octahedral (Al^{VI}) is seen upon hydration-dehydration cycles(30). The ^{29}Si MAS NMR spectra show ill defined peaks (13) suggesting a very heterogeneous incorporation of Si in the framework.

The hydroxyl groups are vibrating in infrared at 3740-3745 cm^{-1} (silanols), 3670-3675 cm^{-1} (P-OH) and 3620 and 3520-3530 cm^{-1} (bridging OH) (9, 24, 41, 45, 46) (table 4). Additionnal bands are sometimes observed (47). The acidity checked by NH_3 TPD (45-47), infrared study of pyridine (21, 46, 47) and of benzene (9) adsorptions (table 5) shows weak or medium strength and a small number of sites.

TABLE 4. Bridging hydroxyl wavenumbers (cm^{-1})(a)

SAPO-n	HF	LF
SAPO-5	3618-3630	3520-3528
SAPO-11	3620	3520-3530
SAPO-34	3625	3600
SAPO-37	3640	3572-3575

(a) see references in the text parts 5.3 to 5.6

5.5   SAPO-34 (STRUCTURE CHABASITE)

It was recently shown that the template-free SAPO-34 is very sensitive to water at room temperature (5, 30, 34, 48, 49) but not above around 400 K (34). The most striking feature is that the effect is reversible upon water desorption (34, 49). It was shown (34) by ^{29}Si MAS NMR that the reversible partial loss of crystallinity is due to the opening and closing of Si-OH-Al bonds as schematized below in the scheme (a). Above around 400 K SAPO-34 is not damaged by water. It is stable up to at least 1173 K.

The hydroxyl groups have been studied by infrared (3, 22, 41, 45, 50) and ^1H MAS NMR (22). Figure 5 gives the infrared spectrum of H SAPO-34 (3). The weak bands at 3745 and 3675 cm^{-1} are assigned to silanol and POH respectively. The two bands usually observed at 3625-3636 and 3600-3605 cm^{-1} are assigned to bridging hydroxyls (3, 22, 50) (table 4). The low frequency one is due to an interaction of the OH groups with framework oxygen in a small cage (22, 50) as this was already proposed in Si-Al zeolites

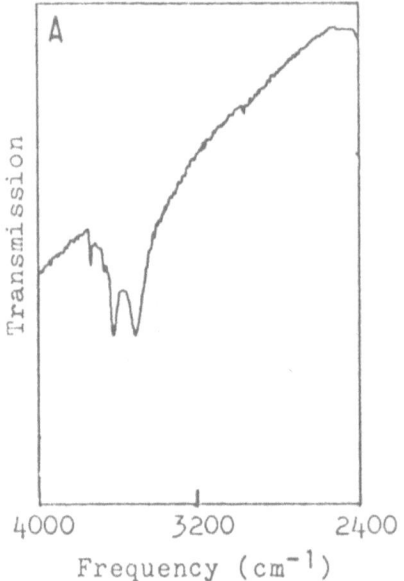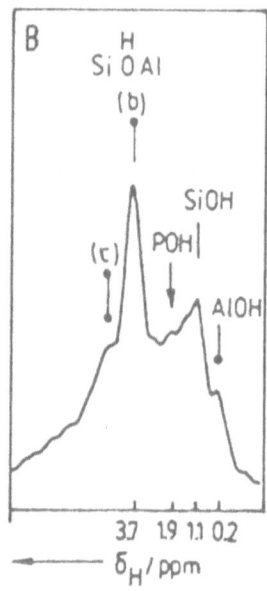

Fig. 4   Hydroxyls groups in SAPO-5. A : Infrared spectrum from ref. (6). B : ^{1}H MAS NMR spectrum from ref. (38) (b and c are bridging hydroxyls).

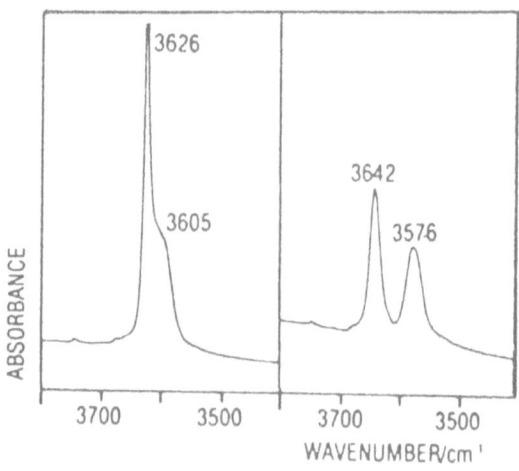

Fig. 5   Infrared spectra in the hydroxyl range for HSAPO-34 (left) and HSAPO-37 (right). From ref. (3).

(51). Due to the small aperture of cages and channels in SAPO-34 (table 3) only a restricted number of probe molecules can be used to check the acidity. Results obtained with NH_3 (22, 45, 52), CH_3I, CD_3CN, N_2O, C_2H_4 (50) and methanol (53) have been described. The adsorption of ammonia removes the bridging OH vibrating at 3620 and 3600 cm^{-1} (45, 54) and the ^1H MAS NMR peak at 3.8 ppm. TPD experiments (45) show two NH_3 desorption peaks at temperatures indicating an acid strength slightly weaker than in HZSM-5. The shift of hydroxyl groups observed upon the adsorption of ethylene and acetonitrile are reported in table 5. They are comparable to those observed with SAPO-5. All these results suggest that HSAPO-34 has an acid strength between those of HZSM-5 and HY.

Compared to other SAPO's presented here (-5, -11, -37) SAPO-34 has few siliceous islands formed during the synthesis as seen from ^{29}Si MAS NMR (5, 34). In the absence of such Si patches one may expect to find a good correlation between the number of Si atoms and that of protons titrated by NH_3. This in fact observed using ^1H MAS NMR (22).

(a)

5.6   SAPO-37 (STRUCTURE FAUJASITE)

SAPO-37, free of template, is damaged when contacted with water at room temperature (20, 30, 31, 35, 49, 55). This irreversible process does not occur above around 350 K (35). A surprising feature is nevertheless a drastic change in the distribution of Al, P and Si atoms at high temperatures between around 800 and 1150 K (17, 18). A solid state transformation, due to the mobility of framework atoms, creates different chemical nanodomains still having the faujasite structure. Besides a pure SAPO-37 phase with isolated Si (figure 1 and 2), Si islands and Si-Al faujasite phases are formed. This implies that some domains of AlPO-37 should exist. These transformations modify greatly the environnement of Si atoms creating Si-OH-Al species at the border of the new siliceous patches, which, as said in part 3.2 have strong acidity.

The hydroxyl groups of HSAPO-37 have been studied mainly by infrared spectroscopy (3, 15, 16, 19, 20, 56).The silanol groups are observed at 3740-3746 cm^{-1}, the P-OH at 3678-3680 cm^{-1} and the bridging Si-OH-Al near 3640 and 3575 cm^{-1} (table 4). As in Si-Al faujasites the high frequency band is assigned to OH groups in the supercage and the low frequency band to hydroxyls in the sodalite cage (19).

The acidity has been studied using ammonia (16, 23, 57) and pyridine TPD (23) and infrared characterization of adsorbed pyridine (19, 20, 58) and benzene (19, 59) (table5). A main feature is the generation of very strong acid sites upon heating above 1073 K (19,

59). The isolated Si (4 Al) species giving a medium acid strength disappear to form Si islands. This increases the number of strong acid sites linked to Si (n Al) with n < 4. After treatment at 1173 K SAPO-37 has sites as strong as those of the ultrastable LZY-82 i. e. quite stronger than in HY (59). This unusual formation of strong acid sites at high temperature explains the simultaneous rise in catalytic cracking activity (60, 61). The heat treatment appears to optimize the acidic and catalytic properties of SAPO-37. It is noteworthy that in Si-Al faujasites the optimization of the catalyst results from Al removal from the framework creating macropores and defects (extraframework phases). In SAPO-37 it arises from the demixing of phases keeping the same faujasite structure and in the formation of Si islands in the framework. A second characteristic property is the appearance of Lewis acidity as the pretreatment temperature rises (19, 23, 60). It is not related to an extraframework phase but rather to the adsorption of basic molecules on distorted AlO_4 tetrahedra (23).

The synthesis of SAPO-37 with high Si content gives rise to Si islands very easily (11, 12, 14, 16, 58). It follows that the number of acid sites tends to decrease as the Si content increases as explained in part 3.1. This effect is verified experimentally (14, 16, 23).

TABLE 5. Shifts Δv_{OH} (cm^{-1}) of acidic hydroxyls upon the interaction with various molecules[a]

Solid	Benzene	C_2H_4	CO	C D_3 CN	Ref
SAPO-5			254-280		39
	325	330		730	7
	250-410				8,9
SAPO-11	120-320				9
SAPO-34		300-325		700-725	50
SAPO-37	320-325				19, 59
HZM-5			310		39
	350	360			7
HY	280-320	300			7
	322				19
			278		39
				700-950	63
SiO$_2$	140		90	295	see 63, 64

(a) interaction at RT (benzene, CD$_3$CN), 300 K (C$_2$H$_4$), 77 K (CO)

6   General characteristics common to the SAPO's

Some general features arise from the studies of Si incorporation : stability at ambient atmosphere, hydroxyl groups and acidity of the various SAPO's considered.

^{29}Si MAS NMR spectra published for samples at increasing Si contents tend to show that SAPO-34 does not form easily siliceous islands while SAPO-5 (8) and even more SAPO-11 (13) contain such patches in many cases.

The stability at ambient atmosphere (30, 33-35, 48, 55) suggests the order of decreasing stability in the presence of water vapour :

SAPO-5 > SAPO-11 > SAPO-34 > SAPO-37

All the SAPO's studied show P-OH groups and two types of bridging Si-OH-Al hydroxyls. The table 4 sums up the range of values observed. The low frequency bands are usually assigned to hydroxyls interacting with framework oxygen in small cages as this was observed in Si-Al zeolites (51). The wavenumbers do not seem to vary significantly, for a given structure, with the Si content (19). This is in contrast with the case of Si-Al zeolites where the OH wavenumbers depend on the H$^+$ level (i.e. Al and metal cation content).

With regards to acidity a first point is that the number of protons cannot be deduced simply from the Si content. The presence of siliceous patches decreases the mean number of H$^+$ per Si. Secondly, the formation of these islands increases the overall acid strength by creating Si (nAl) species with n < 4 at the interface with the SAPO phase.

The acidity studied by NH$_3$ TPD shows broad desorption peaks. The mean acid strength is lower than that of strongly acidic zeolites like HZSM-5. This is in agreement with the results of the shifts of hydroxyl wavenumbers upon adsorption of various compounds (table 5). From the various experiments reported in this paper, SAPO-11 appears to be the material with the weakest acid strength. As discussed in parts 3.1 and 3.2, the acidity (amount and strength) depends not only on the structure type considered but also on the mode of substitution of Si in a hypothetical AlPO$_4$ framework. More work is needed to be able to correlate exactly the acidic properties with the Si location in the lattice.

Some general comparison for the SAPO's may be carried out using catalytic activity in acidic reactions as probes. It is proposed from n-decane hydrocracking that the active sites may be located at three places : SAPO phase (a), siliceous domains (b) and interface between both (c). The activity arises from (a) in SAPO-5, (b) and (c) in SAPO-11 and (a), (b), (c) in SAPO-37 (10). N-octane cracking carried out on SAPO-5 and SAPO-37 shows that when strong acid sites are required, only the centers at the border of Si islands (c case above) are involved (60, 61).

7   Conclusion

In the field of acidic molecular sieves, SAPO's are new (1). What is known on Si-Al zeolites cannot just be extrapolated to these materials. The presence of P introduce not only a different acid strength due to the higher electronegativity of this element but changes in several properties, thermal and hydrothermal stability, hydrolysis of T-O-T bonds, possible mobility of atoms at high temperature giving new chemical phases with the same crystalline structure. The easy formation of siliceous rich islands in some SAPO's during the synthesis cannot be fully explained at the present time. It induces specific acidic properties with regards to the amount and strength of acid sites.

The research in the field of SAPO's brings information which open new areas for the search not only of catalysts, and absorbents but also for materials in view of applications in nanochemistry domains (non linear optics, molecular electronics...).

References

1   B. M. Lok, C. A. Messina, R. L. Patton, R. T. Gajek, T. R. Cannan and E. M. Flanigen, J. Am. Chem. Soc. 106 (1984) 6092.
2   E. M. Flanigen, B. M. Lok, R. L. Patton and S. T. Wilson, in Y. Murakami, A. Iijima and J. W Ward (Editors), Proceed. 7th Inter. Zeol. Conf., Kodansha, Tokyo and Elsevier, Amsterdam, 1986 pp 103.
3   E. M. Flanigen, R. L. Patton and S. T. Wilson, in P. J. Grobet, W. J. Mortier, E. F. Vansant and G. Schulz-Ekloff (Editors), "Innovation in Zeolite Materials Science", Stud. Surf. Sci. Catal. 37 (1988) 13.
4   S. L. Suib, A. M. Winiecki and A. Kostapapas, Langmuir 3 (1987) 483.
5   C. S. Blackwell and R. L. Patton, J. Phys. Chem. 92 (1988) 3965.
6   Qinhua Xu, Aizhen Yan, Shulin Bao and Kaijun Xu in Y. Murakami, A. Iijima and J. Ward (Editors), Proceed. 7th Intern. Zeol. Conf., Kodansha, Tokyo and Elsevier, Amsterdam, 1986 pp 835.
7   S. G. Hedge, P. Ratnasamy, L. M. Kustov and V. B. Kazansky, Zeolites 8 (1988) 137.
8   J. A. Martens, M. Mertens, P. J. Grobet and P. A. Jacobs, in P. J. Grobet, W. J. Mortier, E. F. Vansant and G. Schulz-Ekloff (Editors), "Innovation in Zeolite Materials Science", Stud. Surf. Sci. Catal. 37 (1988) 97.
9   M. Mertens, J. A. Martens, P. J. Grobet and P. A. Jacobs, in D. Barthomeuf, E. G. Derouane, W. Hoelderich (Editors), "Guidelines for the Mastering of Properties of Molecular Sieves", NATO ASI Series, Plenum Press, New York, Series B : Physics vol 221 (1990) 1.
10   J. A. Martens, C. Janssens, P. J.. Grobet, H. K. Beyer and P. A. Jacobs, in P. A. Jacobs and R. A. Van Santen (Editors), "Zeolites : Facts, Figures, Future", Stud. Surf. Sci. Catal. 49 A (1989) 215.
11   J. A. Martens, P. J. Grobet and P. A. Jacobs, J. Catal. 126 (1990) 299.
12   L. Maistriau, N. Dumont, J. B. Nagy, Z. Gabelica and E. G. Derouane, Zeolites 10 (1990) 243.
13   E. Jahn, D. Müller and K. Becker, Zeolites 10 (1990) 151.
14   P. P. Man, M. Briend, M. J. Peltre, A. Lamy, P. Beaunier and D. Barthomeuf, Zeolites 11 (1991) 563.
15   A. F. Ojo, J. Dwyer, J. Dewing, K. Karim, J. Chem. Soc. Faraday Trans. 87 (1991) 2679.
16   M. A. Makarova, A. F. Ojo, K. M. Al-Ghefali, J. Dwyer, in R. Von Ballmoos, J. B. Higgins and M. M. J. Treacy (Editors), Proceed 9th Intern. Zeolite Conf., Butterworth-Heinemann, Boston, 1993, II, pp 259.
17   M. J. Peltre, P. P. Man, M. Briend, M. Derewinski and D. Barthomeuf, Catal. Lett. 16 (1992) 123.
18   M. Derewinski, M. J. Peltre, P. P. Man, M. Briend and D. Barthomeuf, J. Chem. Soc. Faraday Trans. 89 (1993) 1823.
19   S. Dzwigaj, M. Briend, A. Shikholeslami, M. J. Peltre and D. Barthomeuf, Zeolites 10 (1990) 157.

20   L. Sierra de Saldarriaga, C. Saldarriaga and M. E. Davis, J. Am. Chem. Soc. 109 (1987) 2686.

21   N. J. Tapp, N. B. Milestone and D. M. Bibby, in P. J. Grobet, W. J. Mortier, E. F. Vansant and G. Schulz-Ekloff (Editors), "Innovation in Zeolite Materials Science", Stud. Surf. Sci. Catal. 37 (1988) 393.

22   B. Zibrowius, E. Löffler, M. Hunger, Zeolites 12 (1992) 167.

23   M. Briend, M. J. Peltre, A. Lamy, P. P. Man and D. Barthomeuf, J. Catal. 138 (1992) 90.

24   J. Das, V. V. Satyanaryana, D. K. Chakrabarty, S. N. Piramanayagam and S. N. Shringi, J. Chem. Soc. Faraday Trans. 88 (1992) 3255.

25   J. A. Rabo, R. J. Pellet, P. K. Coughlin and E. S. Shamshoum, in H. G. Karge and J. Weitkamp (Editors), "Zeolites as Catalysts, Sorbents and Detergent Builders", Stud. Surf. Sci. Catal. 46 (1989) 1.

26   J. A Rabo, in E. G. Derouane, F. Lemos, C. Naccache and F. R. Ribeiro (Editors) "Zeolite Microporous Solids : Synthesis, Structure and Reactivity", NATO ASI Series, Kluwer, Dordrecht, Series C : vol 352 (1992) 531.

27   R. Carson, E. M. Cooke, J. Dwyer, A. Hinchliffe and P. J. O'Malley in H. G. Karge and J. Weitkamp (Editors), "Zeolites as Catalysts, Sorbents and Detergent Builders", Stud. Surf. Sci. Catal. 46 (1989) 39.

28   M. Briend and D. Barthomeuf, in R. Von Ballmoos, J. B. Higgins and M. M. J. Treacy (Editors), Proceed 9th Intern. Zeol. Conf., Butterworth-Heinemann, Boston 1993, I, pp 635.

29   A. F. Ojo, J. Dwyer, J. Dewing, P. J. O'Malley and A. Nabhan, J. Chem. Soc. Faraday Trans. 88 (1992) 105.

30   M. Goepper, F. Guth, L. Delmotte, J. L. Guth and H. Kessler, in P. A. Jacobs and R. A. Van Santen "Zeolites : Facts, Figures, Future", Stud. Surf. Sci. Catal. 49B (1989) 857.

31   M. J. Peltre, M. Briend, A. Lamy, D. Barthomeuf, F. Taulelle, J. Chem. Soc. Faraday Trans. 86 (1990) 3823.

32   C. R. Theocharis and M. R. Gelsthorpe, in K. K. Unger, J. Rouquerol, K. S. W. Sing and H. Krai (Editors), "Characterization of porous solids", Stud. Surf. Sci. Catal. 39 (1988) 541.

33   R. Khouzami, G. Coudurier, F. Lefebvre, J. C. Vedrine and B. F. Mentzen, Zeolites 10 (1990) 183.

34   R. Vomscheid, M. Briend, M. J. Peltre, P. Massiani, P. P. Man and D. Barthomeuf, J. Chem. Soc., Chemical Comm. 6 (1993) 544.

35   M. Briend, A. Shikholeslami, M. J. Peltre, D. Delafosse and D. Barthomeuf, J. Chem. Soc. Dalton Trans. (1989) 1361.

36   T. Inui, S. Phatanasri and H. Matsuda, J. Chem. Soc., Chem. Comm. (1990) 205.

37   J. M. Thomas, Y. Xu, C. R. A. Catlow and J. W. Couves, Chem. Mater. 3 (1991) 667.

38   H. Pfeifer, D. Freude, J. Kärger, in G. Öhlmann (Editors), "Catalysis and Adsorption by Zeolites", Stud. Surf. Sci. Catal. (1991) 89.

39   L. Kubelkova, S. Beran, J. A. Lercher, Zeolites 9 (1989) 539.

40   Ch. Minchev, V. Kanazirev, V. Mavrodinova, V. Penchev and H. Lechert, in H. G. Karge and J. Weitkamp (Editors), "Zeolites as Catalysts, Sorbents and Detergent Builders", Stud. Surf. Sci. Catal. 46 (1989) 29.

41   E. Löffler, C. Peuker, G. Finger, I. Girnus, E. Jahn, H. L. Zubowa, Z. Chem. 30 (1990) 255.

390

42   L. M. Kustov, S. A. Zubkov, V. B. Kazansky and L. A. Bondar, in P. A. Jacobs (Editors), "Zeolites Chemistry and Catalysis", Stud. Surf. Sci. Catal. (1991) 303.

43   K. J. Chao and L. J. Len, in H. G. Karge and. J. Weitkamp (Editors), "Zeolites as Catalysts, Sorbents and Detergent Builders", Stud. Surf. Sci. Catal. 46 (1989) 19.

44   K. J. Chao, S. P. Sheu, S. H. Chen, J. C. Lin and J. Lievens, in M. L. Ocelli and H. Robson, (Editors) "Synthesis of Microporous Materials, Molecular Sieves", Van Nostrand Reinhold, New York, 1 (1992) 317.

45   K. H. Schnabel, R. Frike, I. Girnus, E. Jahn, E. Löffler, BV. Parlitz and C. Peuker, J. Chem. Soc. Faraday Trans. 87 (1991) 3569.

46   Liu Yang, Yan Aizhen, Xu Qinhua, Appl. Catal. 67 (1991) 169.

47   R. Khouzami, G. Coudurier, B. F. Mentzen and J. C. Vedrine, in P. J. Grobet, W. J. Mortier, E. F. Vansant and G. Schultz-Ekloff (Editors), "Innovation in Zeolite Materials Science", Stud. Surf. Sci. Catal. 37 (1988) 355.

48   R. Vomscheid, M. Briend, J. P. Souron and D. Barthomeuf, in "J. B. Higgins, R. Von Ballmoos and M. M. Treacy (Editors), Abstracts 9th Intern. Zeol. Conf., Butterworth-Heinemann, Montreal (1992) RP 110.

49   Chr. Minchev, Ya. Nienska, V. Valtchev, R. Minkov, T. Tsoncheva, V. Penchev, H. Lechert and M. Hess, in J. B. Higgins, R. Von Ballmoos and M. M. J. Treacy (Editors), Abstracts 9th Intern. Zeol. Conf., Butterworth-Heinemann, Montreal, (1992) RP 104.

50   S. A. Zubkov, L. M. Kustov, V. B. Kazansky, I. Girnus and R. Fricke, J. Chem. Soc. Faraday Trans. 87 (1991) 897.

51   P. A. Jacobs and W. J. Mortier, Zeolites 2 (1982) 226.

52   H. S. Oh and W. Y. Lee, Korean J. Chem. Eng. 9 (1992) 41.

53   S. Hocevar and J. Levec, J. Catal. 135 (1992) 518.

54   R. Vomscheid, M. Briend and D. Barthomeuf, to be published.

55   G. C. Edwards, J. P. Gilson, V. Mc Daniel, U. S. Patent 4 681 864 (1987).

56   A. Corma, V. Fornes, M. J. Franco, F. A. Mocholi and J. Perez-Pariente, in M. L. Occelli (Editor), "Fluid Catalytic Cracking II. Concepts and Design", ACS Symposium Series, ACS Washington 452 (1991) 79

57   I. I. Ivanova, N. Dumont, Z. Gabelica, J. B. Nagy, E. G. Derouane, F. Ghigny, O. E. Ivashkina, E. V. Dmitruk, A. Smirnov and B. V. Romanovsky , in R. Von Balmoos, J. B. Higgins and M. M. J. Treacy (Editors), Proceed 9th Intern. Zeol. Conf. Butterworth-Heinemann, Boston, 1993, II, pp 449.

58   A. Corma, V. Fornes, M. J. Franco, F. V. Melo, J. Perez-Pariente and E. Sastre, in R. Von Ballmoos, J. B. Higgins and M. M. J. Treacy (Editors), Proceed 9th Intern. Zeol. Conf., Butterworth-Heinemann, Boston, 1993, II, pp 343.

59   Bao Lian Su and D. Barthomeuf, J. Catal., 139 (1993) 81.

60   Bao Lian Su, A. Lamy, S. Dzwigaj, M. Briend and D. Barthomeuf, Appl. Catal. 75 (1991) 311

61   M. Briend, M. Derewinski, A. Lamy and D. Barthomeuf, in L. Guczi, F. Solymosi, P. Tétényi (Editors), "New Frontiers in Catalysis", 10th Intern. Cong. Catal., Akadémiai Kiado, Budapest A (1993) 409.

62   W. M. Meier, D. H. Olson, Atlas of Zeolite Structure Types, Butterworth-Heinemann, 1992.

63   N. Echoufi, Thesis Lyon 1992, N. Echoufi, P. Gelin to be published.

64   D. Barthomeuf, in E. G. Derouane, F. Lemos, C. Naccache, F. R. Ribeiro (Editors), "Zeolite Microporous Solids : Synthesis, Structure and Reactivity", NATO ASI Series, Kluwer, Dordrecht, Series C : Vol 352 (1992) 193.

DETERMINATION OF SILICON COORDINATION SPHERES IN SAPO-37 MOLECULAR SIEVE WITH MS-EXCEL

Pascal P. MAN and Jacques FRAISSARD
Laboratoire de Chimie des Surfaces, CNRS URA 1428,
Université Pierre et Marie Curie, 4 Place Jussieu, Tour 55,
75252 Paris Cedex 05, France

ABSTRACT. A program which determines first and second silicon coordination spheres in SAPO-37 is presented; it is written in the macro programming language of MS EXCEL. The algorithm is based on the representation of the crystallographic structure as adjacency lists which are subsequently considered as a data-base. The program takes into account *two constraints*: (1) the validity of Loewenstein's rule, (2) the Si-O-P linkage is not allowed, and *one assumption*: Si atoms form small islands.

1. Introduction

More and more interest is focused on SAPO-37 a silicoaluminophosphate molecular sieve [1-5]. It shows unusual properties attributed to its composition which depends on the way in which it is chemically prepared. It involves a three-dimensional arrangement of SiO_4, PO_4, and AlO_4 tetrahedra connected through shared oxygen atoms and has the faujasite structure. A SAPO-37 can be considered to originate from Si substitution into a hypothetical aluminophosphate AlPO-37 of the same structure in which the Al and P atoms alternate, each surrounded by four oxygens to form $Al(OP)_4$ or $P(OAl)_4$ groups, respectively. Three mechanisms have been proposed: substitution of Si for Al (mechanism 1), substitution of Si for P (mechanism 2), or substitution of 2 Si for one Al and one P (mechanism 3).

Solid-state NMR has been extensively used to determine the various first Si coordination spheres in aluminosilicate zeolites where Si can be surrounded by 0, 1, 2, 3 or 4 Al [6]. Si spectra exhibit a splitting into 5 distinct components corresponding to the number of Al atoms in the four first neighbour sites. Recently it has been shown that the second coordination sphere has also a great effect on the Si chemical shifts [7]. Some results concerning SAPO-37 are well accepted: (a) Loewenstein's rule, established for microporous aluminosilicate materials, is also valid for silicoaluminophosphates; there is no Al-O-Al bond. (b) P is surrounded by four Al atoms (the ^{31}P NMR spectrum is a single line); this means that the Si-O-P linkage is not allowed. The results obtained so far show that mechanisms 2 and 3 are the most probable, while mechanism 1 is unlikely. However, the Si chemical shifts in SAPO-37 do not fall into the same range as those determined in microporous aluminosilicate materials. They can

J. Fraissard and L. Petrakis (eds.), Acidity and Basicity of Solids, 391–402.

be interpreted if we take into account the second coordination sphere, which may contain some P atoms, and if Si atoms form islands [3].

A detailed description of the origin of the protonic acidity in SAPO-37 is presented by D. Barthomeuf in a chapter devoted to SAPO's. The proton centers are the bridging hydroxyls in the bonds Si–O H–Al. As the result, the Si atoms are responsible for the creation of acidity. As far as the Si atoms are isolated, in other words, when the Si environment is Si(4Al)(9P), one negative charge is created at each SiO₄ tetrahedra. However, when Si islands are present, the charge at each Si atom at the border of the islands is less than one, depending on the number of Al neigbours. It is the propose of our program which povides, for a given size of the Si island, the different Si environment, and in particular at the border of a Si island.

Our program determines the various first and second Si coordination spheres in SAPO-37 by taking into account the two constraints (Loewenstein's rule and no Si-O-P linkage) and one assumption (Si atoms form small islands) using Microsoft EXCEL as the programming tool [8, 9]. The proposed method is, of course, applicable to other structures.

2. Principle

In Fig. 1 is represented part of the faujasite structure. T atoms (Al, Si or P) are located at the corners of the polygons with oxygen atoms approximately halfway between them. They are designated by a position number. For clarity, only some values are given. T atoms 1 to 4 constitute the first shell of T atom 0; atoms 5 to 13 the second shell; atoms 14 to 29 the third; atoms 30 to 54 the fourth

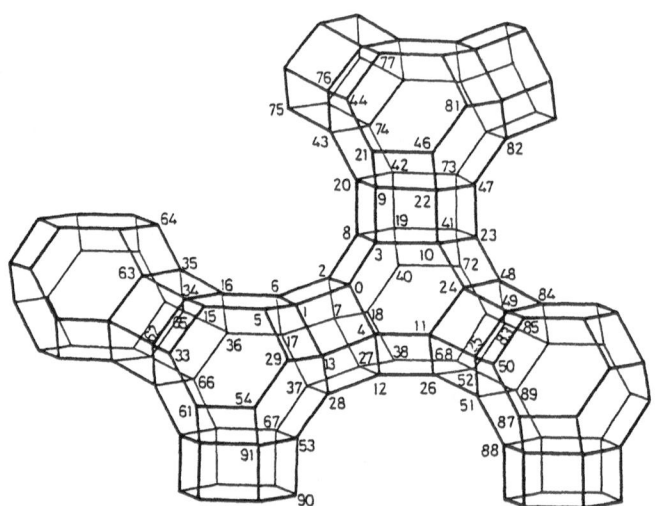

Fig. 1   Part of the faujasite structure and definition of the vertices. For clarity the double 6-membered rings in front of some sodalite cages are not shown.

and atoms 55 to 91 the fifth.

In graph theory terminology [10], polygon corners are vertices and T-O-T linkages are edges. To handle the problem of finding the coordination spheres, the connectivity of the faujasite structure (or any structure) is better represented in a computer by adjacency lists rather than by an adjacency matrix. For each vertex there is an adjacency list formed by a set of nodes. Each node has two fields: vertex and component. The vertex field contains the associated position number and the component field contains the atom located at the vertex (Al, Si or P). The head nodes are sequential, providing easy random access to the adjacency list for any particular vertex. The other nodes in a list represent the adjacent vertices.

Consider a vertex j whose first and second neighbour shells we want to determine: the head node of the (j + 1)th adjacency list contains its vertex number and the associated chemical component. The other nodes of this list correspond to the first neighbour vertices of j from which their adjacency lists can be found. This time the other nodes (except one which is the vertex j itself) of a first neighbour adjacency list correspond to the second neighbour shell of j. The node of this vertex j in all the first neighbour adjacency lists must be idendified in order to avoid it being counted as a second neighbour. Furthermore, among the second neighbour vertices contained in the first neighbour adjacency lists, some appear several times. It is necessary, therefore, to create a flag file which is used to indicate whether or not a second neighbour vertex has been counted.

The need to create adjacency lists and the random access to a specific vertex for substituting one chemical component by another led us to choose a spreadsheet software: Microsoft EXCEL. This tool presents means for searching for a specific item in a data-base. The method consists of creating the adjacency lists in a worksheet. The component fields are then filled with Al or P atoms corresponding to a hypothetical AlPO-37 where vertex 0 contains a P atom (This operation satisfies the first constraint: Loewenstein's rule. The location of a P atom in vertex 0 is suggested by the fact that mechanism 1 is unlikely). Afterwards these lists are transformed as a data-base. The P atom in vertex 0 and some first neighbours (Al atoms) are replaced by Si using search facilities of EXCEL (This way of introducing Si atoms is in agreement with the assumption that Si atoms form islands); then these facilities are used again to apply the second constraint: no Si-O-P linkage. Each time there is a P atom as first neighbour of a Si atom, it is replaced by a Si atom. Finally, the various first and second Si coordination shells forming the island are determined.

3. Program Description

The macro program will determine first and second Si coordination spheres after the replacement of the P atom at vertex 0 and some Al atoms in positions 1 to 4 by Si atoms, by taking into account the two constraints and one assumption. The adjacency lists corresponding to the faujasite structure are represented on the left hand side of the worksheet "Topology" (Fig. 2) and will be called "Topo". There are five columns: C1 to C5 and 55 rows whose head nodes are all the vertices of the four shells of vertex 0. For clarity an empty column is introduced between them. The left cell of a column contains the position number of a atom and the right cell, initially empty, will be filled with A or P

	C1	C2	C3	C4	C5	Temp	AIPO	SAPO		1st neigbour		2nd neigbour		
										Al	Si	Al	Si	P
2	0	1	2	3	4		P		0					
3	1	13	0	5	6		A		1					
4	2	0	6	7	8		A		2					
5	3	0	8	9	10		A		3					
6	4	13	11	12	0		A		4					
7	5	1	14	15	29		P		5					
8	6	1	2	16	17		P		6					
9	7	2	17	18	27		P		7					
10	8	2	3	19	20		P		8					
11	9	3	20	21	22		P		9					
12	10	3	22	23	24		P		10					
13	11	4	24	25	26		P		11					
14	12	4	26	27	28		P		12					
15	13	1	4	28	29		P		13					
16	14	5	30	31	32									
17	15	5	32	33	34									
18	16	6	34	35	36									
19	17	6	7	36	37									
20	18	7	38	39	40									
21	19	8	40	41	42									
22	20	8	9	42	43									
23	21	9	44	45	46									
24	22	9	10	46	47									
25	23	10	41	47	48									
26	24	10	11	48	49									
27	25	11	49	50	51									
28	26	11	12	51	52									
29	27	7	12	37	38									
30	28	13	12	37	53									
31	29	13	5	30	54									
32	30	14	55	56	29									
33	31	14	57	58	59									
34	32	14	15	59	60									
35	33	15	60	61	62									
36	34	15	16	62	63									
37	35	16	63	64	65									
38	36	16	17	65	66									
39	37	17	27	28	67									
40	38	18	27	69	70									
41	39	18	69	70	71									
42	40	18	19	71	72									
43	41	19	23	72	73									
44	42	19	20	73	74									
45	43	20	74	75	76									
46	44	21	76	77	78									
47	45	21	78	79	80									
48	46	21	22	80	81									
49	47	22	23	73	82									
50	48	23	24	83	84									
51	49	24	25	84	85									
52	50	25	85	86	87									
53	51	25	26	88	89									
54	52	26	68	83	89									
55	53	28	67	90	91									
56	54	29	56	61	91									
57														

Fig. 2 Initial state of the worksheet "Topology".

(for Aluminium and Phosphorus, respectively) by the macro program. On the right hand side of this figure is drawn a table in which first and second Si coordination spheres will be found by the macro program. This table will be called "Table". "Topo" and "Table" have to be created by the user.

The listing of the macro program is given in the appendix. The left column contains text values, for use as "variable names" in the program. The central column contains the instructions of the macro program. The right column is reserved for comments about the program [9]. The structure of the macro program is conveniently described in four parts with the command macros (c.m.): *FillAorP*, *SiliconInX*, *NoSi-O-P* and *Configuration*. They have to be used in this order. The c.m. *FillAorP* is used for filling the component field in "Topo" with P and A. The c.m. *SiliconInX* is used for replacing the P atom at vertex 0 and some Al in positions 1 to 4 by Si atoms. It successively prompts the user to introduce the position number of the vertex. The c.m. *NoSi-O-P* applies the second constraint and the c.m. *Configuration* will fill "Table" with the Si coordination spheres.

The flow chart shown in Fig. 3 presents the arrangement of subprograms serving the four c.m. which call other command or function macros (f.m.). The f.m. *PutAP* is called by *FillAorP* five times to fill the component fields with A or P for the five shells of atom 0. The c.m. *EchoOrNot* suppresses screen activities during a program operation, and thus makes the program run faster. The f.m. *FillTopo* locates the adjacency lists in which Si atoms have to be put and calls the c.m. *FillWithS* to fill with an S the component field whose vertex position is given by the calling c.m.. The f.m. *FillTable* puts an S in the columns AlPO or SAPO of "Table" according to the calling instruction. The c.m.

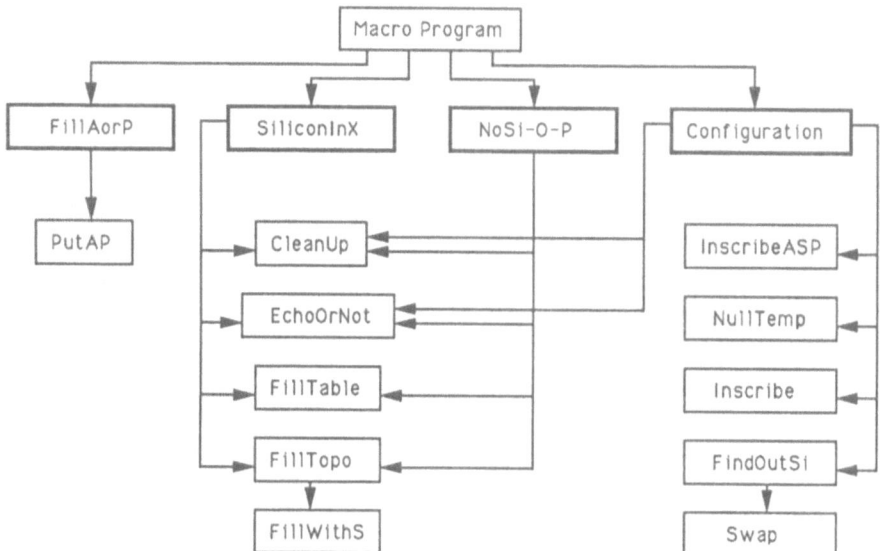

Fig. 3 Flowchart of the macro program.

Topology worksheet — columns C1–C5

Row	C1	C2	C3	C4	C5
1	C1	C2	C3	C4	C5
2	0 S	3 S	1 S	2 S	4 A
3	1 S	13 S	6 S	0 S	5 S
4	2 S	8 S	0 S	6 S	7 S
5	3 S	10 S	0 S	8 S	9 S
6	4 A	13 S	11 P	12 P	0 S
7	5 S	1 S	14 S	15 A	29 A
8	6 S	2 S	1 S	16 A	17 A
9	7 S	2 S	17 A	18 A	27 A
10	8 S	3 S	2 S	19 A	20 A
11	9 S	3 S	20 A	21 A	22 A
12	10 S	3 S	22 A	23 A	24 A
13	11 P	4 A	24 A	25 A	26 A
14	12 P	4 A	26 A	27 A	28 A
15	13 S	1 S	4 A	28 A	29 A
16	14 A	5 S	30 P	31 P	32 P
17	15 A	5 S	32 P	33 P	34 P
18	16 A	6 S	34 P	35 P	36 P
19	17 A	7 S	6 S	36 P	37 P
20	18 A	7 S	38 P	39 P	40 P
21	19 A	8 S	40 P	41 P	42 P
22	20 A	9 S	8 S	42 P	43 P
23	21 A	9 S	44 P	45 P	46 P
24	22 A	10 S	9 S	46 P	47 P
25	23 A	10 S	41 P	47 P	48 P
26	24 A	10 S	11 S	48 P	49 P
27	25 A	11 P	49 P	50 P	51 P
28	26 A	11 P	12 P	51 P	52 P
29	27 A	7 S	12 P	37 P	38 P
30	28 A	13 S	12 P	37 P	53 P
31	29 A	13 S	5 S	30 P	54 P
32	30 P	14 A	55 A	56 A	29 A
33	31 P	14 A	57 A	58 A	59 A
34	32 P	14 A	15 A	59 A	60 A
35	33 P	15 A	60 A	61 A	62 A
36	34 P	15 A	16 A	62 A	63 A
37	35 P	16 A	63 A	64 A	65 A
38	36 P	16 A	17 A	65 A	66 A
39	37 P	17 A	27 A	28 A	67 A
40	38 P	18 A	27 A	69 A	70 A
41	39 P	18 A	69 A	70 A	71 A
42	40 P	18 A	19 A	71 A	72 A
43	41 P	19 A	23 A	72 A	73 A
44	42 P	19 A	20 A	73 A	74 A
45	43 P	20 A	74 A	75 A	76 A
46	44 P	21 A	76 A	77 A	78 A
47	45 P	21 A	78 A	79 A	80 A
48	46 P	21 A	22 A	80 A	81 A
49	47 P	22 A	23 A	73 A	82 A
50	48 P	23 A	24 A	83 A	84 A
51	49 P	24 A	25 A	84 A	85 A
52	50 P	25 A	85 A	86 A	87 A
53	51 P	25 A	26 A	88 A	89 A
54	52 P	26 A	68 A	83 A	89 A
55	53 P	28 A	67 A	90 A	91 A
56	54 P	29 A	56 A	61 A	91 A
57					
58					
59	C1	C2	C3	C4	C5

Temp

AlPO	SAPO		1st neighbour		2nd neighbour		
			Al	Si	Al	Si	P
S	S	0	1	3		7	2
S	S	1		4	7	2	
S	S	2		4	7	2	
S	S	3		4	7	2	
A		4					
P	S	5	3	1		3	6
P	S	6	2	2		5	4
P	S	7	3	1		3	6
P	S	8	2	2		5	4
P	S	9	3	1		3	6
P	S	10	3	1		3	6
P		11					
P		12					
P	S	13	3	1		3	6

Fig. 4 Final state of the worksheet "Topology".

CleanUp deletes the criteria used for searching adjacency lists to be modified. The c.m. *Configuration* calls several other c.m. to fill "Table". The proposed method consists of searching for the column of a given Si in its four first neighbour adjacent lists. Then its position is swapped with the atom located in column C2 so that in column C1 is the first neighbour and in columns C3, C4 and C5 the second neighbours. These operations are performed by the c.m. *FindOutSi* and *Swap*. As mentioned in the previous section, the nine second neighbours located in columns C3 to C5 may appear several times (There are twelve second neighbour vertices in the adjacency lists for a given vertex). The c.m. *Inscribe* and *NullTemp* are designed to count the first and second neighbours. Finally the c.m. *InscribeASP* fills "Table" with the number of different atoms.

4. Results

The macro program was run on a Macintosh SE microcomputer. The version of EXCEL is 2.2. For a Si island consisting of eleven Si atoms, the run-time was about five minutes. An example is given in Fig. 4 where the user replaced the P atom in position 0 and Al atoms in positions 1, 2 and 3 with Si atoms. "Table" shows that P atoms in positions 5 to 10 and 13 are replaced by Si atoms due to the constraint: no Si-O-P linkage. There are four Si environments: Si atom in position 0 is surrounded by one Al and three Si as first neighbours and seven Si and two P as second neighbours, written shortly as Si(1Al,3Si)(7Si,2P). Si atoms in positions 1, 2 and 3 are in the following configuration: Si(0Al,4Si)(7Al, 2Si). Si atoms in positions 5, 7, 9, 10 and 13 are Si(3Al,1Si)(3Si,6P). Finally, Si atoms in positions 6 and 8 are Si(2Al,2Si)(5Si,4P).

5. Acknowledgments

We thank Dr. Barthomeuf for the critical reading of the manuscript.

6. References

[1]   Goepper M., Guth F., Delmotte L., Guth J.L. and Kessler H. (1989) *Stud. Surf. Sci. Catal.* **49**A, 857.
[2]   Maistriau L., Dumont N., B.Nagy J., Gabelica Z. and Derouane E.G. (1990) *Zeolites* **10**, 243.
[3]   Man P.P., Briend M., Peltre M.J., Lamy A., Beaunier P. and Barthomeuf D. (1991) *Zeolites* **11**, 563.
[4]   Martens J.A., Janssens C., Grobet P.J., Beyer H.K. and Jacobs P.A. (1989) *Stud. Surf. Sci. Catal.* **49**A, 215.
[5]   Saldarriaga L.S., Saldarriaga C. and Davis M.E. (1987) *J. Am. Chem. Soc.* **109**, 2686.
[6]   Engelhardt G. and Michel D. (1987) *High Resolution Solid-State NMR of Silicates and Zeolites*, Wiley, New York.
[7]   Melchior M.T. and Newsam J.M. (1989) *Stud. Surf. Sci. Catal.* **49**A, 805.
[8]   Campbell M. (1986) *Excel Macro Library*, Que Corporation, Indianapolis.

[9]   Hergert D. (1986) *Command Performance Microsoft Excel*, Microsoft Press, Washington; *Microsoft Excel with Macros*, Microsoft Press, Washington.

[10] Horowitz E. and Sahni S. (1990) *Fundamentals of Data Structures in Pascal*, Third edition, Computer Science Press, New York.

APPENDIX

	Some parameters	
star1	1 3	
star2	1 3	
kol	4	
AL1b	0	1st Al neighbour counter.
Si1b	0	1st Si neighbour counter.
AL2b	0	2nd Al neighbour counter.
Si2b	0	2nd Si neighbour counter.
P2b	0	2nd P neighbour counter.

Command	EchoOrNot	
message0	="Do you wish to see the screen activity?(y/n)"	Question prompt.
reply0	=INPUT(message0;2;"Echo or No echo")	Get answer.
t f	=MID(reply0;1;1)="y"	Translate to logical.
	=ECHO(tf)	Echo statement.
	=RETURN()	

Function	PutAP	
	=ARGUMENT("X")	First cell number in a shell.
	=ARGUMENT("Y")	Number of atoms in a shell.
	=ARGUMENT("letter1")	Component 1 (A or P).
	=ARGUMENT("letter2")	Component 2 (A or P).
xPri	=X+2	Starting Row of a shell.
xSec	=xPri+Y-1	Ending Row of a shell.
	=SELECT("R"&xPri&"C2:R"&xSec&"C2")	Fill right cells of column C1
	=FORMULA.FILL(letter1)	with letter1.
	=FOR("J";5;14;3)	Fill right cells of columns C2
	=SELECT("R"&xPri&"C"&j&":R"&xSec&"C"&j&"")	to C5 with letter2.
	=FORMULA.FILL(letter2)	
	=NEXT()	
	=RETURN()	

Command	FillAorP	
	Option-Command f	
	=ACTIVATE("TOPOLOGY")	Activate this sheet.
	=PutAP(0;1;"P";"A")	Fill cell no "0" and neighbours.
	=PutAP(1;4;"A";"P")	Fill 1st shell and neighbours.
	=PutAP(5;9;"P";"A")	Fill 2nd shell and neighbours.
	=PutAP(14;16;"A";"P")	Fill 3rd shell and neighbours.
	=PutAP(30;25;"P";"A")	Fill 4th shell and neighbours.
	=SELECT("R1C1:R56C14")	Define the database range...
	=SET.DATABASE()	
	=SELECT("R1C1:RC14")	Select titles of record fields.
	=COPY()	Copy
	=SELECT("R59C1")	Select Row 59 Column 1.
	=PASTE()	Paste titles.
	=RETURN()	

Command	FillWithS	
	=FOR("k";1;5)	Do from C1 to C5.
	=SELECT("RC"&(1+3*(k-1))&"")	Select left cell of column Ck.
what0	=ACTIVE.CELL()	What is the atom cell number?
	=IF(what0=number;GOTO(finish10))	If a cell for Si, branch.
	=NEXT()	Otherwise repeat.
finish10	=FORMULA("S";"RC[1]")	Fill with "S".
	=RETURN()	

Command	CleanUp	
	=SELECT("R60C1:R64C14")	Select criteria range.
	=EDIT.DELETE(2)	Clean criteria.
	=RETURN()	

Function	FillTopo	
	=ARGUMENT("number")	Receive Si atom cell number.
	=FOR("j";1;5)	Define the criterion with
position	=1+3*(j-1)	"=" and the received number.
	=SELECT("R"&(59+j)&"C"&position&"")	
	=FORMULA("="&number&"")	
	=NEXT()	
	=SELECT("R59C1:R64C14")	Select criteria range.
	=SET.CRITERIA()	Set criteria range.
	=DATA.FIND(TRUE)	Find first record.
	=SELECT("RC1")	Select 1st cell of record.
here	=SET.VALUE(star1;ACTIVE.CELL())	Save atom cell number.
	=FillWithS()	Put "S" in cell.
	=DATA.FIND.NEXT()	Find next record.
	=SELECT("RC1")	Select 1st cell of record.
	=SET.VALUE(star2;ACTIVE.CELL())	Save atom cell number.
	=IF(star1<>star2;GOTO(here))	If different cells, branch.
	=RETURN()	Return if same cell.

Function	FillTable	
	=ARGUMENT("number")	Receive atom cell number.
	=ARGUMENT("lonne")	Column number : AlPO or SAPO.
	=FORMULA("S";"R"&(number+3)&"C"&lonne&"")	Fill with "S".
	=RETURN()	

Command	SiliconInX	
	Option-Command x	
	=EchoOrNot()	See screen activity ?
	=ACTIVATE("TOPOLOGY")	Show this sheet.
message1	="Put S in which cell?"	Display message.
numberC	=INPUT(message1;1;"Silicon cell number")	Wait for Si atom cell number.
	=FillTopo(numberC)	Fill Si cell with "S".
	=FillTable(numberC;20)	Fill column AlPO with "S".
	=FillTable(numberC;21)	Fill column SAPO with "S".
message2	="Put another Si?(y/n)"	Display message.
reply2	=INPUT(message2;2;"Continue")	Wait for response.
	=IF(reply2="y";GOTO(message1))	Examine response.
	=RETURN()	Return if no.

Command	NoSi-O-P	
	Option-Command p	
	=EchoOrNot()	See screen activity ?
	=ACTIVATE("TOPOLOGY")	Activate this sheet.
	=FOR("LL";3;16)	Find in column AlPO.
	=SELECT("R"&LL&"C20")	Select atom cell.
	=IF(ACTIVE.CELL()<>"S";GOTO(finish20))	If not "S", repeat.
	=SELECT("RC[2]")	Select next second right cell.
cellOfS	=ACTIVE.CELL()	Get Si atom cell number.
	=FOR("M";5;14;3)	Find from C1 to C5.
	=SELECT("R"&(cellOfS+2)&"C"&M&"")	Select right cell of Cm.
choice	=ACTIVE.CELL()	What does it contain ?
	=IF(OR(choice="S";choice="A");GOTO(finish30))	If "S" or "A", repeat.
	=SELECT("RC[-1]")	Select left cell of Cm.
no1	=ACTIVE.CELL()	Get P cell number.
	=FillTable(no1;21)	Put "S" in column SAPO.
	=FillTopo(no1)	Replace "S" for "P".
finish30	=NEXT()	
finish20	=NEXT()	
	=CleanUp()	Clean criteria.
	=RETURN()	

Fuonction	Swap	
	=ARGUMENT("cellO")	
	=SELECT("RC"&cellO&":RC"&(cellO+1)&"")	Selection in database.
	=COPY()	Copy.
	=SELECT("RC16")	Selection outside database.

	=PASTE()	Paste temporarily.
	=SELECT("RC4:RC5")	Selection in database.
	=COPY()	Copy cells of C2.
	=SELECT("RC"&cellO&"")	Selection in database.
	=PASTE()	Paste definitely.
	=SELECT("RC16:RC17")	Selection outside database.
	=CUT()	Cut.
	=SELECT("RC4")	Selection in database.
	=PASTE()	Paste in C2.
	=RETURN()	

Command	**FindOutSi**	
	=SET.VALUE(kol;1)	Set column Ci counter.
again40	=SET.VALUE(kol;kol+3)	Increment counter.
	=SELECT("RC"&kol&"")	Select a cell.
	=IF(ACTIVE.CELL()<>cellOfS1;GOTO(again40))	If not Si cell number, branch.
	=IF(kol<>4;Swap(kol))	If not C2, swap with C2.
	=RETURN()	Otherwise return.

Command	**Inscribe**	
	=SELECT("RC2")	1st neighbour cell.
elem1	=ACTIVE.CELL()	What does it contain (A or S)?
	=IF(elem1="S";SET.VALUE(Si1b;Si1b+1);SET.VALUE(AL1b;AL1b+1))	Add 1 to counter.
	=FOR("t";8;14;3)	Find from C3 to C5.
	=SELECT("R"&(star1+2)&"C"&t&"")	Select a 2nd neighbour cell.
elem2	=ACTIVE.CELL()	What is it ?
	=SELECT("RC[-1]")	Select next left cell.
cellx	=ACTIVE.CELL()	Get its atom cell number.
	=SELECT("R"&(cellx+2)&"C18")	Select a cell in column Temp.
present	=ACTIVE.CELL()	What does it contain ?
	=IF(present=1;GOTO(finish50))	If there is a "1", branch back.
	=FORMULA("1")	Otherwise fill with "1".
	=IF(AND(elem2<>"A";elem2<>"P");GOTO(finish60))	It is a "S", so branch.
	=IF(elem2="A";SET.VALUE(AL2b;AL2b+1);SET.VALUE(P2b;P2b+1))	It is a "A" or "P", add counter.
	=GOTO(finish50)	Repeat.
finish60	=SET.VALUE(Si2b;Si2b+1)	Add 1 to Si counter.
finish50	=NEXT()	
	=RETURN()	

Command	**NullTemp**	
	=SELECT("R2C18:R56C18")	Select column Temp.
	=FORMULA.FILL("")	Fill with blanks.
	=RETURN()	

Command	**InscribeASP**	
	=SELECT("R"&(cellOfS1+3)&"C23")	Select 4th column of table.
	=IF(AL1b<>0;FORMULA(AL1b);FORMULA(""))	Fill with number of Al or blank
	=SELECT("RC[1]")	Select next right cell.
	=IF(Si1b<>0;FORMULA(Si1b);FORMULA(""))	Fill with number of Si or blank
	=SELECT("RC[1]")	Select next right cell.
	=IF(AL2b<>0;FORMULA(AL2b);FORMULA(""))	Fill with number of Al or blank
	=SELECT("RC[1]")	Select next right cell.
	=IF(Si2b<>0;FORMULA(Si2b);FORMULA(""))	Fill with number of Si or blank
	=SELECT("RC[1]")	Select next right cell.
	=IF(P2b<>0;FORMULA(P2b);FORMULA(""))	Fill with number of P or blank.
	=RETURN()	

Command	**Configuration**	
	Option-Command c	
	=EchoOrNot()	See screen activity ?
	=ACTIVATE("TOPOLOGY")	Activate this sheet.
	=FOR("mC";3;16)	Find in column SAPO the
	=SELECT("R"&mC&"C21")	atom cells containing S.
nucleiS1	=ACTIVE.CELL()	
	=IF(nucleiS1<>"S";GOTO(finish70))	If not a "S", repeat.
	=SELECT("RC[1]")	Get its atom cell number in

cellOfS1	=ACTIVE.CELL()	*3rd column of the table.*
	=SET.VALUE(AL1b;0)	*Initialize counters...*
	=SET.VALUE(Si1b;0)	
	=SET.VALUE(AL2b;0)	
	=SET.VALUE(Si2b;0)	
	=SET.VALUE(P2b;0)	
	=FOR("j";1;4)	*Define the criterion...*
locale	=4+3*(j-1)	
	=SELECT("R"&(59+j)&"C"&locale&"")	
	=FORMULA("="&cellOfS1&"")	
	=NEXT()	
	=SELECT("R59C1:R63C14")	*Set criteria range.*
	=SET.CRITERIA()	
	=DATA.FIND(TRUE)	*Find first record.*
	=SELECT("RC1")	*Select 1st cell of record.*
again70	=SET.VALUE(star1;ACTIVE.CELL())	*Save atom cell number.*
	=FindOutSi()	*Put "S" in C2.*
	=Inscribe()	*Add neighbour counters.*
	=SELECT("R"&(star1+2)&"C1")	*Select 1st cell of record.*
	=DATA.FIND.NEXT()	*Find next record.*
	=SELECT("RC1")	*Select 1st cell of record.*
	=SET.VALUE(star2;ACTIVE.CELL())	*Save atom cell number.*
	=IF(star1<>star2;GOTO(again70))	*If not same record, branch.*
	=NullTemp()	*Otherwise clean Temp.*
	=InscribeASP()	*Fill neighbour columns of table*
finish70	=NEXT()	
	=CleanUp()	*Delete criterion.*
	=RETURN()	

IMPORTANCE OF PREPARATION OF ACIDIC AND BASIC SOLID MATERIALS

KOZO TANABE
Central Research Laboratory
Nippon Shokubai Co., Ltd.
5-8, Nishi Otabi-cho, Suita
Osaka 564, Japan

ABSTRACT. The changes of acid-base properties as well as catalytic activities of solid materials with the changes in the preparative methods and pretreatment conditions are described. The principal factors which control chemically the acid-base properties are discussed.

1. Introduction

Acidic and basic properties of solid materials are greatly influenced by the preparation methods and pretreatment conditions. The large effect of the preparation method is mainly caused by the difference in the kinds of starting materials (precursors), the difference in precipitation procedures, and the difference in the degree of washing precipitates. The effects of the preparation methods and pretreatment conditions on the acid-base property and catalytic activity and selectivity are one or two order of magnitude in some cases. In this paper, the remarkable examples are demonstrated, the importance of preparation of solid materials being emphasized. The reason for the pronounced effect of preparation is given where possible.

2. Effect of Preparation Method

2.1. DIFFERENCE IN STARTING MATERIALS

The surface property and catalytic activity of MgO (I) prepared by evacuating magnesium hydroxide, $Mg(OH)_2$, is different from those of MgO(II) prepared by evacuating magnesium carbonate hydroxide, $4MgCO_3 \cdot Mg(OH)_2 \cdot 5H_2O$.

In the alkylation of phenol with methanol over MgO (I), phenol is selectively alkylated at the ortho position to produce o-cresol and 2,6-xylenol. The orthoselectivity of MgO (I) is higher than 98 percent, while that of MgO (II) is 73 percent [1]. For the isomerization of cis-2-butene, the maximum activity and selectivity (ratio of trans-2-butene

403

J. Fraissard and L. Petrakis (eds.), Acidity and Basicity of Solids, 403–414.

404

to 1-butene) of MgO (I) is about four times higher than those of Mg (II)
as shown in Fig. 1 [1]. The difference in the catalytic action is considered to be due to the difference in regularity of the surface atom
arrangement. The surface structure is less regular or more heterogeneous
on MgO (II) than on MgO(I), which is suggested from the fact that the IR
bands of CO_2 species are broader for MgO (II) than for MgO (I). The
strength of the electrostatic field which can be obtained from the
values in ESR spectra of the adsorbed oxygen species, O_2^-, is also found
to be different [2].

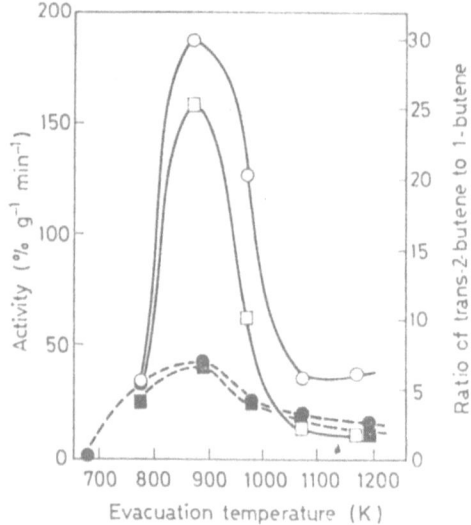

Fig. 1   Activity and selectivity for isomerization of cis-2-butene.
         Activity (○) and selectivity (□) of MgO (I).
         Activity (●) and selectivity (■) of MgO (II).

    A more striking effect was observed in the selectivity of CaO cata-
lysts for the reaction of 3-carene. Double bond isomerization takes
place over CaO prepared from $CaCO_3$, the selectivity for the formation of
2-carene being 96%, while dehydrogenation occurs over CaO prepared from
$Ca(OH)_2$, the selectivity for the formation of p-cymene being 70% [3].

2 - carene   3 - carene   p-cymene

The selectivity of ThO_2 catalysts for the formation of 1-octene from 2-octanol is strongly dependent on the kinds of thorium salts, the preparation method, and the pretreatment as shown in Table 1 [4]. ThO_2 prepared by hydrolysis of the nitrate with ammonia water followed by calcination at 873 K in a hydrogen atmosphere showed high selectivity (99 percent) for 1-octene formation, while the selectivity of ThO_2 prepared by thermal decomposition was 50 percent. The catalyst prepared from the chloride gave a low selectivity (25 percent). Hydrogen treatment makes ThO_2 more active for the dehydration of 2-octanol, whereas oxygen treatment makes it more active for the dehydrogenation. The preferential dehydration with ThO_2 pretreated with hydrogen is due to the acidity increase on the catalyst surface, which is evidenced by poisoning experiments with pyridine [5].

Table 1   Effect of preparation method on selectivity of ThO_2 catalysts for formation of octenes from 2-octanol

Th salts	Pretreatment temp./K	Octenes/mol%		
		1-octene	trans-2-octene	cis-2-octene
Nitrate[a]	H_2, 873	99	trace	trace
Nitrate[b]	O_2, 873	34	36	30
Nitrate[b]	H_2, 873	50	26	24
Oxalate	H_2, 873	72	12	16
Oxalate	O_2, 873	43	27	30
Carbonate	H_2, 873	82	10	8
Carbonate	H_2, 823	86	7	7
Chloride	H_2, 873	25	30	45

a   Precipitation method with NH_4OH
b   Thermal decomposition method

Most metal oxides and complex oxides are usually prepared from their chlorides or sulfates by a precipitation method. In those preparation, about 0.2 - 0.5 wt. percent of anions remain in the precipitates, even if the precipitates are washed thoroughly with distilled water until no anions are detected in the washings with silver nitrate or barium chloride. In particular, residual sulfate anion is not removed by calcination at 773 K for 3 hr. Thus, the effect of anions on acidic and catalytic properties was examined by adding ammonium chloride or sulfate to pure TiO_2, TiO_2 - ZnO and ZnO. The specific surface areas of all oxides increased on the addition of the salts except that of TiO_2 on the addition of NH_4Cl. In Table 2 are shown the effects of the addition of NH_4Cl and $(NH_4)_2SO_4$ on the catalytic activity and selectivity for the isomerization of 1-butene [6]. The addition of $(NH_4)_2SO_4$ increased the activities of TiO_2 and TiO_2 - ZnO, especially the activity of TiO_2. On the other hand, the addition of NH_4Cl decreased the activity of TiO_2 and TiO_2 - ZnO.

Table 2 Effect of the addition of NH_4Cl or $(NH_4)_2SO_4$ on activity and selectivity for isomerization of 1-butene

Catalyst	Additives	Evacuation temperature/K	Activity $min^{-1}g^{-1}$	Ratio of cis/trans	Surface area m^2/g
TiO_2	no	723	2.6	5	42
	no	423	0.9	3	—
TiO_2-ZnO	no	723	4.1	5	37
	no	423	1.1	4	—
ZnO	no	723	0	—	0.4
	no	423	0	—	—
TiO_2	NH_4Cl	723	1.9	5	43
	NH_4Cl	423	1.4	3	—
TiO_2-ZnO	NH_4Cl	723	0	—	76
	NH_4Cl	423	0.5	2	—
ZnO	NH_4Cl	723	0	—	1.1
	NH_4Cl	423	0	—	—
TiO_2	$(NH_4)_2SO_4$	723	240	1	103
	$(NH_4)_2SO_4$	423	520	1	—
TiO_2-ZnO	$(NH_4)_2SO_4$	723	41	2	65
	$(NH_4)_2SO_4$	423	82	2	—
ZnO	$(NH_4)_2SO_4$	723	0.2	3	0.5
	$(NH_4)_2SO_4$	423	0	—	—

Reaction temperature; 423 K. Initial pressure of 1-butene; 13 kPa.
Ratio of cis/trans; the value extrapolated to 0% conversion.

Since the isomerization of 1-butene is known to be catalyzed by acids in the case where the ratio of cis-/trans-2-butene is 1 - 2, the addition of SO_4^{2-} is considered to enhance the acidity of TiO_2 and TiO_2-ZnO. In fact, TiO_2, ZrO_2, and Fe_2O_3 have been confirmed to exhibit superacidity when a small amount of sulfate is added [7, 9].

The application of the above additive effect of sulfate ion to the catalyst for reduction of NO with NH_3 is shown in Table 3 [9]. The increase in NO conversion was attributed to the acidity increase of MoO_3-TiO_2 due to the interaction between TiO_2 and SO_4^{2-} [9].

Table 3 Effect of SO_4^{2-} addition on activity of MoO_3-TiO_2 for NO + NH_3 reaction at 573 K

Amount of SO_4^{2-} wt%	NO conversion %
0	58
0.5	84

2.2. DIFFERENCE IN WASHING PRECIPITATES

When a metal sulfate is used as a starting material, sulfate ion remains in the precipitates and the amount of remaining sulfate ion changes largely depending on the extent of washing precipitates. If one washes the precipitates of $Ti(OH)_4$ and $Ti(OH)_4 \cdot Zn(OH)_2$ mentioned in the foregoing section with hot water, the acidities of TiO_2 and TiO_2-ZnO are remarkably lowered. Washing of precipitates is particularly important in the preparation of the solid materials containing TiO_2, ZrO_2, and Fe_2O_3.

I remember an interesting paper published in J. Catal. about thirty years ago. It was reported that the surface area of ThO_2 prepared by the exactly same precipitation method except the difference in the extent of washing precipitates changed by one order of magnitude.

2.3. DIFFERENCE IN PREPARATION PROCEDURE

Let us compare the acidic property of TiO_2-ZnO prepared by a usual heterogeneous co-precipitation method with that prepared by a homogeneous co-precipitation method. In a usual co-precipitation method where ammonia water is used as a precipitating reagent, the concentration of ammonium hydroxide differs locally during the formation of precipitates, that is, precipitates are formed in a heterogeneous concentration of hydroxide ion. However, if excess of urea is used instead of ammonia, precipitates are formed from a homogeneous solution of hydroxide ion, since urea decomposes to supply ammonium hydroxide at elevated temperatures according to the following equilibria [6].

$$(NH_2)_2CO + H_2O \rightleftharpoons 2 NH_3 + CO_2$$

$$NH_3 + H_2O \rightleftharpoons NH_4OH$$

As shown in Fig. 2, TiO_2-ZnO prepared by the homogeneous method exhibits much larger acidity in the range of $-3 < H_0 \leq +6.8$ and lower acid strength than TiO_2-ZnO prepared by the heterogeneous method [10]. A similar tendency is also observed for TiO_2-SiO_2 [6]. The catalytic activity of TiO_2-SiO_2 prepared by the homogeneous method for isomerization was about two times higher than that of TiO_2-SiO_2 prepared by the heterogeneous method. The difference in the activity is approximately equal to the difference in the acidity in the range of $-3 < H_0 \leq +3.3$, indicating that the acid sites having this acid strength are active sites.

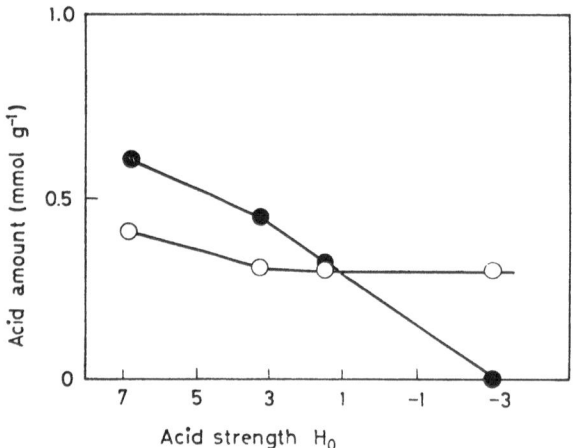

Fig. 2   Acidity and acid strength of TiO_2-ZnO prepared by homogeneous (●) and heterogeneous (○) coprecipitation methods

It might be said from the above results that the heterogeneous precipitation method using aqueous ammonia produces strong acid sites and the homogeneous precipitation method using urea forms a large amount of relatively weak acid sites. Another distinctive feature of the homogeneous precipitation method is that this method allows the preparation of mixed oxides having reproducible acidic properties.

An interesting effect of the preparation method and preparation procedure on the surface area of Al_2O_3, and the activity for the exchange reaction between methane and deuterium is given in Table 4. In this case, the distance between acidic sites and basic sites seems to control the catalytic activity [11].

Table 4 Preparation methods of aluminas and surface areas and catalytic activities of aluminas evacuated at 873 K

Al_2O_3	Preparation		Area	Surface activity[a]
	Starting material	Method	m^2g^{-1}	
I	Isopropoxide	hydrolysis	236	9.9
II	$Al(NO_3)_3$	homogeneous precipitation (urea)	110	4.2
III	$Al(NO_3)_3$	heterogeneous precipitation (ammonia)	156	12.8
IV	$Al(NO_3)_3$	direct thermal decomposition	135	0
V	$Al_2(SO_4)_3$	homogeneous precipitation (urea)	11	0
VI	$Al_2(SO_4)_3$	heterogeneous precipitation (ammonia)	216	3.9
VII	$AlCl_3$	homogeneous precipitation (urea)	147	27.4
VIII	$AlCl_3$	heterogeneous precipitation (ammonia)	248	10.9
IX	$AlCl_3$	direct thermal decomposition	103	1.1
X	$AlBr_3$	homogeneous precipitation (urea)	173	3.8
XI	$NaAlO_2$	heterogeneous precipitation (CO_2)	194	0
XII	high purity α-Al_2O_3 (Sumitomo Chemical Co., Ltd., AKP 20)		5	0
XIII	γ-Al_2O_3 (Nishio Industries CO., Ltd., AE 11)		—	6.5

[a] activity at room temperature for $CH_4- D_2$ exchanges; arbitrary units per unit mass of catalyst.

3. Effect of Pretreatment

Basic sites, reducing sites and the sites active for hydrogenation of olefins appear on the surface of CaO when $Ca(OH)_2$ is heated at different temperatures. Fig. 3 shows how the generation of three kinds of the sites depends on temperature of heat treatment. Model structure of basic sites and reducing sites on CaO is given in Fig. 4 [12]. The sites active for hydrogenation of olefins are the pair-sites of Lewis acid site and base site on CaO. In the case of MgO, the acid-base pair sites suitable for hydrogenation of olefins appear when evacuated at very high temperature (1373 K), where the pair sites are oriented so as to split a hydrogen molecule into a proton and a hydride ion [13].

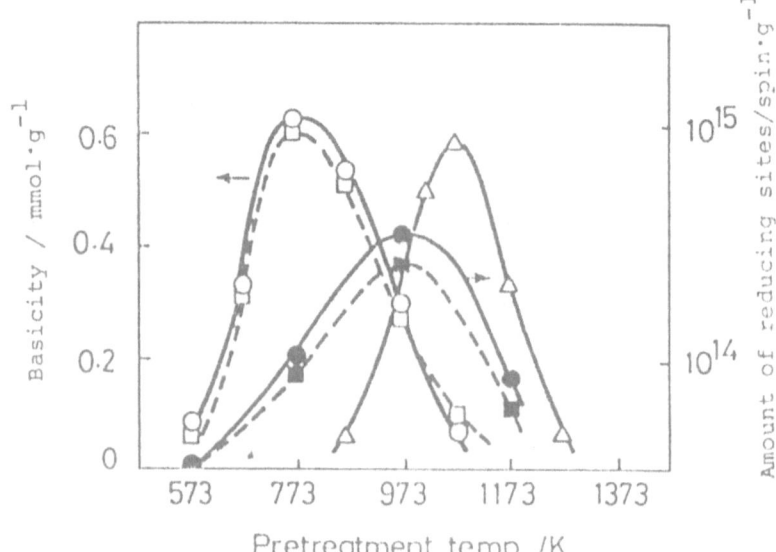

Fig. 3   Change of surface property and catalytic activity of CaO with change of calcination temperature ○:Basicity, □ :Activity for reaction of benzaldehyde, ● :Amount of reducing sites, ■ :Activity for styrene polymerization, △ :Activity for hydrogenation of propylene.

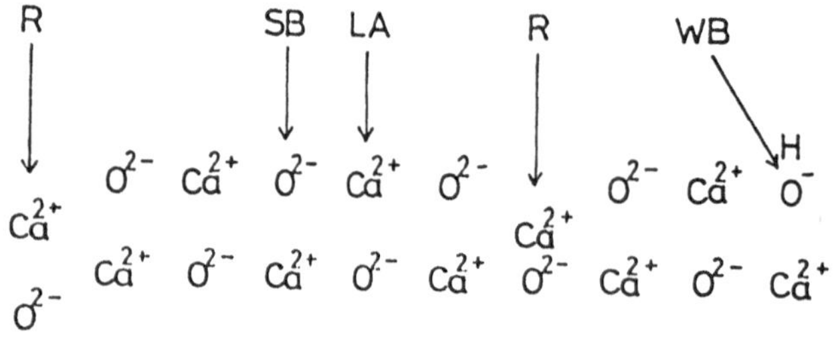

Fig. 4   Surface model of CaO.
SB; strong basic site, WB; weak basic site, LA; Lewis acid site, R; reducing site.

In the cases of SiO_2-TiO_2 [14], and TiO_2-SnO_2 [15], strongly acidic property appears when calcined in air at 773 K, whereas strong reducing property appears when calcined in a vacuum at the same temperature.
The activity and selectivity of TiO_2 and ThO_2 [5] for the isomeriza-tion of 1-butene are found to increase on pretreatment with hydrogen, but decrease on pretreatment with oxygen, while the activity of TiO_2-SiO_2 increases on oxygen pretreatment, but decreases on hydrogen pre-treatment (Table 5) [14]. These phenomena are interpreted to be caused by the appearance and disappearance of acidic sites and reducing sites [14]. It is also known that hydrogen pretreatment increases the acidi-ties of TiO_2-WO_3, TiO_2-NiO, Al_2O_3-NiO, Al_2O_3-CdO, and Al_2O_3-Cr_2O_3, but decreases the acidities of TiO_2-MoO_3, TiO_2-Fe_2O_3, TiO_2-Cr_2O_3, TiO_2-V_2O_5, TiO_2-Co_3O_4, TiO_2-CuO, and Al_2O_3-Fe_2O_3 [16].

Table 5   Changes in the activity and the ratio of cis/trans-2-butene in the isomerization of 1-butene with various treatments and poisonings after evacuation at 123 K

	TiO_2		TiO_2-SiO_2 (9/1)	
	Activity /$\%$ $min^{-1}g^{-1}$	Ratio of cis/trans-2-butene	Activity /$\%$ $min^{-1}g^{-1}$	Ratio of cis/trans-2-butene
Untreated	13.3	6.1	13.2	2.2
Poisoned with NH_3	3.0	9.0	2.8	2.2
Poisoned with CO_2	14.2	6.8	14.9	2.3
Treated with O_2	7.0	2.9	27.7	1.9
Treated with H_2	15.8	11.0	11.2	3.2

Calcium oxides calcined in air at 873 K were catalytically inactive for the isomerization of olefins whose acid strengths are weaker than that of CO_2 because of poisoning of the basic sites (O^{2-}) with CO_2, but found to exhibit an extremely high activity for the isomerization of 1-butene when CO_2 adsorbed on the basic sites was removed by evacuating at 873 K, as shown in Table 6 [17]. Adsorbed states of CO_2 are unidentate and bidentate forms which appear under higher and lower pressure of CO_2, respectively, as shown in Fig. 5 [18]. Since the basic sites are cover-

-ed with CO_2 in both cases, it is necessary to remove CO_2 for the gene-ration of the basic property.

Table 6 Isomerization of 1-Butene over CaO

Catalyst	Catalyst weight/mg	Reaction time/min	Reaction temp/k	Conversion /%
CaO calcined at 873 k in air	140	120	473	0
CaO calcined at 873 k in vacuo	17	20	303	63

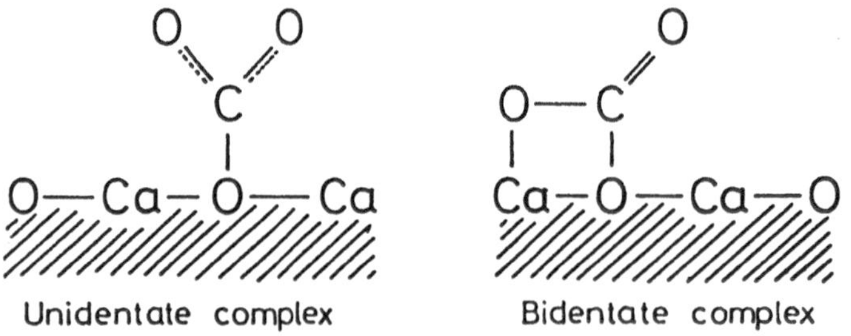

Fig. 5 Adsorbed states of CO_2 on CaO. Unidentate form appears under higher pressure of CO_2. Bidentate form appears under lower pressure of CO_2.

In general, acid-base properties on metal oxides and complex oxides appear by calcination or evacuation at 700 - 900 K. In the case of nio-bium oxide, however, Nb_2O_5 obtained by calcination at 773 K is almost neutral. It has been found recently that hydrated niobium oxide, $Nb_2O_5 \cdot nH_2O$ (niobic acid) which is obtained by calcining the hydroxide at 373 - 573 K shows a high acid strength corresponding to 70% sulfuric acid [19, 20]. The new type solid acid containing considerable amounts of water was reported to show high catalytic selectivities and long catalyst life for the reactions in which water molecules participate or are liberated. As an example, the catalytic behavior of $Nb_2O_5 \cdot nH_2O$ for an esterifica-tion reaction is compared with those of the other solid acids in Table 7 [21].

Table 7 Activity and selectivity for esterification of ethyl alcohol
with acetic acid at 393 k. reaction time; 1 h.

Catalyst	Conversion %	Selectivity %	By-products
$Nb_2O_5 \cdot nH_2O$	72	100	
Resin	38	< 99	
$ZrO_2-SO_4^{2-}$	39	98	$(C_2H_5)_2O$
$Fe_2O_3-SO_4^{2-}$	8	98	$(C_2H_5)_2O$
$TiO_2-SO_4^{2-}$	$95(54)^{a)}$	$99(98)^{a)}$	
$SiO_2-Al_2O_3$	4	99	$(C_2H_5)_2O$, C_2H_4
H-ZSM-5	82	92	$(C_2H_5)_2O$, C_2H_4

a) Reaction time; 2 h.

4. Future Problems

The concentration of a starting material in solution, the valency of
cation and anion of starting material and precipitating reagent, pH in
precipitating solution, the kinds of solvents used, and the aging of
precipitates are known to largely affect the acid-base and catalytic
properties of metal oxides and complex oxides. Although no systematic
study has been made hitherto, such studies will be important not only
for finding new acid-base property, but also for establishing the reproducible preparation method of solid acids and bases.

5. References

1. H. Hattori, K. Shimazu, N. Yoshii, and K. Tanabe, Bull. Chem. Soc.
Jpn., 49, 969 (1976).
2. T. Iizuka, M. Saito, and K. Tanabe, J. Res. Inst. Catal., Hokkaido
Univ., 28, 189 (1981).
3. K. Tanabe, K. Shimazu, and H. Hattori, Chem. Lett., 1975, 507.
4. B. H. Davis, J. Catal., 25, 81 (1972).
5. Y. Imizu, T. Yamaguchi, H. Hattori, and K. Tanabe, Bull. Chem. Soc.
Jpn., 50, 1040 (1977).
6. K. Tanabe, M. Itoh, K. Morishige, and H. Hattori, Preparation of
Catalyst, ed. B. Delmon , P. A. Jacobs, G. Poncelet, Elsevier,
Amsterdam, 1976, p. 65.
7. K. Tanabe, M. Misono, Y. Ono, and H. Hattori, New Solid Acids and
Bases, Kodansha, Tokyo, Elsevier, Amsterdam, 1989, p. 199.
8. K. Tanabe, H. Hattori, and T. Yamaguchi, CRC Review in Surface
Chemistry, 1, 1 (1990).

414

9. S. Okazaki, M. Kumasaka, J. Yoshida, K. Kosaka, and K. Tanabe, Ind. Eng. Chem., Product Research and Development, 20, 301 (1981).
10. K. Tanabe, C. Ishiya, I. Ichikawa, I. Matsuzaki, and H. Hattori, Bull. Chem. Soc. Jpn., 45, 47 (1972).
11. K. Tanabe, M. Utiyama, and H. Hattori, Proc. Symp. Sci. Catalysis and its Application in Ind., Sindri, India, 1979.
12. K. Tanabe, in Acid-Base Catalysis, ed. K. Tanabe et al., Kodansha, Tokyo, VCH, Weinheim, 1989, p. 513.
13. K. Tanabe, in Catalysis by Acids and Bases, ed. B. Imelik et al., Elsevier, Amsterdam, 1985, p. 1.
14. H. Hattori, M. Itoh, and K. Tanabe, J. Catal., 38, 172 (1975).
15. M. Itoh, H. Hattori, and K. Tanabe, J. Catal., 43, 192 (1976).
16. K. Shibata, T. Kiyoura, and K. Tanabe, J. Res. Inst. Catal., Hokkaido Univ., 18, 189 (1970).
17. K. Tanabe, N. Yoshii, and H. Hattori, J. Chem. Soc., Chem. Commun., 464 (1971).
18. Y. Fukuda and K. Tanabe, Bull. Chem. Soc. Jpn., 46, 1616 (1973).
19. K. Tanabe, Catalysis Today, 8, 1 (1990).
20. K. Tanabe, Chemtech, 21, 628 (1991).
21. Z. Chen, T. Iizuka, and K. Tanabe, Chem. Lett., 1085 (1984).

REGULATION OF SURFACE ACIDITY OF OXIDIC CARRIERS USED FOR THE PREPARATION OF SUPPORTED CATALYSTS.

A.LYCOURGHIOTIS
Chemistry Department
Research Institute of Chemical Engineering and High-
-Temperature Chemical Processes
University of Patras
Patras 26 500, Greece

ABSTRACT: The methods developed in the last decade to regulate the surface acidity of the γ-alumina, silica and titania used as carriers for the preparation of supported catalysts are described in the present lecture. Specifically, the change in the impregnation temperature as well as the doping of the carriers with several amounts of Na^+, Li^+ and F^- ions allows a fine regulation to be achieved of the constants related with the protonation-deprotonation equilibria of the surface hydroxyls $(-SOH_2^+ \rightleftharpoons -SOH + H_s^+, -SOH \rightleftharpoons SO^- + H_s^+)$ and thus of the point of zero charge and the surface concentration of the protonated $(-SOH_2^+)$, deprotonated $(-SO^-)$ and neutral (SOH) surface hydroxyls.

Since SOH_2^+ and SOH (SO^-) are responsible for the creation of the depositing sites for the negatively (positively) charged species containing the catalytically active ions, the regulation achieved may be served to maximize the amount of the active species deposited on the support surface by "equilibrium adsorption" from the liquid phase. This, in turn, may result to catalysts with high active surface.

1. Introduction

It is well known that the catalytically active species rarely have more than one of the following properties necessary to be used in industry:
* sufficiently high specific surface area.
* high porosity.
* convenient pore size and particle size distribution.
* high mechanical strength.
* sufficient resistance to sintering, fouling and poisoning.
It is therefore obvious the usefulness to disperse these active species on the surface of a support exhibiting more than one of the mentioned characteristics. The most important oxidic supports used for the preparation of supported catalysts are illustrated in table 1. The deposition of species containing the active element on the surface of

415

J. Fraissard and L. Petrakis (eds.), Acidity and Basicity of Solids, 415–444.
© 1994 Kluwer Academic Publishers.

Table 1. Illustrates the main characteristics of the most important
oxide supports.

Support	SSA/$m^2.g^{-1}$	*OH/nm^2
γ-Al$_2$O$_3$	120	8
SiO$_2$	600	5
TiO$_2$(mixture of anatase and rutile)	50	12

* number of hydroxyl groups per nm^2.

oxidic supports is one of the most critical steps in the preparation of
supported catalysts. This deposition is usually performed by impregnating powder or pellets of the support in an electrolyte solution containing one or more inorganic species of the element to be deposited
(e.g. 1-5). Three types of impregnation are usually followed:
* dry or pore volume impregnation.
* non dry impregnation.
* successive dry impregnations.
In most cases impregnation is followed by drying and calcination. Non-
dry or successive dry impregnations are followed when the amount of the
active element to be deposite cannot be disolved in a volume of the impregnating solution equal to the pore volume of the impregnated support.
    Following the above procedures the deposition of the active element
takes place via three mechanisms:
* by precipitation mainly inside the pores of the support
* by adsorption on the support surface.
* by reaction with the support surface.
The precipitation may be controlled or uncontrolled. The uncontrolled
precipitation, taking place in the step of drying, prevails when the impregnation, usually the pore volume impregnation, is followed by drying.
It results in the formation of large supported crystallites and thus
gives supported catalysts with poor dispersion of the active element.
Therefore, impregnation followed by drying is only used when large
amounts of inexpensive active ions should be deposited on the support
surface. In this case we may increase the active surface by simply increasing the amount of the depositing active species. Controlled
precipitation could be achieved by adding several substances in the impregnating suspension. Relatively small supported crystallites are obtained using "deposition-precipitation" and this new technique is quite
promissing for preparing supported catalysts (e.g. 6-10).
    Extremely small supported crystallites are obtained when adsorption or
chemical reaction is the predominant deposition process. A very simple
way to increase their contribution on the whole deposition is to impreg-
nate the support in a relatively large volume of an electrolyte solution
containing the species to be deposited and to separate the liquid from
the solid phase by filtration after equilibration, under stirring, for

many hours. Following this technique, called "equilibrium adsorption", we may, in effect, increase the relative contribution of adsorption and reaction, which take place in the step of the long-time equilibration, with respect to the uncontrolled precipitation occuring in the step of drying which follows filtration (11-41). An alternative way to maximize deposition by reaction is to use organometallic or carbonyl species to deposit the active element (e.g. 42-46). However, the procedure required for anchoring is usually quite complicated and the extremely high active surface obtained, being rather unstable, usually diminishes considerably upon calcination. These are presumably the main reasons for which grafted catalysts have not yet found industrial applications. In the contrast to that "equilibrimum adsorption" is a very simple technique and in this respect is better compared with anchoring. However, although equilibrium adsorption results to catalysts with high dispersion of active element and therefore is suitable for depositing expensive ions, it has the following weaknesses. It some times results to catalysts with low active surface. This is due to the fact that usually the density of the depositing sites (adsorption or reaction sites) of the industrial oxidic supports is usually low, thus limiting the amount of the species deposited by adsorption or reaction. Therefore, the increase of the density of the depositing sites is necessary when supported catalysts with high active surface have to be prepared by equilibrium adsorption.

The presentation of the methods developed so far for increasing the density of the depositing sites located on the surface of the oxidic supports is the subject of the present lecture.

2. Determination of the surface parameters related with the depositing sites.

Before starting with this presentation it is necessary to discuss the nature of the depositing sites and to describe the methodology followed for determining the surface parameters closely related with these sites.

2.1. THE NATURE OF THE DEPOSITING SITES.

It is well known that the surface of the oxidic supports illustrated in table 1 is fully hydroxylated in electrolyte suspensions. According to the surface ionization model(47) the surface hydroxyls may be protonated or deprotonated. Fig. 1 illustrates the typical surface structures of these supports in electrolyte solutions.

The charging the surface mechanism may be adequately described by the following acid-base equilibria.

$$-SOH_2^+ \overset{K_1^{int}}{\rightleftharpoons} -SOH + H_s^+$$
$$-SOH \overset{K_2^{int}}{\rightleftharpoons} -SO^- + H_s^+ \qquad (1)$$

$$H_s^+ \rightleftharpoons H_b^+$$

SOH, SOH_2^+ and SO^- represent, respectively, the neutral, protonated and deprotonated surface hydroxyls. H_s^+ and H_b^+ represent the hydrogen ions on the surface and in the impregnating solution, respectively. K_1^{int} and K_2^{int} symbolize the acidity constants. The suspension pH at which the

Fig. 1 illustrates the surface structure of the oxidic supports in electrolyte solutions. S: Al, Si, Ti. (a): very low pH, (b): pH near to the point of zero charge (p.z.c.) (c): very high pH.

concentration of the positively charged groups equals the concentration of the negatively charged groups is defined as the point of zero charge (p.z.c). At pH values exceeding the p.z.c. the negatively charged groups on the surface of the oxidic support predominate, while at pH values lower than that corresponding to p.z.c. the positively charged groups are in excess (fig. 1). It is well established that p.z.c. is a surface property depending exclusively on the nature of the oxidic support and that near p.z.c the neutral surface hydroxyls usually predominate. It will be proved later that for simple oxides p.z.c is equal to $[pK_1^{int} + pK_2^{int}]/2$.

As the surface of the oxidic supports is generally charged in aqueous solutions, an electrical double layer is formed between the support surface and the electrolyte solution. Various models have been developed to describe the oxide/solution intrface. The "constant capacitance model" of Stumm and co-workers (48-50), the "diffuse layer model" of Stumm, Huang and Jenkins (51-52), the "Stern model" modified by Westall and Hohl (53), the "site-binding model" of Yates et al (54) and finally the triple layer model developed by Davis et al (55, 56) should be mentioned. A simplified picture of the triple layer model is illustrated in fig. 2. It should be noticed that the SOH_2^+, SOH and SO^- groups are considered to be localized on the surface of the support. On the other hand the centers of the water molecules surrounding the surface of the support particle constitutes the so-called Inner Helmholtz Plane(IHP).

Numerous results have shown that the species containing the catalytically active ions are also deposited on the IHP by replacing water Molecules. (e.g. 36-40). It has been, moreover, found that the negatively charged species, e.g. MoO_4^{2-} ions, are retened on the IHP by two ways: First by electrostatic interactions with the SOH_2^+ groups. Second by reaction with the SOH groups (35, 36, 38, 40). On the other hand it has been found that the positively charged ions such as Co^{2+}, and Ni^{2+} are retened by electrostatic interactions with the SO^- surface groups(39). Thus, it may be concluded that the SOH_2^+ and SOH groups are responsible for the creation of the depositing sites for the negatively charged species whereas the SO^- groups are responsible for the creation

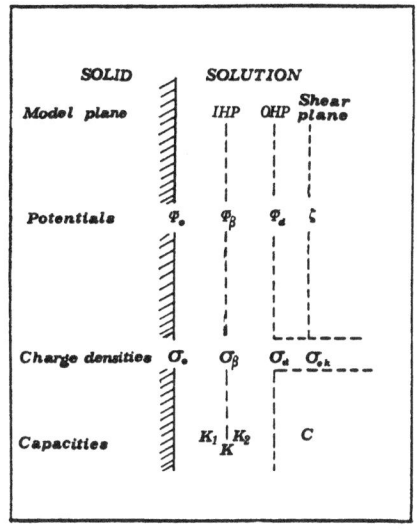

Figure 2. Schematic representation of the triple-layer model.

of the depositing sites for positive species. The above clearly show that the regulation of the concentration of the forementioned surface groups is necessary in order for the concentration of the depositing sites and therefore the active surface to be maximized. As already mentioned the presentation of the methods developed so far for regulating the concentration the forementioned surface groups is the subject of the present lecture.

2.2. DETERMINATION OF p.z.c., pK_1^{int}, pK_2^{int} AND THE CONCENTRATION OF THE CHARGE SURFACE GROUPS.

2.2.1. *Determination of p.z.c. (57-68)*. Although several methods may be used, the potentiometric titrations of the electrolyte suspension containing the support particles is usually employed for determining the title parameters (57-68). These titrations are performed at a given temperature using a thermostated double walled Pyrex vessel with a Perspex lid equipped with holes for electrods and nitrogen gas. [fig. 3]. Titrant solutions are delivered by a microburette and pH is recorded with a recording pH-meter equipped with a combination pH electrode. The titration of the support suspension is performed at three ionic strengths, for instance at 0.1, 0.01 and 0.0001N potassium nitrate usually used as background electrolyte.

A time period of 18-20h is proved necessary to reach a constant value of the suspension pH. Following equilibrium a small quantity (0.1-0.5 ml) of a strong base, usually 0.1N KOH solution, is added to render the support surface negative by deprotonating all surface hydroxyls. After 15-20 min, the new equilibrium value at pH is recorded. This value is noted as initial pH. The suspension is then titrated by adding small aliquots (15-20 ml) of a strong acid, usually nitric acid, and pH is re-

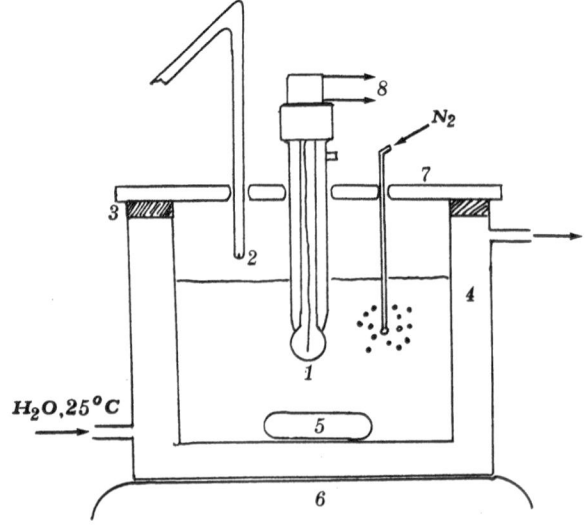

Figure 3.The system used for the potentiometric titrations: (1) combined glass/Ag–AgCl electrode, (2) microburette, (3) Perspex lid, (4) thermostated double walled pyrex vessel, (5) magnetic bar, (6) magnetic stirrer, (7) N_2 supply, (8) to pH-meter.

corded as a function of the volume of titrant added to the suspension. Titrant additions and corresponding pH values of the suspensions are usually recorded every 2 min.

Provided that the ions of the indifferent electrolyte do not hydrolyse or specifically adsorb the only equilibrium taking place in the bulk solutions is

$$H_2O \overset{Kw}{\rightleftharpoons} H_b^+ + OH_b^- \qquad (2)$$

Thus, we write down the following mass balance equation for the hydrogen ions (69)

$$H_{add}^+ = H_{ads}^+ + H_{ac}^+ + H_{H2O}^+, \qquad (3)$$

where H_{add}^+, H_{ads}^+, H_{ac}^+ and H_{H2O}^+ represent, respectively, the hydrogen ions added to the suspension during titration, the hydrogen ions uptaken by the surface $\underline{via}$ equilibria (1), the hydrogen ions accumulated in solution decreasing pH and the hydrogen ions reacted with the OH_b^- through the equilibrium (2). It is rather obvious that H^+_{add}, H^+_{ads}, H^+_{ac} and H^+_{H2O} are respectively equal to $C.\Delta V$, $\{(SOH_2^+)-(SOH_2^+)_{in}+(SO^-)_{in}-(SO^-)\}.WS=\{[(SOH_2^+)-(SO^-)]-[(SOH_2^+)_{in}-(SO^-)_{in}]\}WS=QWS$(where $Q=\{[(SOH_2^+)-(SO^-)]-[(SOH_2^+)_{in}-(SO^-)_{in}]\}$, $[(V+\Delta V).C_{H+}-VC_{H+,in}]$ and $[VC_{OH-,in}-(V+\Delta V)C_{OH-}]$. By C, ΔV, V, (SOH_2^+), (SO^-), W, S, C_{H+} and C_{OH-} we denote, respectively, the concentration of the titrant (mol.dm^{-3}) the titrant volume increment (dm^3), the volume of the suspension (dm^3), the concentration of the SOH_2^+ groups (microequivalents per m^2), the concentration of the SO^- groups (microequivalents per m^2), the weight (g) and the specific surface area (m^2.g^{-1}) of the support dispersed and the concentration of the H$^+$ and

OH⁻ in the solution (mol.dm⁻³). The symbol "in" stands for the initial state. The values of C_H^+, C_{OH}^-, $C_{H}^+{}_{in}$ and $C_{OH}^-{}_{in}$ are related with pH and the activity coefficient for monovalent ions with the following relationships.

$C_H^+ = 10^{-pH}/\gamma_1$,
$C_{OH}^- = 10^{-(pKw-pH)}/\gamma_1$,
$C_{H.in} = 10^{-pH,in}$ and
$C_{OH^-,in} = 10^{-(pKw-pH,in)}/\gamma_1$ (4)

Substituting the terms of eqn(3) with their equals and taking into account the equations (4) we obtain

$Q = \{C.\Delta V - [((V+\Delta V)10^{-(pH)} - 10^{-(pKw-pH)})/\gamma_1 - V(10^{-(pH)in} - 10^{-(pKw-pH)in})/\gamma_1]\}$ /w.S (5)

The values of the activity coefficient for the z-valent ions may be calculated from the ionic strength of suspension, I, using the Davies equation (70)

$$\log \gamma_z = -AZ^2\{[I/(1+I^{1/2})]^{1/2} - 0.3I\}$$ (6)

The value of the parameter A depends on the suspension temperature (70). The pKw values may be determined from the following equation (71)

$$pKw = 447.33/T - 6.0846 + 0.017053\ T$$ (7)

The values of $Q = \{[(SOH_2^+) - (SO^-)] - [(SOH_2^+)_{in} - (SO^-)_{in}]\}$ determined <u>via</u> eqn (5) at each pH are generally dependent on the ionic strength of the suspension. However, at p.z.c. where $[(SOH_2^+) - (SO^-)] = 0$ the value of Q being equal to $-[(SOH_2^+)_{in} - (SO^-)_{in}]$ is constant irrespective of the value of the ionic strength. A typical example is illustrated in fig 4.

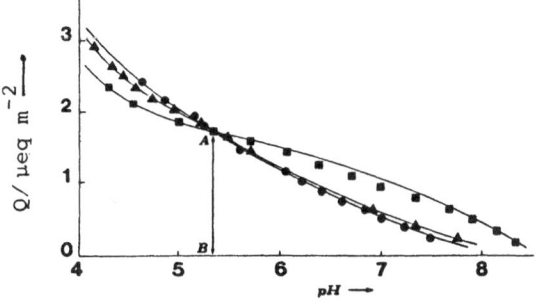

Figure 4. Relative adsorption density Q versus pH for γ-Al$_2$O$_3$.

The appererance of a common intersection point allows the graphical determination of the point of zero charge of a support. Moreover, the y axis could be rearranged so that the intersection point corresponds to a net surface charge equal to zero. This is obtained by substracting the quantity $-[(SOH_2^+)_{in}-(SO^-)_{in}]$ corresponding to the distance AB from each value of the y axis. The curves thus obtained may be transformed to illustate the variation of the surface charge, $\sigma_o=F[(SOH_2^+)-(SO^-)]$, with pH of the suspension. A typical example is illustrated in figure 5.

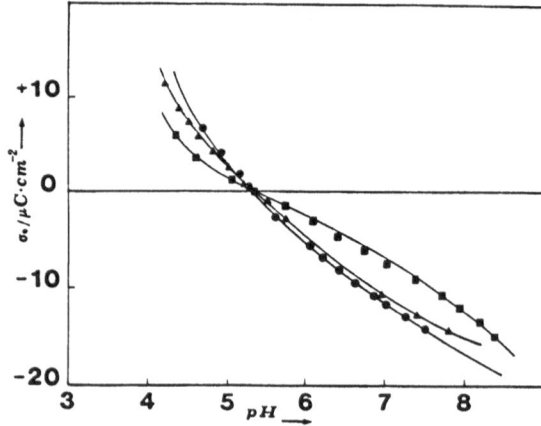

Figure 5. Surface charge σ_o versus pH for γ-Al$_2$O$_3$ (●,0.1; ▲,0.01; ■,0.001 M KNO$_3$).

2.2.2. *Determination of acidity constants K_1^{int} and K_2^{int} (57-68)*. The acidity constants K_1^{int} and K_2^{int} illustrated in the equilibria (1) are defined by

$$K_1^{int} = (H_s^+)\ (SOH)/(SOH_2^+) \tag{8}$$

$$K_2^{int} = (H_s^+)(SO^-)/(SOH) \tag{9}$$

The surface concentration of the protons adsorbed on the solid particles is related to the bulk solution concentration by

$$(H_s^+) = (H_b^+)\exp\,(-\,F.Yo/RT), \tag{10}$$

where F and Yo denote the Faraday constant and the potential on the surface of solid particles, respectively. The fraction of the positively, a_+, and negatively, a_-, charged sites is related respectively with the positive, $+\sigma_o$, and negative, $-\sigma_o$, surface charge and the total concentration of the SOH$_2^+$, SOH and SO$^-$ surface groups, denoted by N$_s$.

$$a_+ = +\sigma_o/N_s = (SOH_2^+)/[(SOH_2^+)+(SOH)] \tag{11}$$

$$a_- = -\sigma_o/N_s = (SO^-)/[(SO^-) + (SOH)] \tag{12}$$

The value of Ns, usually determined by tritium exchange, (72,73), depends on the nature of the support. The $+\sigma o$ and $-\sigma o$ are determined at pH values lower and higher than p.z.c., respectively, where the concentrations of SO$^-$ and SOH$_2^+$ are considered to be negligible, respectively. Combinations of the equations (8), (10) and (11) as well as of the equations (9), (10) and (12) allows the derivation of the equations (13) and (14).

$$pK_1^{int} = pH + \log[a_+/(1-a_+)] + F.Y_o/2.3RT \qquad (13)$$

$$pK_2^{int} = pH - \log[a_-/(1-a_-)] + F.Y_o/2.3RT \qquad (14)$$

Unfortunately the potential, Y_o, on the surface of the support particles is unknown over the pH range with the exception of the p.z.c. where it is zero. Therefore, equations (13) and (14) cannot be used for a direct determination of the surface acidity constants. However, these constants may be determined by extrapolating the curves pH + log[a_+/(1-a_+)] vs a_+ and pH-log[a_-/(1-a_-)] vs a_- at a_+ and a_- near to zero, namely near to p.z.c., where Y_o=o and the term $F.Y_o/2.3RT$ is considered to be negligible. Thus, the intercepts of the above mentioned curves become equal to pK_1^{int} and pK_2^{int}, respectively. The graphical determination of pK_2^{int} for a typical sample is illustrated in fig. 6.

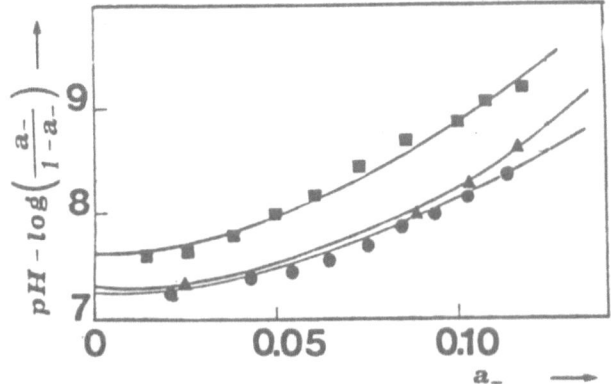

Figure 6. Variation of pH-log(a_-/1-a_-) versus a_-for γ-Al$_2$O$_3$ ($\bullet$,0.1; $\blacktriangle$,0.01; $\blacksquare$,0.001 M KNO$_3$).

2.2.3. *Determination of the concentration of the surface groups SOH$_2^+$, SO$^-$ and SOH (57-68).* The values of N_s (groups.nm^{-2}), specific surface area (SSA, m^2.g^{-1}), solid concentration (SC, g.l^{-1}), capacitance (is an adjustable parameter denoted by C), ionic strength (I), Kw, K_1^{int}, K_2^{int} and a guessed value for (SOH) (mol.l^{-1}) were inserted in a computer program (SURFEQL) and the equilibrium concentration initially as mol.l^{-1}, for SOH, SOH$_2^+$ and SO$^-$ as well as the surface charge σo (Cb.m^{-2}) and surface potential (Yo, volts) may be calculated at each pH. Specifically the values of N_s are used to determine T^{exp}·$_{groups}$, namely the total concentration of surface groups (expressed in mol.l^{-1}) using eqn (15).

$$T^{exp}·_{groups} = N_s. SSA. SC. 10^{18}/6.023.10^{23} \qquad (15)$$

The values of ionic strength are used to calculate the values of the activity coefficient for H^+ ions in order to correct the value of Kw. The guessed value of (SOH) allows the calculation of (SOH_2^+) and (SO^-) at each pH using the equations (8), (9) and (10) and assuming initially that $\log[\exp(-F.Yo/RT)]=1$. Moreover, the mass action equation

$$(OH^-)_b = Kw \ (H^+)_b^{-1} \qquad\qquad (16)$$

allows the detrmination of the $(OH^-)_b$ at each pH. The so calculated values of (SOH_2^+), (SO^-) and (OH^-) are inserted into the mass balance equations

$$(SOH_2^+) + (SOH) + (SO^-) = T^c{}_{groups} \qquad\qquad (17)$$
$$(H^+)+(SOH_2^+)-(SO^-) = T_{H+} \qquad\qquad (18)$$

$$(SOH_2^+)-(SO^-) = T\sigma = C.Yo.SSA.SC/F \qquad\qquad (19)$$

and the calculated total concentration of the surface groups, $T^c{}_{groups}$, the amount of H^+ ions added or removed to suspension in order to reach a given pH value, T_{H+}, the total surface charge $T\sigma(eq.l^{-1})$ and therefore the surface charge, σ, and the surface potential may be calculated for a given pH. The $T^c{}_{groups}$ value is compared with the $T^{exp}{}_{groups}$ value and based on convergence criterion the procedure is automatically repeated by testing various values for (SOH) until $T^c{}_{groups}=T^{exp}{}_{groups}$. The validity of the procedure may be then tested by comparing the values of surface charge determined experimentally at various pH's with the corresponding calculated values.

It should be stressed that the above described procedure for calculating $K_1{}^{int}$, $K_2{}^{int}$, (SOH_2^+), (SO^-) and (SOH) may be followed only in the case where no specific adsorption takes place. In this case, therefore, the constant capacitance model may be employed due to the relatively high values of ionic strength. On the contrary, in the case where an ion is specifically adsorbed the triple layer model must be used rendering the procedure more complicated. Moreover, it should be noticed that the above derscribed procedure may be followed only in the case of simple oxides like γ-alumina and silica. It has been recently proved (63) that this cannot be followed for a mixte oxide like the comercial titania which contain rutile and anatase. However, in this mixte oxide it can be calculated p.z.c. and surface charge. The later parameter may be used to estimate the relative concentration of the positive, negative and neutral hydroxyls in this oxide.

3. Regulation of the concentration of the depositing sites.

Now we shall try to present the methods developed in order to regulate the concentration of the SOH_2^+, SOH and SO^- groups responsible for the creation of the depositing sites for negative and positive species, respectively. The following methods will be discussed:
* the change of pH of the impregnating solution.
* the change of temperature of the impregnating solution.
* the doping of the support.

3.1.CHANGE OF pH OF THE IMPREGNATING SUSPENSION (57-68).

Inspection of the equilibria (1) clearly shows that the simplest way to regulate the concentration of the formentioned groups is to change pH of the impregnating solution. Fig 7 shows the variation with pH of the SOH_2^+, and SO^- groups, for γ-alumina and silica as well as the variation of the surface charge for titania. In agreement with what it is anticipated from equilibria (1) for γ-alumina and silica it can be observed that increase in the pH of the impregnating solution increases the concentration of the negatively charged groups mainly at expenses of the positively charged groups. Similarly, for the titania increase in pH of the impregnating solution decreases the surface charge.

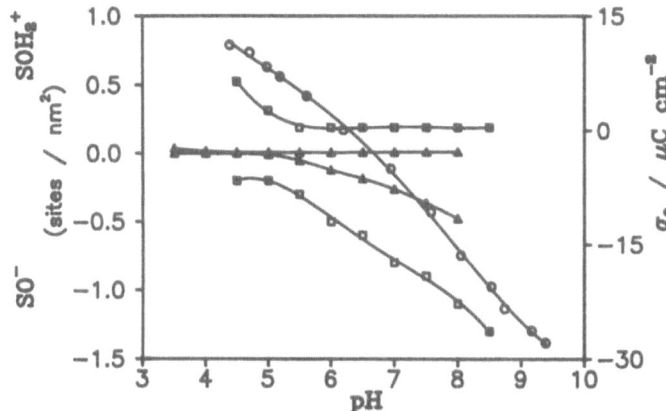

Figure 7.Variation with pH of the SOH_2^+ and SO^- groups for (□) γ-Al$_2$O$_3$ and (△) SiO$_2$ and of surface charge for (○) TiO$_2$ as well.

3.2. CHANGE OF THE IMPREGNATING TEMPERATURE (60, 62-66, 68).

The above method, seemingly very simple, may become problematic in some cases. More specifically, at the pH value where deposition by adsorption or reaction is enhanced, the species to be deposited may be unstable and the support may be partially dissolved. Moreover, deposition by spontaneous precipitation may occur in considerable extent, resulting to low dispersity of the supported phase. The necessity to develop alternative methods for regulating the relative concentrations of the SOH, SOH_2^+ and SO^- groups has been realized some years ago. Inspection of the equilibria (1) shows that at constant pH we could change the concentration of the formentioned groups if we could develop a method for changing the acidity constants K_1^{int} and K_2^{int}. In this point it should be noted that these constants are surface parameters depending exclussively on the nature of the support surface. Therefore, the acidity constants should be related with the point of zero charge which is also a surface parameter. In fact, by combining equations (8), (9) and (10) and taking into account that at p.z.c. $Y_0=0$ and $(SO^-)=(SOH_2^+)$ we may easily obtain

$$(pK_1^{int}+pK_2^{int})/2 = p.z.c. \qquad (20)$$

Equation (20) reveals the fundamental character of p.z.c. Moreover, it shows that the development of a methodology for changing the values of the acidity constants should also be a methodology for regulating p.z.c. Observing again the equilibria (1), the idea to change the values of $K_1{}^{int}$ and $K_2{}^{int}$ and therefore the value of p.z.c. by altering the temperature of the impregnating solution comes quite easily.

3.2.1. *γ-alumina*. We have tested for a first time this simple idea six years ago on γ-alumina (60) and then on silica (60) and titania (63). Figures 8(a) and 8(b) show the plots of $\ln K_1{}^{int}$ and $\ln K_2{}^{int}$ respectively against the inverse temperature of the impregnating suspension for γ-alumina. The quite good Van't Hoff curves achieved show that decrease in the impregnating temperature increases the acidity constants. Therefore, the deprotonation precesses described by equilibria (1) should be exothermic. To determine the thermodynamic parameters for these surface deprotonation we have assumed that the surface of the γ-Al_2O_3 particles may be simulated with a two-dimensional ideal solution. In fact in this case, the chemical potential for each species involved in the equilibria is given by the well known equation (74):

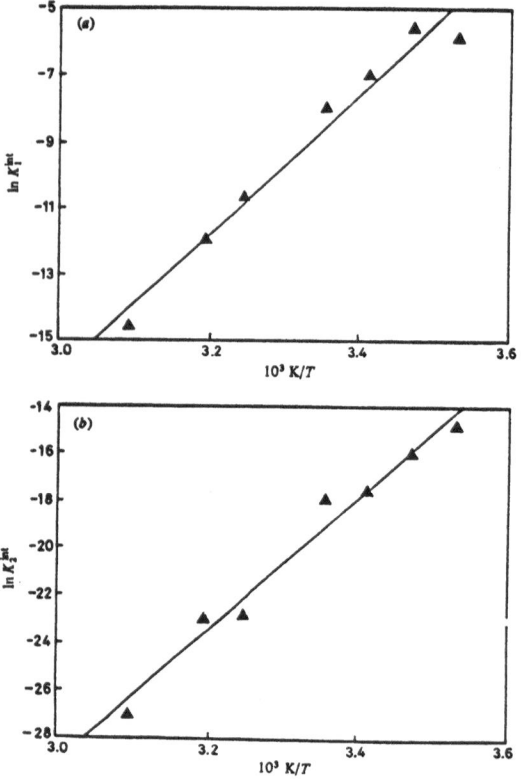

Figure 8. Plots of (a) $\ln K_1{}^{int}$ vs, T^{-1} and (b) $\ln K_2{}^{int}$ vs. T^{-1} obtained for the surface dissociation of the $AlOH_2{}^+$ and AlOH, respectively.

$$\mu_i = \mu_i^\circ - RT \ln C_i \qquad (i = SOH_2^+, \; SO^-, \; SOH \; or \; H_s^+), \qquad (21)$$

where C_i represents the surface concentration (mol.dm^{-2}) of the ith species and μ_i° its standard-state chemical potential taken for $c_i = 1$ mol.dm^{-2}. Starting from the above expression for the chemical potential and following a well established procedure (75) we may easily derive eqns(22) and (23).

$$\ln K_1^{int} = -\Delta H_1^\circ / RT + \Delta S_1^\circ / R \qquad (22)$$

$$\ln K_2^{int} = -\Delta H_2^\circ / RT + \Delta S_2^\circ / R \qquad (23)$$

These equations allowed us to determine the standard-state enthalpies and entropies of the surface deprotonations from the slope and intercept, respectively, of the Van't Hoff curves achieved. The values so achieved are illustrated in table 2. Carefull observation of the table shows that the protonation-deprotonation equilibria in the γ-alumina are related with rather high values of thermodynamic parameters. The negative values achieved for the enthalpy confirmed the highly exothermic character of the deprotonation processes. From the above it may be easily anticipated that increase in the impregnation temperature should Increase pK_1^{int} and pK_2^{int} and therefore p.z.c. Fig 9 shows the fine regulation of p.z.c. with changing of the impregnation temperature in the range 10-50°C. The very good agreement between the calculated from

Table 2. Enthalpies and entropies of the surface processs illustrated in equilibria (1)[a]

i	$\Delta H^\circ_i / kJmol^{-1}$	$\Delta S^\circ_i / JK^{-1}mol^{-1}$
1	-180 ± 10	-660 ± 50
2	-230 ± 20	-940 ± 60

[a] Errors given are the standard deviations.

eqn(20) and the experimentally determined values of p.z.c., corroborates the correctness of the method followed for determining K_1^{int} and K_2^{int}. The practical consequences from a preparative point of view of the regulation obtained for p.z.c. are obvious: a simple rise of temperature from 25 to 50°C is sufficient for an extension of the pH range in which the surface of γ-Al$_2$O$_3$, being positive, may adsorb negatively charged species in considerable extent from 1.00-5.30 to 1.00-8.95. Moreover, a decrease of temperature from 25 to 10°C allows an enlargement of the pH range in which γ-Al$_2$O$_3$ surface, being negative, may adsorb positively charged species by almost 1pH unit.

From the above described influence of the temperature on the value of the acidity constants it is easily anticipated that increase in the impregnating temperature should cause an icrease in the concentration of the positive surface groups as well as a decrease in the concentration of the negative surface groups. This anticipation was confirmed at three

428

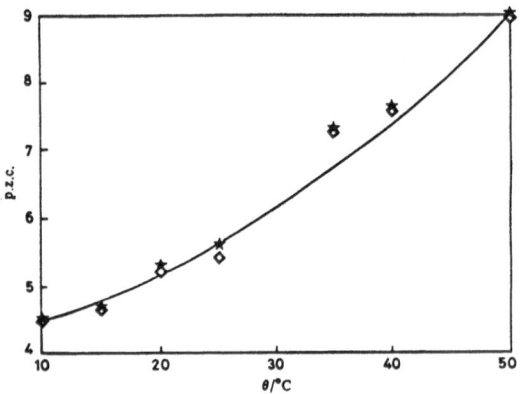

Figure 9. Variation of the experimental and calculated values of p.z.c. with the temperature of γ-Al₂O₃ suspension. (◇Experimental p.z.c., ★ calculated p.z.c.)

ionic strengths over the pH range examined. A typical example is illustrated in fig. 10. An inspection of this figure shows that the variation in the temperature of the impregnating solution is an attractive
method not only for regulating the p.z.c. but also for obtaining a
desired concentration of the protonated and deprotonated surface
hydroxyls of γ-alumina at a given pH and ionic strength. It was,
moreover, found that the influence of the ionic strength on the concentration of the charged surface groups is negligible as compared with
that of temperature.

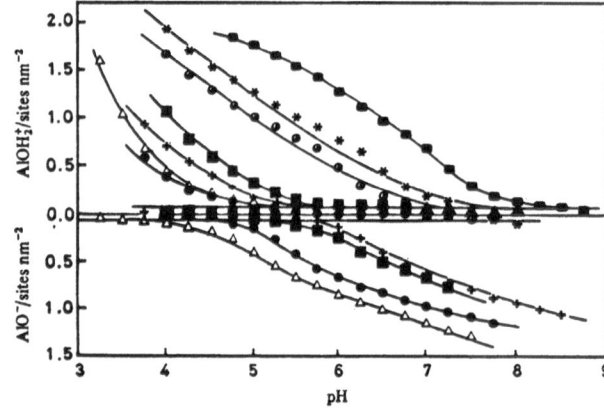

Figure 10. Variation of the concentration of the charged groups (AlOH₂⁺ and AlO⁻) with the temperature of the γ-Al₂O₃ suspension at ionic strength corresponding to 0.1moldm⁻³KNO₃ (△,10°C; ●, 15°C; #,20°C; ■, 25°C; ◐,35°C; ✱,40°C; ■,50°C).

3.2.2. *silica*. With the underlined idea to develop industrial supports
with controlled p.z.c. and concentration of surface groups we extended
the methodology of changing the impregnation temperature to silica (62),
an important carrier used for the preparation of many supported
catalysts. Although this support presents a very high specific surface
area, it has two disadvantages, quite important from the viewpoint of

the preparation of supported catalysts by equilibrium adsorption: (i)
Although the pH range in which SiO_2 is soluble depends on the
preparation/pretreatment conditions, in most cases it is partially dissolved at pH higher than = 5.50, making it difficult to deposit an active ion by adsorption (or surface reaction with neutral hydroxyls) of a
positively charged species beyond this pH. (ii) Since the p.z.c. of this
support is equal to 2.80, only a very narrow pH range is suitable for
the deposition of negative species. The deposition of positive species,
on the other hand, does not present serious problems, because it may be
done over a wide pH range (pH>2.80), though with certain difficulties at
pH higher than ~5.50. Thus, an increase of the p.z.c. of SiO_2 from 2.80
to 5.50 as well as an increase of the concentration of the protonated
surface groups in the pH range 1.00–5.00 would increase the amount of
negative species deposited on this support by equilibrium adsorption in
this pH range and thus the active surface of the resulting supported
catalysts.

In view of the above requirements and the results obtained in
γ-alumina we applied the method of changing the temperature of impregnating suspension to silica (62). Fig's 11a and 11b illustrate the
variation of lnK_1^{int} and lnK_2^{int} with the inverse of the absolute temperature of the impregnating suspension. It can be observed that increase in the impregnating temperature increases the values of K_1^{int} and
K_2^{int} indicating that the deprotonation processes are endothermic in the
case of silica. Following the procedure adopted in the case of γ-alumina

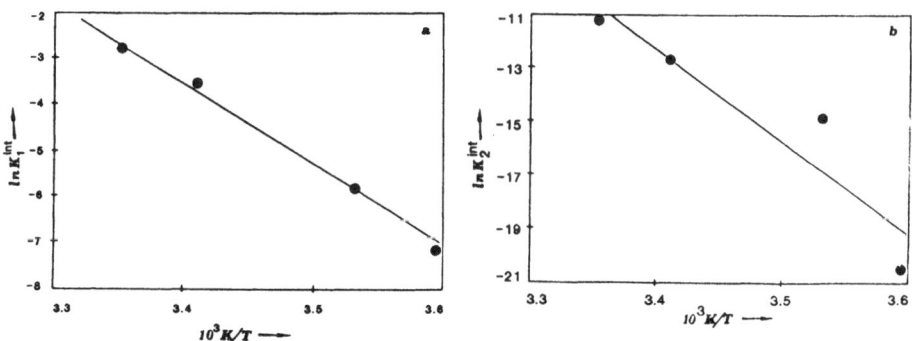

Figure 11. Plots of the lnK_1^{int}(a) and lnK_2^{int}(b) versus 1/T for the
surface dissociation of the $SiOH_2^+$ and $SiOH$ groups, respectively.

we determined the enthalpies and entropies of the surface processes illustrated by equilibria (1). Their values are compiled in table 3. Comparison of the ΔH_1^o with ΔH_2^o value shows that K_2^{int} is more
temperature-sensitive than K_1^{int}. Moreover, comparison of the influence
of temperature on K_1^{int} and K_2^{int} observed in silica with the one observed in γ-alumina shows that the direction of the enthalpy changes
which accompany the deprotonation of the protonated and neutral

Table 3. Enthalpies and Entropies of the Surface Processes Illustrated
in equilibria (1)[a].

i	$\Delta H^{\circ}{}_i / KJmol^{-1}$	$\Delta S^{\circ}{}_i / JK^{-1}mol^{-1}$
1	150±10	490±30
2	290±80	900±300

[a] Errors given are standard deviations.

hydroxyls depends on the nature of the support. From the above one may
anticipate that the p.z.c. should decrease with the temperature of the
impregnating suspension. Fig.12 shows that this is, in effect, the case.

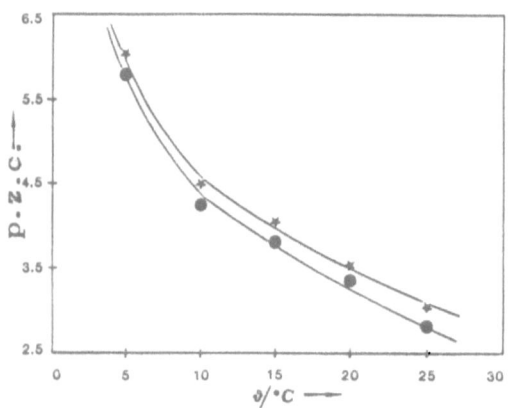

Figure 12. Variation of
experimental (●) and
calculated from
$(pK_1{}^{int}+pK_2{}^{int})/2$ (★)
values of p.z.c. with the
suspension temperature of
the untreated SiO_2.

It can be observed that a precise regulation of the p.z.c. from 2.80 to
5.78 can be achieved by a simple decrease of the suspension temperature
from 25 to 5°C. The practical consequences of this effect are obvious: A
decrease in the temperature of the impregnating suspension from 25 to
5°C allows for an extension of the pH range in which silica can adsorb
negatively charged species almost 3 pH units. The decrease in temperature of the impregnating suspension may have another favorable effect:
It will possibly decrease the surface solubility of silica. As in the
case of γ-alumina the satisfactory agreement obtained between the experimental and calculated values of p.z.c. using the corresponding
values of $K_1{}^{int}$ and $K_2{}^{int}$ corroborates the validity of the method followed for the determination of the surface acidity constants.
   Due to the endothermic character of the deprotonation processes in the
case of silica it is anticipated that the concentration of the
protonated surface hydroxyls should increase whereas the concetration of
the deprotonated surface hydroxyls shoud decrease as temperature
decreases. Inspection of fig's 13 and 14 shows that this is, in effect,
the case. In particular it may be observed that the intensity of both
effects depends on pH.

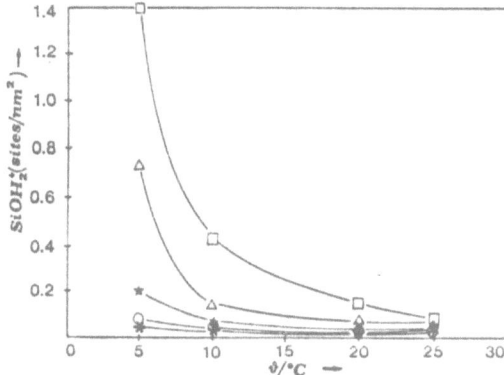

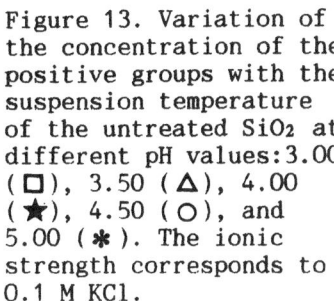

Figure 13. Variation of the concentration of the positive groups with the suspension temperature of the untreated SiO_2 at different pH values: 3.00 ($\square$), 3.50 ($\triangle$), 4.00 ($\bigstar$), 4.50 ($\bigcirc$), and 5.00 ($\ast$). The ionic strength corresponds to 0.1 M KCl.

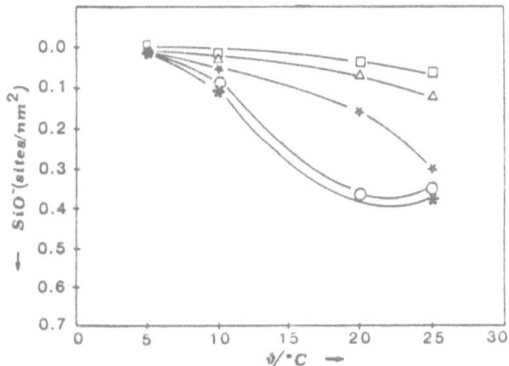

Figure 14. Variation of the concentration of the negative groups with the suspension temperature of the untreated SiO_2 at different pH values: 3.00 ($\square$), 3.50 ($\triangle$), 4.00 ($\bigstar$), 4.50 ($\bigcirc$), and 5.00 ($\ast$). The ionic strength corresponds to 0.1 M KCl.

3.2.3. *Titania*. The method of changing the temperature of the impregnating suspension was finally applied to the commercial titania used in catalysis (63). As already mentioned it is a mixture of two pure oxides, namely anatase and rutile. Therefore, two charging mechanisms, one for each oxide, should be taken into account.

$$
\begin{aligned}
Ti^rOH_2^+ &\xrightleftharpoons{K_1{}^r} Ti^rOH + H_s{}^{r+} \\
Ti^aOH_2^+ &\xrightleftharpoons{K_1{}^a} Ti^aOH + H_s{}^{a+} \\
Ti^rOH &\xrightleftharpoons{K_2{}^r} Ti^rO^- + H_s{}^{r+} \\
Ti^aOH &\xrightleftharpoons{K_2{}^a} Ti^aO^- + H_s{}^{a+} \\
\\
H_s{}^{r+} &\xrightleftharpoons{} H_b{}^+ \\
H_s{}^{a+} &\xrightleftharpoons{} H_b{}^+
\end{aligned}
\tag{24}
$$

The question raised in this point is whether the common intersection point observed in the case of pure oxides which is identified as the point of zero charge is expected to be observed in the case of a mixture of oxides. Before trying to answer this question it is necessary to point out that for simple oxides, like γ-Al_2O_3 and SiO_2, the common intersection point (see for instance fig. 4) is interpreted as follows: At the pH corresponding to this point the ionic strength, namely the concentration of the counterions, has no effect on the value of Q because the concentration of SO^- is equal with the concentration of SOH_2^+. In fact, at this pH the concentrations of SO^- and SOH_2^+ are negligible compared with the concentration of SOH when $K_1^{int} \gg K_2^{int}$. In the case of a mixture of oxides, like titania, it has been established that the surface charge determined experimentally should be equal to that calculated from the weighted sum of the experimentally determined charges of the components, where the weight of a component is determined by its surface area fraction in the mixture (76). Thus, we may write

$$\sigma_o^T = f_r\sigma_o^r + f_a\sigma_o^a, \tag{25}$$

where σ_o^T, σ_o^r, σ_o^a, f_r and f_a represent, respectively the surface charge of the commercial titania, of rutile and anatase as well as the fraction of the total surface area contributed by rutile and anatase. Taking into account that $\sigma_o^r = e^-[(Ti^rOH_2^+)-(Ti^rO^-)]$ and $\sigma_o^a = e^-[(Ti^aOH_2^+)-(Ti^aO^-)]$ we obtain eqn (26) at the p.z.c. of titania, if exist, where $\sigma^T_o = 0$,

$$f_r(Ti^rOH_2^+)-f_r(Ti^rO^-)+f_a(Ti^aOH_2^+)-f_a(Ti^aO^-) = 0 \tag{26}$$

The above equation shows that generally would exist a pH value located in between p.z.c.^{r} and p.z.c.^{a} where $\sigma_o^T = 0$. Assuming that the p.z.c.^{r} is very close to the p.z.c.^{a} and that at pH where $\sigma_o^T = o$ $f_r(Ti^rOH)+f_a(Ti^aOH)$ takes its maximum value we may anticipate the presence of a common intersection point corresponding to the p.z.c. of titania, p.z.c.^{T}. Such a point has been observed corroborating the assumption mentioned. The presence of this intersection point allows the determination of p.z.c.^{T} and its surface charge following the method presented for the simple oxides, like γ-alumina and silica.

However, a major consequence of the fact that titania is actually a mixture of two simple oxides is that it is extremely difficult, though not imposible, to calculate the acidity constants and therefore the concentration of the surface groups. In fact, to the extent of our knowlege there is no way to determine simultaneously, from potentiometric titrations performed only on a mixture of simple oxides, the four acidity constants defined by the following equations.

$$K_1^r = \frac{(Ti^rOH)(H_s^{r+})}{(Ti^rOH_2^+)} \tag{27a}$$

$$K_1^a = \frac{(Ti^aOH)(H_s^{a+})}{(Ti^aOH_2^+)} \tag{27b}$$

$$K_2^r = \frac{(Ti^rO^-)(H_s^{r+})}{(Ti^rOH)} \tag{28a}$$

$$K_2^a = \frac{(Ti^aO^-)(H_s^{a+})}{(Ti^aOH)} \tag{28b}$$

$(H_s{}^{r+})= (H_b{}^+)\exp(-FY_o{}^r/RT)$ (29a) $H_s{}^{a+}=(H_b{}^+)\exp(-FY_o{}^a/RT)$,   (29b)

where $Y_o{}^r$ and $Y_o{}^a$ represent, respectively, the potential on the surface of rutile and anatase.

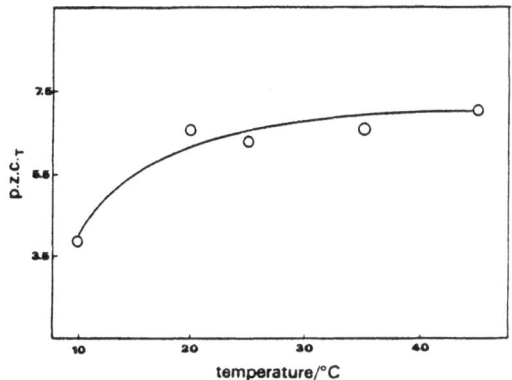

Figure 15. Variation of p.z.c.^{T} with the temperature of TiO$_2$ suspension.

In view of the above considerations it is necessary to limit our study to the influence of the impregnation temperature only on p.z.c.^{T} and $\sigma_o{}^T$. Fig. 15 shows that the p.z.c.^{T} increases with the temperature of the impregnating suspension. It can be observed that increase in this temperature from 10 to 45°C causes an increase in the p.z.c.^{T} from 3.84 to 7.06. Therefore, an extension of the pH range in which negative species can be adsorbed on titania should be expected by increasing the impregnation temperature. The above indicate the exothermic character of the deprotonation processes taken place on the surface of the commercial titania.

Fig. 16 illustrates the $\sigma_o{}^T$ vs pH curves obtained at different temperatures. These curves, though less informative compared with the curves of concentration of charged surface groups against pH, are quite useful for catalyst preparation because increase in the $\sigma_o{}^T$ is expected to bring about increase in the density of sorptive sites and therefore

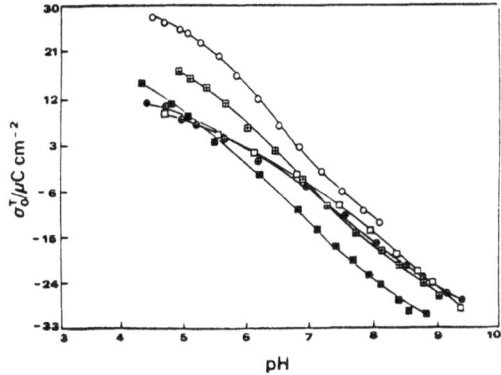

Figure 16. Surface charge of TiO$_2$($\sigma_o{}^T$) vs. suspension pH at different temperatures (°C): ■, 10; □, 20; ⊕, 25; ⊞, 35; ○, 45. Ionic strength: 0.1 mol dm^{-3} KNO$_3$.

in the extent of adsorption of species with opposite charge. It can be observed that an increase in the impregnation temperature causes an increase (decrease) of the positive (negative) surface charge of titania determined at a given pH. Therefore, an increase (decrease) in the extent of adsorption of negative (positive) species is expected by increasing the temperature of the impregnating solution. These are easily explained taking into account the exothermic character of the deprotonation processes mentioned above.

3.3. MODIFICATION OF THE SURFACE OF CARRIERS BY VARIOUS DOPANTS. (57-59, 61, 62, 64, 67, 68).

An alternative method to regulate p.z.c, acidity constants and concentration of surface groups is to modify the support surface using various dopants. A Systematic effort has been started seven years ago (57) though the doping of the support surfaces had been extensively studied in the past in order to regulate the physicochemical characteristics and the catalytic activity of supported catalysts (e.g. 77-81). The effort has been so far focused on the modification of γ-alumina, silica and titania with various amounts of sodium ions and on the modification of γ-alumina with various amounts of lithium and fluoride ions.

3.3.1. *Influence of modification on p.z.c. and surface acidity constants.* Let's starting with the influence of modification on p.z.c. and surface acidity constants. The Na^+, Li^+ and F^- doped specimens studied were prepared from the solid oxides by pore volume impregnation. The ions in the impregnating solutions were provided from the respective solid reagents ($NaNO_3$, $LiNO_3$ and NH_4F), dissolved in triply distilled CO_2-free water without any further purification. The impregnates were dried and then calcined in air. The sodium contained in the undoped γ-alumina originates from the carrier.

Inspection of figs 17 and 18 shows that the modification of the γ-alumina with relatively small amounts of Na^+ and Li^+ ions causes a considerable increase in pK_1^{int}, pK_2^{int} and p.z.c. Further increase in

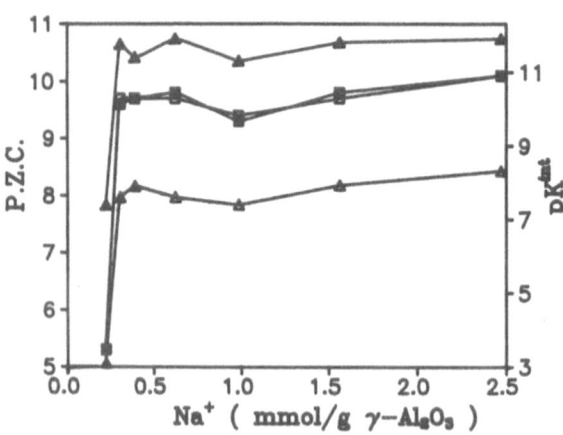

Figure 17. Variation of the experimental and calculated values of p.z.c. as well as of the surface acidity constants (pK_1^{int}, pK_2^{int}) with sodium content for Na^+ doped samples. ($\square$) Experimental p.z.c.,($\blacksquare$)calculated p.z.c. (p.z.c.=($pK_1^{int}+pK_2^{int}/2$) ($\triangle$) pK_1^{int}, ($\blacktriangle$) pK_2^{int}.

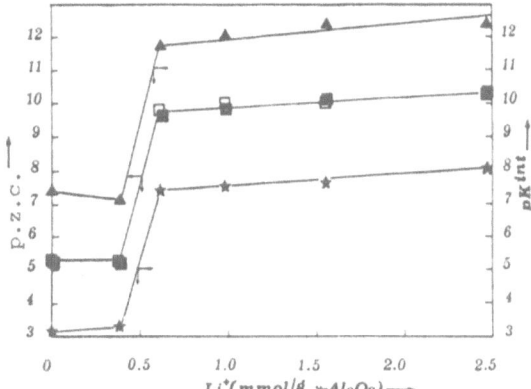

Figure 18. Variation of the experimental and calculated values of p.z.c. as well as of the surface acidity constants (pK_1^{int}, pK_2^{int}) with lithium content for Li^+ doped alumina samples. ($\square$) experimental p.z.c., ($\blacksquare$) calculated p.z.c. (p.z.c.=$pK_1^{int}+pK_2^{int}/2$) ($\bigstar$) pK_1^{int}, ($\blacktriangle$) pK_2^{int}.

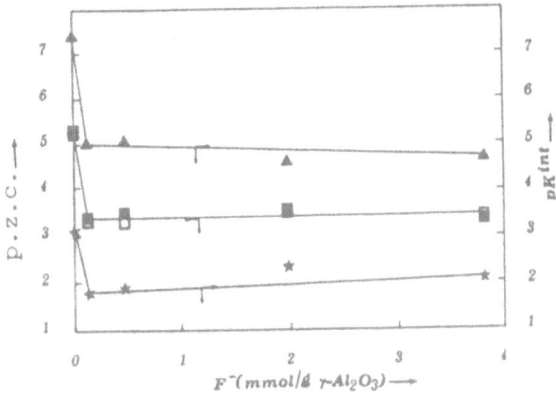

Figure 19. Variation of the experimental and calculated values of p.z.c. as well as of the surface acidity constants (pK_1^{int}, pK_2^{int}) with fluoride content for F^- doped alumina samples. ($\square$) experimental p.z.c., ($\blacksquare$) calculated p.z.c. (p.z.c.=$pK_1^{int}+pK_2^{int}/2$), ($\bigstar$) pK_1^{int}, ($\blacktriangle$) pK_2^{int}.

the Na^+ or Li^+ content does not increases considerably the above parameters. The opposite effects is caused due to the F^- doping (fig. 19). A quite small amount of F^- ions is sufficient to lower considerably p.z.c. as well as pK_1^{int} and pK_2^{int} whereas additional increase in the F^- concentration has practically no effect on these parameters. The sodium doping caused similar effects in the case of silica. In fact, fig. 20 shows that a considerable increase in p.z.c. may be achieved by doping silica with a small amount of sodium ions. On the other hand further increase in the sodium content causes a slight increase in p.z.c. Finally, an increase in p.z.c. from 5.41 to 7.67 was found after the doping of titania with 0.65 mmol Na^+ per g of the support.

436

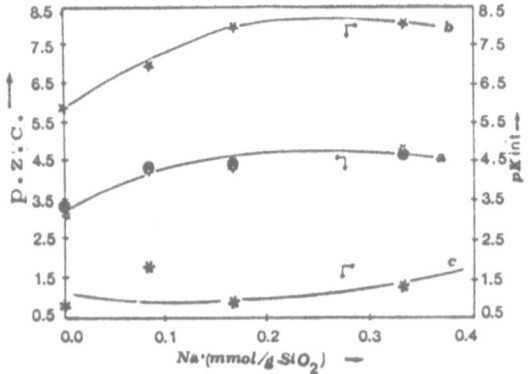

Figure 20. Variation of the experimental and calculated values of p.z.c. as well as of the surface acidity constants (pK₁^{int}, pK₂^{int}) with sodium content for Na⁺ doped silica samples. (☆) experimental p.z.c., (●) calculated p.z.c. (p.z.c.=pK₁^{int}+pK₂^{int}/2), (✳) pK₁^{int}, (★) pK₂^{int}.

The above clearly show that the doping with small amounts of sodium or lithium ions enlarges considerably the pH range where predominate the positive groups and thus the support may adsorb negatively charged species. In the contrast to that the modification with a very small amount of fluoride ions is sufficient for an extension of the pH range where γ-alumina adsorbs positive species to be achieved. It is important to notice the excellent agreement achieved between the experimental and calculated p.z.c. values which again corroborated the validity of the method followed for determining K_1^{int} and K_2^{int}.

3.3.2. *Influence of modification on the concentration of charged surface groups*. Figures 21,22 and 23 show the variation, with pH, of the charged surface groups for γ-alumina doped with various amounts of sodium, lithium and fluoride ions, respectively. It is shown that the sodium and lithium doping generally increases the concentration of the positive groups whereas it decreases the concentration of the negative groups. Both effects promote the adsorption of negatively charged species. On the contrary, modification with fluoride ions povokes a drastic increase in the concentration of the deprotonated surface hydroxyls in pH lower than 5.30 whereas it does not affect the concentration of the protonated surface hydroxyls. Obviously this effect promotes the adsorption of positively charged species.

The above results are generally in line with those achieved concerning the influence of doping on p.z.c. and surface acidity constants. However, carefull observations of the trends achieved for the charged groups (figs. 21, 22 and 23) on one hand and for p.z.c. and acidity constants on the other hand (figs 17, 18 and 19) reveals important differences which should be discussed. The subject could be a part of a more general discussion concerning the mechanism of the sodium, lithium and fluoride action. A convenient way to start the discussion is to recall the model adopted for the deposition state on the γ-alumina surface of sodium, lithium and fluorine. Concerning sodium and lithium ions (77-80) the model adopted involves the following alternatives: (a) Na⁺ and Li⁺ ions diffuse into γ-alumina lattice during calcination. (b) They neutralize the surface AlOH groups forming Al-O$^{δ(-)}$-Na$^{δ(+)}$ and

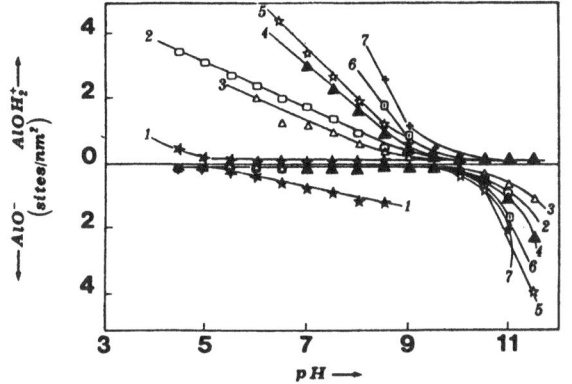

Figure 21.Dependence of the concentration of the charged groups on suspension pH in Na-doped γ-Al$_2$O$_3$,0.1 M KNO$_3$, 25ºC. Numbers 1, 2, 3, 4, 5, 6, 7, on the curves correspond to 0.226, 0.309, 0.392, 0.621, 0.984, 1.560 and 2.470 mmoles of Na per gram of the carrier, respectively.

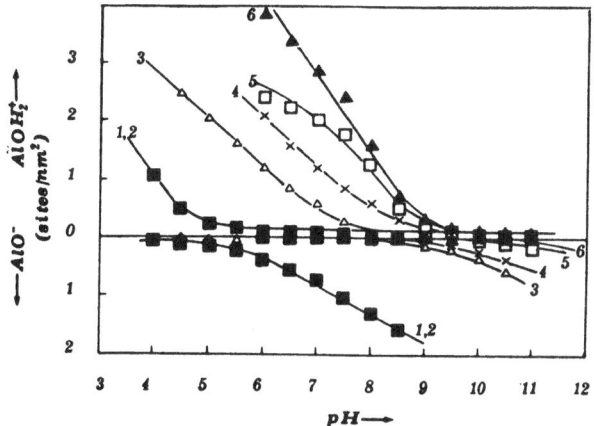

Figure 22. Dependence of the concentration of the charged groups on the pH of suspension for Li-doped γ-Al$_2$O$_3$, 0.1M KNO$_3$, 25ºC. Numbers 1, 2, 3, 4, 5, 6, on the curves correspond to 0.000, 0.392, 0.621, 0.984, 1.560 and 2.470 mmoles of Li per gram of the carrier, respectively.

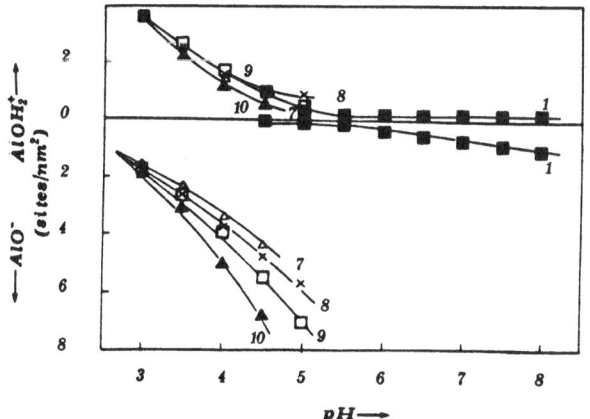

Figure 23. Dependence of the concentration of the charged groups on the pH of suspension for F-doped γ-Al$_2$O$_3$, 0.1M KNO$_3$, 25ºC. Numbers 1, 2, 3, 4, 5, 6, on the curves correspond to 0.000, 0.392, 0.621, 0.984, 1.560 and 2.470 mmoles of Li per gram of the carrier, respectively.

438

Al-$O^{\delta(-)}$-Li$^{\delta(+)}$ groups. (c) They form islands of Na$^+$ and Li$^+$ compounds. In agreement with the literature we assume that the extent of processes (a) and (b) is greater in the case of Li-doped carriers, whereas process (c) is favored in the case of the sodium based system. In both systems the relative extent of process (c) increase at the expense of the relative extents of processes (a) and (b) upon increasing the dopant concentration. It has been suggested that the Al-$O^{\delta(-)}$-Na$^{\delta(+)}$ and Al-$O^{\delta(-)}$-Li$^{\delta(+)}$ groups, predominated in the specimens with low cation content, are responsible for the abrupt increase in pK_1^{int}, pK_2^{int}, p.z.c. and therefore in the concentration of the protonated hydroxyls observed after modification with 0.309 mmolNa$^+$ per g of γ-Al$_2$O$_3$ or with 0.621mmol Li$^+$ per g of γ-Al$_2$O$_3$. The effect of the forementioned groups on the parameters mentioned could be explained assuming that they promote adsorption of hydrogen ions on sites adjacent to Al-OH and Al-O^- groups, facilitating the surface reaction illustrated in the equilibria (1) Fig. 24 illustrates the mechanism for the sodium based system. Taking into account that process (a) is greater in the case of Li$^+$-doped carrier and that the Li$^+$ ions located in the interior of the γ-alumina do not affect its surface we may explain why it is required relatively large ammount of Li$^+$ ions in order to observe the abrupt increase in pK_1^{int}, pK_2^{int}, pzc and therefore in the concentration of the AlOH$_2^+$ groups. The lack of an additional increase in the pK_1^{int}, pK_2^{int}, and pzc, with increasing sodium and lithium content could be attributed to the fact the sodium (lithium) ions in excess of 0.309(0.621) are deposited mainly <u>via</u> the third mechanism. In that case the increase in the concentration of the AlOH$_2^+$ groups with Na$^+$ and Li$^+$ content at a given pH, pK_1^{int} and pK_2^{int} remaining practically the same could be explained as follows: The sodium or lithium ion originated from the islands mentioned above are dissolved and subsequently adsorbed specifically in the IHP of the double layer forming "ion pairs" with the surface AlOH groups and releasing on the surface hydrogen ions [AlOH +Na^+_b--> AlO$^-$...Na$^+$ +H$_s^+$, AlOH+Li^+_b --> AlO$^-$...Li$^+$+H$_s^+$].

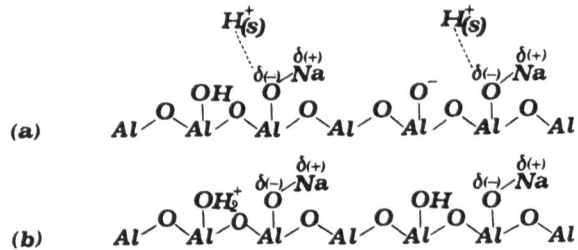

(a)

(b)

Figure 24. Schematic representation of the effect of the Al-O-Na groups on the pK_1^{int}, pK_2^{int}, and p.z.c. values for the specimen containing 0.309 mmol Na$^+$/g of γ-Al$_2$O$_3$.

This assumption is in excellent agreement with the literature (55, 56). The proton released promote the formation of additional protonated surface hydroxyls by reaction with the adjacent neutral surface hydroxyls (see equilibria (1)). In this case it is anticipated an increase in the concentration of the protonated surface hydroxyls with the dissolved Na$^+$ and Li$^+$ ions. Figures 25 and 26 show that this anticipation is experimentally confirmed. However, the nonlinearity observed in the case of Li⁺ doping suggests that the forementioned mechanism is not the only one responsible for the increase in the concentration of the protonated

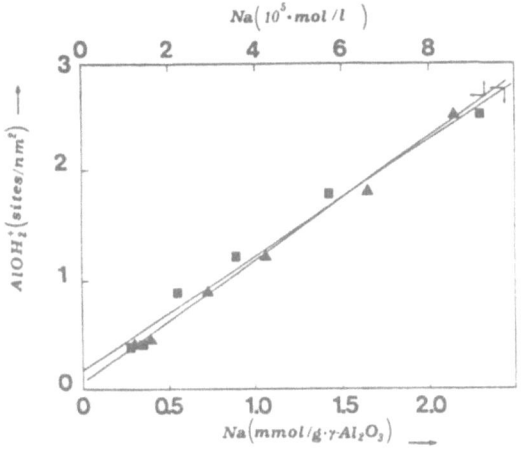

Figure 25. Variation of positively charged groups (AlOH₂⁺) for 0.1 M KNO₃ at pH 8.5 versus (■) the real concentration of sodium on the γ-Al₂O₃ surface and (▲) the concentration of dissolved sodium.

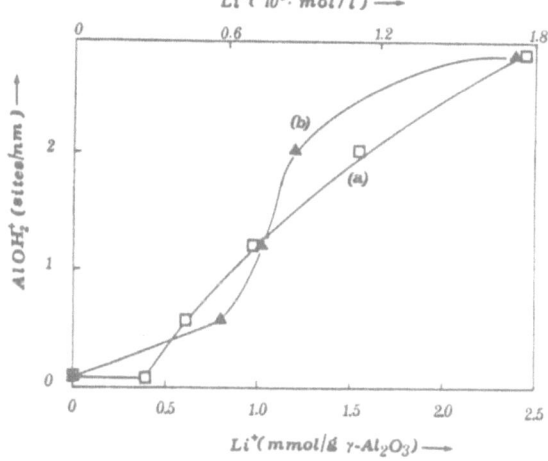

Figure 26. Variation of positively charged groups (AlOH₂⁺) for 0.1 M KNO₃ at pH 7.0 versus (□) the real concentration of lithium on the γ-Al₂O₃ surface and (▲) the concentration of dissolved lithium.

hydroxyls. As, to the fluorine action (figs 19, 23) the sharp increase in the concentration of the AlO⁻ groups observed after modification with 0.125 mmol F⁻ ions per g of γ-alumina may be attributed to the abrupt increase in the values of the surface acidity constants brought about by this amount of F⁻ ions. Moreover, the additional increase in the concentration of the deprotonated surface groups with the F⁻ content in the range 0.125-3.818 mmol F⁻ per g of γ-Al₂O₃, where the values of K_1^{int} and K_2^{int} remain practically constant, could be explained assuming the

440

following mechanism.

$$AlOH + F_b^- \longrightarrow Al^+...F^- + OH^-_s$$
$$AlOH + OH^-_s \longrightarrow AlO^- + H_2O$$

Specifically the dissolved from the surface fluoride ions, F^-_b are specically adsorbed in the IHP of the double layer forming "ion pairs" with Al^+ and releasing on the surface hydroxyl ions OH^-_s. This is in excellent agreement with the literature. The OH^-_s released cause the formation of additional deprotonated surface hydroxyls by reaction with the adjacent neutral surface hydroxyls. From the above it is anticipated an increase in the concentration of the AlO^- groups with the amount of the dissolved F^- ions. (Fig. 27).

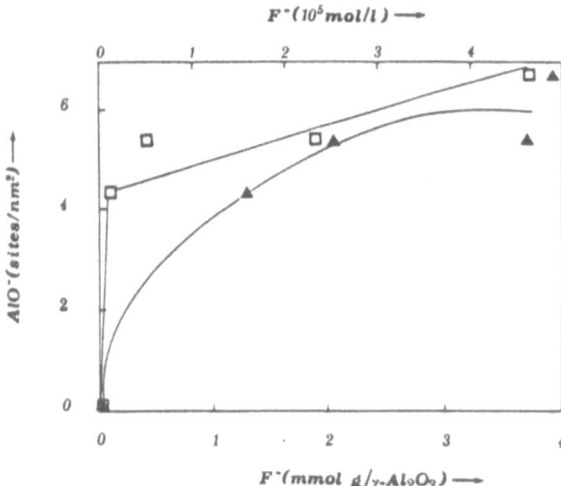

Figure 27. Variation of negatively charged groups (AlO^-) for 0.1 M KNO_3 at pH 4.5 versus (□) the concentration of the surface and surface ion-paired fluoride and (▲) with the concentration of dissolved fluoride.

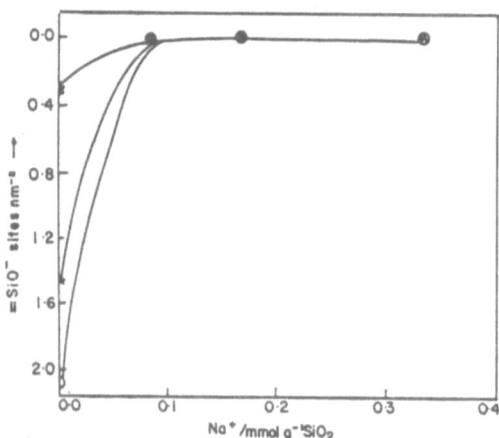

Figure 28. Depedence of the concentration of the negatively charged surface groups on Na-doped SiO_2, on the content of the carrier in Na^+, 0.1M KCl, 25°C. (O) pH 3.50; (★) 4.0; (✳) 4.50.

Finally, returning to the sodium/silica system (fig. 20) we may observe that sodium does not markedly affect the surface dissociation constant pK_1^{int}, related to the positive groups. The value of this parameter oscillates rather randomly about a mean value with the sodium content. Therefore, the change in the $(pK_1^{int}+pK_2^{int})/2=p.z.c.$ should be attributed to the change of the pK_2^{int}. This strongly suggests that sodium doping should promote the protonation of the SiO^- but not of the SiOH groups. The above has been confirmed by observing the dependence of the concentration of the $SiOH_2^+$ and SiO^- groups on the sodium content. In fact, an almost complete disappearence of the SiO^- groups was observed after modification with the minimum Na^+ concentration (Fig. 28). On the contrary insignificant change in the concentration of the neutral surface hydroxyls was observed due to the sodium doping.

References

1. B. Delmon, P.A. Jacobs, and G. Poncelet, eds, "Preparation of catalysts I", Elsevier, Amsterdam, 1976, Stud. Surf. Sci. Catal., Vol. 1.
2. B. Delmon, P.Grange, P.A. Jacobs, and G. Poncelet, eds, "Preparation of Catalysts II", Elsevier, Amsterdam, 1979, Stud. Surf. Sci. Catal., Vol. 3.
3. G. Poncelet, P. Grange, and P.A. Jacobs, eds, "Preparation of Catalysts III", Elsevier, Amsterdam, 1983, Stud. Surf. Sci. Catal., Vol. 16.
4. B. Delmon, P. Grange, P.A. Jacobs and G. Poncelet, eds, "Preparation of Catalysts IV", Elsevier, Amsterdam, 1987, Stud. Surf. Sci. Catal., Vol 31.
5. G. Poncelet, P.A. Jacobs, P. Grange and B. Delmon, eds, "Preparation of Catalysts V", Elsevier, Amsterdam, 1991, Stud. Surf. Sci. Catal., Vol. 63.
6. J.W. Geus in ref. 3 pages 1-33.
7. K.P. de Jong in ref. 5, pages 19-36.
8. L.M. Knijff, P.H. Bolt, R. Van Yperen, A.J. Van Dillen and J.W. Geus in ref. 5, pages 165-174.
9. European Patent Specification 258, 942 (1988) to S.T.R.M-B.V.
10. Netherlands Patent Application 68,1677 (1970) to Stamicarbon.
11. L. Wang, W.K. Hall, J. Catal. 77, 232 (1982).
12. L. Wang, W.K. Hall, J. Catal. 83, 242 (1983).
13. L. Wang, W.K. Hall, J. Catal. 66, 251 (1980).
14. S. Kasztelan, J. Grimblot, J.P. Bonnelle, E. Payen, H. Toulhoat, Y. Jacquin, Applied Catalysis 7, 91 (1983).
15. A. Iannibello, S. Marengo, F. Trifiro, P.L. Villa, in "2nd Int. Sym. on Scientific Bases for the Preparation of Heterogeneous Catalysts", Louvain La Neuve, Belgium, 1978, paper A5.
16. C.V. Caceres, J.L.G. Fierro, A.L. Agudo, M.N. Blanco, H.J. Thomas, J. Catal. 95, 501 (1985).
17. J.A.R. van Veen, H. De Wit, C.A. Emeis, P.A.J.M. Hendriks, J. Catal. 107, 579 (1987).

442

18. L.P. Milova, N.M. Zaidman, L.M.Plyasova, S.V. Ketchik, K.G. Rikhter, Kinetic Katalysis 23, 123 (1982).
19. J.N. Fiedor, A. Proctor, M. Houalla, P.M.A. Sherwood, F.M. Mulcahy, D.M. Hercules, J. Phys. Chem. 96, 10967 (1992).
20. M.J. Fay, A. Proctor, D.P. Hoffman, M. Houalla, D.M. Hercules, Microchimica Acta 109, 281 (1992).
21. J. Sonnemans, P.J. Mars, J. Catal. 31, 209 (1973).
22. C.C. Williams, J.G. Ekerdt, J.M. Jehng, F.D. Hardcastle, I.E. Wachs, J. Phys. Chem. 95, 8791 (1991).
23. T. Machej, J. Haber, A.M. Turek, I.E. Wachs, Applied Catalysis 70, 115 (1991).
24. L. Wang, W.K. Hall, J. Catal. 82, 177 (1983).
25. J.A.R van Veen, P.A.J.M. Hendriks, Polyhedron 5, 75 (1986).
26. J.P. Brunelle, Pure Appl. Chem. 50, 1211 (1978).
27. K.Y.S. Ng, E. Gulari, J. Catal. 92, 340 (1985).
28. K. Segawa, D.S. Kim, Y. Kurusu, I.E. Wachs, in "Proceedings, 9th Int. Con. on Catalysis", Calgary, 1988, M.J. Phillips, M. Ternan, Eds., Chemical Institute of Canada: Ottawa, 1988, p. 1960.
29. D.S. Kim, Y. Kurusu, I.E. Wachs, F.D. Hardcastle, K. Segawa, J. Catal. 120, 325 (1989).
30. D.C. Vermaire, P.C. van Berge, J. Catal. 116, 309 (1989).
31. P.H. Tewari, W. Lee, J. Colloid Interface Sci. 52, 77 (1975).
32. F.M. Mulcahy, M.G. Fay, A. Proctor, M. Houalla, D.M. Hercules, J. Catal. 124, 231 (1990).
33. A. Iannibello, P.C.H. Mitchell, in "2nt Int. Sym. on Scientific Basis for the Preparation of Heterogeneous Catalysts", Louvain La Neuve, Belgium, 1978, paper E2.
34. Cr. Contescu, M.I. Vass, Applied Catalysis 33, 259 (1987).
35. N. Spanos, L. Vordonis, Ch. Kordulis, A. Lycourghiotis, J. Catal. 124, 301 (1990).
36. N. Spanos, L. Vordonis, Ch. Kordulis, P.G. Koutsoukos, A. Lycourghiotis, J. Catal. 124, 315 (1990).
37. L. Vordonis, P.G. Koutsoukos, A. Lycourghiotis, Colloids and Surfaces 50, 353 (1990).
38. N. Spanos, H.K. Matralis, Ch. Kordulis, A. Lycourghiotis, J. Catal. 136, 432 (1992).
39. L. Vordonis, N. Spanos, P.G. Koutsoukos, A. Lycourghiotis, Langmuir 8, 1736 (1992).
40. L. Karakonstantis, Ch. Kordulis, A. Lycourghiotis, Langmuir 8, 1318 (1992).
41. H.J. Thomas, M.N. Blanco, C.V. Caceres, N. Firpo, F.J. Gil Llambias, J.L.G. Fierro, A.L. Agudo, J. Chem. Soc., Faraday Trans. 86(15), 2765 (1990).
42. YU. I. Yermakov: Catal. Rev. Sci. Eng. 13, 77 (1976).
43. YU. I. Yermakov: Adv. in Catalysis 24, 173 (1975).
44. D.H. Ballard : Adv. in Catalysis 23, 263 (1973).
45. J.P. Candlin et al Adv. Chem. Ser. 132, 212 (1974).
46. M.S.Scurrell, "Catalysis" Vol. 2, specialist periodical reports, the Chemical Society, Burlington House, London W1VOBN, 1978, p. 215.

47. T.W. Healy, and L.R. White, Adv. Colloid Interface Sci. **9,** 303 (1978).
48. W. Stumm, H. Hohl and F. Dalang, Croat. Chim. Acta **48,** 491 (1976).
49. H. Hohl and W. Stumm, J. Colloid Interface Sci. **55,** 281 (1976).
50. W. Stumm, R. Kummert, and L. Sigg, Croat. Chim. Acta **53,** 291 (1980).
51. C.P. Huang, and W.J. Stumm, J. Colloid Interface Sci. **43,** 409 (1973).
52. W. Stumm, C.P. Huang, and S.R. Jenkins, Croat. Chim. Acta **42,** 223 (1970).
53. J. Westall, and H. Hohl, Adv. Colloid Interface Sci. **12,** 265 (1980).
54. D. Yates, S. Levine, and T.W. Healy, J. Chem. Soc., Faraday Trans 1 **70,** 1087 (1974).
55. J.A. Davis, R.O. James and J.O. Leckie, J. Colloid Interface Sci. **63,** 480 (1978).
56. R.O. James, J.A. Davis and J.O. Leckie, J. Colloid Interface Sci. **65,** 331 (1978).
57. L. Vordonis, P.G. Koutsoukos and A. Lycourghiotis, J. Chem. Soc., Chem. Commun., 1309 (1984).
58. L. Vordonis, P.G. Koutsoukos and A. Lycourghiotis, J. Catal. **98,** 296 (1986).
59. L. Vordonis, P.G. Koutsoukos and A. Lycourghiotis, J. Catal. **101,** 186 (1986).
60. K. Akratopulu, L. Vordonis and A. Lycourghiotis, J. Chem. Soc., Faraday Trans 1 **82,** 3697 (1986).
61. L. Vordonis, P.G. Koutsoukos and A. Lycourghiotis, Langmuir **2,** 281 (1986).
62. K. Akratopulu, L. Vordonis and A. Lycourghiotis, J. Catal. **109,** 41 (1988).
63. K. Akratopulu, Ch. Kordulis and A. Lycourghiotis, J. Chem. Soc., Faraday Trans 1 **86,** 3437 (1990).
64. L. Vordonis, K. Akratopulu, P.G. Koutsoukos and A. Lycourghiotis in ref. 4 pages 309-321.
65. K. Akratopulu, L. Vordonis and A. Lycourghiotis "Heterogeneous Catalysis" Proc. of the Sixth Inter. Sym. (Bulgarian Academy of Sciences), Sofia, 1987, pages 412-417.
66. K. Akratopulu, L. Vordonis and A. Lycourghiotis "in proc. of the 10th Panhellenic Conference of Chemistry" Greek Chemists Assoc., Patras, 1985, p. 700.
67. L. Vordonis PhD thesis University of Patras, Chemistry Department, Patras, Greece, 1988.
68. K. Akratopulu PhD thesis University of Patras, Chemistry Department, Patras, Greece, 1989.
69. N. Spanos, PhD thesis, University of Patras, Chemistry Department, Patras, Greece 1991
70. C.W. Davies, "Ion Association", Butterworths, London, 1962.
71. R.A. Robinson and R.H. Stokes, "Electrolyte Solutions", Butterworths, London, 1970.

444

72. R. Sprycha, J. Colloid Interface Sci. **96**, 551 (1983).
73. C.P. Huang "The Chemistry of the Aluminum Oxide-Electrolyte Interface" PhD thesis, Harvard University, 1971.
74. G.M. Barrow, "Physical Chemistry" Mc Graw-Hill, New York, 2nd Edition, 1966, pages 612-629.
75. G.M. Barrow, ref. 74, p. 234.
76. J.F. Kuo and T.F. Yen, J. Colloid Interface Sci. **121**, 220 (1989).
77. A. Lycourghiotis, C. Defosse, F. Delannay, F. Lemaitre, and B. Delmon, J. Chem. Soc., Faraday Trans. 1 **76**, 1677 (1980).
78. A. Lycourghiotis, C. Defosse, F. Delannay, and B. Delmon, J. Chem. Soc., Faraday Trans. 1 **76**, 2052 (1980).
79. A. Lycourghiotis, D. Vattis, and Ph. Aroni, Z. Phys. Chem. (N.F.) **121**, 257 (1980).
80. Ch. Kordulis, S. Voliotis, and A. Lycourghiotis, J. Less Common Met. **84**, 187 (1982).

INDUSTRIAL CATALYTIC APPLICATIONS
OF ACIDIC MOLECULAR SIEVES*

P. R. Pujadó, J. A. Rabó, G. J. Antos, and S. A. Gembicki
UOP
Des Plaines, Illinois

ABSTRACT

Over the last three decades, zeolite molecular sieves have become the basic materials for a variety of catalysts used in a number of important commercial applications. Some of the factors that have contributed to the successful development of catalytic applications of molecular sieves are the ability to:

- Introduce strong acidity by the choice of crystal structure, by the modification of chemical composition, or as a result of the type and extent of ion exchange
- Incorporate the catalytic activity of various metals, whether in the crystalline lattice or ion exchanged within the molecular sieve structure
- Institute size and shape selectivity as a result of the regular spatial dimensions of the molecular size channels within the crystalline lattice

This paper reviews the evolution of a number of industrial catalytic applications of molecular sieves with emphasis on the development of new families of molecular sieves with a potential for industrial catalytic applications.

* Originally published in *Catalysis Today*, 13(1992) 113-141, Elsevier Science Publishers, B. V. Amsterdam

J. Fraissard and L. Petrakis (eds.), Acidity and Basicity of Solids, 445–474.

1. INTRODUCTION

Zeolites are widely found in natural deposits. Some of the deposits are believed to have been formed by precipitation from fluids that permeated basaltic rocks. The great bulk of zeolites occurs in sediments and low-grade metamorphic rocks with a grain size near the limit of optical microscopy. In general, the composition of natural zeolites correlates with that of the host rock [1]. Thus, mordenite and other zeolites rich in silicon occur in rock supersaturated with silica, while faujasite, chabazite, gmelinite, and other silicon-poor zeolites occur preferentially in rocks deficient in silica [2].

More than 30 zeolites occur naturally, but only 8 — analcime, chabazite, clinoptilolite, erionite, ferrierite, laumontite, mordenite, and phillipsite — dominate the sedimentary deposits and permit commercial exploitation. Type A zeolite is totally lacking in nature, and faujasite is extremely scarce. Natural zeolites appear to be most suited for applications in which purity is not important or where large tonnages are necessary [2]. Natural zeolites can be used as catalysts after concentration and purification, but the greater efficiency and quality control of synthetic zeolites greatly outweigh any potential cost advantages of natural zeolites.

A new class of industrial materials, molecular sieve zeolites, were first introduced by Union Carbide in 1954 as adsorbents for industrial separations and purifications. Since then, a series of important adsorption, catalytic, and ion-exchange applications have been developed, principally in the petroleum refining and petrochemical industries. The discovery and synthesis of the new zeolites A, X, and Y by Milton and Breck at the Union Carbide laboratories marked the beginning of today's molecular sieve industry and led to the commercial application of zeolites as selective adsorbents and catalysts. The success of molecular sieve zeolites, principally in catalytic applications, has been due to the discovery of the strong acidity of crystalline H-zeolites and to other properties that have been engineered into improvements of existing processes or into the development of new ones [3].

In the late 1960s and early 1970s, highly siliceous zeolites, such as ZSM-5 and Beta, were discovered at the Mobil Oil laboratories. In this same period, the elimination of aluminum in the zeolite framework led to all-silica molecular sieves, as typified by silicalite.

Over the last two or three decades, more than 150 zeolite species have been synthesized. However, only a few have been manufactured in commercial quantities and with a consistent quality and purity. Out of these materials, only a few crystal types are used commercially. The major, large-volume zeolites used commercially in ion exchange, adsorption, and catalysis are A, X, Y, ZSM-5, and mordenite. Tables 1 and 2 show the evolution of zeolites and other molecular sieves as a result of ongoing scientific investigation.

Table 1
Synthetic Molecular Sieve Materials

Time Frame	Materials
1950s and early 1960s	Aluminum-rich zeolites: A, X, Y, and mordenite
1960s and 1970s	Silicon-rich zeolites: ZSM-5
Late 1970s	Silica molecular sieves: silicalite
1980s	• Aluminophosphate-based molecular sieves • Silicon-enriched zeolites: LZ-210 • Framework-substituted molecular sieves • Sulfide molecular sieves • Pillared clays

Table 2
The Evolution of Molecular Sieve Materials

Aluminum-rich zeolites:
• Si/Al from 1 to 1.5: A, X
• Si/Al from ~2 to 5:
  — Natural zeolites: erionite, chabazite, clinoptilolite, mordenite
  — Synthetic zeolites: Y, L, large-pore mordenite, omega

Silicon-rich zeolites:
• Si/Al from ~10 to 100
  — By thermochemical framework modification:
    highly siliceous variants of Y, mordenite, erionite
  — By direct synthesis: ZSM-5

Silica molecular sieves: silicalite
New framework elements: aluminophosphates

Table 3
The Transition in Properties

Changes in:	Si/Al, from 1 to infinity
Result in:	• Stability, from <700° to ~ 1300°C • Surface selectivity, from hydrophilic to hydrophobic • Acidity increasing in intrinsic strength • Cation concentration decreasing • Structure from 4-, 6-, and 8-rings to 5-rings

Table 3 shows the transition in properties of zeolites as the silicon-to-aluminum (Si/Al) ratio increases from aluminum-rich Si/Al = 1 to infinity (silicalite). The property transitions shown are approximate and shown only as a generalized guideline.

The thermal stability of the crystalline lattice of zeolites varies substantially from about 700°C for aluminum-rich zeolites to about 1300°C for silicalite. Aluminum-rich zeolites are unstable in the presence of acids, while silicon-rich zeolites are stable even in concentrated mineral acids. In contrast, silicon-rich zeolites have low stability in basic solutions. Likewise, aluminum-rich zeolites exhibit a highly polar, hydrophilic surface. Silicon-rich zeolites tend to be more nonpolar and hydrophobic (organophilic). The onset of hydrophobicity appears to occur at a Si/Al ratio of about 10.

Zeolites have become the dominant materials in the large-scale adsorptive separation processes, whether in cyclic or in continuous-flow systems, that are used in the refining and petrochemical industries [4]. Molecular sieves have also become the leading contenders for a large number of important catalytic applications, principally in reactions that involve the conversion of hydrocarbons.

Molecular sieves also hold the greatest promise for the development of new catalysts that will enable refiners to face the processing challenges of the coming decade. Table 4 illustrates some of the challenges that petroleum catalysis will face in the 1990s. Other similar summaries can be prepared for the petrochemical industries in the processing of paraffins, olefins, and aromatics, for example, and their derivatives.

Figure 1 illustrates an example of the potential rewards that await those willing to face the challenges. The current use of zeolite catalysts in fluidized catalytic cracking (FCC) units has resulted in an incremental value added of about $1.00 per barrel (bbl) over earlier amorphous catalysts. However, an analysis of the feed composition and product distribution reveals that the reaction selectivity can further be improved to yield approximately an additional $1.00/bbl in value added. Even if $1.00/bbl does not seem significant, it is a huge figure when considering that about seven million barrels are processed every day throughout the world (about five million barrels in the United States and Canada).

Whereas applications of molecular sieves to date have essentially made use only of zeolites, nonzeolitic molecular sieves synthesized in recent years and new ones now being developed may hold great promise for both existing and new catalytic applications. These new molecular sieve materials include metallosilicate crystals with either the aluminum or the silicon substituted by another element, aluminophosphates of a wide range of compositions and crystal structures, and pillared clays, among others. In this paper, the term *zeolite* is reserved for aluminosilicate molecular sieves; all other molecular sieves are identified explicitly as required.

FIGURE 1

VALUE ADDED TO FCC PRODUCTS

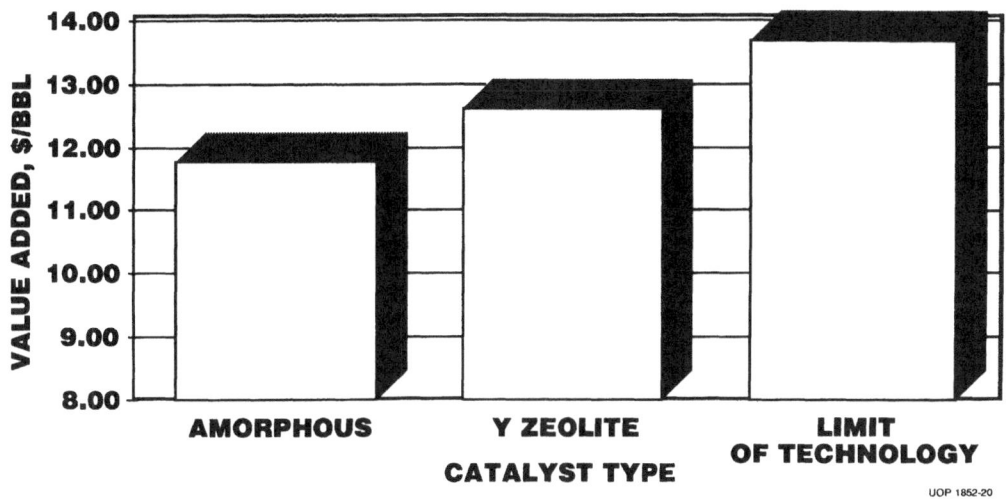

UOP 1852-20
UOP 1910-13

FIGURE 2

CATALYTIC PETROLEUM CONVERSION

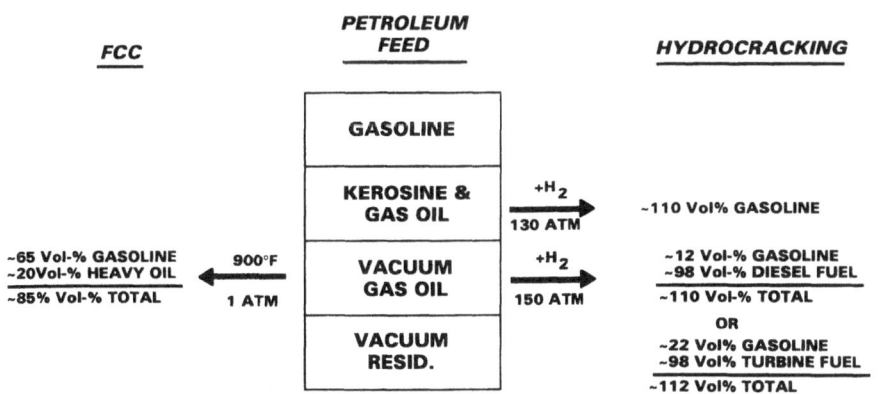

UOP 1910-16

2. INDUSTRIAL APPLICATIONS OF ZEOLITE CATALYSTS

Zeolites are widely used in petroleum refining applications and are finding increasing application in the petrochemical and chemical industries. Zeolites and molecular sieves have made significant inroads in the processing and conversion of paraffins, olefins, and aromatics. Recently, they are also finding applications in oxidation reactions and in organic synthesis involving heteroatoms [5,6].

Catalytic processes in petroleum refining aim at the production of various types of motor fuels. Figure 2 illustrates two typical large-scale conversion processes — FCC and hydrocracking — that are extensively used in petroleum refining and that make use of zeolite catalysts. The conversion of heavier feedstocks into lighter products and the increase in the octane number of the hydrocarbons are typical endeavors faced by petroleum refiners all over the world.

2.1. Fluidized Catalytic Cracking

In FCC applications, all catalysts are now zeolitic in nature. Historically, the use of Y zeolites in FCC applications evolved from rare-earth-exchanged Y zeolite to steam-stabilized Y and then to aluminum-depleted Y and finally to silicon-enriched Y. Some of these zeolites are also partially rare earth exchanged. The development of silicon-enriched Y zeolites, like LZ-210, that has Si/Al ratios of 6 and above have contributed to the enhancement of the thermal and hydrothermal stability of the Y crystal and improved selectivity to high-octane gasoline.

The Y zeolite is the predominant molecular sieve used in this catalytic application. The annual consumption of Y-zeolite-based catalysts in FCC applications exceeds 300,000 metric tons per annum (MTA). The ZSM-5 zeolite, which exhibits selective paraffin cracking characteristics, is also used as a catalyst additive to increment the octane value of the product. For an excellent discussion on the subject, see the monographs "Octane Enhancing Zeolitic FCC Catalysts," by J. Scherzer [7,8].

2.2. Hydrocracking

In hydrocracking, all zeolite-based catalysts, whether for the production of naphtha (gasoline range) or distillate (diesel fuel range) contain Y zeolite [9,10]. The large pore size, three-dimensional pore lattice and corresponding high diffusivity, high acidity, and molecular sieve action of the Y zeolite provide an excellent combination of desirable characteristics for a catalyst in the cracking of large molecules, such as those found in vacuum gas oil. Zeolite materials with these characteristics produce high yields of naphtha fractions that are well suited for further treatment in a reforming step at high conversion and long catalyst life, and low C_1-C_2 production.

The steam-stabilized form of H-Y zeolite containing either base transition metals or palladium has been the principal form of this zeolite used for the production of naphtha.

However, other modifications of this zeolite including magnesium, rare-earth-exchanged species, and silicon-enriched Y zeolites, have also been reported [11-13].

In the production of diesel fuel, amorphous acid catalysts loaded with base transition metals have been in commercial use since the 1960s. In recent years, however, Y zeolite-based catalysts have been successfully introduced that effectively utilize the higher activity and stability of the Y zeolite. In this application, the extensively steamed and aluminum-depleted forms of Y zeolite in combination with base metals have been reported [14-16].

2.3. Catalytic Dewaxing

Normal paraffins are desirable in diesel fuel because of their high cetane index. Likewise, *n*-paraffins display a high viscosity index in lube oils. However, excessive amounts of long-chain *n*-paraffins tend to solidify and thus increase the pour point of both diesel fuel and lube oil products. Catalytic dewaxing is intended to selectively remove a fraction of the longest chain *n*-paraffins to lower the pour point of the product.

Mordenite or ZSM-5 alone or in combination with transition metals have been reported in commercial use for catalytic dewaxing applications [17,18]. Medium-pore ZSM-5 and oval-pore mordenite have been used for catalytic dewaxing based on their strong acidity combined with diffusion control that permits the selective processing of the relatively small size *n*-paraffins. By using these catalysts, a fraction of the long-chain paraffins is cracked and converted to an olefinic, light hydrocarbon product. As a result of this treatment, the diesel or lube oil product displays greatly reduced pour points but has some loss of product yield. The product yield can be recovered if the light olefins thus produced can be used elsewhere, for example, in an alkylation unit.

2.4. Isomerization of Paraffins

Isomerization is another reaction of great importance in the production of high-octane gasoline components because it promotes the formation of highly branched paraffin isomers [19]. Although conventional halided nonzeolitic catalysts still retain a performance edge in this application, mordenite-based zeolitic catalysts have a substantial market share based on their enhanced tolerance of feed impurities and the greater simplicity of their operation.

2.5. Aromatic Alkylation Reactions

Typical applications include the synthesis of ethylbenzene by the alkylation of benzene with ethylene and the synthesis of cumene (isopropylbenzene) by the alkylation of benzene with propylene. Recently, the heterogeneous alkylation of aromatics has been extended to much heavier olefins, as in the production of linear alkylbenzene by the direct alkylation of benzene with heavy, detergent-range (C_{10}-C_{14}) olefins. The alkylation of aromatics is an acid-site catalyzed reaction that requires strong acid sites with low multiple alkylation or disproportionation activity. Also, in the case of cumene, formation of the *n*-propyl isomer is to be avoided. Zeolites such as ZSM-5 and H-Y are most commonly used for these applications.

2.6. Disproportionation and Transalkylation of Aromatics

The disproportionation or transalkylation of aromatics includes the conversion of toluene to xylenes and benzene or the recombination of toluene and C₉ aromatics to maximize the production of xylenes. Although the typical distribution of the xylene isomers obtained by disproportionation is essentially at equilibrium, technology has recently been commercialized by Mobil for the selective production of *p*-xylene by the disproportionation of toluene [20]. Strong acid zeolites like mordenite and ZSM-5 have traditionally been used in this application.

2.7. Isomerization of Xylenes

Zeolites are being used as an additional component for xylene isomerization catalysts, either with ethylbenzene dealkylation capacity, or with capacity for ethylbenzene disproportionation, which is the ability to isomerize ethylbenzene to xylenes. Shape selectivity to *p*-xylene can also be observed with some zeolites and other molecular sieves in isomerization applications. The most common catalytic systems make use of mordenite or ZSM-5 zeolite and its boron-substituted form.

3. METALLOSILICATES AND THEIR APPLICATIONS

The first synthetic molecular sieves were zeolites with only aluminum or silicon as the tetrahedral cations. Several other elements of similar electronic configuration and size can be substituted either for the aluminum or the silicon. However, aluminosilicate compositions remain the most broadly used materials in practical applications, principally because of their high thermal and hydrothermal stability, high acidity, and the low cost of their synthesis [21].

Metallosilicate molecular sieves may exhibit performance characteristics that are superior to those of their zeolite analogues in terms of controlled acidity or the specificity of their catalytic action. However, because aluminosilicates of a given crystalline structure are generally more stable than their metallosilicate homologues, the hydrolysis of framework cations becomes important in the latter. The secondary phase formed from framework cations either by hydrolysis or by reduction may become an important catalytic contributor to desired or undesired products. Thus, the characterization of metallosilicate catalysts requires structural, chemical, and catalytic information. An excellent review of metallosilicates and their characterization is given by R. Szostak [22].

The substitution of zeolite framework elements by others closely related in the periodic table has been successfully accomplished without substantial change in the original hydrothermal synthesis. Thus, gallium can be substituted for Al and germanium for Si. The substitution of other elements may require changes in the synthesis conditions because of either differences in the solution chemistry of the substituent or other requirements needed for crystal formation. The use of organic templates has enabled the substitution of many other elements, such as boron, iron, chromium, cobalt, titanium, zirconium, zinc, beryllium, hafnium, manganese, magnesium, vanadium, and tin, among others. Most of these metallosilicate compositions have been synthesized with ZSM-5 crystal structures. Others are analogues of ZSM-11, ZSM-12, Theta-1, ZSM-34, or Beta.

Of the many compositional and structural varieties described in the literature, the most important practical catalytic applications employ the B-, Ga-, and Ti-substituted forms of the ZSM-5 crystal. Recent publications describe the synthesis and the catalytic application of a variety of metallosilicates [22-24].

3.1. Borosilicate Catalysts

Borosilicates prepared by the substitution of B for the Al in ZSM-5 have been described as practical catalysts for shape-selective acid catalysis [25]. The preferred application is in the processing of C_8 aromatics using isomerization, disproportionation, or transalkylation reactions. The borosilicate catalysts may also contain a metal additive, such as nickel or a noble metal, usually for hydrogen-transfer functions.

3.2. Gallosilicate Catalysts

The gallium-framework-substituted form of ZSM-5 is known to catalyze the aromatization of propane and butane. Also, similar effects are obtained with catalysts compounded from mixtures of gallium oxide and the hydrogen form of ZSM-5 zeolite. These divergent approaches serve to exemplify the importance of characterization studies at each step of catalyst preparation to understand the species that is actually catalyzing the reactions. In this case, these studies point to a zeolite modified by gallium oxide as the likely active agent. The gallium oxide is extremely well dispersed and both affects the acidity of the ZSM-5 zeolite and provides hydrogenation-dehydrogenation functionality. This bifunctional behavior can also be achieved (to differing efficiencies) with other zeolites that have strong acid sites (Y, Beta) and with other metal functions (zinc, platinum).

3.2.1. UOP Cyclar Process

A commercial application of such an aromatization reaction is in the UOP Cyclar process. Discovered by BP and jointly developed with UOP, the Cyclar process uses a single catalyst system to convert propane and butanes into aromatic hydrocarbons (BTX) at high selectivity [26]. The catalyst consists of a zeolite with a non-noble metal promoter. The zeolite component provides the oligomerization and aromatization acid functions, and the non-noble metal additive provides the dehydrogenation function [27].

The aromatization of propane and butanes is a thermodynamically feasible endothermic reaction (35 to 45 kcal/mol). Aromatization in the Cyclar process includes a series of distinct reaction steps: dehydrogenation, olefin oligomerization, cyclization, hydrogen transfer, alkylation, dealkylation, and cracking. A simplified diagram showing the basic Cyclar reaction mechanism is shown in Figure 3. Light paraffins are dehydrogenated to the corresponding olefins in the initial, rate-limiting, reaction step. Once formed, the highly reactive olefins rapidly oligomerize to intermediates with higher molecular weight. These intermediates then rapidly cyclicize to naphthenes, which are rapidly dehydrogenated to aromatics. The shape selectivity of the catalyst limits the size of the naphthenic rings to single-ring compounds and also limits the production of coke precursors, such as polynuclear aromatic compounds.

A side reaction that the intermediates can undergo is hydrocracking to form light paraffins. Cracking reactions result in a net loss in yield because methane and ethane are unreactive at Cyclar process conditions, although the aromatization of ethane is feasible at more severe conditions. The presence of a dehydrogenation function in the catalyst greatly increases the dehydrogenation rate relative to the cracking rates, and thus the net aromatics yield is increased. The result of the nearly complete conversion of the oligomer is an aromatic product with virtually no impurities. The individual BTX components can be recovered from this aromatic product at high purities by simple direct fractionation without extraction.

Other reactions that occur over the Cyclar catalyst are isomerization, dealkylation, and transalkylation of the aromatic species formed by the main reaction mechanism. The transalkylation reactions result in a distribution of benzene and alkylbenzenes close to the equilibrium composition, depending on charge stock and process conditions. Slightly more benzene is produced from propane than from butane-rich feedstocks [27].

The aromatization of pentanes can be accomplished if processed separately at milder operating conditions. If mixed with propane and butanes and operated at the usual conditions of the Cyclar process, pentanes tend to crack to propane and ethane and thus yield a product distribution akin to that from propane aromatization.

The Cyclar process was successfully commercialized in 1990 in a unit built by BP in their facilities in Grangemouth, Scotland. Although the unit was designed for operation with the original catalyst formulation, DHCD-2, a new high-activity catalyst, DHCD-4, introduced commercially in 1991 has vastly exceeded the original performance representations. Figure 4 illustrates a typical flow diagram of one Cyclar unit. The process itself has been described elsewhere [27,28].

3.2.2. Cyclar Process Economics

Catalyst performance and investment costs are influenced by the choice of operating pressure. Two different extreme pressures are usually considered for the economic analysis of commercial designs, depending on site-specific needs for minimum investment or maximum yield. An example is a typical Cyclar unit with a capacity for 12,000 BPSD of mixed propane-butanes feed, roughly corresponding to about 350,000 MTA of feed (320,000 to about 375,000 MTA depending on feed composition, from 100% propane to

FIGURE 3
CYCLAR REACTION PATHWAYS

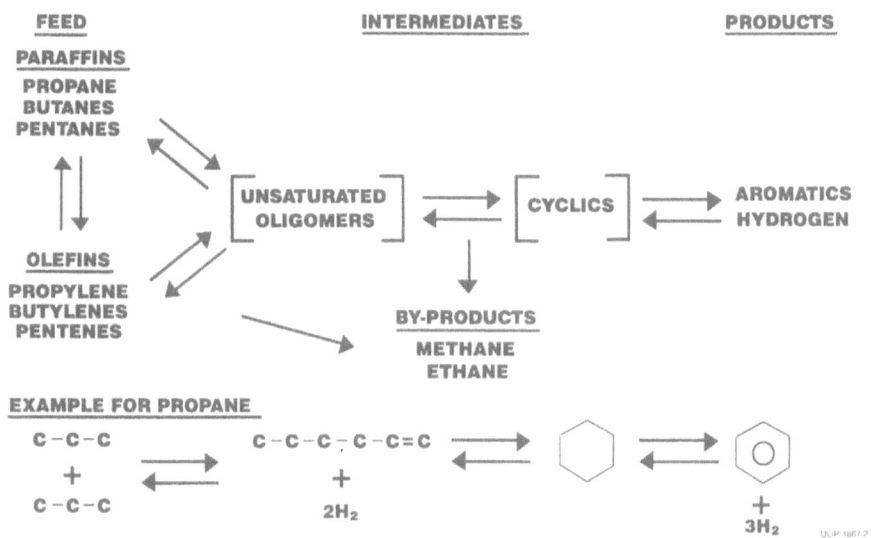

FIGURE 4
UOP-BP CYCLAR PROCESS FOR LPG AROMATIZATION

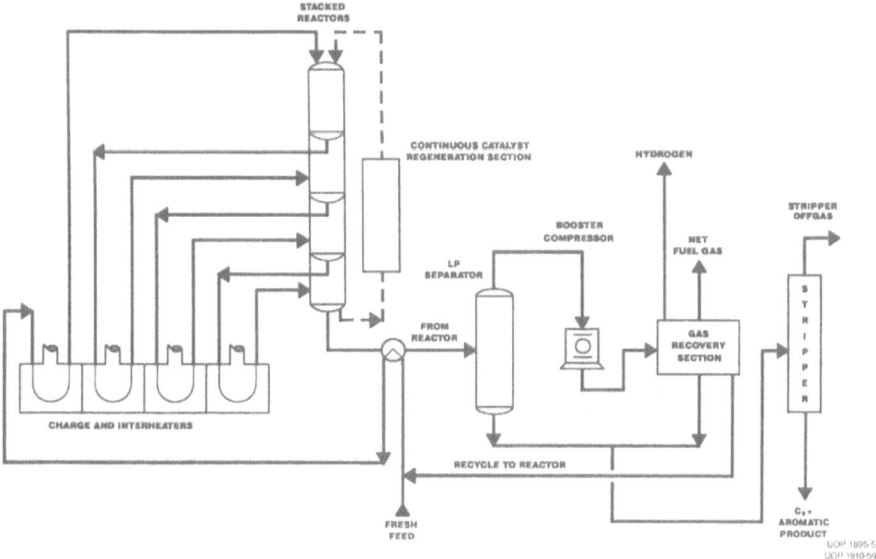

100% butanes). Both high-pressure (minimum-investment) and low-pressure (maximum-yield) modes have been considered. The minimum-investment mode reduces the estimated erected cost by 25% and the catalyst requirements by 50% relative to the maximum-yield mode. The economics are based on the use of the latest DHCD-4 catalyst now in commercial operation.

Figure 5 illustrates the simple payback as a function of the price differential between the liquefied petroleum gas (LPG) feed and the aromatics product. The simple payback was calculated using the estimated erected cost within battery limits and a 30% allowance for offsites for a unit located in the U.S. Gulf Coast and built in the second quarter of 1991. The choice of operating pressure has relatively little economic impact as far as the simple payback is concerned. The minimum-investment option can be particularly attractive for remote locations where low-priced feedstock can be made available.

The aromatization of light paraffins produces significant amounts of the coproduct hydrogen. Hydrogen valuation has therefore a pronounced effect on plant economics, as illustrated in Figure 6 for the high-pressure mode. The increasing need for efficient hydrogen management to meet the demand for lighter and cleaner fuels in oil refineries provides an opportunity to install hydrogen-generating technology, such as the Cyclar process, as an alternative to a conventional hydrogen plant.

The Cyclar process is particularly attractive in locations where the LPG feedstock is available at reasonable prices and where hydrogen is in demand. For example, the payback for a typical 12,000 BPSD installation is less than 2.5 years if the mixed C_3-C_4 feed has a value of $135/MT, and the aromatics are sold at petrochemical value with the hydrogen value at $650/MT. If the value of the hydrogen coproduct is reduced to $300/MT, the payback increases to three years in this case.

3.3. Titanosilicate Catalysts

Titanium-substituted silicalite can be prepared by hydrothermal synthesis with a homogeneous distribution of Ti ions in the crystal. The Ti ions seem to be all Ti^{4+}; ESR shows the absence of Ti^{3+}. Also, all Ti^{4+} are surrounded by four Si ions, and thus, the catalytic site is a single Ti^{4+} ion [29].

Titanium silicalite has been reported to be an industrial catalyst for the oxidation of olefins or aromatics. The oxidant is hydrogen peroxide (H_2O_2). For example, Ti silicalite catalyzes the epoxidation of both olefins and diolefins with high yields and high epoxide selectivity (>98% for propylene). The reaction can be conducted with dilute H_2O_2 solution in contrast to other catalysts that require nearly pure H_2O_2 as oxidant.

Glycol monomethyl ethers can be synthesized in one step from the mixture of olefins, methanol, and H_2O_2. Ether selectivity is >95%. Aldehyde and ketone derivatives are also obtained from primary or secondary alcohols in high purity [29].

FIGURE 5

SIMPLE PAYBACK FOR MINIMUM INVESTMENT vs. MAXIMUM YIELD

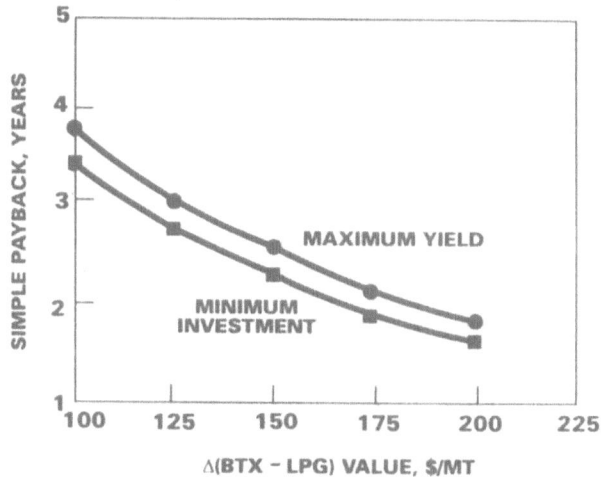

BASIS: 12,000 BPSD FEED RATE, $120/MT FUEL GAS VALUE, $650/MT HYDROGEN VALUE

UOP 1910 51

FIGURE 6

HYDROGEN PRICE SENSITIVITY FOR THE MINIMUM-INVESTMENT MODE

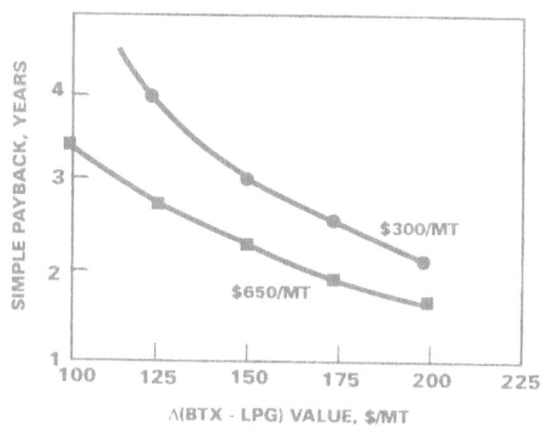

BASIS: 12,000 BPSD FEED RATE, $120/MT FUEL GAS VALUE

UOP 1910-52

Table 4
Challenges for Hydrocarbon Processing Catalysts in the 1990s

- Changing feedstocks
  - Heavier feeds
  - Higher sulfur and metals content
- Alternative fuels, CHy, syngas
- Changing fuel product slate
  - More motor fuel
  - Less fuel oil
- Changing motor fuel specifications and environmental issues
  - Gasoline reformulation
  - Desulfurization of motor fuels
  - Reduction of SO_x and NO_x emissions
  - Use of noncorrosive, inert catalysts
- Provide economic raw materials for petrochemicals

Table 5
Catalytic Oxidation with Titanosilicalite

Reactants	Products
$R\text{-}C_6H_5 + H_2O_2$	$R\text{-}C_6H_4\text{-}OH + H_2O$
$R\text{-}C_6H_4\text{-}OH + H_2O_2$	$R\text{-}C_6H_3\text{-}(OH)_2 + H_2O$
$R\text{-}CH{=}CH\text{-}R + H_2O_2$	$R\text{-}CH\text{-}CH\text{-}R + H_2O$ $\underset{O}{\diagdown\diagup}$
$R\text{-}CH{=}CH\text{-}CH{=}CH\text{-}R + H_2O_2$	$R\text{-}CH\text{-}CH\text{-}CH{=}CH\text{-}R + H_2O$ $\underset{O}{\diagdown\diagup}$
$R\text{-}CH{=}CH\text{-}R + CH_3OH + H_2O_2$	$R\text{-}CH(OH)\text{-}C(OCH_3)\text{-}R + H_2O$
$R\text{-}CH_2OH + H_2O_2$	$R\text{-}CHO + 2H_2O$
$R\text{-}CHOH\text{-}R + H_2O_2$	$R\text{-}CO\text{-}R + 2H_2O$

Titanium silicalite is unique in catalyzing the oxidation of aromatics with H_2O_2 at high selectivity. For example, benzene can be oxidized to phenol, and phenol can be further hydroxylated to a mixture of catechol and hydroquinone. A listing of known oxidation reactions catalyzed with Ti silicalite is shown in Table 5. A restricted transition state selectivity [30] appears to be effective because of the specific pore structure of the catalyst.

4. ALUMINOPHOSPHATE MOLECULAR SIEVES

This section is abstracted from a publication by E. M. Flanigen, et al. [23].

4.1. Compositions

In the late 1970s, the Union Carbide research laboratory (now part of UOP) set out to explore new framework compositions of oxides outside the range of the then-known aluminosilicate zeolites and silica molecular sieves. In 1982, a major discovery of a new class of aluminophosphate molecular sieves was reported by Wilson, et al. [31]. By 1986, some 13 elements had been reported that could be incorporated into the alum-inophosphate framework: lithium, beryllium, boron, magnesium, silicon, titanium, manganese, iron, cobalt, zinc, gallium, germanium, and arsenic [31]. This new generation of molecular sieves consists of more than two dozen structures and about 200 compositions. New additions are being reported on a regular basis.

Aluminophosphate molecular sieves exhibit a number of novel crystalline structures as well as topologic analogues of those found in zeolites. Pore sizes range from very small (~0.3 nm) to very large (~2 nm). Table 6 illustrates a variety of morphologies, compositions, and pore sizes in the aluminophosphate family.

4.1.1. Aluminophosphates

Aluminophosphate ($AlPO_4$) molecular sieves, which were first reported in 1982, include the first very large pore material, VPI-5, reported more recently by Davis, et al. [21, 23]. The VPI-5 structure has a unidimensional channel of an 18-membered ring with a free pore diameter of 1.25 nm. The $AlPO_4$-8, $AlPO_4$-11, $AlPO_4$-36, $AlPO_4$-41, and others, alternatively, display characteristic oval shapes. For example, $AlPO_4$-8 is a large 14-membered ring with one-dimensional channels that are slightly elliptical (0.87 x 0.79 nm) (Figure 7). Channel walls can be smooth, as in $AlPO_4$-5 and $AlPO_4$-31, or display annular side pockets, as in $AlPO_4$-36.

The $AlPO_4$ molecular sieves have a univariant framework composition with Al/P = 1, a high degree of structural diversity, and a wide variety of pore sizes and volumes. They are neutral frameworks and, therefore, have no ion-exchange capacity. Their surface selectivity is mildly hydrophilic. Some members of the series (although not the large pore material) exhibit excellent thermal and hydrothermal stability, up to 1000°C and 600°C, respectively.

Table 6
Selected Structures and Compositions
in Binary, Ternary, and Quaternary Systems

Structure Type	Molecular Sieves			Metals in MeAPSO
	AlPO₄	SAPO	MeAPSO	
Large Pore:				
5	X	X	X	Co, Fe, Mg, Mn, Zn
36	—	—	X	Co, Mg, Mn, Zn
37	—	X	—	
40	—	X	—	
46	—	—	X	Co, Fe, Mg, Mn, Zn
Intermediate Pore:				
11	X	X	X	Co, Fe, Mg, Mn, Zn
31	X	X	X	Co, Fe, Mg, Mn, Zn
41	—	X	—	
Small Pore:				
41	X	—	—	—
17	X	X	X	Co
34	—	X	X	Co, Fe, Mg, Mn, Zn
44	—	X	X	Co, Fe, Mg, Mn, Zn
47	—	—	X	Co, Mg, Mn, Zn
Very Small Pore:				
20	X	X	X	Co, Fe, Mg, Mn, Zn

Table 7
Vapor-Phase Propylene Oligomerization with SAPO Catalysts

Molecular Sieve	SAPO-5	SAPO-11	SAPO-31	SAPO-34	LZ-105
Pore Size, Å	8	6	7	4.3	6
Run Temp., °F	700	700	700	700	703
Pressure, psig	25	25	50	25	25
Propylene, WHSV	0.98	0.94	1.04	0.53	0.90
Time On-Stream, hr	4.3	4.2	5.5	2.33	3.5
Propylene Conversion, %	0	86.3	76.2	41.6	81.6
C₅+ Selectivity, wt-%	—	77.0	82.7	19.5	37.2

FIGURE 7

STRUCTURE OF AlPO$_4$-8 AND VPI-5

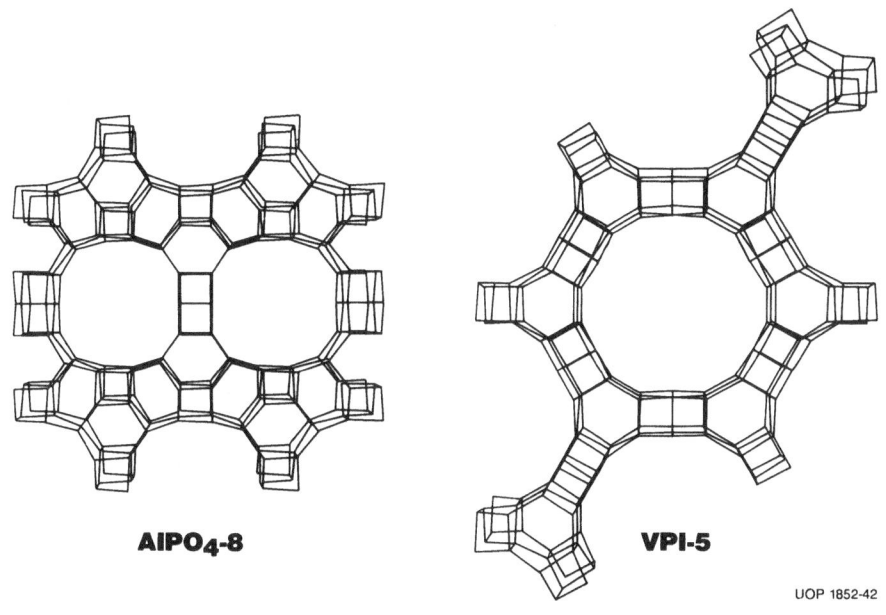

AlPO$_4$-8   VPI-5

UOP 1852-42

FIGURE 8

SCHEMATIC REPRESENTATION OF THE RELATIONSHIPS IN THE ALUMINOPHOSPHATE-BASED MOLECULAR SIEVES

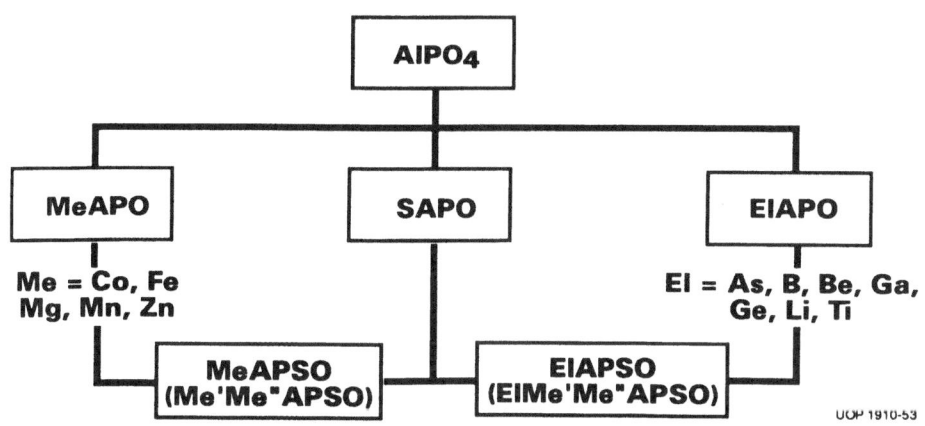

UOP 1910-53

4.1.2. Silicoaluminophosphates

Silicoaluminophosphates (SAPO) constitute the next family of new molecular sieves. Structures again range from totally novel to analogues of existing zeolites or AlPO$_4$ materials [31,33]. For example, Si-VPI-5, a SAPO analogue of the 18-membered ring VPI-5, is known. There are also a number of large-pore 12-membered ring structures including SAPO-37; medium-pore structures in the 0.6 to 0.65 nm range; and small-pore structures with pores of 0.35 to 0.45 nm, including SAPO-34. A pure silica analogue of the SAPO-5 structure, SSZ-24 is also known [34].

Because of the substitution of silicon for hypothetical phosphorus sites, SAPOs have negatively charged frameworks with cation exchange properties and weak-to-mild acid properties. They also exhibit excellent thermal and hydrothermal stability.

4.1.3. Metal Aluminophosphates

The framework of the metal aluminophosphates (MeAPO) family contains a metal, Al, and P. The known metal species include divalent forms of Co, Fe, Mg, Mn, and Zn and the trivalent Fe. The MeAPOs exhibit great structural diversity and even more extensive compositional variation. Many of their structures have never been observed in zeolites, but some are topologically related to zeolites or to AlPO$_4$ materials.

The MeAPOs can be regarded as hypothetical AlPO$_4$ frameworks that have undergone substitution of metal for Al. The result is a negative or neutral framework for bivalent and trivalent metals. Like SAPOs, the negatively charged MeAPO frameworks possess ion exchange properties and Brönsted acid sites. The thermal and hydrothermal stability of MeAPOs is somewhat less than those of the AlPO$_4$ and SAPO series.

4.1.4. Metal Silicoaluminophosphates

The metal silicoaluminophosphates (MeAPSO) family further extends the structural diversity and compositional variation found in the SAPO and MeAPO molecular sieves. These quaternary frameworks contain a metal, Al, P, and Si as framework species. Structural types include topologies observed in the binary AlPO$_4$ materials, ternary SAPOs, and MeAPOs as well as novel structures, such as MeAPSO-46, with a 0.7 nm pore size [35]. Quinary and senary framework compositions that contain Al, P, Si and additional combinations of divalent metals have also been synthesized.

In an additional series (ElAPO and ElAPSO), the elements Li, Be, B, Ga, Ge, As, and Ti have been incorporated into the AlPO$_4$ framework. Figure 8 shows the compositional relationships among the aluminophosphate-based families of molecular sieves.

Most recently, a molecular sieve with extra large pores has been reported by Eastermann, et al. [36]. These molecular sieves have a gallium phosphate composition. Called cloverite, this molecular sieve displays a cloverleaf-like pore opening with about a

2 nm diameter (Figure 9). Metallophosphate molecular sieves including zinc-beryllium phosphates constitute a new collection of materials now being developed [37].

4.1.5. Sulfide Molecular Sieves

Whereas all crystalline molecular sieves and microporous crystals had hitherto been based on oxide frameworks, an important new direction for molecular sieves is provided by the recent discovery of sulfide-based molecular sieves [37]. The first publication on these template-containing materials described a whole family containing germanium, tin, and several other metals. The crystal structures are all new and include a number of unique framework structures. These materials may have potential catalytic applications in the processing of sulfur-containing feedstocks in which the feed sulfur may help to stabilize the composition of the sulfide molecular sieve. The series is recently expanded to include antimony sulfides [39].

4.2. Catalytic Fingerprinting of Aluminophosphate-Type Molecular Sieves

Aluminophosphate-based molecular sieves have acidic and functional activity and exhibit a broad spectrum of potential or actual catalytic applications in the conversion of hydrocarbons.

4.2.1. Reactions of Olefins

The SAPOs are effective for the conversion of light olefins to gasoline-range products. Typical results are reproduced in Table 7, which shows conversion and selectivity data for propylene oligomerization to liquid products. These data were obtained using large-, medium-, and small-pore SAPOs and the medium-pore reference LZ-105 zeolite.

The large-pore SAPO-5, with unidirectional nonintersecting channels [24], was inactive for oligomerization by the time the first sample had been taken. Deactivation appeared to be the result of rapid catalyst coking. The small-pore SAPO-34, with a crystal structure analogous to chabazite, was also ineffective for propylene oligomerization. In contrast, the medium-pore SAPO-11 and SAPO-31 exhibited significant oligomerization activity and selectivity to liquid products. The medium-pore SAPO-11 consists of unidirectional, nonintersecting channels formed by 10-membered rings [24]. By way of reference, the medium-pore LZ-105 zeolite exhibited high propylene conversion but poor selectivity. The LZ-105 zeolite is structurally similar to ZSM-5 but is prepared without an organic template. The liquid product from the LZ-105 zeolite was highly aromatic, whereas that from the SAPOs was predominantly olefinic. The higher olefinic selectivity of the SAPOs can be attributed to their reduced acid strength, which prevents the occurrence of hydride shift reactions [40].

Table 8 shows similar results for the conversion of 1-hexene over a number of aluminophosphate-based molecular sieves. Of particular interest is the coke resistance of the medium-pore SAPO-11. Spatial constraints within the SAPO-11 crystal structure inhibit the formation of coke precursors. Similar resistance can be observed in

Table 8
Reactions of 1-Hexene over Aluminophosphate-Based Molecular Sieves

Run Conditions:
Run Temperature - 650°F
Pressure - 40 psig
WHSV - 5.5 hr^{-1}

Molecular Sieve	SAPO-5	SAPO-11	FAPO-11	MnAPO-11	SAPO-31	FAPO-31	MnAPSO-31	LZ-105
Pore Size, Å	8	6	6	6	7	7	7	6
Total Conversion, %	85.1	84.5	90.1	89.6	86.0	89.1	94.5	93.7
Selectivities:								
Double-Bond Isomerization, %	79.1	46.2	22.1	28.0	82.2	42.5	43.2	2.4
Skeletal Isomerization, %	10.9	41.9	70.6	63.6	14.3	52.8	44.6	12.2
Oligomerization, %	5.6	4.3	2.4	0.9	1.7	0.9	4.1	55.1
Cracking, %	3.2	3.3	1.5	1.9	0.9	1.3	2.6	25.6

FIGURE 9

STRUCTURE OF CLOVERITE

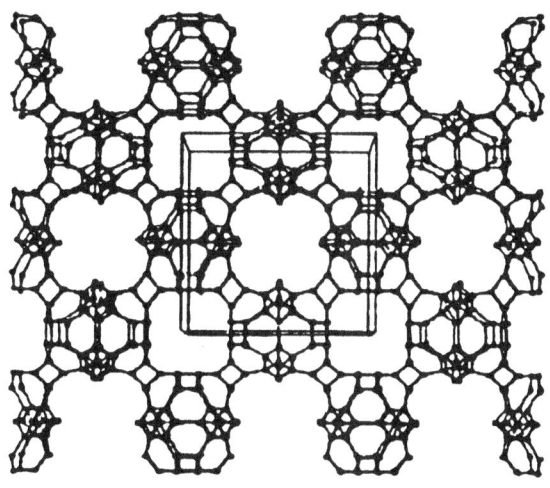

UOP 1903-9

Table 9
***m*-Xylene Reactions Catalyzed by Aluminophosphate-Based Molecular Sieves**

Run Conditions:
Run Temperature - 1000°F
Pressure - 100 psig
WHSV - 5.6 hr^{-1}

Molecular Sieve	Pore Size, nm	Pore Volume, cc/g	*m*-Xylene Disproportionation, % Conversion
SAPO-5	0.8	0.31	20.7
MAPO-5	0.8	0.31	22.4
MAPO-36	0.8	0.31	12.6
SAPO-31	0.65	0.17	15.1
SAPO-41	0.6	0.22	7.1
SAPO-11	0.6	0.16	3.2
MAPO-11	0.6	0.16	1.8

medium-pore ZSM-type molecular sieves. The incorporation of transition elements into the crystal structure can greatly enhance a particular catalytic feature. Thus, FAPO-11 (FeAPO-11) and MnAPO-11 show much improved skeletal isomerization relative to that of SAPO-11. Similar trends can be observed for FAPO-31 and MnAPO-31 relative to SAPO-31.

4.2.2. Reactions of Paraffins

The dehydrocyclization of paraffins represents an important class of reactions in the catalytic reforming of naphthas for the production of high-octane gasolines. Zeolitic acids such as mordenite and ZSM-5 can be substituted for the usual acidity derived from chlorided alumina used in catalytic reforming. However, the enhanced octane observed with these materials is often accompanied by increased cracking and reduced yields.

Catalysts consisting of Pt on large-pore SAPO-5 are less active than reference alumina-supported platinum catalysts but appear to have similar selectivity. Catalysts prepared with Pt on small-pore SAPO-34 are active for hexane conversion but have lower selectivities. The Pt-loaded medium-pore SAPO-11 and SAPO-41 exhibit excellent hexane conversion and selectivity. The results are in agreement with those observed for 1-hexene isomerization: medium-pore SAPOs show significantly reduced cracking activity and the maintenance of high skeletal isomerization activity.

4.2.3. Reactions of Aromatics

The isomerization of C_8 aromatics to produce p-xylene is another area of commercial interest. Also of interest for the production of p-xylene are the alkylation of toluene with methanol and the disproportionation of toluene. Medium-pore SAPO materials are distinguished by their low disproportionation activity and their high selectivity for toluene methylation reactions. This activity-selectivity pattern results in lower benzene losses and improved xylene yields.

The isomerization of C_8 aromatics, toluene disproportionation, and toluene trans-alkylation with C_9 aromatics are all important commercial routes toward the production of mixed xylenes. The p-xylene can be recovered at high purities from C_8 aromatic streams, which result, at least in part, from the isomerization of mixtures of xylenes and ethylbenzene. The p-xylene can also be obtained at relatively high concentrations either by the shape-selective disproportionation of toluene (now in commercial operation) or by the shape-selective alkylation of toluene with methanol (not yet commercialized). Various zeolite compositions are being used or have been proposed for these applications [41,42]; nonzeolitic molecular sieves also offer promise in these areas.

4.2.3.1. Without Hydrogen

Aluminophosphate molecular sieves have been used for the isomerization of C_8 aromatics either alone or in the presence of Pt and hydrogen. In a screening test, a series of large- and medium-pore molecular sieves were evaluated for catalytic activity in the isomerization of *m*-xylene at about 540°C in the absence of Pt and hydrogen. Under these conditions, all molecular sieves evaluated yielded virtually complete isomerization to an equilibrium distribution of xylene isomers. However, the disproportionation activity of toluene and trimethylbenzenes does vary significantly. Table 9 shows disproportionation activity as a function of molecular sieve pore size. Large-pore molecular sieves are more active for toluene disproportionation.

Several medium-pore molecular sieves were also tested for catalytic performance with an ethylbenzene and *m*-xylene feed. No added metals or hydrogen were used, although several of these aluminophosphate-based molecular sieves did contain transition metals as framework constituents. The results of this study are summarized in Table 10. Surprisingly, MeAPSO-31 yields *p*-xylene concentrations in excess of equilibrium that are not apparent when only SAPO-31 is used. Obviously, one interpretation can be that the substituting framework cations (Mn^{2+} or Co^{2+}) are larger than the silicon or aluminum cations and result in shape-selective restrictions in the molecular sieve channels. However, gravimetric adsorption studies indicate that MeAPSO-31s have pore sizes and volumes comparable to those of SAPO-31, which does not appear to be *p*-selective. Furthermore, SAPO-11, with only slightly smaller pore sizes, does not exhibit the level of *p*-selectivity seen with MeAPSO-31. Either chemical interaction or the internal distribution of acid sites must play a role to explain the enhanced selectivity of MeAPSO-31.

4.2.3.2. With Hydrogen

Catalysts consisting of selected SAPOs and supported Pt have been evaluated in the presence of hydrogen for bifunctional C_8 aromatic isomerization with a feed that contained 17% ethylbenzene and 83% *m*-xylene (reformate gasoline) and with a feed that contained 40% ethylbenzene and 60% *m*-xylene (pyrolysis gasoline). Both large-pore SAPO-5 and medium-pore SAPO-11 promote near complete xylene isomerization, and SAPO-5 achieves higher ethylbenzene conversion. However, with the low ethylbenzene feed, the large-pore molecular sieve incurred nearly 22% xylene losses as a result of disproportionation and the production of noncyclic cracked byproducts. In contrast, SAPO-11 actually produced 2.1% more xylenes than were originally present in the feed. With SAPO-11, disproportionation activity is reduced and almost no paraffin products result from ring opening.

The beneficial effect of the medium-pore SAPO is even more dramatic for feed with 40% ethylbenzene content. With SAPO-11 at 23.6% ethylbenzene conversion, a nearly 13% increase in xylene yield was observed. This increase is equivalent to an ethylbenzene conversion selectivity of 75% to xylene isomers. The remaining conversion is accounted for by disproportionation reactions, but essentially no ring opening is observed.

4.3. Initial Industrial Applications

Medium-pore aluminophosphate-based molecular sieves are active and selective catalysts for a variety of important hydrocarbon conversion reactions. As acid catalysts, they promote olefin isomerization and oligomerization, but they are less effective at the

Table 10
Selective Reactions of Aromatics

Run Conditions:
Run Temperature - 800°F
Pressure - 100 psig
WHSV - 5.6 1/hr
Feed - 17% EB, 83% *m*-xylene

Molecular Sieve	*para*/ortho Xylene Ratio	% of Equilibrium Concentration	% Disproportionation	
			Xylenes	Ethylbenzene
SAPO-11	0.97	96	9.9	23.2
SAPO-11 (Al⁺³ Exchanged)	1.15	102	6.6	20.1
SAPO-11 (Steam Treated)	1.52	63	0.2	6.0
SAPO-31	0.78	100	31.6	56.3
SAPO-31 (Steam Treated)	1.56	102	1.7	10.4
MnAPSO-11	0.88	104	5.0	23.3
CoAPSO-11	1.81	91	0.0	7.7
MnAPSO-31	3.59	120	0.0	17.7
CoAPSO-13	3.06	111	1.7	23.0
LZ-105	0.99	100	23.6	58.5

Table 11
Conversion of Methanol to Olefins

Methanol to Ethylene Conversion, 100%		Methanol to Propylene Conversion, 100%	
Product Yield, Wt-%		Product Yield, Wt-%	
Ethylene	53.0	Ethylene	31.2
Propylene	27.1	Propylene	53.0
Butenes	6.4	Butenes	7.8
Pentenes	3.5	Pentenes	2.0
Total Olefins	90%	Total Olefins	93%

competing hydride transfer and cracking reactions. In aromatics reactions, they are effective for skeletal isomerization but show low ethyl group disproportionation activity. As acid components of bifunctional catalysts, they are selective for paraffin and cycloparaffin isomerization with low cracking activity.

In general, medium-pore SAPOs are considerably less active than LZ-105 and other zeolites for olefin, paraffin, and aromatic conversions when compared at the same temperatures, but they are more selective for olefin and paraffin isomerization when evaluated at comparable conversions. Medium-pore aluminophosphates have remarkable resistance to coking and can thus remain stable and active over long periods.

The introduction of transition metals into framework positions enhances the activity and selectivity for olefin oligomerization relative to plain SAPOs. The MeAPOs and MeAPSOs are also more selective for C_8 aromatic rearrangement reactions, an effect that cannot be attributed solely to improved shape selectivity. The enhanced selectivity may be due to uniquely located acid sites in the transition-metal-containing molecular sieve combined with unexpected spatial effects. It may also be due to a ligand or electronic effect of the transition metal that affects the transition states in aromatic isomerization and disproportionation reactions.

Some of the most promising applications are enumerated below. A few of these applications are expected to become commercial in the near future.

4.3.1. Catalytic Dewaxing

Because of characteristically high yields in reactions with paraffins and olefins, catalytic dewaxing is a typical application in which aluminophosphate molecular sieves should prove advantageous. Indeed, medium-pore aluminophosphate compositions show outstanding selectivity in the catalytic dewaxing of heavy petroleum products [43]. Whereas conventional catalytic dewaxing zeolites typically crack the paraffins to lighter olefinic materials, aluminophosphates isomerize the long-chain n-paraffins to nonwaxy hydrocarbons of similar molecular weight. The result is an incremental yield advantage and high product quality.

4.3.2. *p*-Xylene Production

Certain aluminophosphate catalysts show high selectivity, particularly high xylene yields in the conversion of ethylbenzene-containing C_8 aromatic mixtures, high catalytic activity, and outstanding resistance to coke formation. These characteristics provide opportunity for commercial application as catalysts in the production of *p*-xylene from C_8 aromatics.

4.3.3. Olefin Isomerization

The low coking tendency of medium-pore aluminophosphate molecular sieves has resulted in the first viable approach to the skeletal isomerization of olefins, such as the isomerization of n-butenes to isobutene.

4.3.4. Conversion of Methanol to Light Olefins

Although ZSM-5 zeolites have been proposed for this application, more recent work has shown that nonzeolitic molecular sieves, like SAPO-34, can produce ethylene and propylene at high selectivity and still remain active (coke-free) over long periods of time (Table 11). The selectivity to ethylene reportedly can be enhanced by the use of MeAPSO compositions when the metal consists of either Ni or Co.

4.3.5. Alkylation of Aromatics with Methanol

Methyl-substituted aromatic compounds can be produced by the alkylation of aromatics with methanol. In particular, p-xylene can be selectively produced by the shape-selective alkylation of toluene with methanol over a medium-pore aluminophosphate catalyst.

4.3.6. Octane-Enhancing Catalyst Additives

Medium-pore aluminophosphate molecular sieves show high selectivity toward isomerization over cracking when processing paraffins and olefins. This characteristic may be beneficially used in octane-boosting applications, both in FCC and in hydrocracking, using the aluminophosphates as catalyst additives. These applications are similar to the use of ZSM-5 in process applications [44-47].

5. LAYERED STRUCTURES

Clays have traditionally been used for the alkylation of residual olefins in aromatic streams for the control of the bromine index. Recent years have brought increased interest in new materials for the catalysis of feedstock molecules with a high molecular weight and for the separation of large molecules. Although the recent discovery of aluminophosphate-based molecular sieves has led to significant increases in the maximum pore size (1.2 nm and larger), even larger pores are desired for the treatment of large molecules in heavy petroleum feedstocks. Pillared clays and other layered structures can provide such large-pore networks [48]. Thermal and hydrothermal stabilities are key properties that affect the utility of a material in these applications and have so far limited successful applications.

Clays and pillared clays have acidities and pore sizes that are intermediate between those of amorphous aluminosilicates and zeolites. Some clays of particular interest are: hectorite, montmorillonite, saponite, and beidellite. Not all clays are readily available; some require fairly specialized synthesis.

Because pillared clays offer the capability of developing well-defined gallery pores, they may display some molecular sieve character. The pore sizes are defined by the chosen pillaring agent. Clays are usually pillared by ion-exchanging metal cations between the clay layers. To achieve thermal stability and large pore sizes, polymeric oxycations have been used as a prop or pillar to keep the silicate layers permanently separated. In this way, a porous network is obtained. For example, a montmorillonite that has been pillared with

Al_{13} Keggin ion [49] has a BET surface area between 200 and 400 m²/g and an interlamellar distance (gallery) of 0.9 nm.

The great diversity of clay and pillar combinations offers a growing, technologically unexplored frontier. Thermally stable pillars have been prepared with Al, Ti, Zr, Cr, Si, and Fe and mixtures thereof. The use of these pillaring agents also offers the possibility of incorporating a metal function, which can play a catalytic role within the clay pore. The synthesis of modified clays with larger gallery spacing and with improved thermal and hydrothermal stabilities is expected to lead to valuable materials in the catalytic processing of heavy feedstocks.

Recently, a pillared clay with a free gallery spacing of 1.8 nm was reported using rare earth and aluminum oxide units to provide the pillar. This large-pore pillared clay is claimed to have excellent thermal and hydrothermal stability and so can be used for applications as demanding as the FCC process.

6. FUTURE OPPORTUNITIES FOR MOLECULAR SIEVE CATALYSTS

In the coming years, a broad spectrum of new catalytic materials that offer a wider range of pore size, chemical composition, acidity, adsorptivity, and catalytic behavior will be essential to meet the needs for higher processing flexibility and for the production of environmentally desirable fuels. New catalytic materials are also needed to improve the process economics in the chemical and petrochemical industries and to make needed progress in environmental control. These improvements will require catalysts with higher activity and selectivity and hence new and improved materials. The ability to use new catalytic materials with a wide range of well-controlled properties to tailor reaction pathways, such as acid site strength and concentration, solvent effects, coke resistance, controlled molecular ingress and egress, and specific metal site interaction, will offer new opportunities in obtaining optimal selectivity and increased efficiency. The need for improved catalysts is anticipated for both existing and new process applications.

Considerable progress has been made recently in bringing advanced surface science techniques to the characterization of metals, and even oxides, at the atomic scale. These methods applied to catalytic sites are expected to help to reveal the true chemistry of catalytic reaction steps. They will also contribute to the development of custom-designed catalysts at an accelerated rate.

The remarkable diversity in structure and chemistry of the new and modified molecular sieves and other crystalline microporous materials offers a nearly unlimited number of parameters to tailor catalytic properties. As a result, catalysts can be developed with unique combinations of acidity and shape selectivity that are not achievable with ordinary zeolites or other current catalysts. The capability of these materials to carry out selective and, sometimes, novel chemical reactions will play a prime role in the catalytic specificity required in the future. These catalysts will be the key in the development of advanced processes to meet the environmental, feedstock, product, and economic requirements of the 21st century.

472

7. REFERENCES

1. D. S. Coombs, A. J. Ellis, W. S. Fyfe, and A. M. Taylor, Geochim. Cosmochim. Acta, 17 (1959) 53.
2. J. V. Smith, Origin and Structure of Zeolites, in Zeolite Chemistry and Analysis, J. A. Rabó (ed.), ACS Monograph 171, American Chemical Society, New York City, 1976.
3. E. M. Flanigen, Molecular Sieve Zeolite Technology: The First Twenty-Five Years, in Zeolites: Science and Technology, F. R. Ribeiro, A. E. Rodrigues, L. D. Rollmann, and C. Naccache (eds.), NATO ASI Series, Series E: Applied Sciences, No. 80, Martinus Nijhoff Publishers, The Hague, 1984.
4. B. McCulloch, J. A. Johnson, and A. R. Oroskar, Zeolites for Petrochemical Separation, AIChE 1991 Spring National Meeting, Houston, TX, Apr. 7-11, 1991.
5. J. W. Ward, Molecular Sieve Catalysis, in Applied Industrial Catalysis, Vol. 3, B. E. Leach (ed.), Academic Press, 1984.
6. W. Holderich, M. Hesse, F. Naümann, Angew. Chem. Int. Ed., 27 (1988) 226.
7. J. Scherzer, Octane-Enhancing Zeolitic FCC Catalysts: Scientific and Technical Aspects. Marcel Dekker, Inc., New York City, Chemical Industries Series, Vol. 42, 1990.
8. J. Scherzer, Octane-Enhancing Zeolitic FCC Catalysts: Scientific and Technical Aspects. Catalysis Reviews, Science and Engineering, A.T. Bell and J.J. Carberry (eds.), 31 (1989).
9. A. P. Lamourelle, M. E. Reno, and G. J. Thompson, Hydrocracking for High Quality Distillates, NPRA Annual Meeting, San Antonio, TX, 1991.
10. S. E. George, and R. M. Foley, Hydrocracking Catalyst Applications: Criterion/-Zeolyst Approach, NPRA Annual Meeting, San Antonio, TX, 1991.
11. D. F. Besi, U.S. Patent 4,735,928 (1988).
12. J. W. Ward, U.S. Patent 4,429,053 (1984).
13. J. W. Ward, Preparation of Catalysts III, G. Ponceloi, editors, Elsevier, 1983. p. 597.
14. D. F. Best, U.S. Patent 4,897,178 (1990).
15. J. W. Ward, U.S. Patent 4,419,271 (1983).
16. R. D. Bezman and J. A. Rabo, U.S. Patent 4,401,556 (1983).
17. J. D. Hargrove, G. J. Elkes, A. H. Richardson BP CAT Dewaxing - Experience in Commercial Operation, NPRA National Fuels and Lubricants Meeting, Houston, TX, Nov. 9-10, 1978.
18. D. J. O'Rear and B. K. Lok, Ind. Eng. Chem. Res. 30 (1991) 1100.
19. M. E. Reno, R. S. Haizmann, B. H. Johnson, P. P. Piotrowski, and A. S. Zarchy, Improved Profits with Paraffin Isomerization Innovations, UOP Technology Conference Series, UOP, 1990.
20. Y. Y. Huang, M. P. Nicoletti, and R. A. Sailor, DeWitt 1990 Petrochemical Review, Houston, TX, Mar. 27-29, 1990.
21. D. W. Breck, Zeolite Molecular Sieves: Structure, Chemistry, and Use, John Wiley & Sons, Inc., New York City, 1974.
22. R. Szostak, Molecular Sieves: Principles of Synthesis and Identification, Van Norstrand Reinhold, New York City, 1989.

23. E. M. Flanigen, R. L. Patton, and S. T. Wilson, Structural, Synthetic and Physicochemical Concepts in Aluminophosphate-Based Molecular Sieves, Stud. Surf. Sci. Catal., 37 (Innovation in Zeol. Mater. Sci.), 1990, 13-27.

24. J. A. Rabó, and G. J. Gajda, Catal. Rev. Sci. Eng., 31 (1989-1990) 385-430.

25. M. R. Klotz, U.S. Patent 4,269,813, Standard Oil Co. (Indiana) (1981).

26. M.T. Barlow, U.S. Patent 4,761,511, BP Co. (1988) and M. T. Barlow, EP Appl. 84302178.8, BP Co., (1984).

27. T. Imai, J. A. Kocal, and C. D. Gosling, Proceedings of lst Tokyo Conf. on Adv. Cat. Sci. and Tech. (1990) and C. D. Gosling, F. P. Wilcher, and J. A. Kocal, in JPI Petr. Ref. Conf., Tokyo, Oct. 18-19, 1990.

28. P. C. Doolan, and P. R. Pujadó, Make Aromatics from LPG, Hydrocarbon Processing, 68 (1989) 72-76.

29. G. Perego, G. Bellussi, C. Corno, M. Tara, F. Buonomo, and A. Esposito, in Proc. 7th Intern. Zeolite Conf., Y. Murakami, A. Iijima, and J. Ward (eds.), Kodansha Ltd., Tokyo and Elsevier, Amsterdam, pp. Aug. 17-22, 1986, 129-136.

30. D. K. Simmons, R. Szostak, P. K. Agrawal, and T. L. Thomas, J. of Catal., 106 (1987) 287-291.

31. S. T. Wilson, B. M. Lok, C. A. Messina, T. R. Cannan, and E. M. Flanigen, J. Am. Chem. Soc., 104 (1982) 1146-1147.

32. E. M. Flanigen, B. M. Lok, R. L. Patton, and S. T. Wilson, New Developments in Zeolite Science and Technology, in Proc. 7th Intern. Zeolite Conf, Y. Murakami, A. Iijima, and J. Ward (eds.), Kodansha Ltd., Tokyo, and Elsevier, Amsterdam, Aug. 17-22, 1986, 236-245.

33. M. E. Davis, C. Saldarriaga, C. Montes, J. Garcés, and C. Crowder, Nature (London), 331, (1988) 698-699.

34. R. A. Van Nordstrand, D. S. Santilli, and S. I. Zones, in Perspect. Mol. Sieve Sci, W. H. Flank and T. E. Whyte (eds.), ACS Symp. Ser. 368 (1988) 236-245.

35. J. M. Bennett and B. K. Marcus, in Innovation Zeolite Mater. Sci., P. J. Grobet, W. J. Mortier, E. F. Vansant, and G. Schulz-Ekloff (eds.), Stud. Surf. Sci. Catal., Elsevier, Amsterdam, 37 (1988) 269-279.

36. M. Eastermann, L. B. McCusker, C. Baerlocher, A. Merrouche, and A. Kessler, A Synthetic Gallophosphate Molecular Sieve with a 20. Tetrahedral-Atom Pore Opening, Nature, 352 (1991) 320.

37. T. E. Gier and G. D. Stucky, Nature 349 (1991) 508.

38. R. L. Bedard, S. T. Wilson, L. D. Vail, J. M. Bennett, and E. M. Flanigen, in Zeolites, Facts, Figures, Future, P. A. Jacobs and R. A. Van Santen, R. A. (eds.), Elsevier, Amsterdam, 1989.

39. J. B. Parise, Science 251 (1991) 293.

40. J. A. Rabó, New Applications of Molecular Sieve Catalysts, in NATO Advance Study Institue (Zeolite Microporous Solids: Synthesis, Structure, and Reactivity), Alcabideche, Portugal, May 13-25, 1991.

41. W. O. Haag and D. H. Olson, U.S. Patent 3,856,871 (1974).

42. W. C. Carr, L. M. Polinski, S. G. Hindin, and J. L. Kosco, U.S. Patent 4,128,591 (1978).

43. F. P. Gortsema, A. R. Springer, G. N. Long, U.S. Patent 4,880,760, UOP (1989).

44. R. J. Pellet, P. K. Coughlin, G. N. Long, and J. A. Rabó, U.S. Patent 4,859,314, UOP (1989).

45. R. J. Pellet, P. K. Coughlin, M. T. Staniulis, J. A. Rabó, and G. N. Long, U.S. Patent 4,791,083, UOP (1988).

46. F. P. Gortsema, R. J. Pellet, A. R. Springer, J. A. Rabó, and G. N. Long, U.S. Patent 4,913,799, UOP (1990).

47. F. P. Gortsema, A. R. Springer, J. A. Rabó, and G. N. Long, U.S. Patent 4,818,739, UOP (1989).

48. D. E. W. Vaughan, Catalysis Today 2 (1988) 187.

49. T. J. Pinnavaia, M. S. Tzou, S. D. Landau, and R. H. Raythatha, J. Molec. Catal., 27 (1984) 195.

APPLICATION OF SOLID ACIDS AND BASES TO VARIOUS FIELDS

KOZO TANABE
Central Research Laboratory
Nippon Shokubai Co., Ltd.
5-8, Nishi Otabi-cho, Suita,
Osaka 564, Japan

ABSTRACT. The application of solid acids and bases to various fields other than catalysis is described. The fields cover those of gas sensors, adsorbents, pressure sensitive recording paper, cosmetic pigments, and others.

1. Introduction

Solid acids and bases have been extensively applied as catalysts for diversified reactions. Recently, some of those are being applied to various fields of gas sensors, removal or separation of certain compounds in gas mixtures and solutions, pressure sensitive recording paper, cosmetic pigments, etc. In this paper, the examples of the application are demonstrated and the principles in the application discussed.

2. Gas Sensors

Gas sensors are devices which detect or monitor a desired component or components in gases, and quantitatively convert the information to an electric signal. The component(s) may be combustible gases, oxygen, water (humidity), poisonous gases, etc.

For any gas sensor of either a semiconductor-type or catalytic combustion-type, sufficient sensitivity, accuracy, selectivity, reproducibility, life, and stability are required, and the sensing mechanism starts by adsorption or reactive adsorption of the gas component on the surface of sensor materials. Metal oxides such as SnO_2, Al_2O_3, Fe_2O_3, ZnO and mixed oxides are widely used as sensor materials. Therefore, the sensitivity and selectivity depend to a considerable extent on the acid-base interactions between the gas component and the surface of metal oxides. However, so far there have been very few studies which explicitly take into account the acid-base properties of the surface of sensor materials.

It has been reported that α-Fe_2O_3 promoted by sulfate ion exhibited

475

J. Fraissard and L. Petrakis (eds.), Acidity and Basicity of Solids, 475–484.
© 1994 Kluwer Academic Publishers.

higher sensitivity to combusible gases than α -Fe$_2$O$_3$ alone [1]. The addition of tetravalent metal ions such as Ti^{4+}, Zr^{4+}, and Sn^{4+} further enhanced the sensitivity, as shown in Fig. 1.

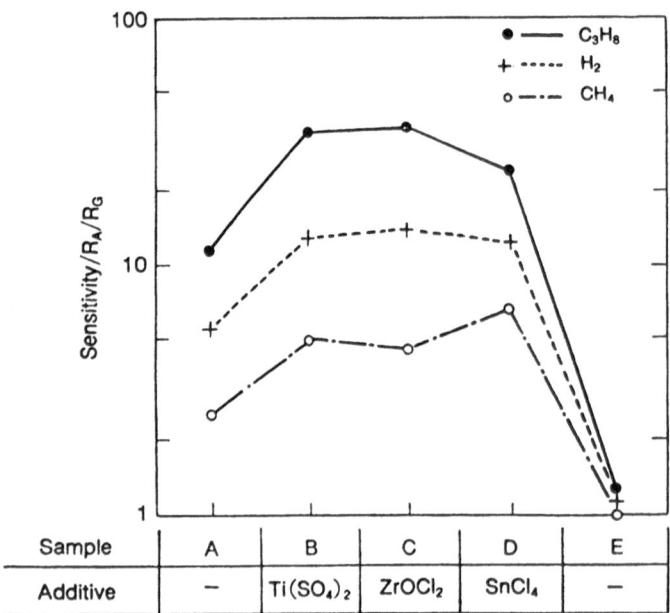

Fig. 1   Effects of sulfate ion and additives on gas sensitivity.
R$_A$; resistance in clean air, R$_G$; resistance in air containing 0.5 vol % of the gas.   Starting Fe salts were ferric sulfate for A — D and ferric nitrate for E.

Sulfate ion is contained in samples A, B, C, and D, but not in sample E. Thus, the role of sulfate ion is thought to stabilize fine particles and maintain a high surface area, by suppressing the crystal growth [2]. The acidity of Fe$_2$O$_3$ increases a little on the addition of Ti^{4+} or Sn^{4+}, but not on the addition of Zr^{4+} [3].   However, Fe$_2$O$_3$ containing a small amount of SO$_4^{2-}$ is known to be a very strong acid (one of solid super-acids), the acid sites being considered to generate by the formation of a complex as shown in Fig. 2, where the central Fe ion acts as a Lewis acid, whose acid strength is greatly enhanced by the inductive effect of S = O having high double bond nature [4, 5].   The high gas sensor's sensitivity of samples A, B, C, and D could be explained by their high acidity, the little difference in their sensitivities being due to the difference in the additive effect of Ti^{4+}, Zr^{4+}, and Sn^{4+}.

Lewis acid site

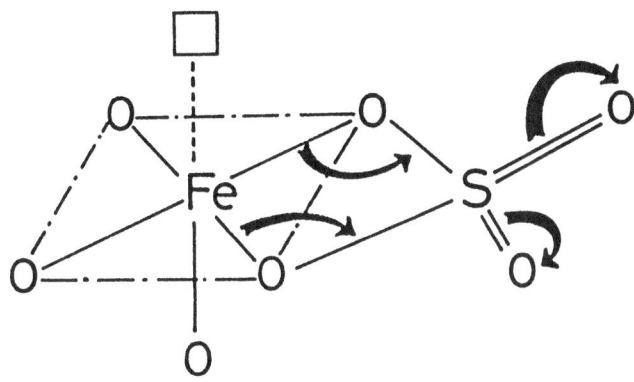

Fig. 2 Model structure of a superacid, $Fe_2O_3 - SO_4^{2-}$.

Humidity sensors such as $MgCr_2O_4-TiO_2$, $MgO-ZrO_2$, and thin film of Al_2O_3 detect changes of electric properties (conductance or capacitance) caused by hydroxylation, adsorption of water and/or condensation of it in micropores. Therefore, in addition to the pore-size distribution, the affinity of the surface to water vapor (hydrophilic or hydrophobic) also play important roles in the sensing function. For example, the sensitivity, the resistivity-$vs.$-humidity correlation and response of Cs, NH_4, and alkyl ammonium salts of 12-tungstophosphate varied closely corresponding to those properties [6].

3. Adsorbents

Solid acids and bases are widely used as adsorbents for the removal or separation of certain compounds in gas mixtures or solutions. Some typical examples will be described below.

Zeolites selectively adsorb molecules and ions with the aid of shape-selective micropores and/or acid sites or the electric field created by metal-ions exchanged in the pore. Shape selectivity is governed by the relative diameter of the adsorbate molecule and pore-opening (window) which is principally determined by the number of oxide ions involved in the window and secondarily by the metal-ion present in the wall of the window. An example of precise control of shape-selective adsorption is shown in Fig. 3 [7]. As the extent of exchange of Na^+ with Ca^{2+} increases and slightly widens the window, at a certain level of exchange n-butane starts to adsorb in the pores of A zeolite.

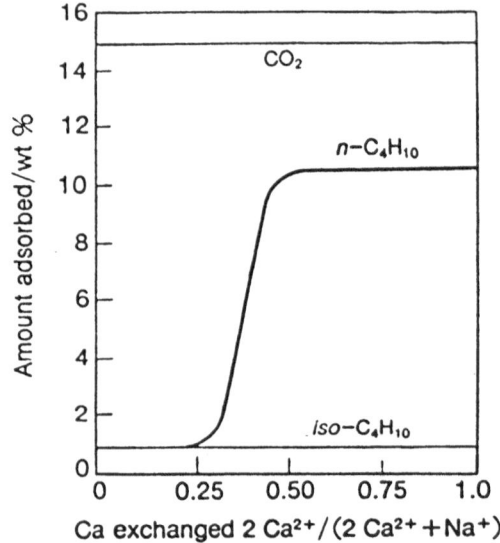

Fig. 3   Precise control of the shape-selective adsorption of A zeolite
by ion-exchange.

The hydrophilic surface of zeolite changes to hydrophobic with increasing silica/alumina ratio. This property can be also used for selective
adsorption. Examples of practical applications of zeolites as adsorbents are listed below.

Zeolites	Adsorption processes
A	Drying of ethylene and butadiene (3A), drying and removal of CO_2 from natural gas (4A), O_2 separation from air (5A), recovery of n-paraffins from naphtha and kerosene (5A), purification of monosilane (NaZnA)
ZSM-5	Dewaxing (selective cracking of n-paraffins)
Mordenite	Separation of p-xylene from a mixture of xylene isomers and ethylbenzene

Reactive adsorbents which contain alkali-earth metal oxides (MO) are
added to the FCC process to remove SO_x by the following reactions.

$$SO_3 \ + \ MO \ \longrightarrow \ MSO_4 \qquad\qquad (1)$$

$$MSO_4 \ + \ 4H_2 \ \longrightarrow \ MO \ + \ H_2S \ + \ 3H_2O \qquad (2)$$

Hydroxyapaptite, $Ca_{10}(OH)_2(PO_4)_6$ has acidic Ca^{2+} sites (C sites) and basic sites (P sites, one site surrounded by four PO_4^{3-}) of moderate strength on its surface. It can be used as a stationary phase of liquid chromatography for the separation of biopolymers of 10^4 - 10^9 Daltons such as enzymes, antibody, and other proteins, as shown in Fig. 4 [8]. C sites interact with anionic parts of biopolymers, carboxylate and phosphate, as well as phosphate ion in buffer solution, and P sites with cations and cationic parts.

TiO_2 is used to remove radioactive Co^{2+} ions [9] and SiO_2 to remove radioactive RuO_4 in exhausts of nuclear reactor.

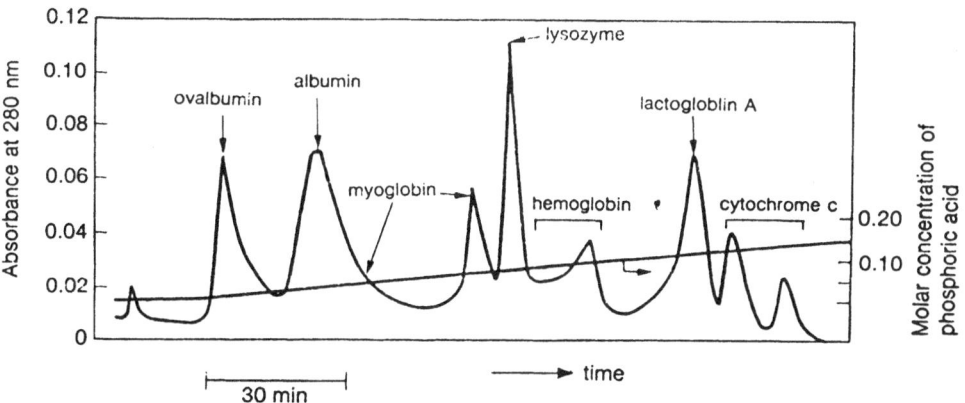

Fig. 4 Chromatographic separation of a mixture of proteins with a hydroxyapatite column.

4.   Pressure Sensitive Recording Paper (Carbonless Paper)

4.1. PRINCIPLE

Solid acids are utilized as one component of pressure sensitive recording paper. The principle of the carbonless paper is indicator color change by adsorption on solid acids. The model construction of carbonless paper consists of a set of two types of paper as schematically shown in Fig. 5. The top sheet is coated with indicators encapsulated in microcapsules [10] and the bottom sheet is coated with solid acids. Applying pressure to the top sheet destroys the microcapsule, and the indicator is released and transferred to the solid acids on the bottom sheet to be adsorbed.

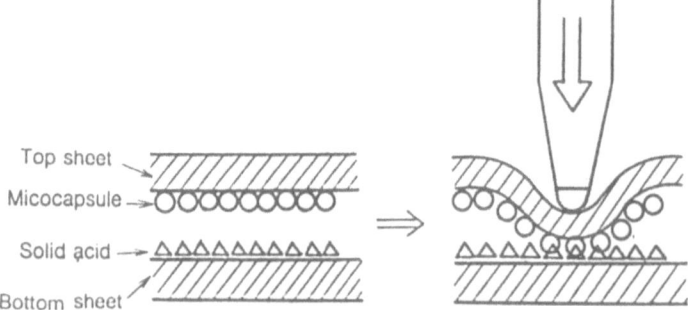

Fig. 5 Pressure sensitive paper.
        Indicator contained in microcapsules changes color when
        adsorbed on solid acid.

4.2. TYPES OF SOLID ACID

The solid acids utilized for carbonless paper must meet the following
requierments:

a) Colorless or white
b) Large adsorption capacity
c) Fast and strong coloring
d) High resistance to weather and light
e) Nontoxicity

Both organic and inorganic solid acids are used. As organic acids,
Brönsted acids such as condensed compounds of para-substituted phenol
with formaldehyde [11 - 13] and Lewis acids such as zinc salts of the
above compounds and those of aromatic carboxyl acids are used. As inorganic acids, montmorillonite clays such as bentonite, fuller's earth,
kaoline, chinaclay, and their chemically modified materials are used.
    Any kind of indicator can be used provided that it changes colors when
adsorbed on solid acids. Selecting the appropriate indicators makes it
possible to obtain the desired color. The types of coloring materials
in practical use are as follows: phthalides such as crystal violet
lactone, acyl leucomethylene blues such as benzoyl leucomethylene blue,
and fluorane derivatives. In particular, fluorane derivatives give desired color of yellow, red, and black by changing the substitutional
groups.

4.3. PREPARATION OF THE PAPER

The key technique which makes it possible to commercially produce carbonless paper is microcapsulation of the indicators. The indicators are

dissolved in proper solvents, and the solutions are encapsulated into microcapsules made of gelatins and Arabian rubbers. The size of the microcapsules is in the range 1 - 5 μ m. Aqueous solutions containing the microcapsules are then painted onto the paper.

SEM photographs of the microcapsules coated on the top sheet of paper and the solid acids on the bottom sheet are shown in Figs. 6 and 7.

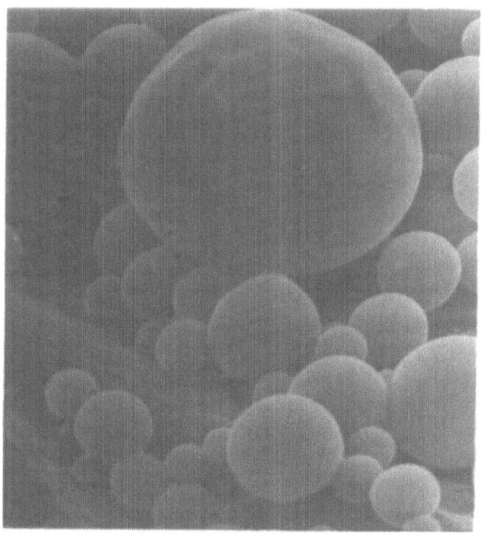

Fig. 6   SEM image of microcapsules coated on top sheet, x2400

5. Cosmetic Pigments

Clays and metal oxides and salts such as talc, kaolinite, mica, zinc oxide, titanium oxide, iron oxide, hydrated chromium oxide, cobalt blue, ultramarine blue, calcium carbonate, barium sulfate, etc. are widely used as pigments for cosmetics. Since these pigments possess acidic and basic surface properties, and hence catalytic activity, cosmetic products containing perfumes, oils, and medicaments as the other components occasionally deteriorate as time elapses by decomposition of the components due to catalytic action of the pigments, markedly depreciating the value of the cosmetic. In the case of perfumes, the isomerization of 2-pinene [14, 15], the reaction of d-limonene oxide [16, 17], and the dehydrogenation of d-limonene to p-cymene [18] are known to take place over solid acids and bases. As a result, the odor or the quality of the cosmetic changes.

It is essential to deactivate the catalytic activity of such pigments,

482

Fig. 7 SEM image of solid acid coated on bottom sheet, x2400

and measurement of their catalytic activity is important to maintain the
stability of the cosmetics containing them. Fukui et al. measured the
catalytic activities of various pigments for the dehydration and the dehydrogenation of isopropanol which are known to be catalyzed by acid
sites and by both acid and base sites, respectively [19]. As shown in
Table 1, the pigments used in cosmetics can be classified into four
groups by the reaction products of isopropanol: group 1 pigments which
mainly produce propylene by dehydration, group 2 pigments which mainly
produce acetone by dehydrogenation, group 3 forming both acetone and
propylene by dehydrogenation and dehydration, and those of group 4 which
do not decompose isopropanol. It was found that the catalytic activities
of pigment groups 1 - 3 decreased markedly by treatment with basic sodium
carbonate or acidic acetic acid [19].

In the reaction between pigments and linalool, which is a common component of perfumes, pigments such as black iron oxide and hydrated chromium oxide which showed low conversion rates formed only ocimene and
myrcene, while limonene and terpinolene were also formed over pigments
such as silica and mica with moderate conversion rates [20]. With pigments such as prussian blue and red iron oxide, which showed high conversion rates, alloocimene, terpinene, and p-cymene were also formed
[20]. p-Cymene is the cause of one of the most unpleasant odors in de-
teriorated cosmetics. The decomposition of perfumes can be controlled
by treating pigments with suitable acidic and/or basic materials. Thus,
the change in odor and hence the deterioration of the quality of cosme-

tics is prevented.

Table 1 Decomposition of isopropyl alcohol over various pigments

Pigment	Selectivity	
	Propylene/%	Acetone/%
Group 1		
Talc	8.8	0
Kaolinite	99.5	0
Titanium dioxide (A)	19.2	0
Titanium dioxide (R)	33.8	trace
Ultramarine blue	2.7	0
Cobalt blue	1.0	0
Titanium-coated mica	5.7	0
Group 2		
Zinc oxide	0.9	32.1
Group 3		
Black iron oxide (M)	27.0	33.4
Red iron oxide (H)	3.8	20.9
Hydrated chromium oxide	5.2	17.9
Group 4		
Mica	0	0
Barium sulfate	0	0
Calcium carbonate	0	0
Silica	0	0
Synthesized pigment G	0	0

Solid acids and bases of fine particles on which suitable indicators were adsorbed as monolayers to give the conjugate acid and base forms are considered to have a possibility of use as cosmetic pigments. Mixed oxides of Fe_2O_3 with ZnO, etc. which show beautiful colors like pink depending on the composition and preparation method [21] are also promising as cosmetic pigments.

6. Others

The measurement method of the basicity of solids has been applied to the classification of the slags of iron and steel [21]. A quality of a paint containing titanium oxide is said to change depending on the chemical property (acid-base property) of titanium oxide. Agricultural medicines and fertilizers are known to decompose on soil and the decomposition rate changes depending on acidity and basicity of soil. It is said that the mode of life of some insects living in one kind of soil is

484

different from those living in another kind of soil, suggesting a possibility of the influence of acidity and basicity of soil.

7. References

1. Y. Nakatani, M. Sakai, and M. Matsusaka, Jpn. J. Appl. Phys., 22,
   912 (1983).
2. Y. Nakatani and M. Matsusaka, Jpn. J. Appl. Phys., 21, 758 (1982).
3. Y. Nagase, H. Hattori, and K. Tanabe, unpublished data.
4. T. Yamaguchi, T. Jin, and K. Tanabe, J. Phys. Chem., 90, 3148
   (1986).
5. K. Tanabe, M. Misono, Y. Ono, and H. Hattori, New Solid Acids and
   Bases, Kodansha, Tokyo, Elsevier, Amsterdam, 1989, p. 202-204.
6. N. Mizuno, K. Inumaru, and M. Misono, Hyomen Kagaku, 10, 21 (1989).
7. T. L. Thomas, 6th World Petr. Congr., Frankfurt, Part 3, No. 16
   (1963).
8. T. Kawasaki, Ceramics Japan, 20, 195 (1985) (in Japanese).
9. F. Kawamura and K. Funabashi, ibid., 20, 203 (1985) (in Japanese).
10. U. S. Patent, 2,711,375 (1955).
11. Jpn. Patent (Toku-Kou-Sho) 40-9,309.
12. Mitsui Toatsu Chemicals, Inc. (M. Asano, K. Hasegawa, Y. Tamura,
    and Y. Ohno), U. S. Patent, 4,574,110 (1985).
13. Mitsui Toatsu Chemicals, Inc. (K. Yamaguchi, Y. Tanabe, M. Asano,
    and A. Yamaguchi), U. S. Patent, 5,023,366 (1991).
14. R. Ohnishi, K. Tanabe, and S. Morikawa, T. Nishizaki, Bull. Chem.
    Soc. Jpn., 47, 571 (1974).
15. K. Tanabe, R. Ohnishi, and K. Arata, Terpene Chemistry, (ed. J.
    Verghese) Tata McGraw, New Delhi, 1981.
16. K. Arata, S. Akutagawa, and K. Tanabe, J. Catal., 41, 173 (1976).
17. K. Arata and K. Tanabe, Catal. Rev., 25, 365 (1983).
18. H. Pines, J. A. Vesely, and V. N. Ipatieff, J. Am. Chem. Soc., 77,
    347 (1955).
19. H. Fukui, T. Saito, M. Tanaka, and S. Ohta, Cosmetics & Toiletries,
    96, 37 (1981).
20. H. Fukui, R. Namba, M. Tanaka, M. Nakano, and S. Fukushima, J. Soc.
    Cosmet. Chem., 38, 385 (1987).
21. H. Shima and K. Tanabe, unpublished results.
22. K. Niwa, S. Kado, and H. Kuki, Report of 19th Committee of Japan
    Soc. Promotion Sci., No. 6673 (1962).

CHARACTERIZATION OF BRÖNSTED ACIDITY OF SOLIDS BY RIGID-LATTICE BROAD-LINE ^{1}H-NMR. PART 2 : APPLICATIONS

C. DOREMIEUX-MORIN
Laboratoire de Chimie des Surfaces
Associé au CNRS - URA1428
Université P. et M. Curie
Casier 196, Tour 55
4, place Jussieu
F 75252 PARIS CEDEX 05

ABSTRACT. The acid strength of solids is investigated using simulation of ^{1}H broad-line NMR spectra, recorded at 4 K. Two types of results are presented : (i) a scale of intrinsic Brönsted acidity, with the condition that water molecules interact with as many acidic sites ; (ii) evidence for an increase in Brönsted acidity by Lewis acid sites.

1. Introduction

A method to determine the strength of Brönsted acidity of solids is described in Part 1 of this contribution. Broad-line ^{1}H NMR allows measurement of the concentrations of the oxy-protonated species formed by the interaction of water molecules with acid sites denoted SOH. These species are hydroxonium ions and water molecules hydrogen-bonded with the Brönsted sites, $H_2O...HOS$. The concentrations of non-interacting SOH groups and/or other OH groups as well as H_2O are also determined. Two kinds of application are presented in this second part of our contribution.

Results obtained on HY zeolite samples are shown first. One of the sample has nearly no framework defects and gives values of the concentrations which can be related to its specific acidity (only Brönsted type). It is denoted ND, for non-dealuminated (1,2). The other HY sample has been partially dealuminated using an aqueous solution of ammonium hexafluorosilicate. It is denoted D and contains a few Lewis acid sites. Comparison of results obtained for ND and D shows a large difference in the number of hydroxonium ions formed for relatively large water adsorption ratios (3,4). This difference is attributed to a synergy of Lewis and Brönsted acid sites in the presence of water.

Results obtained on ND HY zeolite are then compared with those obtained on other well characterized solids : silanol-type zeolite SOH group defects (5), superficial SOH groups of amorphous titanium oxide (6), dihydrated dihydrogeno-aluminium-tripolyphosphate $H_2AlP_3O_{10}.2H_2O$ (7), dihydrated antimonic acid $H_2Sb_4O_{11}.2H_2O$ (8) and commercial nafion. On the basis of this comparison, tests to measure the acid strengths of solids are proposed.

485

J. Fraissard and L. Petrakis (eds.), Acidity and Basicity of Solids, 485-500.

The measured sample composition at 4 K seems the equilibrated one at room temperature (5): the cooling of the sample in the helium dewar can be considered as a quenching.

2. Comparison of a non-dealuminated (ND) and partially dealuminated (D) HY zeolite samples

As described in Part 1, prior to NMR experiments the samples are "shallow-bed" pretreated, then, in turn, water is adsorbed, the sample is homogenized and, finally, sealed.

The characteristic properties of the samples are given in Table 1. The Si / Al ratio, per number, is 2.4 for ND and 4.4 for D. Sample ND still contains some Na^+ cations so that the number of SOH is 47.7 per unit cell (uc). D contains 35.8 SOH per uc, negligible quantities of sodium and 1.5 silanol per uc. A significant difference between the two samples is that D, contrarily to ND, contains a small number of Lewis acid sites (3,4). These sites have been identified in ^1H MAS experiments by a signal attributable to water molecules coordinated to them. This signal is clearly visible when the adsorbed water molecules are not numerous (4 per uc as example).

2.1. SAMPLE ND HY ZEOLITE

As an example, three experimental and simulated spectra of ND sample with different concentrations of adsorbed water are shown in Figure 1. Weighted contributions of the different magnetic configurations used are also shown.

The species concentrations obtained from the simulations are reported in Table 2 and the distance parameters in Table 3. In agreement with Sauer's statement and the explanation given in Part 1 (9), some hydroxonium ion contributions use an isosceles configuration of spins. Let $n_i(H_2O)$ be the number of adsorbed water molecule per uc and $n_i(OH)$ the number of initial SOH; the results show that as long as $n_i(H_2O) < n_i(OH)$ there are no remaining "free" water molecules (Figure 1, 40 H_2O/uc). All H_2O molecules interact with SOH to give H_3O^+ and $H_2O \cdots HOS$; some SOH groups remain isolated. When $n_i(H_2O) = n_i(OH)$, no SOH group nor "free" water molecule remains. In this case, 20% of the protons initially in OH groups are in the H_3O^+ form and the others are involved in hydrogen bonds. For concentrations where $n_i(H_2O) > n_i(OH)$ the spectra include also a "free" water molecule contribution but no more "free" SOH groups (Table 2, Figure 1, 134 H_2O/uc).

Therefore, the results of the simulation of broad-line spectra can be expressed by the complete interaction :

$$x \text{ SOH} + y \text{ } H_2O \rightarrow \inf(x,y) \text{ } H_2O \cdots HOS + /x-y/ \text{ SOH or } H_2O \quad [1]$$

and the equilibrium :

$$H_2O \cdots HOS \rightleftharpoons H_3O^+ + SO^- . \quad [2]$$

Let us take concentrations as activities. This is a poor approximation (10) but it is not possible to know activities. If k is the equilibrium constant of equation [2] in the direction left to right, the hydroxonium ion concentration per unit cell is given by :

$$n(H_3O^+) = (-k + (k^2 + 4 k \text{ } n_i(H_2O))^{1/2})/2 \quad \text{for } n_i(H_2O) \leq n_i(OH)$$
$$n(H_3O^+) = (-k + (k^2 + 4 k \text{ } n_i(OH))^{1/2})/2 \quad \text{for } n_i(H_2O) > n_i(OH),$$

where $n_i(OH)$ is constant and equal to 47.7. According to these equations, in the first case $n(H_3O^+)$ increases with $n_i(H_2O)$. In the second case $n(H_3O^+)$ remains constant, as is approximately observed (2) (Table 2, Figure 2A).

2.2. SAMPLE D HY ZEOLITE

It is known from a previous study on HZSM-5 zeolites using ^1H NMR, both broad-line at 4K and high resolution MAS, that zeolite silanol OH groups do not interact with water (5).

In order to simulate the broad-line spectra of D HY zeolite more precisely we have introduced an additional constraint when $n_i(H_2O)$ < 140 : we include *a priori* into the fit the distinct contribution of silanol OH groups of the pretreated sample (about 1.5 / uc, Table 1), known not to interact with water molecules. For this zeolite D Figure 2B shows that, after remaining constant in the region where $n_i(OH)$ < $n_i(H_2O)$ < $2 n_i(OH)$, the number of H_3O^+ again increases markedly with $n_i(H_2O)$. To summarize, in the horizontal section the dissociation coefficient of the $H_2O \cdots HOS$ complex is about 0.20. This value is similar to that obtained for sample ND. However, the dissociation coefficient of sample D increases to about 0.45 per initial acid OH group when $n_i(H_2O)$ / $n_i(OH)$ attains 4.5. The number of hydroxonium ions in the sample is then about 1.5 times larger in D than in ND with the same degree of hydration.

2.3. DISCUSSION

A discussion of the difference observed for the plots of $n(H_3O^+)$ against $n_i(H_2O)$ for ND and D HY samples is to be published (4). The Lewis sites are assumed to be Al atoms still partially bonded to the framework but with at least one of their four bonds broken. A synergy of Lewis and Brönsted acid sites is considered, which, in presence of water molecules gives the increase in $n(H_3O^+)$ proposed in Figure 3. It is assumed that there are both an increase in the $SOH \cdots OH_2$ dissociation coefficient and a subsequent dissociation of water molecules coordinated to the Lewis acid sites.

3. Comparison of the Brönsted intrinsic acidity of different solids. Establishement of a scale of acidity

We propose now to compare the results obtained using the same technique of broad-line ^1H NMR on solids whose acid strength can be quite different. In order to normalize this comparison, we chose to compare the species obtained (nature and concentration) for **samples where the number of water molecules is equal to the number of OH groups being tested.**

3.1. SCALE OF ACIDITY BASED ONLY ON DISSOCIATION COEFFICIENT

First, hydroxonium ion concentrations can be compared (Table 4).

The ionization coefficient for HY zeolite when $n_i(H_2O)$ = $n_i(OH)$ is 0.2, corresponding to the formula $(H_3O^+)_{9.7}(H_2O \cdots HO)_{38.5}Na_{7.8}Al_{56}Si_{136}O_{345.5}$. For the dihydrated antimonic acid $H_2Sb_4O_{11}.2H_2O$, the ionization coefficient attains 0.4 (8) expressed as

488

$(H_3O^+)_{0.80}(H_2O\cdots HO)_{1.20}Sb_4O_{9.80}$. The simulation of the spectrum requires only two distinct magnetic configurations (Figure 4). Experiments have also been performed on commercial nafion. The sample was first dehydrated under the same conditions as for zeolites except that the temperature was kept lower (390 K). After rehydration, though the sensitivity is low, the ionization coefficient is about 1.

These results can be compared to those obtained on less acidic solids. No interaction has been observed between zeolite silanol OH groups and water molecules as already mentioned (5). We consider that these SOH groups are not acid. Amorphous titanium oxide covered with as many water molecules as OH groups gives no hydroxonium ions (6). In the same way, no H_3O^+ is formed in dihydrated dihydrogeno-aluminium-tripolyphosphate (7). Therefore, basing an acidity scale on the ionization coefficient gives only limited results :

zeolite (SiOH.$1H_2O$)
TiOH.$1H_2O$ < HY zeolite SiOH(Al).$1H_2O$ < $H_2Sb_4O_{11}$.$2H_2O$ < nafion.
$H_2AlP_3O_{10}$.$2H_2O$

3.2. SCALE OF ACIDITY BASED ONLY ON THE HYDROGEN BOND STRENGTH

However, broad-line ^1H NMR gives us the opportunity to improve the classification of the samples relative to their acid strength by comparing the hydrogen-bonded complexes that their SOH groups form with water molecules. If it is assumed that the three H atoms of $H_2O\cdots HOS$ are located at the apices of an isosceles triangle (C_{2v} symmetry (Figure 5)), the NMR results give the two following H H distances (see Part 1) : (i) r, the intra-water molecule distance, known with a simulation accuracy of 2 pm ; (ii) r', the mean value of those between the H atom of SOH and each of the H atoms of H_2O. The simulation accuracy is then 5 pm. The value of r' gives an estimate of the hydrogen bond strength : the smaller r', the stronger the hydrogen bond (Table 4). As an example, we give a more usual description of the hydrogen bond strength in terms of the O O distance, calculated from r'and r, assuming that OH distances are 100 pm and that the $H_2O...HOS$ group symmetry is C_{2v}. The accuracy on O O distances is, of course, less than on r'.

Let us consider again the results obtained with amorphous titanium oxide covered with as many water molecules as SOH groups (6). All H atoms belong to isosceles groups. Parameter r' is 265 pm which, with the above assumptions, means that the O O distance is about 285 pm : the hydrogen bonds are weak. The formula obtained for dihydrated dihydrogeno-aluminium-tripolyphosphate is $(H_2O\cdots HO)_2AlP_3O_8$ (7). Only one three-spin magnetic configuration was required for a good simulation (Figure 6), showing that the H atoms are really at the apices of an isosceles triangle. Parameter r' is now 215 pm which indicates a very short O O distance of 242 pm. The hydrogen bonds are then strong. We can infer that the SOH groups of dihydrated dihydrogeno-aluminium-tripolyphosphate are stronger acids than those at the surface of amorphous titanium oxide. In HY zeolite and in dihydrated antimonic acid all the initial OH groups not transformed into hydroxonium ions are hydrogen-bonded to water molecules. The comparison of these $H_2O\cdots HOS$ groups is also meaningful. The r' values are 232 and 221 pm for HY zeolite and antimonic acid, respectively, though the r values are the same (about 163 pm). Therefore, the O O distance is larger (259 pm) in the zeolite than in the antimonic acid (247 pm) : the order of HY zeolite and antimonic acid in an acidity scale based on the hydrogen-bond strength is the same as that given by the ionization coefficient. Ranging the samples versus the strength of the hydrogen-bonds gives:

zeolite $SiOH.1H_2O$ < $TiOH.1H_2O$ on amorphous titanium oxide < HY zeolite $SiOH(Al).1H_2O$ < $H_2Sb_4O_{11}.2H_2O$ ≤ $H_2AlP_3O_{10}.2H_2O$.

3.3. SCALE OF ACIDITY BASED ON BOTH DISSOCIATION COEFFICIENT AND HYDROGEN BOND STRENGTH

By analogy with the usual expression for acidity in aqueous media we propose that the ionization coefficient be regarded as the principal criterion for the acid strength. When the classification that it gives differs from that based on the hydrogen bond length the order following the ionization coefficient will be adopted. However, the strength of the hydrogen bonds will be used when no ionization occurs. With these proposals the complete order of the samples is :

zeolite $SiOH.1H_2O$ < $TiOH.1H_2O$ on amorphous titanium oxide < $H_2AlP_3O_{10}.2H_2O$ < HY zeolite $SiOH(Al).1H_2O$ < $H_2Sb_4O_{11}.2H_2O$ < nafion.

These results are summarized in Table 4. We assume that the difference between the hydrogen bonds in $H_2AlP_3O_{10}.2H_2O$ (r' = 215 pm, O O = 242 pm), which contains no hydroxonium ion, and HY zeolite (r' = 232 pm, O O = 259 pm), which is partly ionized, is due to a relaxation of the hydrogen bonds of the complex when ionization occurs.

For samples containing both hydroxonium ions and $H_2O\cdots HOS$ it would be possible to consider that two kinds of acid sites coexist, the stronger of them leading to complete ionization and the weaker giving $H_2O\cdots HOS$ species only. However, two reasons can be proposed against this interpretation. First, there is, a priori, no reason to consider two types of SOH groups in some of the samples studied and, moreover, the ionization coefficient found shows no visible relationship with the sample stoichiometries. Secondly, in cases of interaction between water molecules and OH groups , the results can always be interpreted in terms of the following complete first step of the interaction :

$SOH + H_2O \rightarrow H_2O\cdots HOS$.

4. Conclusion

The concentrations and the internal H H distances of the oxy-protonated species formed by interaction of water molecules with SOH Brönsted acid sites of solids are obtained by simulation of broad-line ^{1}H NMR spectra recorded at a temperature low enough to rule out motion of the species (in practice, 4 K). The results allow us to define a partial new scale of Brönsted acidity (limited by the properties of water as a base).

To establish this scale, results are compared for samples containing as many water molecules as SOH groups to be tested. First, these SOH groups are qualitatively classified according to the species formed by the interaction, which are the following :

- SOH groups for which there is complete ionization, giving hydroxonium ions and SO⁻ species ; they are the strongest acid sites which can be tested with water. To test them quantitatively a weaker base than water is necessary. H_2S would probably be convenient from the NMR standpoint because, like H_2O, it contains only two H atoms.

- SOH groups which give hydroxonium ions and hydrogen-bonded $H_2O\cdots HOS$ groups are strong acid sites, the strength of which is characterized by the ionization coefficient, which is the number of hydroxonium ions formed per SOH.

- SOH groups which give only hydrogen-bonded $H_2O\cdots HOS$ species are weak to moderate acid sites. The strength of the hydrogen bond gives a quantitative measure of the acid strength ;

- SOH groups of samples which give no interaction with water ; in the present scale they are assumed to have no acid properties.

The ^1H broad-line NMR method makes it possible to study also the already proposed "synergy" between Lewis and Brönsted acid sites when they coexist in a sample. This synergy is assumed to be the origin of increasing hydroxonium ion / SOH ratio when the water molecule concentration is larger than that of SOH groups.

Acknowledgments

I am grateful to H. Arribart, P. Batamack, M. A. Enriquez, J. Fraissard, D. Freude, Y. Piffard, J. Sanz , R. Vincent and F. d'Yvoire for their contributions to the above results.

References

(1)   P. Batamack, C. Dorémieux-Morin, R. Vincent and J. Fraissard, *Chem. Phys. Let.*, 1 9 9 1, *180*, 545.

(2)   P. Batamack, C. Dorémieux-Morin, and J. Fraissard, *J. Chim. Phys.*, 1 9 9 2, *89*, 423.

(3)   P. Batamack, C. Dorémieux-Morin, and J. Fraissard, *Catalysis Letters.*, 1 9 9 1, *11*, 119.

(4)   P. Batamack, C. Dorémieux-Morin and J. Fraissard, to be published.

(5)   P. Batamack, C. Dorémieux-Morin, J. Fraissard and D. Freude, *J.Phys. Chem.*, 1 9 9 1, *95*, 3790.

(6)   C. Dorémieux-Morin, M. A. Enriquez, J. Sanz and J. Fraissard, *J. Colloid Interf. Sci.*, 1 9 8 3, *95*, 502.

(7)   C. Dorémieux-Morin, *J. Magn. Res.*, 1 9 7 6, *21*, 419.

(8)   H. Arribart, Y. Piffard and C. Dorémieux-Morin, *Solid State Ionics*, 1 9 8 2, *7*, 91.

(9)   J. Sauer, H. Horn, M. Häser and R. Ahlrichs, *Chem. Phys. Letters*, 1 9 9 0, *173*, 26.

(1 0)   D. Barthomeuf, *C. R. Acad. Sc. Paris*, 1 9 7 8, *C286*, 181.

Table 1
Characteristics of the non-dealuminated (ND) and dealuminated (D) HY zeolites

Sample	Preparation	Si / Al (NMR)	hexacoordin. Al (NMR)	Na+ per unit cell	OH per unit cell of "anhydrous" zeolite	
					ZOH	Silanols
ND	"Y64" from UOP	2.4	no	15 % of the ions	47.7	0
D	(NH4)2SiF6	4.4	no	negligible	35.8	1.5

Number of adsorbed water molecules	Number of H_3O^+ ions	Number of $H_2O \cdot \cdot H_a OS$ groups	Number of SOH_a groups	Number of H_2O molecules
3.0±0.5	0.9±0.3 *	2.3±0.4	44.0±1.1	0
9.3±1.1	2.2±0.4 *	8.2±0.4	35.1±1.4	0
19.0±1.9	5.7±0.6 *	14.3±0.6	25.7±1.7	0
40.0±2.0	8.1±0.8 *	31.1±0.8	10.2±2.4	0
	8.1±0.8 **	31.1±0.8	10.2±2.4	0
48.5±2.4	10.6±1.0 *	37.6±1.0	0	0
	8.7±1.0 **	39.5±1.0	0	0
129.0±6.5	10.2±2.1 *	37.7±2.0	0	81.0±3.1
	10.2±2.1**	37.7±2.0	0	81.0±3.1
134.0±6.5	11.5±2.1 *	36.8±2.1	0	85.2±3.1
	11.5±2.1**	36.8±2.1	0	85.2±3.1
142±7.1	10.5±3.0 *	37.0±4.2	0	94.5±7.9
	10.5±3.0**	37.0±4.2	0	94.5±7.9

Table 2 : Number of oxygen-protonated species groups per unit cell of HY zeolite containing initially 47.7±1.9 SOH_a groups per unit cell (all acidic), after adsorption of the number of water molecules per unit cell given in the first column. The accuracy is that for the simulation using the proposed model; * and ** indicate that the hydroxonium ion protons have equilateral or isosceles symmetry, respectively.

Number of adsorbed water molecules per uc	Hydroxonium ions			$H_2O\cdots H_aOS$ groups			SOH_a groups		H_2O molecules	
	r±2	r'±5	X±5	r±2	r'±5	X±5	r±5	X±5	r±2	X±5
3.0± 0.5	166		255*	165	239	240	287	287		
							and			
							Lorent.	1.1		
9.3± 1.1	166		255*	162	235	235	300	300		
19.0±1.9	166		248*	158	231	269	270	270		
40.0± 2.0	166		248*	160	234	253	Lorent.	1.05		
	158	169	248**	160	234	253	Lorent.	1.05		
48.5± 2.4	165		220*	161	235	259				
	158	169	248**	162	232	252				
129.0±6.5	165		225*	159	246	250			159	200
	156	167	225**	159	247	250			159	200
134.0±6.5	163		240*	159	237	240			159	200
	156	167	235**	159	237	240			159	200
142.0±7.1	161		255*	155	232	231			158	204
	155	167	250**	155	232	232			158	206

Table 3 : Distance parameters in pm used for the simulations of the HY zeolite spectra; * and ** indicate that the hydroxonium ion protons have equilateral or isosceles symmetry, respectively. The absorption for the SOH groups is sometimes described using the derivative of a Lorenzian function whose parameter is the half-width (in 10^{-4} T) of the signal at half-height.

Order of increasing Brönsted acid strength	Sample	Number of H_3O^+ per initial OH : ionization coefficient	Number of $H_2O...HO$ per initial OH	Value of r (pm)	Value of r' (pm)	O O distance (pm) from r and r', assuming C_{2v} symmetry for $H_2O...HO$
1	Zeolite silanols	0	0			
2	Superficial OH groups on amorphous TiO_2	0	1	143	265	285
3	$H_2AlP_3O_{10}\cdot 2H_2O$	0	1	163	216	242
4	Zeolite SiO(H)Al	0.2	0.8	162	232	259
5	$H_2Sb_4O_{11}\cdot 2H_2O$	0.4	0.6	164	221	247
6	Nafion	1	0			

Table 4 : Acidity scale using broad-line ^1H NMR : the condition is that samples contain one water molecule per OH group being tested.

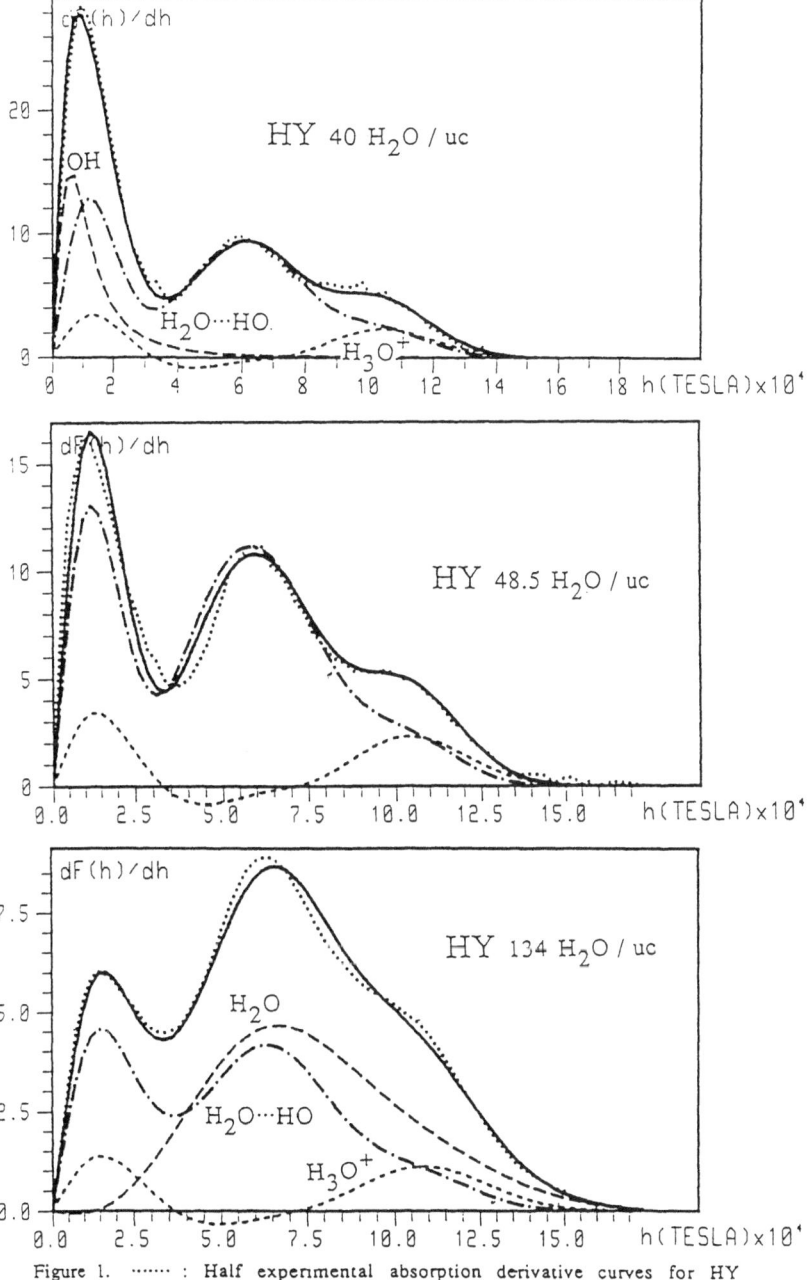

Figure 1. ········· : Half experimental absorption derivative curves for HY zeolite loaded with water molecules ; ——— : fitted spectrum; ---- : weighted contribution of H_3O^+ ions; ·—··— : weighted contribution of $H_2O...HO$ groups; — — — — : weighted contribution of : (i) OH groups without interaction with water (for 40 H_2O/uc) ; (ii) water molecules without interaction with OH groups (for 134 H_2O/uc).

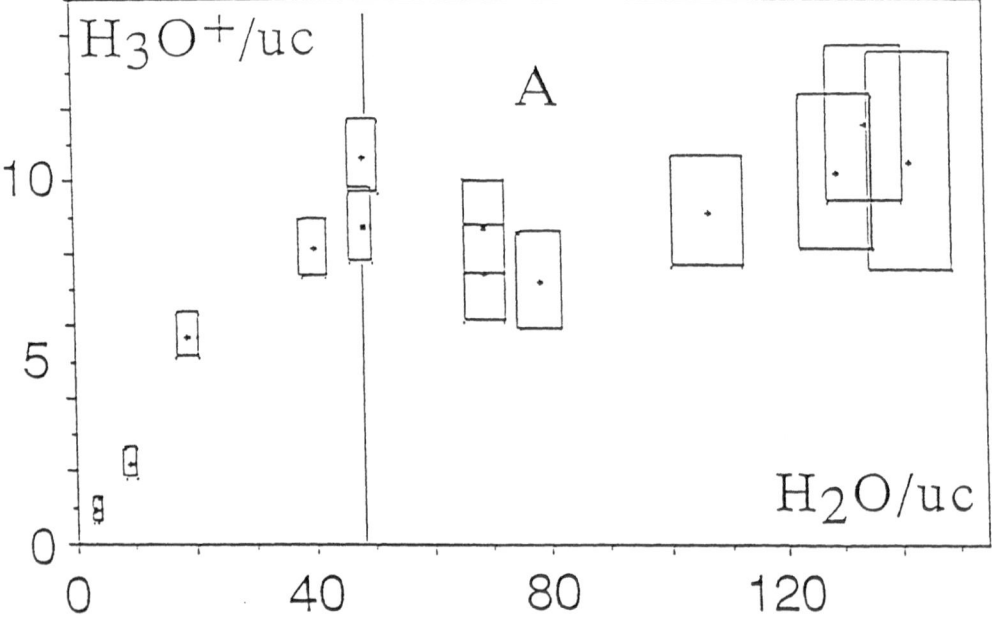

Figure 2A Number of hydroxonium ions formed per unit cell versus the number of water molecules adsorbed per unit cell in a HY zeolite without framework defects denoted ND (non dealuminated). The straigth line corresponds to a number of water molecules equal to the number of SOH groups.

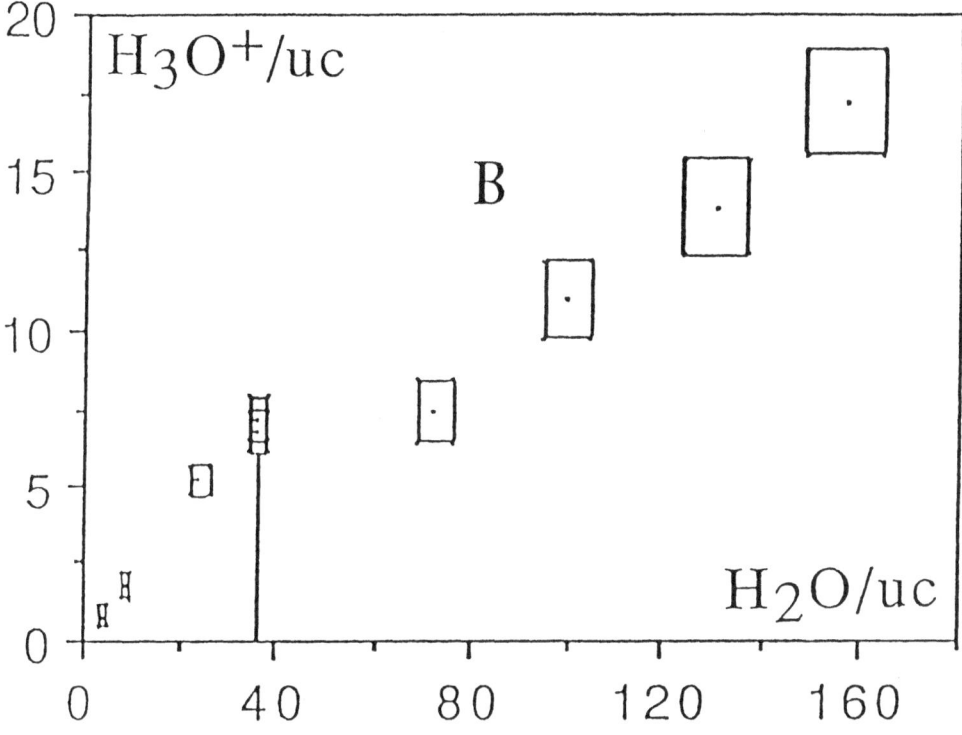

Figure 2B Number of hydroxonium ions formed per unit cell versus the number of water molecules adsorbed per unit cell in a HY zeolite partially dealuminated denoted D (dealuminated). The straigth line corresponds to a number of water molecules equal to the number of SOH groups.

498

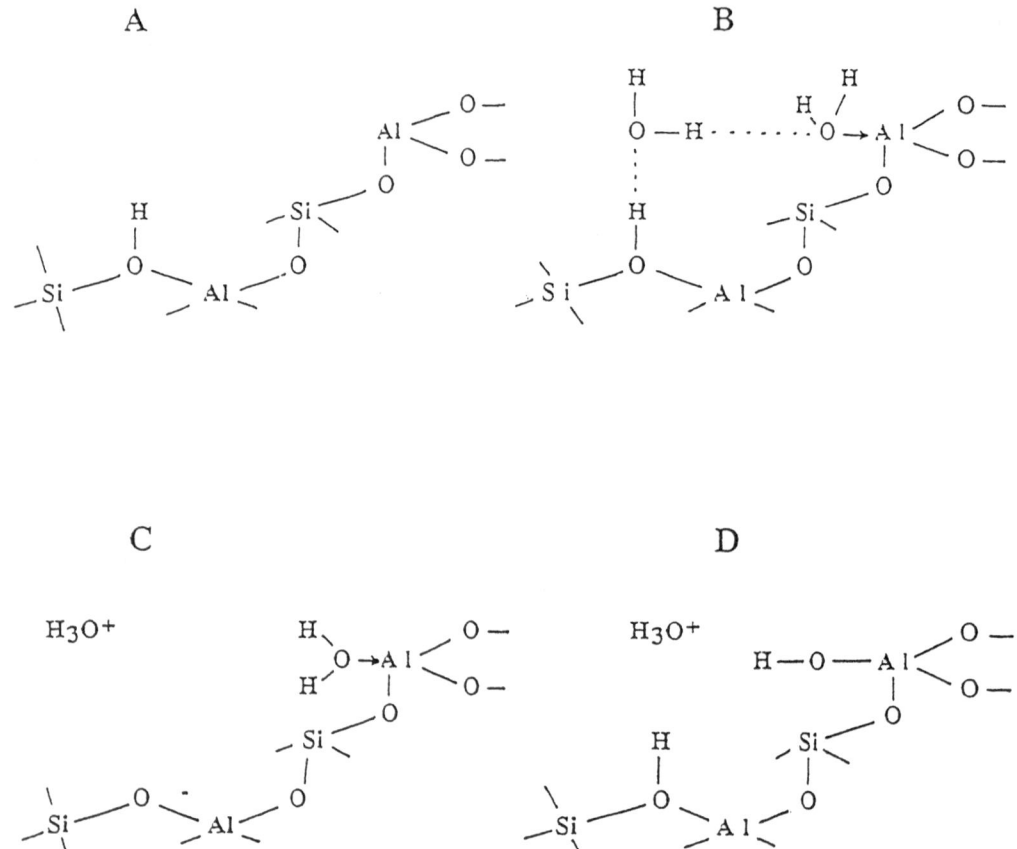

Figure 3   Proposed process for the increasing in the number of hydroxonium formed in the D HY zeolite when the number of water molecules is larger than the number of SOH.

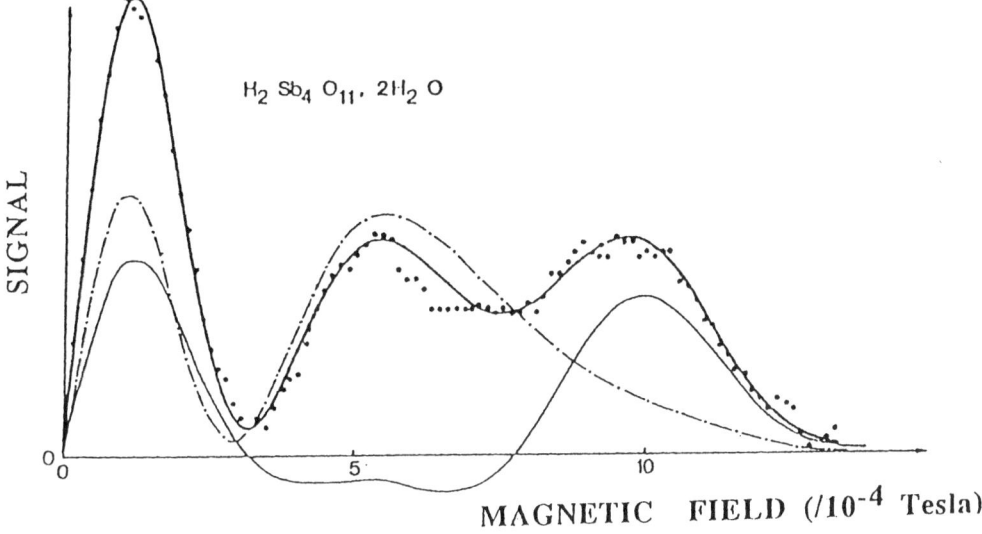

Figure 4   NMR spectra for the antimonic acid $H_2Sb_4O_{11}.2H_2O$: ...   experimental spectrum ; ▬▬, simulated spectrum ; - · - · -, weighted contribution of SOH groups hydrogen-bonded with water ; ⌐, weighted contribution of hydroxonium ions.

500

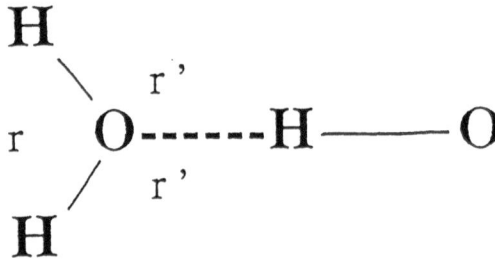

Figure 5   Assumed C2v symmetry for each SOH group hydrogen-bonded to a water molecule.

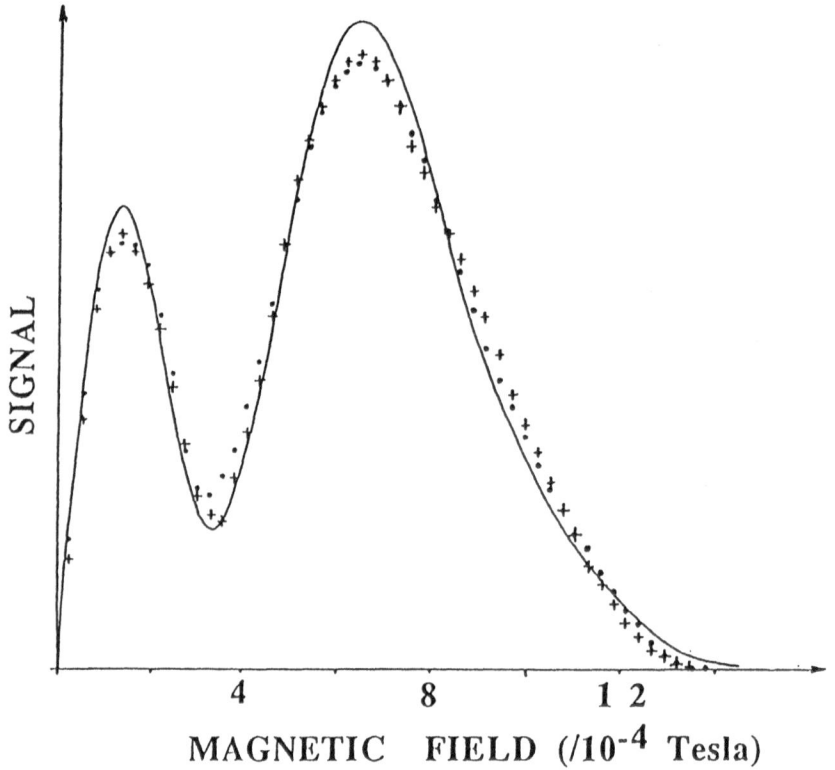

Figure 6   NMR spectra for $H_2AlP_3O_{10}.2H_2O$: . . . and +++ experimental spectra ;   ,
simulated spectrum assuming each SOH group to be hydrogen-bonded with a water
molecule.

CLOSING REMARKS

Jule A. Rabo
U.O.P
Des Plaines, Il 60017

Acid catalysis is the most important chemical aspect of many important petroleum and petrochemical technologies. While acid catalysis grew out of successful use of solution catalysis, recent trends resulting from environmental considerations give preference to heterogenous acid catalysis. Heterogenous acid catalysis received great boost in recent decades by the discovery of zeolite catalysis and by its amazing success in achieving commercial applications in hydrocarbon processing technologies. Another important category of solid acids is the sulfonated resins, including perfluorinated polymers, displaying acid strengths from strong to super acidity. At present, there are important areas of hydrocarbon conversion technology that are limited to homogenous catalysis (isoparaffin alkylation) while others are limited to heterogenous catalysis (shape selective rearrangements). For these reasons, the best future strategy is to combine our knowledge in molecular chemistry gained in homogenous catalysis and with both sulfonated resins and zeolites, to synthesize superior, environmentally inert solid acid catalysts for future applications.

The understanding of the chemistry of acid catalysis in solution systems made great advances in the last decades by effective use of NMR spectroscopy as well as by extensive application of theory in molecular modeling, to define the reaction path including reaction intermediates. This progress was made possible by the current state of science in describing chemical behavior of molecules in a predictive manner. Heterogenous acid catalysis has also shown great progress in science, and even more in the development of new industrial technologies. However, in spite of this progress, our knowledge here has less predictive power relative to solution catalysis. The reason is that the surface of solids is less defined and less understood. As a result, the chemistry and structure of acid catalytic sites is not known in most cases, and the chemical/structural models used to conceptualize the site of acid catalyzed reactions is not based on adequate scientific evidence. Therefore, the conceptual models are usually too simplistic or too inaccurate to be useful for theoretical modeling or as basis in the evolution of superior catalysts. Since the chemistry/structure of active sites is not well known, our knowledge of catalytic intermediates is even less understood in heterogenous acid catalysis.

For these reasons, the most effective steps toward progress in acid catalysis are those leading to increased predictive power of present science. Here, the following steps are recommended in catalyst synthesis, characterization and in chemical theory :

Catalyst Synthesis :

- Continue exploration of highly structured catalysts such as zeolites, pillared clays and alike ;
- Explore development of acid catalyst membranes suitable to combine acid

501

J. Fraissard and L. Petrakis (eds.), Acidity and Basicity of Solids, 501 502.
© 1994 Kluwer Academic Publishers.

catalysis and product separation steps.
- Develop solid acid catalysts with controlled interaction between protic and Lewis acid sites at desirable acid strengths.

Catalyst Characterization :

- Define the whole acid site (H^+ and Lewis acid) with a precision suitable for theoretical modeling.
- Define reaction intermediates.
- Develop fast spectroscopy for the detection of reactive, short-lived reaction intermediates.

Theoretical Modeling :

- Reconcile local and long-range effects affecting acid site chemistry in solids.
- Evaluate models including interacting protic and Lewis acid sites.
- Apply theoretical modeling as functional part of the catalyst evolution cycle : catalyst conception - synthesis - characterization - catalyst testing - theoretical modeling - catalyst improvement.

504